CONSTRUCTION LITIGATION: REPRESENTING THE CONTRACTOR

CONSTRUCTION LITIGATION: REPRESENTING THE CONTRACTOR

ROBERT F. CUSHMAN
JOHN D. CARTER
ALAN SILVERMAN
Editors

Coopers & Lybrand Edition

Wiley Law Publications
JOHN WILEY & SONS
New York · Chichester · Brisbane · Toronto · Singapore

Library of Congress Cataloging-in-Publication Data

Main entry under title:

Construction litigation.

(Wiley law publications)
Includes index.
1. Building—Contracts and specification—United States. 2. Construction industry—Law and legislation—United States. 3. Liability (Law)—United States. 4. Liability for building accidents—United States. 5. Actions and defenses—United States. I. Cushman, Robert Frank, 1931- . II. Carter, John D. 1946- . III. Silverman, Alan. IV. Series.
KF902.C626 1986 343.73'0786 24 85-20369
ISBN 0-471-84592-2 (Coopers & Lybrand special edition)

Printed in the United States of America

10 9 8 7 6 5 4 3 2 1

FOREWORD

I am pleased to present this special Coopers & Lybrand edition of *Construction Litigation: Representing the Contractor.*

Because of Coopers & Lybrand's extensive experience in the construction industry, we were asked to write the chapters on "Pricing the Loss," "Tax Considerations of Construction Litigation," and "The Use of Experts to Contain the Cost of Construction Litigation."

As accountants and consultants to contractors, we have tried to explain how to decrease the likelihood and reduce the cost of potential litigation. We have also made suggestions on how to be better prepared if litigation becomes unavoidable.

With our offices throughout the United States and around the world, we are equipped to deal with all the financial aspects of the construction industry, on projects large and small.

I trust that businessmen, lawyers and accountants will find this book a useful reference.

San Francisco, California
October 1985

ALAN SILVERMAN
Coopers & Lybrand

PREFACE

The construction industry has always been and always will be highly litigious.

The double-digit inflation of the 1970s brought about "newfangled" methods of designing and constructing such as fast-track, phased construction, value engineering, and construction management, each of which brought to the litigation arena its own set of legal relationships and problems. The cutback in public and private work brought about by the recession of the 1980s not only caused many contractors, subcontractors, and material suppliers to go out of business, but created fierce competition, tight budgets, and a profound unwillingness to contribute to the solution of site problems.

It is true that there is a movement to develop alternative methods of dispute resolution, Unfortunately, litigation remains the dominant method of resolving construction claims and disputes. Perhaps the primary reason for this is that the bare elements of cooperation necessary to stimulate an alternative dispute resolution approach are lacking by the time the parties are forced to turn outside for a solution to their problem.

Major construction litigation continues to be fueled by new third-party pressures which upset the balance of the contractual bargain struck between the owner and its contractor, and this disruption has a ripple effect down to the subcontractors to the suppliers. To use the utility industry and "heavy construction" as an example, there is active intervention in licensing and rate proceedings by activist public utility commissions, by concerned shareholders who see earnings eroding, by takeover specialists, by management prudency audits, by regulatory roadblocks, and so forth. Across the board we see the development of these and other outside pressures, which stimulate owners to squeeze every possible dollar out of their contractors, and to reject claims whenever possible.

We have invited to participate as co-authors in this endeavor a distinguished group of experienced construction litigators and claim specialists to share their "in the trenches" experience, information, and strategies and to show how contractors, subcontractors, and material suppliers can best avoid costly and time-consuming litigation, or, if necessary, best protect themselves and champion their interests if litigation becomes unavoidable.

October 1985

ROBERT F. CUSHMAN
Philadelphia, Pennsylvania

JOHN D. CARTER
San Francisco, California

ALAN SILVERMAN
San Francisco, California

ABOUT THE EDITORS

Robert F. Cushman is a partner in the national law firm of Pepper, Hamilton & Scheetz and a recognized specialist and lecturer on all phases of real estate and construction law. He serves as legal counsel to numerous trade associations, construction, development, and bonding companies.

Mr. Cushman is the editor and co-author of *Doing Business In America, The Dow Jones Business Insurance Handbook, The Businessman's Guide to Construction and High Tech Real Estate,* published by Dow Jones-Irwin; *The McGraw-Hill Construction Business Handbook, The Construction Industry Formbook,* McGraw-Hill; *Avoiding Liability in Architecture, Design and Construction, Construction Litigation: Representing the Owner, The John Wiley Handbook on Managing Real Estate in the 1980s, Handling Property and Casualty Claims* and *Handling Fidelity and Surety Claims*, published by John Wiley & Sons.

Mr. Cushman, who is a member of the Bar of the Commonwealth of Pennsylvania and who is admitted to practice before the Supreme Court of the United States and the United States Claims Court, has served as Executive Vice President and General Counsel to the Construction Industry Foundation, as well as Regional Chairman of the Public Contract Law Section of the American Bar Association. He is a member of the International Association of Insurance Counsel.

John D. Carter is a Chief Litigation Counsel with the Legal Department of the Bechtel Group of Companies, responsible for management of major claims and all litigation.

Carter joined Bechtel in January 1982. Prior to that, he practiced with the firm of Thelen, Marrin, Johnson & Bridges, where he specialized in litigation as a partner in the firm. His experience included several major construction and engineering trials for a wide variety of the leading firms in the industry.

He received a bachelor's degree in history from Stanford University and a law degree from Harvard University.

Alan Silverman is a partner in the international accounting and consulting firm of Coopers & Lybrand. He is national director of the firm's Litigation Services practice, which provides financial analysis and expert testimony for trial lawyers in the areas of litigation, construction claims, and bankruptcy.

Mr. Silverman is chairman of the American Bar Association's subcommittee on the Use of Computers in Litigation, and is a member of the Litigation Services committee of the California Society of Certified Public Accountants. He has been a visiting lecturer at Columbia Law School and the University of London. He was formerly the director of Litigation Analysis for IBM.

SUMMARY CONTENTS

DETAILED CONTENTS

CHAPTER 1

THE CONTRACTOR CONTEMPLATING LITIGATION AND ITS ALTERNATIVES: AN OVERVIEW

Overton A. Currie, James E. Stephenson, and Philip E. Beck

Overton A. Currie, senior partner of the Atlanta law firm of Smith, Currie & Hancock, is head of the firm's 40-lawyer construction law department. Mr. Currie has served the American Bar Association as National Chairman of the Public Contract Section, National Chairman of the Committee on Subcontracts, and National Chairman of the Litigation Section. He earned a Master of Law degree from Yale University as valedictorian, and remained on the Yale Law School faculty as a research associate until returning to private practice in 1959. His experience includes both private and public projects across the nation and abroad.

James E. Stephenson is a partner in the Atlanta law firm of Smith, Currie & Hancock. He is a graduate of Yale University and a 1976 graduate of Duke University School of Law, where he was a member of the Duke Law Journal. He has been involved primarily in construction litigation and other aspects of construction law since joining Smith, Currie & Hancock in 1976. Mr. Stephenson has represented a wide range of clients, including most of the participants in the construction process—owners, general contractors, subcontractors, sureties, and construction managers.

Philip E. Beck is an associate with the Atlanta law firm of Smith, Currie & Hancock, which he joined upon graduation from the University of Tennessee College of Law in the spring of 1981. He also holds a Master of Business Administration degree from the University of Tennessee, where he earned an undergraduate business degree as well. Mr. Beck is a member of the legal honorary society Order of the Coif, and the business honorary fraternity Beta Gamma Sigma. Mr. Beck is a member of Smith, Currie & Hancock's construction litigation department.

THE CONSTRUCTION ENVIRONMENT

§ 1.1 The Construction Business

In many ways, the construction business is like any other business. It is the process of assembling various resources—labor, materials, and expertise—and combining them to create a valuable product—a hospital, a school, an airport, an office building, an industrial plant, or some other facility or structure. With few exceptions, the long-term objective of a construction company, like most businesses, is to earn a profit. There is nothing dishonorable about such an objective; it is the basis of most economic systems and provides incentive for undertakings that benefit and advance the whole of human society.

The construction business is also unique in many ways. The profit margin—the percentage difference between revenues and costs—is very small relative to the risk exposure. Good business and money managers analyze investments in terms of a natural tradeoff between risk and return. The construction industry as a whole is characterized by much lower dollars-profit per dollars-at-risk than almost any other type of business. The construction business requires large capital outlays, yet is also labor-intensive. On almost every construction project, it is possible to lose much more money than it is possible to make. It is not unusual for a job that constitutes only 5 percent of a large contractor's business to have the potential to destroy 50 percent or more of the company's net worth.

§ 1.2 —Risk Factors

The time-sensitivity of construction work is one factor contributing to its risk. Time is money, especially in the construction industry. Construction projects typically extend over long periods of time and there are many time-related costs that increase in proportion to any increases in the performance period. The cyclical nature of the construction business also increases risk. While many costs are variable and can be avoided during periods of low construction activity, many capital costs are not, and must be met even though the amount of available work is insufficient to support them. The predominance of performance and payment surety bonds in the construction industry is evidence of the high-risk nature of the industry, and distinguishes it from most others. The construction business is undoubtedly exciting, but it is truly an excitement born of risk.

§ 1.3 —Risk Management

There are profits to be realized in the construction business, but to be successful over the long term requires effective management skills and techniques. Management has been defined as "the function of getting things done through people and directing the efforts of individuals towards a common objective."[1] Only through good, sound management can a contractor marshal the integration of labor, materials, and expertise to construct a building or structure so as to accomplish the business's objective: realization of a profit.

§ 1.4 Construction Disputes

Most writings and programs on the topic of construction law and litigation begin with a discussion of the various factors which combine to make the construction industry the arena for a great many disputes. If high risk is the salt of the construction industry, controversy and disputes are the pepper. This is true for a number of reasons. First of all, the construction industry is the largest industry in the United States. However, the construction industry breeds a large number of disputes even for its size, not because the participants are unusually litigious by nature, but rather because of inherent characteristics of the industry.

One of the principal factors leading to construction disputes is the unavoidable complexity of the industry. A multitude of factors affects the respective rights, responsibilities, and liabilities of the various parties involved in the construction process. The technical and changing nature of construction adds to this complexity.

The large number of separate, yet interdependent, parties involved in a typical construction project is also a factor in the generation of disputes. Unlike many

---

[1] Haimann, Professional Management Theory and Practice 1 (Houghton Mifflin Co., 1962).

businesses that are characterized by two-party transactions, a construction project will likely involve an owner, one or more prime contractors, a number of different subcontractors and suppliers, an architect, one or more engineers, a number of sureties and insurance carriers, a construction lender, and perhaps a construction manager or various other consultants. The complex nature of construction and the number of parties involved also create communication difficulties which can themselves lead to disputes.

The time-related characteristics of the construction business may also generate disputes. These include the protracted performance period of construction, over which a multitude of different but interrelated activities are performed, and the time-sensitivity of the costs involved.

Finally, the many different contract structures by which the various parties may align themselves create disputes by altering the rights and liabilities of the parties, often without their realization. For example, an owner may:

1. Employ a single general contractor, who in turn employs and coordinates various subcontractors
2. Contract directly with the various trade contractors, establishing multiple prime contracts, and expressly or implicitly assume the right and responsibility to coordinate the work of these various prime contractors itself
3. Employ multiple prime contractors and, in addition, employ a construction manager to assist in the coordination of their work
4. Enter into a design-build contract, contracting with a single entity to perform design and construction services, rather than employing a separate architect/engineer
5. Devise some hybrid of the foregoing, or fashion some different role, relationship, and contract structure.

§ 1.5 —Types of Disputes

Most construction disputes that arise can be grouped into three broad categories: (1) disputes pertaining to the quantity of the work, (2) disputes pertaining to the quality of the work, and (3) disputes pertaining to the method (including the time frame) of performing the work. Construction disputes, like other types of disputes, generally arise because the parties have different and conflicting interests and expectations. When an owner provides a construction design to a contractor, he imposes certain quantity, quality, method, scheduling, and sequencing restraints upon the contractor. Many owners, conscious of their substantial cash investment in the project, view these guidelines as minimum standards below which the contractor cannot trespass, but above which the contractor should aspire. The natural tradeoff between cost and time on the one hand and quality on the other breeds disputes.

§ 1.6 –Judge or Party

The architect is established under many contracts as the neutral referee of disputes between the owner and contractor. However, the architect has his or her own set of interests and expectations, which may conflict with those of the owner and/or the contractor. When the architect who designed the building also supervises its construction, he or she is often charged with responsibility for insuring adherence to the design, but not with responsibility for the timely completion of the project. The architect, therefore, has a natural tendency to demand quality, often at the expense of time and cost.

§ 1.7 Need for Effective Risk, Dispute, and Litigation Management

The demanding nature of the construction business and the prevalence of disputes in the industry require a contractor, in order to be successful, to be a successful dispute manager. This entails being an effective manager of risk, to prevent and minimize disputes; an effective manager of disputes, once they arise; and an effective manager of the various instrumentalities of dispute resolution, including negotiation, arbitration, and litigation.

An old proverb states, "Agree, for the law is costly."[2] The potential costs of disputes are enormous. Through effective contract negotiation and administration, a contractor can minimize the likelihood of disputes and save itself enormous costs. When disputes do arise, as they inevitably will, the contractor must effectively manage the resolution of the dispute—be it by litigation, arbitration or negotiation—to ensure that the cure is not worse than the ailment. Any experienced lawyer can cite examples of where litigation costs have exceeded the amount in controversy in a given case. This is usually due to poor dispute resolution management.

Litigation and its alternatives should be viewed as means of accomplishing a certain objective, and should be subjected to the same type of cost-benefit analysis as any other business undertaking. If the anticipated benefits to be derived from litigation do not exceed the anticipated costs of litigation, it should not be undertaken. Benefits can, of course, include noneconomic and nonimmediate benefits. Sometimes a litigant will consciously choose to pursue a cause, with little regard for the cost, on principle. Most construction disputes, however, are purely economic disputes, and should be evaluated purely in economic terms.

In evaluating the costs of litigation and its alternatives, one should not lose sight of a number of indirect or hidden costs. One is the time value of money. It is a rudimentary concept of finance that a dollar today is worth more than a dollar tomorrow. It is the reason banks are willing to pay interest on deposits and can charge interest on loans. When litigation takes years to be completed, the opportunity costs of not having capital to invest in other projects and ventures can become

[2] Herbert, Jacula Prudentum (1651).

enormous. Another sometimes overlooked cost of litigation is the management time and resources which must be devoted to the matter rather than to the pursuit of other profit-generating endeavors.

The numerous parties and long time periods involved, the volume of paper generated, the necessity of using experts, the necessity of protracted discovery, and the length of trials and appeals (due to the complexity and size of construction disputes) all combine to make construction disputes more costly to litigate, both in terms of time and of money, than many other types of disputes. The best way to streamline, minimize, or avoid litigation and arbitration is to implement a management system that avoids many potential disputes, identifies and documents them when they occur, and generates and preserves documentation and evidence of the supporting facts. By doing this, the contractor not only minimizes the possibility of disputes, but enhances its position for a favorable resolution of disputes when they do occur. Recognition, preparation, presentation, defense, and resolution of construction claims should all be performed in accordance with the same management standards applied to operational decisions.

RISK MANAGEMENT

§ 1.8 Parties and Relationships

This chapter has touched upon the numerous parties involved in the construction process, and subsequent chapters of this book address in greater detail the relationship of the contractor with these various parties. A brief introduction to these parties and relationships is, however, appropriate here.

§ 1.9 —Owners: Public and Private

The *owner* is primarily a purchaser of materials and services: it pays for the construction and derives the benefit of the product. It is important for the contractor to know the owner it is dealing with and looking to for payment, specifically, the owner's ability to pay, expertise, and reputation. Much of the construction work performed, even in the United States, is performed for the benefit of public owners such as the federal government or state and local governments. A prudent contractor adjusts its construction administration methods depending upon the nature of the owner. In addition to differences in the contract documents, there are varying interests, philosophies, and legal rights and responsibilities.

Dispute resolution mechanisms also vary for the different types of owners. Disputes under private contracts are generally resolved, if not by agreement, through arbitration, if the contract so provides, and otherwise through litigation. Construction contracts with public bodies, on the other hand, frequently refer disputes to special tribunals and procedures. For example, under the Contract Disputes Act of 1978,[3] disputes arising under federal government construction contracts are to be resolved through proceedings before one of the various agency boards of contract appeals or the United States Claims Court. Some states and municipalities also have special disputes procedures or requirements.

Contracting with governmental bodies also raises the issue of sovereign immunity, which is an ancient doctrine prohibiting or limiting suits against the government. Today, sovereign immunity is not a bar to most contract suits against governmental bodies, but it can in some cases bar or limit recovery, and, therefore, should not be ignored. In short, the nature of the owner must be taken into account by both the contractor and the contractor's attorney, and both contractor and attorney must recognize and respond to the distinctions.

§ 1.10 –Architect/Engineers

Under most contract structures, the *architect/engineer* is a design professional who is not in contractual privity (does not have a direct contractual relationship) with either the prime contractor(s) or the subcontractors. This limits the ability of a contractor to seek recourse against the architect/engineer for damages flowing from improper design, inspection, contract interpretation, or contract administration. The *Spearin* doctrine[4] is applied in federal government construction contract disputes and under the law of many states to provide the contractor a remedy for defective design against the owner, with whom it is in contractual privity. The *Spearin* doctrine provides that the owner, in providing the contractor the construction design, impliedly warrants that the plans and specifications are reasonably free from defects.

In addition, while the contractor may not have a direct contractual cause of action against the architect/engineer, it may be able to recover directly against the design professional under one or more of the following legal theories:

1. *Negligence* on the part of the design professional
2. The commission of an *intentional tort* by the design professional
3. The existence of a *third-party beneficiary* relationship whereby the contractor can recover damages resulting from the design professional's breach of its contract with the owner.

The contractor's decision as to whether to seek recovery from the owner or directly from the design professional is an extremely important one and should depend on a number of factors. These include the law governing the dispute, the nature of the claim, the solvency of the owner and the design professional, the terms

---

[3] 41 U.S.C. § 601 *et seq.*

[4] The ''*Spearin* doctrine'' originates from the 1918 United States Supreme Court decision in United States v. Spearin, 248 U.S. 132, 136 (1918).

of the construction contract, the terms of the design professional's contract, and the forum in which the contractor would prefer to resolve the dispute.

§ 1.11 —Construction Managers

In recent years, it has become popular with many owners to contract directly with the various trade contractors and employ a *construction manager* to fulfill many of the functions traditionally performed by the general contractor. Contractually, this results in the existence of multiple prime contractors who may or may not be in contractual privity with the construction manager charged with the duty to coordinate and schedule their work. In the pure construction management scenario, where the prime contractors are not in contractual privity with the construction manager, the ability of a contractor to seek recourse directly against the construction manager for the construction manager's failure to properly schedule or coordinate the work is limited in the same way as is the contractor's ability to proceed directly against the architect (discussed in **§ 1.10**). Generally speaking, when an owner chooses to contract directly with multiple trade contractors, the owner assumes the obligation to coordinate the work of the various contractors, and cannot divest itself of that responsibility by employing a construction manager.[5]

The decision of whether to bring an action directly against the construction manager on a negligence, intentional tort, or third-party beneficiary contract theory, or to seek recovery from the owner, is an important strategic decision and should depend on several factors. Construction managers are sometimes grouped into two categories, distinguished by the background and experience of the firm providing the construction management services. The *professional construction manager* is a construction manager who is a design professional by background and training. The *contractor-construction manager* is usually a general contractor performing in this role. Which of these two categories the construction manager in question falls into should have a bearing on the contractor's litigation strategies; for instance, the professional construction manager is more likely to carry insurance and, therefore, be capable of paying a judgment.

Other factors include the nature and origin of the action; the relative financial strengths of the construction manager and the owner; the relative experience of the construction manager and the owner; whether the contractor desires to do business in the future with either the construction manager or the owner, or otherwise maintain a good relationship with one or the other; the relative culpability

---

[5] *See* Eric A. Carlstrom Constr. Co. v. Independence School Dist. No. 77, 256 N.W.2d 479 (Minn. 1977); Stehlin-Muller-Henes Co. v. City of Bridgeport, 117 A. 811 (Conn. 1922). This obligation is analogous to the general contractor's obligation to coordinate subcontractors. *See* Johnson v. Fenestra, Inc., 305 F.2d 179 (3d Cir. 1962); United States v. Citizens & S. Nat'l Bank of Atlanta, GA., 367 F.2d 473 (4th Cir. 1966); Great Lakes Constr. Co. v. Republic Creosoting Co., 139 F.2d 456 (8th Cir. 1943); Melwin Constr. v. Stonewall Constr. Co., 113 A.2d 108 (D.C. Mun. App. 1955).

of the construction manager and the owner; and the specific facts of the particular situation, such as contract terms.

§ 1.12 —General Contractors, Subcontractors, and Suppliers

The relationship between a general contractor and its subcontractors and suppliers is a very important one, which is in many ways analogous to the owner-general contractor relationship. The general contractor and its subcontractors are often either powerful allies or powerful adversaries. The general contractor must not lose sight of the fact that the owner is contractually looking to it for timely and proper performance of the entire contract.

General contractors often make the mistake of building a case against a subcontractor that the owner can then use against the general contractor. Correspondence between the general contractor and its subcontractors, in which accusations are made against each other, often provide powerful ammunition for a capable construction litigator representing the owner when disputes between the owner and the contractor develop. This is not to say that a subcontractor should never be critical of its general contractor or that a general contractor should never be critical of its subcontractors; sometimes a critical or demanding letter may be necessary to force another party to adhere to its obligations. Rather, the parties should be aware of the legal ramifications of their actions and weigh those in making their decisions.

§ 1.13 —Sureties

Often, on a construction project, both the general contractor and the various subcontractors will be required to provide payment and performance bonds. This introduces another group of parties, with their own rights and obligations, to the project: the *construction sureties*.

Only in rare situations is it in the best interest of the contractor (the principal) to initiate litigation against its own surety. This is because the surety will in most cases have indemnity rights against the contractor, and in the case of small corporations, against the owners of the corporation. Thus, suing one's surety is almost, in effect, suing oneself. There are, more frequently, situations where it may be in the best interest of a general contractor to bring suit against a subcontractor's surety, or in the interest of a subcontractor or supplier to bring suit against the general contractor's surety. The typical situation where this occurs is when the principal defaults on its contract and is insolvent.

On federal construction projects and most other public work, the general contractor is required to provide performance and payment bonds to serve as a substitute for the lien rights which subcontractors and suppliers would normally have on a private project but do not have on public work.[6] A general contractor often has the option, however, of requiring or not requiring subcontractors to provide performance and payment bonds, and should carefully evaluate whether the cost of doing so is warranted under the particular circumstances.

§ 1.14 —Construction Lenders

In some situations, a contractor may be entitled to seek legal recourse directly against the construction lender who is financing the project. Such action should be seriously considered if there is concern about the solvency of the owner. With regard to construction financing, the contractor must remember that it will lose important leverage if it allows progress payments to fall behind performance progress to a point where it would be cheaper for the owner and lender to terminate it and employ another contractor than to pay the existing contractor to complete.

§ 1.15 Legal Parameters

There are a number of legal rights and obligations that exist separately and independently of those expressly created and assumed in the contract documents. Some of these can be altered or shifted in the contract documents; some cannot. The contractor must be aware of these and take them into account in bidding and performing its work.

§ 1.16 —Building Codes, Ordinances, and Regulations

Even though not specifically referenced in the contract documents, contractors are generally required to comply with the local laws of the jurisdiction in which the project is performed. One issue which commonly arises is that of who is responsible when a building is constructed in accordance with the design but does not conform to local building and safety codes. The question usually boils down to whether the contractor acted reasonably in relying upon the design professional to take local building and safety codes into account. As a rule of thumb, if the contractor has substantial experience in the area where the project is being built, it is not reasonable for it to rely entirely on the design professional in this regard. If the contractor is not familiar with the area, such reliance may be deemed

[6] The "Miller Act," 40 U.S.C. § 270(a)–(f), imposes such requirements on federal work. A number of states have enacted "Little Miller Acts"; *see* Lifschitz, *Little Miller Acts*, Construction Briefings No. 8 (Aug. 1984).

reasonable. A contractor should, however, fully apprise itself of local codes in order to avoid disuptes of this type.

In addition to local building and safety codes, a number of other laws and ordinances may affect the contractor and require its attention and compliance. Among these are licensing requirements, registration requirements, and tax requirements. The failure of a contractor to comply with these can result in substantial penalties and may even bar its ability to enforce the obligations of others.[7]

When a contractor contracts with a federal, state, or local governmental body to provide construction services, it impliedly submits itself to a myriad of public regulations and requirements. These can include safety regulations, wage standards, minority and female employment requirements, reporting and recordkeeping requirement, notice requirements, and others. These, too, should be taken into account by the contractor and should govern its actions and decisions.

§ 1.17 —Trade Usage and Implied Obligations

When the express language of a contract does not unambiguously resolve a given dispute or controversy, courts, boards, and arbitration panels often look to trade custom and usage and the parties' prior dealings to interpret the intent of the contract. While the parties are free to establish by contract rights and obligations totally different from prior dealings and trade custom, they must be careful to do so expressly and unambiguously.

There are a number of legal rights and obligations that arise from a contract even though not expressly stated therein. For example, a party to a contract impliedly promises not to do anything to hinder or impede the other party's performance.[8] This basic principle becomes the basis of the contractor's right, under appropriate circumstances, to recover delay damages and other forms of damages. The *Spearin* doctrine, referred to in **§ 1.10**, is another example of an implied obligation that arises from the contract without being expressed in words. The owner's obligation to coordinate multiple prime contractors, and the general contractor's obligation to coordinate the various subcontractors are other examples of important implied rights and obligations. Finally, there are various implied warranties that may arise to impose obligations on the contractor, including implied warranties regarding the quality of materials and workmanship, and, in the case of residential construction in some states, the implied warranty of habitability.

Aside from express and implied contractual rights and obligations, the contractor has certain other rights and obligations, the breach of which can give rise to a right of recovery in tort. A contractor has an obligation to exercise reasonable

[7] *See, e.g.*, Sample v. Morgan, 319 S.E.2d 607 (N.C. 1984); Allan S. Meade & Assocs., Inc. v. McGarry, 315 S.E.2d 69 (N.C. Ct. App. 1984).

[8] *See* Gulf M.&O. Ry. Co. v. Illinois Central R.R. Co., 128 F. Supp. 311 (N.D. Ala. 1954), *aff'd*, 225 F.2d 816 (5th Cir. 1955); S. Williston, A Treatise on the Law of Contracts §§ 1296, 1316 (3d ed. 1961).

care in the performance of the work and, in many states, has a right to expect the exercise of reasonable care by the architect, the construction manager, the owner, and others. The contractor has an obligation to refrain from committing various intentional torts and a right to be free from intentional torts of others. These tort rights and obligations arise independent of any contract and are fundamentally similar to those imposed on anyone driving a car or walking down the street.

§ 1.18 Contract Documents–Beginning

Like a marriage license, the contract documents are essential, but only a beginning. The goal of the contract documents should be to define as clearly and unambiguously as possible the various rights and obligations of the parties. Some contractors bemoan the fact that the day of the handshake agreement is past. This is true, however, not solely due to any lack of trustworthiness on the part of modern-day participants, but because of the increased complexity of construction.

In the contract documents, the parties should strive to anticipate potential problems, to allocate the risks of such problems, and to provide for the efficient resolution of any disputes which may arise. This applies not only to the owner/general contractor contract but also to the general contractor/subcontractor contract. Moreover, the contractor must be careful to coordinate its subcontracts with the general contract to ensure consistency and desired result.

§ 1.19 –Uniform Commercial Code Applicability

The Uniform Commercial Code (UCC) was devised as a recommended code of laws governing commercial conduct. Its substantial adoption by all 50 states except Louisiana resulted in a system of relatively uniform state laws governing frequently recurring forms of business transactions, including the sale of goods, the use of negotiable instruments and investment securities, bulk transfers, and financing arrangements involving security interests.

Article Two of the UCC's nine articles governs transactions involving the sale of movable goods. Since the creation of the UCC, a debate has raged as to whether construction contracts, which typically involve the sale of services primarily and goods only secondarily, are covered by Article Two. The various states which have addressed the issue are split, with some holding that Article Two of the UCC does apply to construction contracts.[9] In addition to those courts holding that the UCC is directly applicable, a number of other decisions have held that its principles should be applied to construction contracts by analogy, thus yielding the

---

[9] *See, e.g.*, Pittsburgh-Des Moines Steel Co. v. Brookhaven Manor Water Co., 532 F.2d 572 (7th Cir. 1976); Bonebrake v. Corx, 499 F.2d 951 (8th Cir. 1974); Omaha Pollution Control Corp. v. Carver-Greenfield Corp., 413 F. Supp. 1069 (D. Nev. 1976); Port City Constr. Co. v. Henderson, 266 So. 2d 896 (Ala. Ct. App. 1972).

same result.[10] In addition, supply contracts involving solely the sale of movable goods fall squarely within the scope of UCC Article Two.

The application of UCC Article Two, directly or by analogy, has several ramifications. Section 2-204(1) provides that a contract "may be made in any manner sufficient to show agreement, including conduct by both parties which recognizes the existence of such a contract." Section 2-204(3) provides that a contract may be enforceable "even though one or more terms are left open." Both of these provisions represent major modifications of common law.

Perhaps the most important effect of the applicability of the UCC is the application of § 2-207, commonly referred to as the "battle of forms." Under traditional contract law, a "mirror image" rule was applied requiring, as a prerequisite to contract formation, an acceptance strictly conforming to the terms of the offer. Under UCC § 2-207, an expression of acceptance can operate as an acceptance of a contract "even though it states terms additional to or different from those offered or agreed upon, unless acceptance is expressly made conditional on assent to the additional or different terms." Section 2-207 goes on to describe when those additional or changed terms become part of the contract and when they do not.

§ 1.20 —Basic Provisions

There are a number of contract provisions basic, if not essential, to all construction contracts. These include payment provisions, provisions defining the scope of the work, provisions governing changes in the scope of the work, provisions establishing requirements and procedures for termination of the contract, and provisions establishing procedures for the resolution of disputes arising under the contract.

The *changes clause* is unique, yet essential, to construction contracts. Unlike most contracts, where the scope of the undertaking can be firmly established at the outset, most owners desire the flexibility to make changes in the scope of the work as it progresses. In the absence of a changes clause, the scope of the work could only be changed by mutual modification of the contract agreed to by both parties. Under the changes clause, a procedure is established whereby the owner may unilaterally order a change in the work, with the contractor's compensation and performance time adjusted accordingly.

A general contractor, in establishing the terms for its subcontracts, must incorporate *flow-down* provisions, binding the subcontractors to the same obligations, liabilities, and procedures to which the general contractor has bound itself in its contract with the owner. Critical provisions to be addressed in this way include:

1. Payment provisions (so that the general contractor is not bound to pay the subcontractor before the general contractor has been paid for that work)

---

[10] *See, e.g.*, Padbloc Co. v. United States, 161 Ct. Cl. 369 (1963); Redman Dev. Corp. v. Piedmont Heating & Air Conditioning Co., 197 S.E.2d 167 (Ga. Ct. App. 1974); Aced v. Hobbs-Sesack Plumbing Co., 55 Cal. 2d 573, 360 P.2d 897, 12 Cal. Rptr. 257 (1961).

2. Provisions for changes to the scope of the work (so that the general contractor is not obligated to grant the subcontractor a change order and additional compensation for which the general contractor is not compensated by the owner, unless the change merely involves shifting work within the original scope of the general contract from one subcontractor to another)
3. Default and termination provisions (so that, in the event the general contractor is terminated, the general contractor has the power to terminate the subcontract)
4. Quality standards provisions
5. Suspension of work provisions
6. Special damages provisions (e.g., no damages for delay clauses)
7. Provisions establishing disputes resolution procedures.

The general contractor should also be careful to ensure that all the work included within the scope of the general contract is included within the scope of one of the subcontractors' work, unless the general contractor intends to perform and incur the costs of that work itself.

§ 1.21 —Exculpatory and Special Clauses

In addition to the basic contract provisions, an owner or general contractor may desire to include one or more exculpatory or special provisions in the contract. An *exculpatory clause* is a clause barring or limiting some type of liability.

A common example of an exculpatory clause (which may appear either in a general contract or in a subcontract) is the *no damages for delay* clause. Such a clause attempts to limit the owner's or general contractor's liability to the general contractor/subcontractor for damages resulting from construction delays. Such a clause may be broadly worded to apply to all delays, or may apply only to delays of a specific nature. It may purport to deny any relief for delays, or may only limit the liability in some fashion. As a general rule, such provisions, when agreed to by both parties to the contract, are enforceable. However, the courts construe such provisions very narrowly and have developed many exceptions to the general rule of enforceability.[11] Although the contractor may often recover delay damages despite the presence of such a clause, the contractor should weigh very carefully the consequences of such a clause before entering a contract containing one. Exculpatory clauses are basically a means of shifting risk allocation, and, presumably, there is a cost associated with that risk. The contractor should, therefore, demand a higher contract price when such a clause is included.

Another type of risk-allocation provision is a provision pertaining to unanticipated subsurface or latent conditions. Modern federal government construction contracts

---

[11] *See* 74 A.L.R. 3d 187 (1976); Gontar, *The Enforceability of "No Damage for Delay" Clauses in Construction Contracts*, 28 Loy. L. Rev. 129 (1982).

include a clause making the owner liable for the costs of any unanticipated and unusual subsurface conditions, thus placing that risk upon the owner. In the absence of such a clause, the owner would be responsible for the costs of unanticipated conditions if the conditions were misrepresented in the contract documents, but the contractor might bear the cost of any conditions which were merely unanticipated. The federal government's *differing site condition* clause places responsibility for both types of unanticipated conditions on the owner. By doing this the government, in theory, encourages lower bids by eliminating this risk factor. A third possible type of provision pertaining to such conditions is an exculpatory clause expressly making the contractor liable for any unanticipated or unexpected subsurface or latent conditions. Again, such exculpatory clauses are narrowly construed, but may bar contractor recovery.

§ 1.22 —Modifications

As discussed in **§ 1.20**, the presence of a changes clause eliminates the need for a mutual contract modification to effect changes in the scope of the work. The changes clause does not, however, prevent the parties from modifying the terms of the contract by mutual agreement. Contractors should not lose sight of the fact that a postcontract modification has the same power and effect as the preprinted contract form signed at the execution of the contract, and, in fact, overrides any conflicting provisions in the original contract. In addition, contractual rights and obligations can sometimes be modified or waived by the parties' actions or course of conduct.

DISPUTE MANAGEMENT

§ 1.23 Types of Disputes and Their Causes

Disputes may arise before, during, or after construction, for a variety of reasons. Contractors who are aware of areas of potential disputes and the usual causes of such disagreements will be better able to recognize, avoid, and resolve them.

§ 1.24 —Bid Disputes

It is not unusual for disputes to arise at the very outset of the construction process: the bidding stage. Most public construction work is required by law to be awarded on the basis of secret, competitive bidding to the lowest responsive and responsible bidder. Often disputes arise as to who is the lowest responsive and responsible bidder, resulting in a bid protest. There are two most important words of advice to be offered to a bidder desiring to protest a proposed award: act quickly. After the contract award is announced, a bidder desiring to protest a proposed award has roughly 10 percent of the rights and likelihood of success that it had before the announcement.

Bid mistakes can lead to another type of dispute. Such disputes usually result when the low bidder realizes that it made an error in compiling its bid and desires to modify or withdraw the bid.

A final type of dispute that may arise during the bidding process is when a subcontractor, on whose bid the general contractor relied in preparing and submitting its bid to the owner, refuses to honor the bid. Under appropriate circumstances, many states hold that a general contractor can recover damages from the prospective subcontractor on the theory of promissory estoppel, but other states refuse to impose liability in such situations.[12]

§ 1.25 —Performance Disputes

Of the disputes arising during performance of a construction contract, without question the most common are disputes involving (1) extra work and (2) delays and disruption.

Whether a contractor has performed extra work beyond the original scope of the contract, entitling it to additional compensation under the changes clause, depends on a comparison of the scope and nature of work specified in the contract and the work actually performed. Extra work may result from changes needed to remedy defects in the design documents; directives by the owner or the owner's architect or representative; instructions, explanations, or interpretations of the contract by the owner or architect requiring work beyond the reasonable contemplation of the contract documents; unanticipated subsurface conditions; improper rejection or unreasonable inspection of the work; or various other acts and omissions of the owner or its representative.

Delays and disruptions may arise from any number of causes. As mentioned previously, the basis of the right to recover delay damages is the implied obligation on the part of the owner to not hinder or impede the contractor's performance. For the contractor to recover the costs associated with a delay from the owner, the delay must be both *excusable* and *compensable* under the terms of the contract.

---

[12] For cases allowing an action against the subcontractor in such circumstances, *see, e.g.*, Allen M. Campbell Co. v. Virginia Metal Indus., 708 F.2d 930 (4th Cir. 1983); Montgomery Indus. Int'l, Inc. v. Thomas Constr. Co., 620 F.2d 91 (5th Cir. 1980); Drennan v. Star Paving Co., 333 P.2d 757 (Cal. 1958); C.E. Frazier Constr. Co. v. Campbell Roofing & Metal Works, Inc., 373 So. 2d 1036 (Miss. 1979); E.A. Coronis Assocs. v. M. Gordon Constr. Co., 216 A.2d 246 (N.J. 1966). *Contra: see, e.g.*, Southeastern Sales & Serv. Co. v. T.T. Watson, Inc., 172 So. 2d 239 (Fla. Dist. Ct. App. 1965).

An *excusable delay* is a delay for which the contractor is entitled to an extension of the contract performance time. Under American Institute of Architects Form Contract Document A201, delay resulting from "any act or neglect of the Owner or the Architect, or by any employee of either, or by any separate contractor employed by the Owner, or by changes ordered in the Work, or by labor disputes, fire, unusual delay in transportation, adverse weather conditions not reasonably anticipatable, unavoidable casualties, or any causes beyond the Contractor's control" are excusable delays.[13] Generally speaking, nonexcusable delays are delays which could and should have been avoided by the contractor and delays resulting from risks assumed by the contractor, such as those resulting from anticipated conditions such as normal bad weather.

Compensable delays, a subset of excusable delays, are those delays which result from unanticipated acts or omissions chargeable to the owner or someone for whom the owner is responsible. This will generally include the architect, separate contractors, and employees and agents of the owner. Delay claims are probably the most litigated of all construction-related claims, because of the prevalence of construction delays, the difficulty in precisely assessing and allocating responsibility for those delays, and the enormous economic impact of construction delays due to the highly time-sensitive nature of construction costs. To understand, evaluate, and present construction delay claims requires a high level of expertise, often including the use of sophisticated scheduling and analysis techniques.

Compensable construction delays may arise from a number of different causes, including the owner's failure to timely deliver owner-furnished materials,[14] the owner's failure to have the construction site prepared on time for the contractor's performance,[15] the owner's failure to provide timely and sufficient access to the construction site,[16] the owner's furnishing of defective contract documents prepared by the architect,[17] the owner's failure to make timely payment,[18] the owner's failure to properly coordinate the work of parallel prime contractors,[19] the owner's failure to give written orders for extra work,[20] and the owner's failure, through its architect, to approve shop drawings within a reasonable period of time.[21]

---

[13] AIA Form A201, General Condition 8.3.1 (1976 ed.).

[14] *See* Brown v. East Carolina R.R. Co., 70 S.E. 625 (N.C. 1911).

[15] *See In re* Roberts Constr. Co., 111 N.W.2d 767 (Neb. 1961).

[16] *See* Blinderman Constr. Co. v. United States, 695 F.2d 552 (Fed. Cir. 1982); Higgins v. City of Fillmore, 639 P.2d 192 (Utah 1981).

[17] *See* Pathman Constr. Co., A.S.B.C.A. No. 22343, 81-1 B.C.A. (CCH) ¶ 15010 (1981); Dewey Jordan, Inc. v. Maryland-National Capitol Park & Planning Comm'n, 265 A.2d 892 (Md. Ct. Spec. App. 1970).

[18] *See* Seretto v. Rockland, S.T.&O.H. Ry. Co., 63 A. 651 (Me. 1906).

[19] *See* Eric A. Carlstrom Constr. Co. v. Independence School Dist. No. 77, 256 N.W.2d 479 (Minn. 1977); Stehlin-Muller-Henes Co. v. City of Bridgeport, 117 A. 811 (Conn. 1922).

[20] *See* Baltimore v. Clarke, 97 A. 911 (Md. Ct. Spec. App. 1916).

[21] *See* Langvin v. United States, 100 Ct. Cl. 15 (1943).

§ 1.26 Dispute Recognition

Early recognition of potential claims and disputes is extremely important. It allows the contractor to (1) comply with any contract prerequisites for relief, (2) involve experts, including attorneys, at an early stage when they can often be of the most assistance, and (3) begin documentation and preservation of important facts and evidence. Most construction-related disputes are controlled by the facts and by the contract. Therefore, in order to recognize potential claims and disputes early on, the contractor's key contract administration personnel, the people closest to the facts, must be intimately familiar with the contract documents.

For example, only by knowing the scope of the original contract can the contractor assess whether it is performing work beyond the scope of the contract, entitling it to additional compensation. A lack of familiarity with the contract documents can result in the contractor losing the entitlement to additional compensation for extra work due to the (1) failure to give timely notice as required by the contract, (2) failure to preserve evidence necessary to prove the facts, or (3) being considered to have performed the work as a ''volunteer'' or under an implied or express agreement to perform it without additional compensation.

The necessity of informed decisions also extends beyond the contract documents to a need for understanding of legal principles not expressed in the contract. For example, the authors are aware of projects where the contractor or construction manager consciously failed to submit or revise construction schedules on the mistaken belief that to do so would forgive the other party to the contract for past delays, or otherwise adversely impact certain rights and obligations. This can result in a failure to effectively administer the project and delays to its completion. While the contractor can usually read and understand the express provisions of the contract, understanding and appreciation of many of the legal and technical ramifications of the contractor's actions and decisions require consultation with experts.

§ 1.27 Preservation of Evidence and Claims

It is imperative that the contractor, once it recognizes the potential for a claim or dispute, begin to document, and thereby preserve a written or tangible record of, the facts necessary to prevail. For one thing, the contractor should endeavor to comply with any contractual, statutory, or other notice requirements. Most construction contracts purport to require written directives or notice as a prerequisite to any recovery for extra work, delays, unanticipated subsurface conditions, and other types of potential claims. It is a good idea for a contractor to maintain a checklist of all notice requirements to ensure compliance. While contractors can often prevail despite noncompliance with written notice requirements, such noncompliance is an unnecessary hurdle that sometimes cannot be overcome.

In addition to satisfying contractual and statutory prerequisites to the submission of claims and disputes, the contractor should thoroughly document the facts from which the claim or dispute arose. Among the most common and useful types of project documentation are job site logs or diaries, photographs, conversation memoranda, cost accounting records, contract correspondence, superintendents' and foremen's daily reports, and internal memoranda, including memoranda to the file. It is possible to implement a routine reporting and recordkeeping system utilizing fairly simple standard forms and procedures. A good construction documentation system actually serves a two-fold purpose: to ensure adequate project monitoring and control, and to ensure a complete and accurate record of job conditions, problems, and disputes and their effects. It is extremely difficult, in litigation years after an event has occurred, to recreate and prove the facts supporting the contractor's position without good documentation. The construction schedule is another piece of documentation that serves both as a very useful project monitoring and control tool and as a strong measure to prevent, preserve, present, defend, and prove claims.

§ 1.28 Use of Experts

Contractors often do not consider employing expert engineers, attorneys, or consultants until long after the facts giving rise to their involvment occur. In so doing, these contractors fail to make the most effective use of these resources. Experts, when employed during the construction project, can:

1. Assist in identifying and resolving technical and complex problems and disputes
2. Suggest ways of mitigating damages
3. Recommend ways of preserving evidence
4. Advise the contractor as to how to optimize its legal and factual position
5. Through contributing to an informed decisionmaking atmosphere, increase the chances of early resolution and decrease the need for litigation.

§ 1.29 Managing Dispute Resolution—Settlement

Quick and satisfactory settlement of claims and disputes is often essential to the continued success, and sometimes to the survival, of a contractor. Management time consumed by the pursuit of claims, while profitable business opportunities languish, compounds the red ink flowing from the original claim. In short, the contractor should settle or resolve disputes by agreement when it is possible to achieve a fair and reasonable result by doing so. Thus, it is important that the contractor utilize effective settlement and negotiation techniques and be able to appraise the costs and benefits of the alternatives for resolving given disputes.

§ 1.30 —Settlement Techniques

The best way to achieve a successful settlement is through early and thorough preparation. A well-prepared, well-documented, accurate, and thorough claim presentation goes a long way toward contributing to resolution of the claim or dispute. Such a presentation usually should be written and should contain a discussion of the factual basis of the claim, the costs associated with the claim, and the legal principles and contract provisions upon which the claim is based. A well-prepared claim document is an indication to the other side that the contractor is serious about the claim, that it believes in the merits of the claim, and that it is willing, if necessary, to go to trial. A successful negotiated settlement of the claim is far more likely if the other party is persuaded that the contractor is prepared to, and will, try the case if necessary. In addition, the chances of inconsistent positions or statements being asserted are diminished when the contractor prepares thoroughly and early.

In presenting a claim, the contractor should not lose sight of the human dynamics involved. The role of the experienced construction litigator is to make the complex appear simple and the dry interesting. The case should be reduced to its common denominator and the best theory of the case should be emphasized, in lieu of using a shotgun approach. It is important that the contractor also maintain credibility by not overstating the case or by "crying wolf" when no claims exist. The approach of many is to grossly overstate a claim in the hope of achieving settlement somewhere in the middle. Such a strategy, at best, only destroys credibility, and in some cases may subject the contractor to liability.

§ 1.31 —Negotiation Strategies

Once the well-prepared claim or defense has been effectively presented to the other party, the stage is set for effective negotiation. Negotiation is both an art and a science, requiring skill and experience. There have been many books and articles written on the subject, some of which are quite useful.[22] First of all, it should be noted that the well-prepared claim or defense takes into account the necessity for negotiation by providing bargaining room. In preparing the claim, the contractor should exercise sound judgment, but that judgment should be as favorable to the contractor as reasonableness, truth, and fair dealing allow.

In this limited space, all of the rules and suggestions regarding negotiation that have been developed from years of experience and analysis cannot be stated. A

---

[22] *See,* for example, Nierenberg, The Art of Negotiating (Hawthorne 1968); Hermann, Better Settlements Through Leverage (Lawyer's Co-op. Publishing Co. 1965); Coffin, The Negotiator (Barnes & Noble 1976); McDonald, Principles and Techniques of Negotiation (Southwestern Legal Found. Inst. on Gov't Contracts, CCH, 1963); Armed Services Procurement Regulations Manual for Contract Pricing 13–18 (ASPM No. 1, 1969); *Preparing and Settling Construction Claims*, Construction Briefings, No. 12 (Dec. 1983).

few basic principles, however, deserve mention. Modern-day thinking favors a win/win approach to negotiation: the negotiator should strive for a resolution that both sides will view as at least a partial victory. A skillful negotiator will never try to embarrass or belittle the opponent, but will allow the adversary to save face and feel good about the resolution. A good negotiator will define the objectives prior to entering negotiations, and be cognizant of on what points he or she must prevail to be successful and on what points he or she can concede or compromise. The good negotiator will identify the other party's objectives and needs and will, in fact, restate their objectives and positions to them to convey an understanding and appreciation of their position and to identify common ground.

A claim or defense should be expressed in human terms that everyone can understand and appreciate. The way points are expressed can be very important. For example, the contractor is more likely to persuade the other party of the justness of his claim if its costs are expressed in terms of ''benefits'' received by the other party rather than ''damages'' incurred by the contractor. Finally, the successful negotiator must be innovative, up to and including devising an easy method for the other party to pay the contractor money or otherwise do what needs to be done.

The United States Army Corps of Engineers has published a manual for use by its own representatives in negotiating contract claims. That book is helpful and insightful with regard to negotiation techniques:

> *Establishing Negotiation Strategy.* Before entering into formal negotiations with the contractor, the negotiator must establish exactly what his objectives are and how they may best be attained. Clear-cut decisions must be made beforehand as to which objectives (i) cannot be compromised under any circumstances, (ii) can be compromised and to what extent and in exchange for what, and (iii) merely represent an antidote to anticipated ''pie-in-the-sky'' demands by the contractor. The latter, of course, can be quickly and easily jettisoned as soon as the contractor shows a reciprocal willingness to abandon his extreme demands. Whenever possible, alternative objectives should also be established.[23]
>
> * * *
>
> As a general rule, the counteroffer should be a total offer. It should also be the lowest possible offer that is justifiable and one that will permit logical and orderly negotiations of a mutually acceptable price.[24]
>
> * * *
>
> Within certain limits, at least, negotiation is a matter of ''horse trading.'' The extent to which the negotiator or the contractor makes concessions without getting anything in return depends upon the relative bargaining positions of the parties.[25]

---

[23] Department of the Army, Office of the Chief of Engineers, Construction Contract Negotiation ch. 3, p. 26 (1966).

[24] *Id.* 38.

[25] *Id.* 42.

The negotiator should emphasize the case's strong points, yet acknowledge its problems and weaknesses, explaining them forthrightly. A good negotiator will also carefully consider the location of the negotiating sessions, the timing of the sessions, and the composition of the negotiating team.

§ 1.32 —Arbitration versus Litigation

When settlement negotiations fail to accomplish a full resolution of a claim or dispute, the parties must turn to third parties to resolve them. The traditional means of dispute resolution is litigation before a court of law. Alternatives to litigation, however, exist. As one alternative, the parties can usually contract to submit disputes and claims to arbitration for resolution. Today, arbitration is the rule rather than the exception for construction industry contract disputes. Arbitration has come to be viewed by many as preferable to litigation for construction-related disputes for a number of reasons.

Arbitration is often thought to yield a quicker and less expensive resolution of disputes. While this can be true, it is not always so. Because arbitrators and tribunal administrators often do not exercise the same power as a judge regarding scheduling, time restraints, evidentiary limitations, and other matters, arbitration often requires the parties to exercise more self-control and mutual restraint.

Arbitration normally provides a more technically knowledgeable decisionmaker than would be the case with a judge and jury. A popular arbitration panel composition is one attorney, one contractor, and one architect/engineer. Such a panel provides expertise in each of these three disciplines and is equipped to grasp much more readily than could a judge or jury the often technical and complex facts arising in a construction dispute.

Normally, arbitration provides a much more private forum for the resolution of disputes than does a court of law. This may be an important consideration for governmental bodies and many contractors who do not desire to air their grievances and problems publicly.

Arbitrations are usually much more informal proceedings than a courtroom trial. This can make the contractor more comfortable and contribute to a full and relaxed presentation of the facts. One way in which this informality manifests itself is in the arbitrator's ability to ask questions of the witnesses, a right a juror does not have. This provides a means for ensuring that the decisionmakers understand the presentation and that the parties address the points the decisionmakers want addressed.

Arbitration provides more flexibility than litigation with regard to scheduling the time and location of the hearings. Generally, in arbitration, the parties are free to agree to the most mutually convenient location and time for hearings, and hearings can be conducted other than during normal court hours and days, if desired.

The Greek philosopher Aristotle once said: "The arbitrator looks to what is equitable, the judge to what is law; and it was for this purpose that arbitration was introduced, namely, that equity might avail.''[26] While it is doubtful that Aristotle foresaw modern-day construction arbitration, this statement is still applicable. If a party has a very strong case equitably, but is impaired by legal technicalities, it is more likely to prevail in an arbitration proceeding than in a court of law. The converse is, of course, also true.

Although the parties are free to agree otherwise either in their arbitration agreement or subsequently, generally arbitration does not afford the parties the same broad discovery rights that they would have in proceedings before a court of law. This can reduce costs, but can also lead to ''trial by ambush,'' such as existed in the court system before modern-day discovery practice. Whether this lack of discovery is a net benefit or a net detriment depends upon the circumstances of the particular case.

One limitation of arbitration arises in the situation where multiple parties are involved in a given dispute. Because of the contractual prerequisite to arbitration, arbitrators normally do not have the power to join in one arbitration all the parties that are involved. For example, the American Institute of Architects Form Contract General Conditions A201 precludes the contractor from joining the architect as a party to an arbitration between the contractor and the owner.[27] The contractor also may be unable to include in one proceeding subcontractors, suppliers, sureties, lenders, and other parties with potential liability in the matter, as could be done in a court of law.

While a procedure does exist for petitioning a court of law to set aside an arbitration award, a party unhappy with an arbitration decision is much less likely to have that decision overturned than it would to have a court decision reversed. To vacate an arbitration award, the complaining party usually must be able to demonstrate that there was fraud on the part of the arbitrators, that the arbitrators failed to follow proper procedures, or that the decision was beyond the scope of the arbitrators' power. A disappointed party may not complain that the arbitrators failed to understand and apply the correct legal principles. This increases the risk of arbitration, but decreases the time and cost associated with achieving a final resolution of the dispute.

§ 1.33 —Other Alternatives

Court congestion, the high costs of litigation, and a number of other factors have caused people in recent years to look for alternative means of dispute resolution in addition to arbitration. These methods include nonbinding mediation, binding mediation, and mini-trials. Some companies have even implemented a system whereby they try their case before a jury of their own employees prior to an actual trial, as a means of evaluating their case and the desirablility of settlement. One or more of these alternative dispute resolution procedures may be appropriate, and more desirable than litigation or arbitration, for a given set of circumstances.

---

[26] Aristotle, Rhetoric, Book 1, ch. 13.

[27] AIA Document A201, General Condition 7.9.1 (1976 ed.).

LITIGATION MANAGEMENT

§ 1.34 Litigation Objectives

Litigation of a construction case is perhaps the most challenging of all forms of litigation, because of the complexity of most construction disputes. Unlike personal injury cases, which may involve a traffic accident that occurred at one specific time at one location, construction disputes often involve a mass of details and a number of "wrongs" that did not occur at any one discernable time. In addition, any one given act or omission is likely to have a ripple effect across time, upon many different activities and events. The large quantity of parties, facts, documents, and time involved make construction cases perhaps most analogous to complex antitrust cases, but nevertheless unique.

Trial lawyers are often taught to begin preparation for trial by writing their closing argument, and then working backwards from there. This may seem strange until one remembers that an efficient traveller does not embark upon a journey until he or she knows where he or she wants the journey to end. This same advice is also useful throughout the construction process. Decisions such as what to include in the contract documents, what types of job records and documentation should be maintained, and what experts and consultants should be engaged should all be governed, in part, by what the contractor will need in the event disputes arise. A solid strategy is essential both to construction administration and to litigation, and good organization is the key. Both in construction administration and in litigation management, the contractor should define its objectives and then identify the means of achieving those objectives.

§ 1.35 Selection of Parties, Remedies, and Forum

Subsequent chapters of this book discuss the various possible avenues of recovery a contractor can pursue when it finds itself in a claim situation. The choice of which avenue to pursue should receive careful consideration with the assistance of expert advice. A contractor should be discriminating in its selection of allies and adversaries, remedies to be pursued, and the forum in which to pursue them. Factors such as the role played by each of the various parties, the terms of their contracts, their financial solvency, their reputation, the presence or absence of insurance or bonding, the personalities involved, the procedural, evidentiary, and discovery implications, the governing legal principle, and other dispute dynamics should all be considered before the contractor commits itself by forming alliances and declaring war. A defense unique to a particular owner, such as sovereign immunity, may make it desirable to institute an action directly against the architect or construction manager. Under other circumstances, it may be more profitable to sue the owner and seek the assistance of the construction manager or architect. As mentioned in **§ 1.12**, litigation with a subcontractor can jeopardize the general contractor's position with the owner, but may be appropriate under a given set of circumstances.

In addition to making decisions concerning the parties, and as a part of that process, the contractor must also choose the remedies and legal theories to be pursued and the forum in which to pursue them. Harsh contract terms limiting or barring recovery, or failure to comply with notice requirements, may make it desirable to seek rescission of the contract and recovery under the theory of quantum meruit—unjust enrichment for the reasonable value of the work performed. The facts of a particular case may make it desirable to seek injunctive relief. The circumstances may cause the choice of parties to control the choice of remedy and forum, or a strong preference for a particular forum may control the other decisions.

§ 1.36 Effective Litigation Strategy

When a contractor finds itself in litigation, it should apply the same sound management techniques and standards that it applies in making other decisions regarding investments of the company's time and money. No step or action should be taken in litigation merely because it seems the next logical step. The contractor and its attorney must avoid tunnel vision, and at each step evaluate all options, weighing the costs and benefits of each.

The need for sound managerial decisionmaking begins at the initial pleading stage. Pleadings can usually be as vague and general or as specific as the pleader likes. Different circumstances call for different actions. A well-prepared, specific complaint, accompanied by a request for production of relevant documents and detailed interrogatories can have the same effect on a defendant as a well-prepared claim brief; it shows that the plaintiff is serious and prepared to aggressively pursue resolution of the dispute. Effective use of various pretrial motions and procedures can also reduce the cost of litigation and enhance the chances of success.

Pretrial discovery procedures can be immensely beneficial or wastefully expensive, depending upon whether they are effectively employed. The most common discovery techniques used in construction litigation are requests for production of documents, interrogatories, and depositions. A lesser-used tool, but one that can be extremely useful, is the request for admissions of fact. The extremely large number of documents, facts, people, and locations involved in any given construction dispute makes construction discovery very expensive, but also very important. Effective discovery increases the possibility of an acceptable settlement while at the same time preparing the contractor for a persuasive and successful trial presentation.

In addition to the construction attorney, it is usually necessary to engage other experts to assist in the preparation and presentation of a construction claim. The selection, management, preparation, and presentation of expert opinion testimony and associated demonstrative evidence is perhaps the most important aspect of effective construction litigation. The quality most needed in an expert witness is the ability to persuasively present the facts and opinions supporting the contractor's case. To be persuasive, the expert witness must be able to gather and analyze vast quantities of details and present them in a clear, simple, and understandable fashion. For most cases, this requires technical expertise, analytical ability, communicative skills, and a mastery of demonstrative evidence, visual aids, and other means of persuasive presentation. An expert can be the most knowledgeable person in the world on a given subject, but if he or she cannot persuade the factfinder of the correctness of the position, that expert is of little value as a witness. Such an expert may be very useful in a consulting or advisory role during construction and litigation, but the role of testifying should by assigned to experts with strong communication skills.

§ 1.37 Effective Trial Presentation

There is an adage that "knowing the truth and proving it are two different things." The construction trial lawyer must be able to marshal a multitude of facts and present them in a simple, coherent fashion that will persuade the factfinder that its client is entitled to the relief sought. The most effective evidence that can be introduced at trial is the live testimony of a good witness. Nonlawyers often speak of whether there is "proof" or "evidence" of a particular fact, referring to documentation or other tangible evidence. The observations and recollections of a human witness, stated in court, are themselves legal evidence and proof of the existence of facts. Yet, in construction cases, probably more so than almost any other type of litigation, job documentation, prepared at the time of the facts and events it records, is almost essential to a persuasive case presentation.

This chapter, in § **1.36**, touched upon the use of demonstrative evidence. This is a very effective means of simplifying and assisting the factfinder in comprehending complex construction issues. Demonstrative evidence can take the form of graphs, charts, models, diagrams, photographs, slides, movies, material samples, or virtually anything that helps the factfinders understand the facts. The use of demonstrative evidence in personal injury cases has gained much fame or notoriety (depending on one's perspective) in recent years. Such evidence can, if properly prepared and presented, be even more useful in complex construction litigation.

Finally, effective litigation preparation and presentation includes calculation and proof of damages. Proof of damages is the area where many construction cases are lost; pricing the loss and presenting it in a persuasive fashion are essential elements of the construction case. A contractor can present a compelling case for entitlement, but if it cannot quantify and prove its damages, it has accomplished nothing more than a moral victory. Because of the complex nature of construction costs, proof of construction damages normally requires evaluation of extensive cost documentation maintained during the project.

To recover its costs, the contractor must establish (1) that it actually incurred the costs, (2) that the costs bear a legally sufficient cause/effect relationship to the act or omission of which it is complaining, and (3) that it has acted reasonably to mitigate the damages. The required relationship between the cause and the cost varies depending on the legal theory under which the contractor is seeking to recover. In an action in contract, the damages must flow from the act or omission and be of such a nature that it was reasonably foreseeable at the time of entering the contract that such damages would result from any such breach. When recovery is sought under negligence or a tort theory, the test applied is whether the act or omission was the *proximate cause* of the damages. Obviously, the measure of damages to be applied, and the contractor's ability to prove the requisite facts, are additional factors to be taken into consideration at the outset in determining whom to sue and on what theory.

§ 1.38 Appeal

Without belaboring the point, most construction cases are decided on the facts. Trial of a construction case does not normally involve complex, esoteric legal issues, but rather an organized, persuasive presentation of complex facts to fit a limited number of fairly basic legal principles. Because of this, construction cases are rarely reversed on appeal. Appellate courts generally accept the trial court's findings of fact and only explore whether there has been an error of law. The typical length and complexity of construction cases also make appeals less fruitful than in other areas of the law. It is not unusual, however, for the losing party to appeal a construction decision because of the enormous amount of money that may be at stake. The main effect of such appeals is usually to delay the ultimate resolution of the dispute and payment of the money found due. It is sometimes in the contractor's interest to compromise a favorable judgment to avoid an appeal and the time associated with it; however, when available, postjudgment interest largely diminishes the leverage of the losing party in this regard.

§ 1.39 Conclusion

Construction litigation is a specialized, complex, and sometimes expensive undertaking. This chapter demonstrates that, while it is usually in the contractor's best interest to avoid litigation, there are times when the costs associated with avoiding litigation are greater than the costs of pursuing it. Skillful and conscientious management, prior to and during construction, as well as during litigation, are necessary to minimize the costs of litigation and optimize its benefits. The following chapters of this book address various avenues the contractor may pursue to enforce legal rights and obligations. This chapter has sought to stress the need to carefully consider which of those avenues should be followed and to prepare oneself for the journey.

CHAPTER 2

BID PROTESTS AND BID MISTAKES

David Buoncristiani and Donald R. Fitzgerald

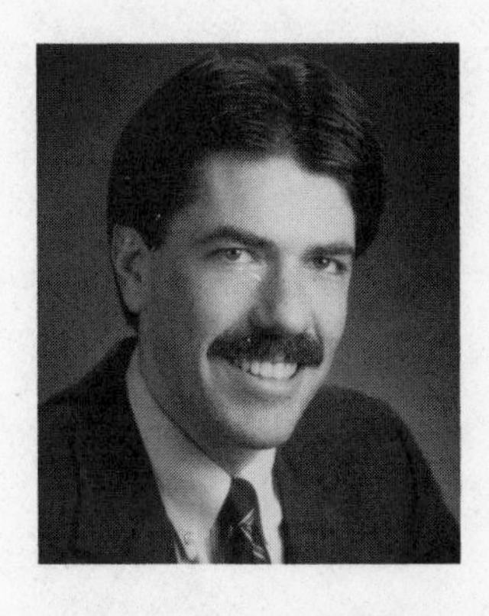

David Buoncristiani is a partner in the San Francisco office of Thelen, Marrin, Johnson & Bridges. He is a 1971 graduate of Hastings College of the Law and has been primarily involved in construction industry related problems, including all facets of the competitive bidding process. He has handled numerous bid protest proceedings, both on the federal and state level. He is a member of the ABA sections on Public Contracts and Litigation and the California AGC Legal Advisory Committee.

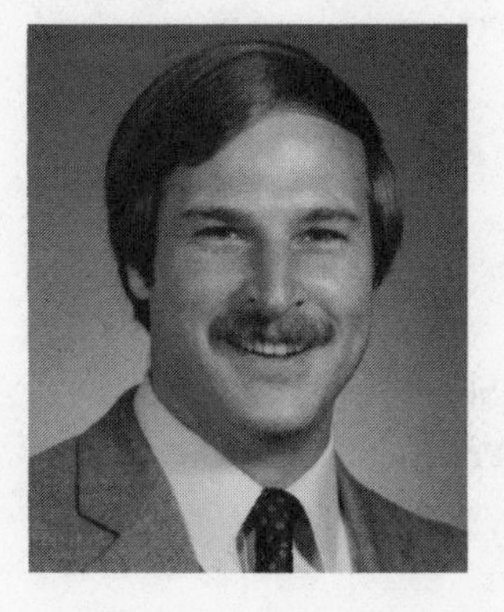

Donald R. Fitzgerald is an associate at the law firm of Thelen, Marrin, Johnson & Bridges in San Francisco. A 1974 graduate of the University of Oklahoma, he first wrote grant proposals with the university's Bureau of Water and Environmental Resources Research while in graduate school. After earning a Master's Degree in 1977, he worked in state government for four years, where he managed environmental program grants for state agencies. He earned his J.D. from Duke University in 1984.

§ 2.1 Introduction

With the present level of competition in the construction industry, all contractors should be aware of the procedures available to challenge a proposed award to another bidder or to defend against such a challenge if one is the apparent low bidder. The purpose of this chapter is to outline generally the procedures available and the rules which apply to the award of competitively bid construction work for federal, state, and local contracts. Because there are also instances where the apparent low bidder may itself desire to be relieved of the obligations of performing the work in accordance with its bid, this chapter also briefly discusses the criteria for a material mistake in a bid and the relief which is then available.

BACKGROUND

§ 2.2 Overview

In the United States, most contracts for the construction of public facilities are awarded through competitive bidding. Federal, state, and local agencies typically require competitive bidding for public works projects, because these public owners desire an objective and impartial method of selecting a contractor to construct the project at the lowest possible price. The statutes and regulations dealing with competitive bidding on public construction contracts are intended to implement this goal.

To insure that the awards will be based on objective criteria, the laws governing public contracting typically provide that *interested persons* who believe a government agency has acted to frustrate the bidding process may obtain a determination of the propriety of the award by means of a *bid protest*. Although competitive bidding is common throughout the United States, these protest procedures are established by different government agencies and vary considerably. This chapter presents an overview of the competitive bidding process in the United States and describes the central concepts and rules applicable to bid protests. Because California has codified the rules for many of the subjects encountered in protests, some of its procedures have been used as generally representative examples.

One preliminary word of caution. Of necessity this discussion cannot be an exhaustive treatise on all legal principles applicable to the award of competitively bid construction contracts. Specific protests are dependent on the terms of the contract documents, the controlling statutes and case law, and the nature of the awarding authority. Our intent is to provide the reader with a general familiarity with the applicable legal concepts. Such an overview is needed because, except for direct federal government contracts regulated by the General Accounting Office (GAO), there is no clearly defined set of procedures or rules common to the various levels of government. As a result, the procedures for pursuing bid protests depend to a great extent on the rules enforced by the particular awarding agency and, in many cases, the source of the funding. Consequently, the analysis is divided into four sections, addressing the procedures and rules employed by various types of public awarding authorities, and briefly addressing the rules applicable to private owners. **Sections 2.3–2.11** provide the necessary background for this analysis with a general discussion of the fundamental concepts of competitive bidding.

§ 2.3 The Competitive Bidding Process

Despite the variety of bid protest procedures, the general competitive bidding process does not vary greatly. The public owner must first advertise the work to be undertaken. This solicitation, often referred to as a *request for proposal* or an *invitation for bids* (IFB), describes the work to be accomplished and establishes a specific date and time by which all the bids are to be received and subsequently opened at a public opening. The bids are opened and reviewed, and the contract is then awarded to the *lowest responsive, responsible bidder*. Every public agency is subject to rules and regulations governing each of these activities for its major contracts.

The bidder who is awarded the contract must have submitted the lowest dollar bid, have been determined to be responsive to the IFB, and be a responsible bidder. As discussed in more detail in **§ 2.7**, the term *responsive* refers to the bidder's conformance with the material elements of the solicitation and is determined at the time of bid opening. The requirement for a *responsible* bidder addresses the trustworthiness, quality, fitness, and capacity of the bidder to perform the work and is often determined after bid opening.

Once the awarding agency announces an intent to award the contract to the bidder determined to be the lowest responsive, responsible bidder, bid protests commonly arise in two situations. The first involves the bidder who submits a lower total price for completion of the contract work (apparent low bidder) but who is determined by the owner to be nonresponsive because of a failure to comply with a material portion of the solicitation. The apparent low bidder may protest any award to a higher bidder on the ground that any deviation from the solicitation was a minor irregularity that did not disqualify the bid nor justify awarding the contract to the next highest bidder.

The more common situation arises when a higher bidder seeks to disqualify one or more bidders whose proposals are lower than the protestor's on the grounds that the lower bidders are not responsive or not responsible. In this situation, the protestor requests the owner to award the contract to the protestor instead of to the bidders whose bids are monetarily lower.

§ 2.4 General Considerations for All Bid Protests

Before submitting a bid protest to the awarding authority, a potential protestor should consider the following factors. The first and most obvious consideration is that the protesting party must have sufficient legal grounds for challenging the award. This requires the protestor to analyze the invitation for bids (IFB), the applicable statutory rules, and substantive case law in the jurisdiction and to satisfy itself that there are material violations of the competitive bidding process. All protests result in some disruption of the contracting process, and frivolous protests without substantive grounds may result in the protestor being penalized.[1]

[1] *See* 40 C.F.R. § 35.939(a) (1982).

Second, the would-be protestor must determine whether it has standing to challenge the action of the awarding agency; that is, whether it is a proper party to pursue a protest.

Third (and contemporaneously with the first two determinations), the protestor must consider the appropriate procedure for filing the bid protest. In all jurisdictions, immediate written notice is critical to preserving the right to protest the action of the awarding agency.[2] In many cases, the statutory deadlines provided for bid protests are extremely short and, if not met, preclude even a meritorious protest.

Once these determinations are made, it is essential that the protestor examine the specific relief available if the bid protest is successful. In some instances, even though the protestor's position is legally sound, the expenditure of considerable time and effort may not be justified if the potential result is limited to a reprimand of the awarding agency by the responsible authority. For example, on the federal level, the GAO is reluctant to overturn an award already made despite the ability of the protester to establish an impropriety.

§ 2.5 General Requirements of Bids

The competitive bidding statutes are designed to allow maximum competition for the award of public works contracts and to prevent collusion and fraud. The competitive bidding process achieves these goals by allowing all those parties interested in performing a given construction project to evaluate the requirements of the project and to submit the lowest price for which they are willing to undertake the described work. From these bids, the government agency or owner is, theoretically, able to select the proper low bid on any given project.

In selecting the lowest bid, the owner does not look solely at the dollar amount of the bid. Both state and federal regulations require that the awarding agency evaluate the bids within certain designated parameters to determine the *low bidder*. The process of employing these parameters to determine which bidder receives the contract requires that the bidders and the owner fulfill their respective obligations. These obligations are established by the rules and regulations applicable in each jurisdiction, but certain rules regarding the respective obligations of the owner and bidder are common to all jurisdictions.

The owner must prepare a complete and accurate set of contract documents fully descriptive of the work to be performed and must make them available to all parties responding to the invitation for bids (IFB). The clarity and accuracy of the contract documents are of utmost importance. If material portions of the documents are vague or susceptible of more than one rational interpretation, the owner will receive bids that are not comparable, thereby frustrating the purpose of competitive bidding and forcing rejection of all bids. The owner must also disclose the

---

[2] *See* 4 C.F.R. § 21.2(a) (1984), limiting bid protests filed with the GAO after bid opening to within 10 days of actual or constructive knowledge of the basis of the protest.

specific procedure it intends to use in evaluating the submitted bids. Any such procedure must be fair and objective.

The contractors are also subject to general procedural requirements which they must follow in order to have their bid considered. Some of the requirements concern the form of the bid and are seemingly routine, such as requiring the contractor to sign all the necessary forms. However, such requirements are strictly enforced. Although the omission of a signature may be merely an oversight and the bid might otherwise be acceptable, failure to sign the bid documents where required generally renders the entire bid void.

The contractor's bid must further address all material aspects of the owner's solicitation, including all addenda issued prior to bid. The bidders may not make any material omissions or include any qualifications except where specifically requested to do so. The bidder is also required to state clearly the dollar value of the bid and to indicate correctly all pricing, whether it be lump-sum, unit price, or some other specific pricing mechanism prescribed by the bid documents.

The contractor should also consider any specialized statutes or provisions in the solicitation which could affect the responsiveness of the bid or the owner's determination of responsibility. These provisions include, but are not limited to:

1. Specialized licensing requirements
2. Subcontractor listing requirements
3. Subcontractor or supplier qualification requirements (i.e., five years minimum experience)
4. Minority business enterprise goals or commitments.

Once the owner has prepared an IFB that is sufficient, and the contractors have submitted completed bids, the owner opens and examines each of the bids. The contract must then be awarded to the lowest responsive, responsible bidder.

§ 2.6 Contract Award Criteria

Determination of the lowest bid is normally an arithmetical computation based on the indicated pricing. The more complicated concepts are the responsiveness of the bid and the responsibility of the bidder. Responsiveness is determined at bid opening; a nonresponsive bid cannot legally be accepted by the owner. However, a bidder can establish responsibility after bid opening and be awarded the contract, even though questions of responsibility existed at bid opening.

§ 2.7 —Responsiveness

The goal of the competitive bidding process is to obtain from each bidder a complete, definitive, unqualified offer which the awarding authority need only accept to create a binding contract to perform all work described in the solicitation. For a bid to be *responsive*, it must conform to the material terms of the solicitation at the time of the bid opening and permit the owner simply to accept the offer. Thus, the contractor's bid may not contain omissions or ambiguities regarding price, quality, quantity, or time of performance, nor will the contractor be permitted to modify its bid to cure such deficiencies after bid opening. Omissions or ambiguities of this type afford the contractor a competitive advantage over other bidders because the owner cannot simply accept the bid and create a binding contract for *all* material elements of the bid. If an owner is required to deviate from the terms of the solicitation, the contractor has the option of accepting or refusing an award, an advantage not shared by other bidders whose offers are unqualified. All rules and regulations on competitive bidding require that nonresponsive bids be disregarded.

However, minor irregularities in a bid may not render it nonresponsive. A *minor irregularity* is a deviation from the solicitation requirements that can be waived by the owner because it does not affect price, quality, quantity, or time, and does not afford the bidder a competitive advantage. The basic test of whether a deviation is a minor irregularity or a fatal defect is whether the bidder is required to perform all of the work described in the solicitation or whether it could escape performance because the bid omits a material element, and thus does not constitute an unqualified offer. For example, a bidder's failure to sign the bid or commit to a precise dollar value for the work could prevent the owner from compelling the bidder's performance on the terms mandated by the invitation for bids (IFB).

However, failure to include in a bid certain information requested by the bid documents, such as the identity of equipment suppliers, will normally be considered either a minor irregularity or a condition of responsibility which can be cured after the bid opening, and will not disqualify the bid unless the IFB clearly makes such information a condition of responsiveness. This is because the owner can accept the bid even without such information, and require the contractor to meet all material conditions of the bid; the bidder obtains no competitive advantage over others who included such information.

§ 2.8 —Responsibility

The requirement that the award must be made to the lowest *responsible* bidder mandates that the contractor whose bid is accepted be capable of performing the work. Commonly, the test of contractor responsibility is measured by such factors as the trustworthiness, quality, fitness, and capacity of the contracting entity.[3] This evaluation is often undertaken after the opening of the bids and the contractor may furnish post-bid-opening information to facilitate a favorable determination.

[3] City of Inglewood-Los Angeles County Civic Center Auth. v. Superior Court, 7 Cal. 3d 861, 500 P.2d 601, 103 Cal. Rptr. 689 (1972).

As the foregoing demonstrates, it is often difficult to determine whether the failure to include requested information renders a bid nonresponsive, and therefore void, or simply constitutes a minor irregularity or condition of responsibility, which can be waived by the awarding authority or addressed after bid opening. For this reason, the terms of the invitation for bids (IFB) must be carefully examined in any protests. The language of an IFB may make certain information required for the bid a condition of responsiveness, even though, without such language, the failure to supply such information would merely be considered an irregularity which would not prevent award of the contract. Subcontractor or supplier listing laws and minority business enterprise requirements provide good examples of IFB conditions that can be either conditions of responsiveness or conditions of responsibility.

§ 2.9 —Subcontractor Listing Requirements

Subcontractor listing laws are codified in several states for the purpose of precluding bid shopping.[4] *Bid shopping* occurs when the contractor, who based its bid on quotations from certain subcontractors or suppliers and who has been awarded the contract, is allowed to disregard those subcontractors and suppliers and obtain lower pricing from different subcontractors. The practice is perceived to be detrimental to the public interest because, although the contractor obtains the benefit of a price reduction, the public obtains no similar benefit and potentially receives a lower quality product. Invitations for bids (IFBs) sometimes require contracts to identify material suppliers for the same reason.

Some states seek to prevent bid shopping by statute. For example, California has enacted legislation to preclude bid shopping which requires that subcontractors whose bids are in excess of one-half of one percent of the total bid price be identified by each general contractor in its bid.[5] The applicable statutes further prescribe an exclusive method for substituting subcontractors which, if not followed, authorizes the owner to assess certain penalties and cancel the contract, and entitles the subcontractor to legal redress against the offending contractor. Failure to comply with such listing requirements can provide a basis for a bid protest either on the grounds of nonresponsiveness or on the basis that the bidder is not responsible.

Even in the absence of anti-bid-shopping statutes, IFBs frequently include mandatory language such as "the contractor 'shall' [or 'must'] comply with prescribed subcontractor or supplier listing requirements." When an IFB contains this language, the bidder must determine whether the failure to list subcontractors will render the bid nonresponsive and lead to rejection of the bid. Although the use of mandatory language seems to indicate that a failure to list will void the bid, such is not always the case.

---

[4] *See* Cal. Gov't Code § 4100 *et seq.*

[5] *Id.* § 4104.

In fact, the effect of an IFB's inclusion of specific listing requirements has been the source of a substantial number of bid protests on Environmental Protection Agency-funded projects. As discussed further in **§§ 2.17–2.21**, the EPA has developed a sophisticated set of protest procedures and has a reported body of bid protest precedents applicable to EPA-funded projects.

In the course of administering its projects, the EPA found that many subcontractor or supplier listing requirements were being improperly used by the recipients of the funds (grantees) to select or reject bidders. The EPA determined that, in cases where the use of terms such as "must" and "shall" in the language of the IFB made it appear that such listing requirements were mandatory, grantees did not uniformly enforce such requirements. If the grantee was satisfied with a low bidder who failed to comply with the listing requirements, the grantee would simply decide that such requirements were a condition of responsibility that did not render the bid void. In other cases, where the grantee favored the second low bidder, and the low bidder failed to comply with the listing requirements, the grantee would interpret the listing requirements as conditions of responsiveness and invalidate the low bid.

Since rejection of the low bidder necessarily meant that the project would require higher funding levels, the EPA adopted a regulation which provided that a grantee could only disqualify an otherwise low bid because of a failure to comply with a subcontractor or supplier listing law if the IFB was clear and unequivocal on two points: first, that the subcontractor or supplier listing requirement was a condition of responsiveness, and second, that a failure to comply with such a condition would render the bid nonresponsive and result in rejection.[6] After the adoption of this rule, the EPA promulgated regulations simplifying the evaluation of the procurement processes of the grantee agencies.[7] While these regulations did not include the rule requiring clear and unequivocal notice of the condition of responsiveness, EPA administrative decisions have consistently held that the standards reflected in this requirement are matters of EPA policy which apply to EPA-funded projects regardless of whether the protest arises under 40 C.F.R. § 33.00 *et seq.* or 40 C.F.R. § 35.00 *et seq.*[8]

There are a number of cases where, although the IFB utilized terms such as "shall" and "must," the EPA determined that the grantee had failed to comply with its regulation and refused to disqualify a low bidder who had not fully complied with the terms of the listing requirements.[9] The EPA administrative determinations on these and other issues provide an excellent source of information regarding the proper considerations in evaluating bid protests.

---

[6] 40 C.F.R. § 35.938-4(h)(6).

[7] *Id.* § 33.00 *et seq.*

[8] City of Los Angeles, Project No. C-06-1205-110 (Protest of Advanco Contractors, Inc.) (EPA Region IX, June 6, 1983).

[9] *See* City of Statesville, N.C., Grant No. C370395-02 (Protest of DPS Contractors, Inc.) (EPA Region IV, Nov. 17, 1982).

§ 2.10 —Minority Business Enterprise Requirements

A second area which has generated a significant number of protests on the federal, federally-funded, and state and local level has been the determination of whether a bidder has fully complied with the minority business enterprise (MBE) requirements of an invitation for bids (IFB). More specifically, since many federal and state contracts have MBE requirements, it is incumbent upon the bidder to determine whether the information submitted at the time of bid regarding such programs are conditions of responsiveness or of responsibility. Again, the issue hinges largely on the language of the IFB. The Environmental Protection Agency (EPA) follows essentially the same standard for determining whether the MBE requirements of an IFB have been made a matter of responsiveness by the grantee agency as it does for listing requirements. In order for a grantee agency to make the MBE requirements a matter of responsiveness, the IFB must clearly establish that compliance with the MBE requirements to be included with the bid documents is a matter of responsiveness.[10] When bidding on an EPA-funded grant project, it is important to pay particular attention to the terms in the solicitation describing the MBE requirements and to determine whether the MBE requirements of the bid are a matter of responsiveness.

Suffice it to say that there are both administrative determinations and federal case law which establish that the MBE requirements of a particular IFB can be either conditions of responsiveness, which render a nonconforming bid invalid, or conditions of responsibility, which permit a bidder to submit postopening information to substantiate compliance.

§ 2.11 General Grounds for Protest

Although an owner retains the right to reject all bids, this right is infrequently exercised unless the bid protests are so numerous or the solicitation is so confusing that the only reasonable option is to rebid the project. However, it can also be invoked for other legitimate reasons, such as where even the lowest bid exceeds the amount of funding awarded the project.

If the owner does not reject all bids, but proposes an award of the contract to the bidder determined by the owner to be the lowest responsive, responsible bidder, a proper party may protest the award of the contract. **Sections 2.12–2.27** examine the general rules for bid protests in four contexts: direct federal contracts, federally funded grant projects, state and local contracts, and private contracts. Regardless of the category of owner or the source of funding, bid protests are normally based on one or more of the following grounds:

[10] *See* Albert Lea, (Protest of Orvedahl Construction, Inc.) (EPA Region V, Sept. 18, 1980); City of Fort Wayne, (Protest of Davis Excavating, Inc.) (EPA Region V, Mar. 24, 1981).

1. Improprieties in the solicitation procedure, including the failure to properly advertise the solicitation or providing bid information to one bidder but not others[11]
2. Defects in the specifications
3. Material deviations from the invitation for bids in the bid of the contractor to whom the award is proposed to be made
4. Deficiencies in the bid evaluation process.

The following discussion highlights the proper grounds for protest, the procedure to follow, and the available relief in the four identified contexts.

FEDERAL CONTRACTS

§ 2.12 General

This section addresses bid protests on direct federal contracts. The rules governing bid protests of federal contract awards are well defined and differ from those applicable to federally funded grant projects. Protests of federally funded grant projects awarded through state and local government agencies are addressed in **§§ 2.22–2.26**.

§ 2.13 Grounds for Protests

If the construction project is funded directly by an appropriation from the federal government, protest procedures are the responsibility of the General Accounting Office (GAO) and are contained in Title 4 of the Code of Federal Regulations.[12] Virtually any deficiency in the specifications or proposed award is a proper subject for a bid protest. Where the protest is directed to the proposed award, it can encompass any material failure to comply with the provisions of the solicitation. In general, this type of protest will necessarily be determined on a case-by-case basis only after review of the bids of other contractors following bid opening.

Deficiencies in the evaluation process may also be sufficient grounds for sustaining a bid protest. For example, the improper determination of responsiveness or responsibility by the federal government has been held to be sufficient grounds

---

[11] 48 C.F.R. § 14.2 *et seq.* (1983); Comp. Gen. Dec. B-167382 (Oct. 16, 1969).

[12] 4 C.F.R. § 21 *et seq.* (1984).

for maintaining a successful bid protest.[13] In addition, if a federal agency improperly rejects all bids, interested parties may successfully protest.[14]

Despite numerous attempts in Congress to include listing requirements on federal projects, none have passed. Consequently, on the federal level at least, subcontractor listing is not a protest issue. At one time, the General Services Administration required listing of subcontractors on contracts which met certain minimum requirements. These requirements were eliminated in 1984 and replaced with a rule providing that all GSA solicitations are to contain language instructing bidders that they must only subcontract with firms *not* listed on the Consolidated List of Debarred, Suspended, and Ineligible Contractors without specific authorization.[15]

§ 2.14 Standing to Protest

Under the federal regulations, any "interested party" may protest the award or proposed award of a federal contract. An *interested party* is generally someone with an economic interest at stake in the outcome of the competitive bidding process.[16] Normally, an interested party is a contractor, a supplier, a contractors' association, or a labor union. At least one decision has held that the legal concept of an interested party does not extend to a bidder who is not the next-lowest bidder because it is not eligible for the award even if the bid protest is sustained.[17]

§ 2.15 Protest Procedure

It is proper to protest either to the contracting officer or directly to the General Accounting Office (GAO). Although the GAO encourages initial review by the awarding agency, it may be strategically wise to submit a bid protest directly to the GAO because an awarding agency may be particularly defensive of its award decision.

Protests to the GAO must be *received* within 10 days of the denial of a protest submitted to the contracting officer or the date on which the protestors knew or should have known of the basis for the protest. This deadline is crucial for obtaining review of the proposed award by the GAO, although it may consider a late

---

[13] Brown & Son Elec. Co. v. United States, 173 Ct. Cl. 465 (1963).

[14] *See* F.A.R. § 14.404-1(c) (Sept. 19, 1983). 48 C.F.R. § 14.404-1(c) (Oct. 1, 1984). The Federal Acquisition Regulations contain the policy and procedure guidance for acquisitions by federal agencies, and are codified at 48 C.F.R. pt. 1 *et seq.*

[15] 41 C.F.R. §§ 5-2.202-51, 5-2.202-52; 49 Fed. Reg. 5755 (Feb. 15, 1984).

[16] Association of Data Processing Serv. Orgs. v. Camp, 397 U.S. 150 (1970); 4 C.F.R. § 20.1(a).

[17] Comp. Gen. Dec. B-210851 (Apr. 16, 1983).

protest for good cause or where the protest raises a significant issue.[18] A protest directed to the contracting officer should be made pursuant to the awarding agency's specific regulations governing bid protests.

The protest should be in writing and should set forth all relevant facts supporting the protest. No specific form is required, but the initial protest must include the name and address of the protestor, contract number, a statement of the grounds, a specific request for ruling, and in general be concise, logical, and direct.[19]

The federal protest procedures were modified in 1984 pursuant to the Competition in Contracting Act of 1984.[20] The Act required the GAO to promulgate new rules for bid protests. The new rules are published at 4 C.F.R. part 21, and are designed to provide the framework for the consideration of bid protests by the General Accounting Office.[21]

The present rules require an awarding agency to file a complete report on the protest with the GAO within 25 days of the date of notice of the protest.[22] The protestor is required to respond to the agency's report within seven days after receipt. This response is mandatory: failure of a protestor to file comments, a statement requesting that the decision be based on the existing record, or a request for an extension will result in dismissal of the protest. This seven-day limit may be extended if the circumstances warrant but the rules provide that extensions are to be considered ''exceptional'' and are to be granted only sparingly.[23]

Several new rules have also been promulgated with respect to withholding an award or suspension of contract performance pending a resolution of a protest filed with the GAO. Section 3553(c) and (d) of the Act provide that, in those instances when the GAO receives a protest, it is to provide notice of the protest to the contracting agency. The contracting agency may not award a contract in the protested matter during the time the protest is pending unless the awarding agency determines that there are ''urgent and compelling circumstances significantly affecting interests of the United States [that] will not permit waiting for the General Accounting Office decision.''[24] A new section has been added to the regulations providing that the GAO will not consider subcontractor protests unless the subcontract is ''by or for the government.''[25]

Parties to a bid protest often desire a hearing or conference for the purpose of confronting other parties to the protest. The GAO has the discretion to grant such a request by a protesting or interested party.[26] In addition, the GAO itself may

---

[18] 4 C.F.R. § 21.2(c).

[19] *Id.* § 21.1(c).

[20] 31 U.S.C. §§ 3551-3556 (1984).

[21] 4 C.F.R. § 21 *et seq.* (1984).

[22] *Id.* § 21.3(c).

[23] *Id.* § 21.3(e).

[24] *Id.* § 21.4(a).

[25] *Id.* § 21.3(f)(10) .

[26] *Id.* § 21 5(a).

request that a conference be held at any time during the protest proceedings if it determines that such a conference is needed to clarify material issues.[27]

Normally, the GAO is required to issue a decision on the protest within 90 days from the date the protest is filed.[28] The new rules also provide for an *express option* where the time for the protest procedure is shortened. The express option may be requested by a protestor, the contracting agency, or an interested party, and must be filed in writing and received by the GAO no later than three days after the protest is filed. The GAO has the sole discretion to employ the express option in those cases suitable for resolution within 45 days. Use of the express option reduces the time for the contracting agency to file a report with the GAO to 10 days from the receipt of notice that the express option will be used, and limits the comment period on the agency report to five days after receipt of the report. Under this option, the GAO retains its discretion to arrange a conference on any material issue, but it is required to issue its decision on the bid protest within 45 days from the date the protest is filed.[29]

§ 2.16 Relief Available

The relief available to a protestor depends upon whether the award has been made. On occasion, all bids may be rejected. Rejection of all bids is not favored but may occur under certain circumstances. For example, the Federal Acquisition Regulations provide that all bids may be rejected and the invitation cancelled when the contracting officer determines that the solicitation is somehow defective.[30] Sufficient justification includes: inadequate or ambiguous specifications; the supplies or services procured are no longer needed; all acceptable bids received are at unreasonable prices; the bids received were not arrived at independently, were collusive, or were submitted in bad faith; or for other reasons where cancellation is clearly in the best interest of the government.

The General Accounting Office (GAO) may overrule a contracting officer's decision to reject all bids.[31] Normally, however, the GAO will not interfere where the contracting officer exercises that authority, absent a clear showing of abuse of discretion.[32]

If the protest is determined favorably to the protestor prior to award, the protestor may be awarded the contract. However, even if the contract was initially awarded to an improper party, the GAO retains the power to direct cancellation of the

---

[27] *Id.* § 21.5(d).

[28] *Id.* § 21.7(a).

[29] *See id.* § 21.8.

[30] F.A.R. § 14.404-1 (Sept. 19, 1983).

[31] 39 Comp. Gen. 396 (Nov. 25, 1959).

[32] Comp. Gen. Dec. B-159287 (Jul. 26, 1966).

contract. This remedy is rarely employed because the awardee who is the subject of the cancellation may seek breach of contract damages in the Claims Court.[33] Instead of cancellation, the GAO will often terminate the contract for convenience. The GAO's refusal to cancel the contract or terminate the award does not prevent the frustrated bidder from recovering bid preparation costs.[34] Bidders can recover bid preparation costs only if the contracting agency has unreasonably excluded the protestor from the procurement. The recovery of these costs is subject to the exception that, if the GAO recommends that the contract be awarded to the protestor and the protestor receives the award, costs are not recoverable.

Relief is also available through the courts. One method for protesting a contract award to another bidder is the writ of mandamus. Although not frequently employed in this context, it has been held that mandamus may issue to prevent an award if the plaintiff can establish that the proposed award is improper and will cause grave and irreparable injury.[35]

Injunctive relief may be obtained where the award by the contracting officer is found to be without any rational basis. However, the burden of the protesting party is substantial and, in the absence of a showing that the award is clearly arbitrary and capricious or in violation of the statute or regulation, injunctive relief is not available in the federal courts.[36]

The rules governing selection of the proper forum for bringing a suit for injunctive relief have recently changed. The Federal Court Improvement Act of 1982 (FCIA) created the United States Claims Court and vested it with the trial jurisdiction of the former Court of Claims.[37] The FCIA also expanded the jurisdiction of the new court over that of the former Court of Claims by giving it exclusive jurisdiction to grant declaratory judgments and equitable relief on contract claims.[38] The Claims Court's equitable jurisdiction was limited by the Act to claims brought *before* the contract is awarded.[39]

A party desiring to protest the solicitation procedures of a contracting agency and enjoin award of the contract must file a claim with the Claims Court prior to the contract award. The fact that the protesting party has filed a protest with the contracting agency prior to award does not satisfy the requirement for filing a claim with the *court* before an award is made.[40] Where the protesting party desires to bring a suit for equitable relief after the award of a contract, it must bring the

---

[33] Warren Bros. Roads Co. v. United States, 173 Ct. Cl. 714 (1965).

[34] *See* Keco Indus., Inc. v. United States, 428 F.2d 1233 (Ct. Cl. 1970).

[35] Noce v. Morgan, 106 F.2d 746 (8th Cir. 1939).

[36] Wheelabrator Corp. v. Chafee, 455 F.2d 1306 (D.C. Cir. 1971).

[37] 28 U.S.C. § 1491.

[38] *Id.* § 1491(a)(3). For an extensive discussion of the legislative History of the Act, *see* United States v. John C. Grimberg Co., 702 F.2d 1362, 1369-72 (Fed. Cir. 1983).

[39] 28 U.S.C. § 1491(a)(3).

[40] United States v. John C. Grimberg Co., 702 F.2d 1362 (Fed. Cir. 1983).

action in the federal district court. The FCIA does not disturb the district courts' jurisdiction to consider both pre-and-post-award government contract disputes.[41]

According to the provisions of the Contract Disputes Act of 1978, the former Court of Claims and the Board of Contract Appeals for each agency had concurrent jurisdiction to review appeals from decisions of the contracting officer.[42] The FCIA explicitly eliminates the equitable jurisdiction of the Boards of Contract Appeals over bid protests—a significant change in the prior law.

Under the present statutory scheme, a protestor desiring to obtain equitable or declaratory relief may bring an action prior to award of the contract in either the Claims Court or the federal district court. If the contract has been awarded, the protestor is limited to bringing the action in the federal district court. If the protestor's appeal is not equitable in nature, then it may be brought either to the district court, the Claims Court, or the Board of Contract Appeals.

FEDERALLY FUNDED GRANT PROJECTS

§ 2.17 General

The rules governing federally funded grant projects differ from those applicable to direct federal contracts. The distinction is founded on the differences in the two procurement processes. Direct federal public works contracts originate in a federal agency and are awarded by that agency directly to the contractor. In contrast, federally funded grant projects involve federal monies given as a grant to a local or state agency (grantee) that is then responsible for awarding the contract. The additional step of channeling the funds through a grantee subjects the protestor to additional legal rules governing the bidding process.

The substantive and procedural rules of the state or local jurisdiction are normally applied in conjunction with applicable federal regulations. The result is that not only must the contractor act pursuant to the regulations promulgated to govern the award of contracts by the grantor (the federal agency), it must also follow the state and local regulations applicable to the grantee.

Although many federal agencies fund grants to state and local entities for the construction of public works projects, in the last 15 years a substantial number of public works construction projects have been funded by grants from the Environmental Protection Agency (EPA). Not surprisingly, the EPA has developed elaborate bid protest procedures, which are often employed by other federal agencies and therefore serve as a model for discussing the legal requirements of bid protests on federally funded grant projects. These legal requirements are applied through decisions of the EPA regional administrators which are published and collected

---

[41] Coco Bros. Inc. v. Pierce, 741 F.2d 675 (3d Cir. 1984).

[42] Contract Disputes Act of 1978, 41 U.S.C. §§ 601-613 (1982).

in annual volumes. The published decisions have significant value as legal precedent and are available for review at all 10 EPA regional offices. For these reasons, this discussion focuses on the EPA protest procedures.

§ 2.18 Grounds for Protests

A contractor may look either to the state and local competitive bidding law or to the federal agencies' regulations to determine if sufficient grounds exist for a bid protest regarding a federally funded grant project. The EPA grant regulations[43] incorporate state and local law into the rules governing bid protests on EPA grants. The contractor desiring to protest must first determine if there has been a violation of a state or local regulation. If no such violation is found, the contractor may still have sufficient grounds to protest under the EPA regulations. In situations where state and local criteria do not provide for fair treatment of bidders or do not provide for basic principles of fairness employed in the federal procurement system, the federal rules will be followed in a bid protest.[44]

§ 2.19 Standing to Protest

Any party with a ''direct financial interest adversely affected'' by a procurement action has standing to file a protest with the EPA. This may include subcontractors or manufacturers of proprietary products who are excluded from the solicitation. The criterion is similar to the federal standing requirements for interested parties used on direct federal contracts.[45]

§ 2.20 Protest Procedure

A bid protest on a federal grant project must initially be presented to the grantee agency, and may be reviewed on appeal by the grantor agency. In general, the grantee agency is the proper authority to hear protests over awards for public works projects supported by federal grants. The protest procedures are generally the same as the bid protest procedures required by the state or local agency for nonfederal contract protests. The grantee is initially charged with the responsibility for determination of the bid protest. Normally, it is also the proper authority to conduct a hearing or otherwise to allow the interested parties to submit documents outlining their respective positions. The grantee agency must also decide questions of

[43] 40 C.F.R. § 33 *et seq.*

[44] Comp. Gen. Dec. B-180278 (Oct. 17, 1975).

[45] See § **2.14**.

local law such as subcontractor or supplier listing requirements. Such determinations based on local law will not be disturbed by the EPA on appeal unless they are contrary to established EPA policies.

The EPA regulations establish definite time limits for filing of a bid protest.[46] A protest relating to an allegedly deficient specification must be brought prior to bid opening and within one week after the protesting party should have been aware of the basis of the protest. Protests against the award of a bid must be in writing directed to the grantee within seven days of the bid opening.

If the grantee agency rules unfavorably on the protest, a protest appeal may be taken to the appropriate EPA official within seven days of the date of receipt of the decision. This initial appeal should be in the form of a written complaint filed with the assistant general counsel for grants or the office of regional counsel,[47] or in the form of a petition for review by the regional administrator.[48]

The EPA award official will permit all interested parties to present written submissions outlining their respective positions on the subject protest. This submission must include all legal and factual contentions relevant to the appeal. The award official also may hold a hearing or telephone conference on the protest appeal and will generally issue a decision within a two-week period. The award official's determination constitutes final agency action.[49]

§ 2.21 Relief Available

Judicial review of the regional administrator's decision is available. However, in most cases it is unlikely to result in cancellation of the project or award of the contract to the protesting party. The grantee agency's decision and the determination of the EPA regional administrator will be overturned only if the protestor can establish that it is arbitrary, capricious, an abuse of discretion, or otherwise in violation of the law.[50]

Although review of the regional administrator's decision by the General Accounting Office (GAO) was previously available, the GAO will no longer review the propriety of contract awards made by grantee agencies upon the request of contract bidders. The general standard of review of a grantee agency's decision was whether it had a ''rational basis'' in light of the fundamental principles of competitive bidding, but as of January 17, 1985, the GAO will no longer consider such protests.[51] As in the case of direct federal contracts, bid protestors whose protests are determined favorably can generally recover bid preparation costs.

---

[46] *See, e.g.*, 40 C.F.R. § 35.939(b)(1) (1978).

[47] *Id.* § 33.1105 *et seq.*

[48] *Id.* § 35.939 *et seq.*, for grants subject to these rules.

[49] *Id.* §§ 33.1145(c) , (g), 35.939(e)(3).

[50] *See* Gruman Ecosys. Corp. v. Gainesville-Alachua, 402 F. Supp. 582 (N.D. Fla. 1975).

[51] *See* Comp. Gen. Dec. B-185790 (Jul. 9, 1976).

STATE AND LOCAL CONTRACTS

§ 2.22 General

Although states and most local public entities are subject to competitive bidding statutes, they are not typically required to formally adopt a protest procedure. In some states, such as Oregon and Montana, the state chapters of the Associated General Contractors have formally published uniform guidelines for protest procedures, but most states do not have uniform protest procedures. As a result, a potential protestor must contact the state or individual agency to determine whether any formalized protest procedures exist.

Even if no formalized procedure has been adopted, applicable statutes and case law will normally require that some procedure be available to fairly test an owner's award of a contract. Although the procedures may vary, they will generally be required to comply with a fundamental fairness standard. The decision of the awarding authority will usually be reviewable by some judicial procedure on a relatively narrow standard of review which will, at a minimum, ensure that the hearing procedure and the ultimate award were appropriate.

§ 2.23 Grounds for Protests

The grounds for protests to state and local agencies are substantially the same as those for federal and federally funded contracts.[52] The concepts of low, responsive, and responsible bidders remain essentially the same and are tested by the same standards.[53]

Although some states employ the concept of "best bidder," which vests the awarding authority with substantial discretion in determining the entity to whom the contract will be awarded,[54] most states, like California, have rejected any kind of "relative superiority" test. That is, the agency may not compare the qualifications of a low bidder with those of other bidders unless the comparison is done in accordance with a disclosed standard which is itself fair and impartial.

§ 2.24 Standing to Protest

The requirements for standing to protest the award of a competitively bid project may vary considerably from state to state. The *interested party* requirement

---

[52] See **§§ 2.12-2.21**.

[53] *See, e.g.*, Cal. Gov't Code §§ 14807, 25454, 37902. The California statute does not use the term "responsive," but the courts have accepted a responsiveness test.

[54] *See* State *ex rel* Fisher Constr. Co. v. Linzell Constr. Co., 101 Ohio App. 219 (1955).

employed in federal bid protest procedures does not necessarily apply to the state and local government agencies who hear bid protests. For example, the California statute provides that any bidder who submitted a bid on a project may initiate a protest on the grounds that it is the lowest responsible bidder meeting the specifications.[55] This requirement differs significantly from the federal requirement in that a bidder who is not the next-lowest bidder may still have standing to bring a protest of the proposed award. Conceivably, the highest bidder has standing to protest if it establishes that the other bidders are not responsive or not responsible.

A person seeking relief from a proposed award by a writ of mandamus (mandate) to restrain a public official's award of a contract is subject to a different standing requirement in most states. Frequently, standing to oppose the award of a contract through a mandamus action is much broader, because the mandamus action to restrain a public official's discretion is available to any taxpayer who is a citizen of the state.[56] In those jurisdictions where this broad standing requirement exists, persons without an economic interest in the award of the contract may still be able to challenge the proposed award with a mandamus proceeding.

§ 2.25 Protest Procedure

If no specific protest procedure has been adopted by the awarding authority, a contractor should follow a procedure similar to that outlined for more formalized bid protests. That is, to commence a protest, a contractor should do the following:

1. Upon determination that a protest should be pursued, the contractor should initiate the protest in writing as soon as reasonably possible after the grounds for the protest have been determined. In all instances, a contractor should not wait more than seven days from the date of bid opening to notify the awarding authority of its intent to pursue a protest.
2. The contractor should request a hearing or some method of presenting its position regarding the protest. In this regard, counsel for the entity involved should be contacted to determine what rules have been adopted by the entity in the past for such hearings.
3. The contractor should prepare and deliver to the awarding authority a written memorandum outlining the basis of its protest and the reason why its protest should be sustained. This memorandum should be delivered either with the original protest letter or as soon as reasonably practical thereafter.

---

[55] Cal. Pub. Contract Code § 10306.

[56] *See* Baldwin-Lima-Hamilton Corp. v. Superior Court, 208 Cal. App. 2d 803 (1962).

§ 2.26 Relief Available

If the protest is favorably determined prior to award of the contract, the protesting party can successfully obtain an award of the contract. If, however, an award is made prior to determination of the protest, or the determination is unfavorable, most states will permit a protestor to file a suit for declaratory relief and may allow the court to entertain a request to restrain the award or, after award, the performance of the contract. Actions of this type often require the posting of a bond to cover damages sustained by the awarding authority as a result of the delay, and normally require the protestor to sustain a heavy burden of proof that performance of the contract by the apparent low bidder will result in irreparable injury or was an abuse of discretion on the part of the awarding authority.

In California and in other states, a protestor who receives an unfavorable determination from the awarding authority may also file for a writ of administrative mandamus. If the awarding authority conducts the hearing in accordance with certain requirements, an unsuccessful protestor may challenge the decision on the grounds that it was improper and unlawful. Significantly, the reviewing court will only consider whether there is some support in the record for the agency's determination rather than deciding whether the court would have reached the same decision. However, the remedy has the benefit of being relatively quick and normally does not require posting of a bond. Some authority exists at the state level for allowing recovery of bid preparation costs if the bid protest is determined favorably.

PRIVATE CONTRACTS

§ 2.27 Bidding on Private Contracts

As the foregoing indicates, a bidder under a public contract can, in proper cases, require the awarding authority to award it the contract, because the controlling statutes mandate an award to the lowest responsive, responsible bidder. No such right exists in the private sector. A private owner's invitation for bids for construction projects normally reserves the owner's right to accept any of the bids. This reservation allows the private owner to reject the low bid and contract with a higher bidder for the construction. The basic principles of contract law establish that no contractual relationship exists until the bid is accepted. Consequently, a disappointed low bidder usually has no remedy against the private owner for rejection of a low bid.[57]

---

[57] Universal By-Prods., Inc. v. City of Modesto, 43 Cal. App. 3d 145 (1974).

If the private owner fails to reserve the right of rejection in the invitation for bids and uses language in the solicitation establishing a commitment to award the contract to the lowest responsible bidder, then the disappointed bidder may have a remedy in contract law founded on the theory of promissory estoppel.[58] If the owner rejects the low bidder, or all bids, in bad faith or in furtherance of a scheme to obtain estimates from the bidders at no cost, the disappointed bidder may recover bidding expenses.[59]

BID MISTAKES

§ 2.28 General

There may be occasions when a bidder desires to be relieved of the obligation to perform the work. Protest procedures are not available for this purpose, but other actions are appropriate. A bidder may obtain relief in some circumstances if the bid submitted in response to the solicitation contains a *material mistake*. The type of relief available and the particular circumstances in which relief may be granted will first depend upon whether the bidding procedures are subject to state or federal law and whether the owner is public or private. Accordingly, this analysis of bid mistakes is divided into two separate discussions of the particular rules applicable in state and federal jurisdictions.

However, before examining particular rules governing bid mistakes, it is important at the outset to classify the kind of mistake. The following outline suggests a classification for evaluation of different types of bid mistakes:

1. Unilateral versus bilateral or mutual
2. Clerical (mistake of fact, mistake in filling out the bid) versus judgmental (error in judgment; carelessness in inspecting the site or in reading the plans or specifications)
3. Material versus inconsequential
4. Nonnegligent (excusable; honest; caused by ordinary negligence) versus culpable (caused by gross negligence or neglect of legal duty).

In general, relief is only available for material, excusable mistakes which are clerical rather than judgmental.

---

[58] *See* Swinerton & Walberg Co. v. Inglewood, 40 Cal. App. 3d 98 (1974).

[59] *See id.*, Milton v. Hudson Sales Corp., 152 Cal. App. 2d 418 (1957).

§ 2.29 Remedies for Bid Mistakes under State Law

Several states have statutes governing bid mistake remedies. California's statutes have evolved over the last 75 years and provide a good example of the principles affecting bid mistakes for other jurisdictions.[60]

If a bid contains a mistake that is a product of a clerical error and is material, a bidder may obtain a release from the mistaken bid by following the statutory procedures. Written notice specifying the mistake in detail must be given to the awarding authority within five days after bid opening. The statutes also require that, in any suit brought for a bid mistake, the summons must be served on the state or public entity within 90 days after bid opening.

The mistaken bidder has the burden of proving that the mistake was material, clerical in nature, and not the result of an error of judgment. A mistake is *material* if it is significant in contrast to the total value of the contract and would lead a reasonable person to conclude that the contractor would not willingly perform the work at the price.

The law in California does not fully recognize relief for a mistaken bid founded on general equitable principles. However, remedies for mistakes at common law may apply where the owner is not one of several state entities expressly covered under the statutes. In these situations, relief from a mistaken bid may be had under a contract theory of rescission or reformation. According to a well-established line of authority, where a mutual mistake of the parties is the basis for the formation of a contract, rescission of the contract will be allowed.[61] Thus, a bid submitted to an owner where both bidder and owner are mistaken as to a material element of the work can be rescinded.

Rescission may also be available for unilateral mistakes in certain situations. Equitable principles may support an action for rescission of a mistaken bid submitted because of a clerical mistake where the circumstances make the enforcement of the mistaken bid unconscionable.[62] Circumstances important to such a determination include: the inadvertance of the error; the owner's knowledge of the error; the owner's opportunity to readvertise; and equitable considerations based on the differential in bid price and contract cost.[63]

A contractor who seeks to rescind an erroneous low bid that is the result of a unilateral mistake of fact must meet certain conditions.[64] In general, the bidder must provide prompt notice of the material mistake of fact and its desire to

---

[60] The procedures governing bid mistakes in California appear in Cal. Pub. Contract Code §§ 10,200-10,205. The Public Contract Code applies only to contracts with most (but not all) agencies in California. *See* Cal. Educ. Code §§ 90,200 and 90,205 for rules applicable to mistakes in contracts awarded through the California State University System and California Colleges.

[61] *See* C. Kaufman, 3 Corbin on Contracts § 608 (1984 Supp.).

[62] M.F. Kemper Constr. v. City of Los Angeles, 37 Cal. 2d 696 (1951).

[63] *Id.* 702, 703.

[64] *See id.*; White v. Berenda Mesa Water Dist., 7 Cal. App. 3d 894 (1970).

rescind the bid. In addition, the bidder must establish that it has not neglected any legal duty and must provide proof to the awarding agency that enforcement of the contract based upon the erroneous bid would be unconscionable. The bidder will be required to reimburse the agency for any value received so that the agency's position is the same as it would have been had the bid not been withdrawn.

A contractor who submits an erroneous bid may recover the bidder's security which would normally be forfeited due to the bidder's failure to execute the contract. In order to recover the security, the bidder must establish: that the mistake made was material to the intended amount of the bid; that it provided notice to the agency within five days of bid opening, describing in detail how the bid mistake occurred; and that it was a mistake in the preparation of the bid not due to an error in judgment or carelessness.

Reformation is available only for mutual mistakes. If the bidder makes a bid mistake and eventually performs the contract after notifying the owner of the error, the bidder may not later reform the contract after performance to secure the higher payment amount.[65]

The statutory provisions of the Public Contract Code and other applicable statutes are the exclusive remedies in California for mistakes in bids submitted to agencies covered by the bidding statutes.[66] Failure to comply with the statutory notice requirements is sufficient ground for denial of recovery. All the California statutes governing bid mistakes expressly provide that no change shall be made in a bid because of mistake. This is in contrast to federal law, where corrections may be allowed. If the bidder is not relieved of the bid, it will be held to the bid if selected as the lowest bidder. The statutes uniformally provide that a bidder who obtains relief for a bid mistake cannot rebid in response to the same solicitation.

§ 2.30 Jurisdictions Following Equitable Principles

Absent a specific statute for relief for material mistakes, a contractor must rely on traditional theories available in the jurisdiction for such relief. Many state courts that have addressed the problem of bid mistakes follow the rule that equity can allow withdrawal of a material clerical mistake if the bidder acts promptly to seek relief.[67] Several courts deny relief for any kind of mistake, citing various theories such as avoiding possible fraud or preserving the sanctity of competitive bidding; other courts rely on statutes providing that bids may not be withdrawn after opening. Some rely on provisions in the invitations for bids to establish that no relief

[65] Lemoge Elec. v. County of San Mateo, 46 Cal. 2d 659 (1956).

[66] A&A Elec., Inc. v. City of King, 54 Cal. App. 3d 457 (1976).

[67] *See* Boise Jr. College Dist. v. Mattefs Constr. Co., 450 P.2d 604 (Idaho 1969); Smith & Lowe Constr. Co. v. Herrera, 79 N.M. 239, 441 P.2d 197 (1968); Rushlight Auto Sprinkler Co. v. City of Portland, 219 P.2d 732 (Or. 1950); cases collected in 52 A.L.R.2d 792-814 (1957); 70 A.L.R.2d 1370, 1375-77 (1960).

will be granted for mistakes. Irrespective of what procedure is available, it is vital to determine the nature of the error and to advise the owner as early as possible.[68]

§ 2.31 Federal Law on Bid Mistakes

Federal law requires the contracting officer to examine all bids submitted in response to an invitation for bids (IFB) to determine if mistakes are present.[69] Two types of mistakes are most frequently identified by the contracting officer: those clerical mistakes which are apparent from the face of the bid; and those mistakes which are errors of assumption or judgment not readily apparent from examining the bid. The basis for the contracting officer's examination is often the appearance of a large discrepancy between the mistaken bid and other bids or the government's estimate.

When a contracting officer discovers what it believes to be a mistaken bid, it must request verification.[70] The contracting officer is obligated to notify the bidder of the information that leads the officer to believe the bid is suspect and must allow sufficient time for the bidder to respond to the request for verification.[71]

Additional federal rules for bid mistakes apply depending on when the mistake is discovered. If the mistake is discovered prior to the award, the contracting officer may allow withdrawal of the bid for an honest mistake. That is, to qualify for withdrawal, the bidder must establish that reasonable grounds exist for the mistake. This principle is based on the theory that a contracting officer cannot accept a bid which it knows or should have known contains an error.

In exceptional cases the contracting officer may allow correction of a bid. If a bidder can establish the manner in which the error occurred and its actual intended price, the bid may be corrected provided that the correction does not displace another bidder. The contracting officer may allow correction of a bid even if it displaces another bidder in those instances where both the mistake and what the contractor intended to bid are obvious on the face of the bid documents.

After an award of a contract, reformation of the contract is available to correct mutual mistakes. Reformation is not available for unilateral mistakes unless the contracting officer, prior to acceptance of the bid, had actual or constructive notice of the probability that an error existed in the bid. If this occurs, the General Accounting Office or the court may allow cancellation of the contract or adjustment of the contract price, but may not adjust the price so that it exceeds the price of

---

[68] *See* Triple A Contractors Inc. v. Rural Water Dist., 603 P.2d 184 (Kan. 1979); City of Newport News v. Doyle & Russell Inc., 179 S.E.2d 493 (Va. 1971); Board of Educ. v. Sever-Williams Co., 258 N.E.2d 605 (Ohio 1970); Modany v. State Pub. School Bldg. Auth., 208 A.2d 276 (Pa. 1965); A.J. Colella, Inc. v. County of Allegheny, 137 A.2d 265 (Pa. 1957).

[69] *See* Comp. Gen. Dec. B-180329 (Oct. 1, 1974); Comp. Gen. Dec. B-182700 (Dec. 23, 1974).

[70] *See* 43 Comp. Gen. 327 (1963).

[71] *See* 44 Comp. Gen. 383 (1965); 54 Comp. Gen. 545 (1974).

the next lowest bidder. If the contract has been substantially performed, the bidder may be able to obtain a quantum meruit or quantum valebant recovery subject to the limitation that it may not exceed the price of the next lowest bidder.

§ 2.32 Private Owners

The rules governing mistakes in bids submitted to private owners vary greatly from those applicable to government contracts. A bidder may withdraw the bid at any time prior to acceptance. A bid may be withdrawn after opening of the bids unless the doctrine of promissory estoppel applies. Consequently, a bidder may withdraw its bid before acceptance for any kind of mistake or for any reason (or for no reason at all). This rule clearly differs from that applicable to public contracts, where the bidder is permitted withdrawal for any reason or without reason prior to bid opening, but is not allowed to withdraw once the bids are open unless it is able to establish that relief should be available because of a mistake in the bid.

Once a private owner has accepted the bid, rescission of the contract is ordinarily not available due to unilateral mistake. However, even in the private sector, some courts have allowed bidders on private contracts to rescind their bids where a clerical error was not discovered until after the contract was signed.[72]

§ 2.33 Conclusion

The foregoing presents a summary of the procedures available to bidders to initiate or defend against protests and to obtain relief in cases of bid mistakes. As the text indicates, the rules are in many cases informal and, in any event, are subject to constant change. Care should be taken to consult appropriate sources and to determine the applicable law before initiating either proceeding.

---

[72] *See* Geremia v. Boyarsky, 140 A. 749 (Conn. 1928).

CHAPTER 3

SUING THE PRIVATE OWNER

Robert M. McLeod and Jesse B. Grove, III*

Robert M. McLeod is a senior partner in Thelen, Marrin, Johnson & Bridges where since 1953 he has represented contractors, subcontractors, owners, engineers, architects, shipbuilders, and others in all phases of design and construction law. Mr. McLeod has contributed to numerous publications on construction law and arbitration and has spoken at many national and international conferences. He is Vice-Chairman of the International Construction Contracts Committee of the International Bar Association and a member of the Forum Committee on the Construction Industry and the Committee on Construction Litigation of the Litigation Section of the American Bar Association.

Jesse B. Grove, III is a trial lawyer with substantial experience in all phases of construction claims on major projects such as dams, tunnels, powerhouses, wastewater treatment plants, highways, bridges, and municipal, industrial, and highrise buildings; antitrust claims involving predatory pricing, tying, attempted monopolization, price discrimination, and mergers; and insurance disputes, including builder's risk and computer lease indemnity claims.

AREAS OF DISPUTE

* Milton C. Butler, Esq., of Thelen, Marrin, Johnson & Bridges assisted in the preparation of this chapter.

AREAS OF DISPUTE

§ 3.1 Introduction

Although the action in the title of this chapter is suing, perhaps as many, if not more, construction disputes are resolved by arbitration,[1] mediation, or other processes[2] agreed upon by the parties to the dispute as are resolved by court

---

[1] In 1983, 2,675 construction arbitration claims with a value of $447.7 million were filed with the American Arbitration Association. Engineering News Record, May 24, 1984, at 53, col. 2.

[2] The Associated General Contractors of California recently introduced a three-tiered dispute resolution procedure: first, informal mediation. Second, if mediation fails to produce a resolution, a mini-trial is held before a decisionmaker from each party and a neutral advisor who gives a nonbinding, written opinion if the decisionmakers cannot reach agreement. The third tier, if necessary, is binding arbitration. San Francisco Att'y, Feb./Mar. 1985, at 15, col. 1.

A *claims review board* successfully and promptly resolved disputes between the owner and the contractor on some very large construction projects such as the second bore of the Eisenhower Tunnel in Colorado and the dam of the El Cajon hydroelectric project in Honduras, C.A. The board is composed of two members, one appointed by each party to the contract, and a third member selected by the first two members. The board meets at the job site at regular intervals and considers and decides any unresolved claims of either party as the job progresses. A decision of the board is binding only if neither party overrides it within a specified time after it is issued. If a decision is overridden, it is admissible in evidence in any subsequent legal or arbitral proceedings. D.C. Davies, *A Review of the Origin, Functions and Operation of the Claims Review Board, Proyecto Hidroelectrico El Cajon* (Dec. 10, 1984) (unpublished paper prepared for the World Bank).

action. This chapter discusses resolving disputes with the private owner[3] regardless of the forum in which the resolution takes place.

In the past, it was the contractor or someone injured on the project who usually sued the owner. More recently, subcontractors have successfully sued the owner. In *Berkel & Co. v. Providence Hospital*,[4] the subcontractor, Berkel, sued the owner and its architect on tort theories to recover what it had expended on unaccepted piles. Count One was based on a breach of duty in directing pile installation. The owner's defense of lack of privity was disposed of by the court by reference to cases which have held that a duty may arise from a social relationship as well as from a contractual relationship. The court quoted Professor Prosser: "[B]y entering into a contract with A, the defendant may place himself in such a relation toward B that the law will impose upon him an obligation, sounding in tort and not in contract, to act in such way that B will not be injured."[5]

The hospital argued that even if privity were not a defense, the facts showed that no duty was owed to Berkel. The court said:

> In deciding whether to impose a duty in a construction context, the trial court should analyze six factors:
>
> "(1) [T]he extent to which the transaction was intended to affect the other person; (2) the foreseeability of harm to him; (3) the degree of certainty that he suffered injury; (4) the closeness of the connection between the defendant's conduct and the injury; (5) the moral blame attached to such conduct; and (6) the policy of preventing future harm." . . . Under this standard, Providence clearly owes Berkel a duty to act reasonably in directing and approving pile construction work.[6]

§ 3.2 Changes

The contract is the common denominator in the construction industry. The construction contract is an anomaly in the law of contracts in that it is often made without a meeting of the minds (e.g., competitively bid public contracts) and usually contains a changes clause which permits one party to unilaterally change the other party's obligations. This detour around the normal requirement of mutual agreement for an effective modification of a contract has a long history and is now a commonplace.

The earliest approach to a changes clause[7] in this country appears in an 1818 United States Army contract with Asa Waters for 10,000 muskets: "Should any

---

[3] In most situations, the courts apply the same rules to the construction of contracts between private parties as they do to those between a private party and a governmental body. Accordingly, cases in which the owner is a public entity have been cited when the rule announced is equally applicable to both types of owners.

[4] 454 So. 2d 496 (Ala. 1984).

[5] *Id.* 502.

[6] *Id.* 502-03 (citations omitted).

[7] R. Nash, Jr., Government Contract Changes 13 (1975).

alterations . . . be decreed by the Ordinance Department, the said Asa Waters will be entitled to compensation for any extra expenses occasioned by such alterations.'' There was no provision in the contract which gave the Ordinance Department the right to decree alterations in the musket specifications; apparently the parties assumed that this was an inherent right of the government.

There are probably more disputes arising from the changes clause than from any other provision of the construction contract. Litigation relating to the changes clause in government contracts was more frequent than that relating to any other clause.[8] In 1984, only appeals involving the default clause exceeded in number those relating to the changes clause.[9]

Normally, the contractor is required to proceed with the changed work even though the parties have not agreed on the price of the change. However, this is not always the case. *Coleman Engineering Co. v. North American Aviation, Inc.*[10] involved the changes clause in a $527,000 purchase order contract for the engineering and manufacturing of trailers for a missile which read: "Buyer [North American] reserves the right to . . . make changes. . . . In such event there will be made an equitable adjustment in the price and time of performance mutually satisfactory to Buyer and Seller . . . .''

The buyer made a change which increased costs of performance at least $257,000. The parties attempted but were unable to agree upon an equitable adjustment. The contractor (Coleman) refused to perform further and successfully sued North American. On appeal, the court affirmed and said:

> . . . if the subsequent changes are minor or of not great magnitude the contractor must perform and obtain a subsequent judicial determination as to the price of the changes. However, where the changes are of great magnitude in relation to the entire contract, the contractor must negotiate in good faith to settle the price . . . and where he has done so, he is not required to continue performance in the absence of an agreement as to the price.[11]

The change order should increase or decrease the contract price by the difference between what it would have cost to perform the unchanged work and what it cost to perform the changed work. In *S.N. Nielsen Co. v. United States,*[12] the contractor mistakenly bid only $22,564.32 for installing new underground ducts for cables because nothing was included in the bid for materials. The value of that work was $60,690; it was deleted by change order and the changed work (above-ground poles added) had a value of $19,180. The government issued a change order reducing the contract price by $41,510. The contractor contended that the $19,180 should have been subtracted from the $22,564.32 and only the difference of $3,384.32

[8] ''In fiscal year 1974, over 31% of the cases before the [Armed Services Board of Contract Appeals] were based on the Changes clause.'' *Id.* 4.

[9] 253 default clause cases as opposed to 243 changes clause cases. 26 Gov't Contractor ¶ 324 (Nov. 12, 1984).

[10] 65 Cal. 2d 396, 55 Cal. Rptr. 1 (1966).

[11] *Id.* 406, 408.

[12] 141 Ct. Cl. 793 (1958).

deducted from the contract price. The court rejected this contention, which would have used the price adjustment as a means to correct the contractor's mistake, pointing out that the contractor's loss under the change order was the same as it would have been if the change order had not been issued.

Typically, the changes clause limits the owner's right to make changes to those which are within "the scope of the work" required under the contract. Changes which are beyond the scope of the work are characterized as *cardinal changes* and have been held to be breaches of contract.[13] A cardinal change agreed to by the contractor constitutes an effective modification of the contract by agreement, not under the changes clause.

In *Hensler v. City of Los Angeles,*[14] the court observed that a "contract entered into by a governmental body and an individual is governed by the same rules which apply to the construction of contracts between private persons."[15] The contract in that case gave the owner (the city) the right to make changes "as may be considered necessary or desirable to complete fully and acceptably the proposed construction in a satisfactory manner."[16] The owner issued a change order deleting portions of the work. As a result of the deletions, the project was unusable and the deleted work was not performed until several months later when the owner let a contract for the deleted work to another contractor.

Although it did not use the term *cardinal change,* the court affirmed a judgment for the contractor on the ground that the deletion of work constituted a breach of contract. The court said that the right to make changes "cannot be used . . . for the purpose of legitimatizing the deletion of so integral a part of the work as to leave the improvement in an unfinished condition and still insulate the city from liability."[17]

§ 3.3 —Notice of Changes

Many construction contracts contain one or more provisions which make the giving of notice of a change which is not the subject of a written change order a condition precedent to the contractor's right to payment for performing the changed work. This requirement may be expressly or impliedly waived by the owner. Implied waivers have been found where the owner considers and acts on the claim for additional compensation on its merits,[18] where the owner had actual knowledge of the change so that notice would have served no useful purpose,[19] and where

[13] Boomer v. Abbett, 121 Cal. App. 2d 449, 263 P.2d 476 (1953); General Contracting Co. v. United States, 84 Ct. Cl. 570 (1937); Saddler v. United States, 287 F.2d 411 (Ct. Cl. 1961).

[14] 124 Cal. App. 2d 71, 268 P.2d 12 (1954).

[15] *Id.* 78, 268 P.2d at 18.

[16] *Id.*

[17] *Id.* 80, 268 P.2d at 19.

[18] George A. Fuller Co. v. United States, 104 Ct. Cl. 176, 218 (1945).

[19] Central Mechanical Constr., ASBCA Nos. 29431-29433, 85-2 B.C.A. (CCH) ¶ 18,061 (1985).

the owner was not prejudiced by the contractor's not giving notice.[20] However, absent proof of waiver, notice provisions will normally be enforced.[21]

Most changes clauses require that change orders be in writing. This requirement may be waived but the waiver must be clearly established, particularly in public contracts. In *Nether Providence Township School Authority v. Thomas M. Durkin & Sons, Inc.*,[22] the contract to construct a school building provided: "No change in the contract shall be made without the written approval of the Board. A request for any change must be in writing."[23]

The board's president authorized its secretary to write a letter to the contractor acknowledging the disagreement as to who should bear the cost of added work and recommending continuation of the work with resolution of any problem at a later time. The contractor completed the job and billed the board for the extra work. The trial court held that the secretary's letter was a waiver of the requirement of writing. The Supreme Court reversed, however, saying that waiver could only be accomplished by formal written action of the board or ratification of the extra work claim by resolution of the board. The secretary's letter was not the action of the full board.

§ 3.4 —Constructive Changes

The legal fiction of *constructive change* developed in federal government construction contracts in order to permit contractors to be paid under the changes clause even though no change order had been issued. The Boards of Contract Appeals and later the Court of Claims held that where a contractor's performance requirements were changed by some action or inaction by the government which amounted to a direction to the contractor, there had been a constructive change compensable under the changes clause.

An example of a constructive change is acceleration. For instance, a contractor experiences an excusable delay which has the effect of extending the contract completion date, but the government insists on completion in accordance with the original schedule. The government's conduct amounts to a constructive acceleration order.

The constructive change doctrine is gaining wider acceptance and has been applied in state and local government and private contracts. In *Weeshoff Construction Co. v. Los Angeles County Flood Control District*,[24] the court said: "Although . . . the district refused to explicitly name the procedure it was ordering, it clarified, by example, the material it was then requiring when it [used] temporary pavement. Since this procedure has been found to constitute a change in the original contract, it is clear that the district, by its conduct, exerted an intentional attempt

---

[20] Charles H. Siever, PSBCA No. 460, 80-2 B.C.A. (CCH) ¶ 14,740 (1980).

[21] State v. Omega Painting, Inc., 463 N.E.2d 287 (Ind. Ct. App. 1984).

[22] 476 A.2d 904 (Pa. 1984).

[23] *Id.* 906.

[24] 88 Cal. App. 3d 579, 152 Cal. Rptr. 19 (1979).

to affect [sic] a contractual change without complying with the change order provision . . . [T]he district's performance . . . itself, affected [sic] a change of contractual terms.''[25]

Some other examples of constructive changes are:

1. The owner requires performance of work which is beyond the scope of the work required by the contract. The cost of performing such work is compensable
2. The owner supplies defective design specifications, compliance with which is impossible. Increased costs of performance resulting from attempting to comply are compensable
3. The owner requires performance in accordance with only one of two or more optional methods of performance specified in the contract. The increase in cost of performance over the cost of performance pursuant to the least expensive option is compensable
4. The owner demands higher standards of inspection or workmanship or applies tests different from those specified in the contract. Increases in the contractor's performance costs due to use of such standards are compensable
5. The owner requires work in accordance with an incorrect interpretation of the specifications. The increase in cost over the cost of the work done in accordance with the correct interpretation is compensable.

§ 3.5 Scope of Work

Most disputes about the scope of the work occur because of inaccurate, incomplete, or ambiguous contract provisions and details in the specifications and plans. For example, in *Ronald Adams, Contractor, Inc. v. State,*[26] the contractor performed a unit price contract in accordance with the original plans and specifications, but the quantity of excavation work performed was 79.9 percent less than estimated and the embankment work was 31.58 percent less than estimated. The contract provided that in instances where there was a difference between the final and estimated quantities, the contractor was to be paid the original unit price ''except as provided in Section 40,'' which read:

> The owner reserves and shall have the right to make such alterations in the work . . . as may increase or decrease the originally awarded contract quantities, provided that the aggregate of such alterations does not change the total contract cost or the total cost of any major contract item by more than 25 percent. . . . Should the aggregate amount of altered work exceed the 25 percent limitation . . . such excess altered work shall be covered by supplemental agreement.[27]

---

[25] *Id.* 590, 591, 152 Cal. Rptr. at 24, 25.

[26] 457 So. 2d 778 (La. Ct. App. 1984).

[27] *Id.* 780.

In affirming a judgment for the contractor, the court rejected the state's contention that "alterations in the work" meant merely alterations in the plans or design.

The court also rejected the state's argument that the contractor was entitled to additional compensation for only the portion of the alterations which exceeded the 25 percent limitation. The court said: "In the very least, it must be said the phrase 'such excess altered work' . . . is not definitive . . . The meaning being . . . ambiguous it must be interpreted against the entity confecting the contract, i.e., the State."[28]

The contractor in *Acchione & Canuso, Inc. v. Commonwealth*[29] entered into a contract to perform 13,131 lineal feet of trenching at a unit price per lineal foot. The trenching was to be performed on earth, sidewalk, and highway, and the contractor estimated the respective costs per lineal foot for the three types of trenching as $13, $23, and $52. Based on estimated quantities, the contractor's composite bid price was $24 per lineal foot. After the contract was entered into, the state added 17,433 lineal feet of trenching. The quantity of the least expensive trenching rose about 50 percent and the quantity of the most expensive trenching rose over 300 percent.

Prior to bid the contractor had questioned the estimated quantities and had been directed by the owner's engineers to assume that 50 percent of the existing conduit would be reusable and thus would require no new trenching. The 17,433 lineal feet of trenching were "required due to the omission by the designer and the consultant in the plans and in the Contract of a contingency item . . . to cover the installation of conduit where reusable conduit does not check out to be usable. The corresponding trenching item was also not revised accordingly. The complete item for trenching on the interconnect system was also omitted by the designer and the consultant."[30]

The court held that the increased expenses incurred by the contractor were a direct result of the contractor's reliance on the misrepresentation of the owner's engineers. That representation was arbitrary and/or a gross mistake and was therefore actionable.

§ 3.6 Changed Physical Conditions

Many of today's construction contracts contain clauses which deal with the rights and obligations of the parties if unforeseen physical conditions are encountered. The federal government's clause is entitled "Differing Site Conditions" and refers to "(1) subsurface or latent physical conditions . . . differing materially from those indicated in this contract, or (2) unknown physical conditions . . . of an unusual nature, differing materially from those ordinarily encountered and generally recognized as inhering in work of the character provided for in this contract."

---

[28] *Id.* 781.

[29] 461 A.2d 765 (Pa. 1983).

[30] *Id.* 767.

The corresponding clause in American Institute of Architects Form A201 is called "Concealed Conditions" and provides for "concealed conditions encountered . . . below the surface of the ground or . . . concealed or unknown conditions in an existing structure . . . at variance with the conditions indicated by the Contract Documents . . . ."[31] The National Society of Professional Engineers Form 1910-8 has a clause entitled "Unforeseen Physical Conditions" which relates to "subsurface or latent physical conditions at the site or in an existing structure differing materially from those indicated or referred to in the Contract Documents . . . ."[32]

The Court of Claims has said that the purpose of the *changed conditions* clause is to take at least some of the gamble of subsurface conditions out of bidding. The clause shifts the risk of encountering an adverse subsurface condition to the owner and encourages the contractor not to include contingency amounts therefor in its bid.[33]

After award of a contract to Empire Construction,[34] the government supplied as-built drawings which did not show a two-inch lateral line coming off the main gas line and continuing below a concrete slab which had to be removed by Empire. During removal of the slab, the two-inch line was ruptured. In sustaining the contractor's appeal, the board said:

> Although the as-built drawings relied upon by appellant to give the locations of existing gas, steam and heat lines were not a part of the original contract drawing package, they were Government drawings furnished to appellant in the course of its performance by the Government representatives at Mather Air Force Base. As such they related to conditions "indicated in the contract" as contemplated by the "Differing Site Conditions" clause. . . . [A]ppellant has demonstrated that a latent condition existed, that it was not shown on the drawings, and that the failure to so show it was the proximate cause of the gas line rupture.[35]

In *P.J. Maffei Building Wrecking Corp. v. United States,*[36] the invitation for bids for demolition work contained the following language: "Some drawings of some of the existing conditions are available for examination at the New York City Parks Department's Administration Building. . . . These drawings are for information only and will not be part of the contract documents. The quantity, quality, completeness, accuracy and availability of these drawings are not guaranteed."[37]

The contractor used the drawings in arriving at its estimate of salvageable steel which turned out to be about 20 percent more than was actually obtained. The court sustained the denial of Maffei's claim of having encountered changed conditions, saying:

---

[31] AIA Form A201, General Conditions (1976).

[32] NSPE Form 1910-8 (1978).

[33] Foster Constr. C.A. & Williams Bros. Co. v. United States, 435 F.2d 873, 887 (Ct. Cl. 1970).

[34] Empire Constr. Co., Inc., ASBCA No. 27540, 84-3 B.C.A. (CCH) ¶ 17,531 (1984).

[35] *Id.* at 87,309.

[36] 732 F.2d 913 (Fed. Cir. 1984).

[37] *Id.* 917.

> [W]e agree with the Board that the contract documents did *not* "indicate" the amount of steel recoverable from the Pavilion, within the meaning of the Differing Site Conditions clause. We also decide (as did the Claims Court) that Maffei failed to show that the Government misrepresented conditions at the Pavilion or that it possessed pertinent information regarding the Pavilion steel structure which it concealed.[38]

A different problem with respect to prebid information which is labelled as not part of the contract documents arose in *McNamara Corp. Ltd. v. State of California.*[39] In that case, the decisions of the owner's engineer that the contractor had not encountered changed conditions were held to lack finality because, in reaching his decisions, the engineer had considered representations made in prebid geological data which were expressly stated not to be in the contract. The court said:

> Since the prebid geological data is not among the representations of the site conditions made in the contract, we agree with the trial court that the engineer erroneously considered this data in ruling on the changed conditions questions. Therefore, for purposes of the changed conditions claims, the engineer's decisions were not entitled to finality.

Other contract clauses shift the risk of encountering unanticipated subsurface conditions to the contractor. In a New Jersey case,[40] the court considered a clause entitled "Article 1.2.12 Subsurface Conditions" which read as follows:

> It is the obligation of the Bidder to make his own investigations of subsurface conditions prior to submitting his Proposal. Borings, test excavations and other subsurface investigations, if any, made by the Engineer prior to the construction of the project, the records of which may be available to bidders, are made for use as a guide for design. Said borings, test excavations and other subsurface investigations are not warranted to show the actual subsurface conditions. The Contractor agrees that he will make no claims against the State, if in carrying out the Project he finds that the actual conditions encountered do not conform to those indicated by said borings, test excavations and other subsurface investigations.
>
> Any estimate or estimates of quantities shown on the Plans or in the form of proposal, based on said borings, test excavations and other subsurface investigations, are in no way warranted to indicate the true quantities. The Contractor agrees that he will make no claims against the State, if the actual quantity or quantities do not conform to the estimated quantity or quantities . . . .

The court said:

> In Sasso . . . we found that article 1.2.12 was a specific exculpatory clause focusing directly on subsurface conditions and the contract language was ". . . straightforward, unambiguous and categorical . . . in placing responsibility for subsurface investigations on the contractor." *Id.* at 489, 414 A.2d 603. Thus we concluded that

[38] *Id.* 916 (emphasis by the court).

[39] Cal. App. No. 38595 (1st Dist. Feb. 28, 1977) (unpublished).

[40] Ell-Dorer Contracting Co. v. State, 484 A.2d 356 (N.J. Super. Ct. App. Div. 1984).

under the terms of the contract the State's representations were merely gratuitous and the reliance by the contractor was at his own peril.

Golomore and *Sasso* stand for the proposition that when the State makes false representations it will be liable for damages resulting from them despite a general disclaimer of liability for inaccurate representations. However, if the disclaimer is sufficiently specific or if the statements only purport to be the results of tests rather than being actual conditions or descriptions of actual conditions, then the contractor cannot recover.[41]

§ 3.7 Delay

Most construction contracts have provisions dealing with delays to the work. **Section 3.10** discusses *no damage for delay clauses,* which provide that a contractor's only remedy for delay is an extension of the contract completion date. *Termination clauses,* which give one or both parties the right to terminate performance of the work in the event of certain delays, are discussed in **§ 3.16.** Other types of delay clauses and delay situations are dealt with in this section.

The highway contract involved in *J.A. Tobin Construction v. State Highway Commission*[42] was executed on July 12, 1968 and provided:

> The Contractor is advised that the Commission and the Kansas City Power & Light Company have an agreement executed November 7, 1966, for the relocation of the utility facilities which are on the right of way . . . . The Kansas City Power & Light Company . . . has advised the Commission that such work . . . would be completed by October 1, 1968 . . . . No additional compensation will be allowed for any delay, inconvenience, or added expense to the Contractor resulting from any restriction of operations which may be necessary to protect utility facilities prior to their removal from the right of way.

The utility lines were not relocated until May of 1969 and both parties agreed that as a result Tobin lost the equivalent of one year's work and suffered increased costs due to the delay. The court of appeals affirmed a judgment for Tobin based on a breach of the express written warranty to relocate the utility lines by October 1, 1968.

In *Tranco Industries, Inc.*,[43] a painting contractor notified the government that all contract work was complete before June 1, 1981. The government did not inspect the work until October 13, 1981 and offered no justification for the four-month delay in inspection. In sustaining the contractor's appeal from a default termination, the board said that "[t]he Government's failure to reject the work within a reasonable time after its completion constituted an acceptance of the work."[44]

---

[41] *Id.* 360.

[42] 680 S.W.2d 183 (Mo. Ct. App. 1984).

[43] ASBCA No. 26955, 83-1 B.C.A. (CCH) ¶ 16,414 (1983).

[44] *Id.* at 81,655.

In *Continental Heller Corp.*,[45] the government was late in furnishing an environmental report which was a prerequisite to the issuance of a building permit. The lack of a building permit forced the contractor to adopt unanticipated construction procedures which created inefficiencies in and delays to the work. The board held that the government's conduct constituted a constructive change entitling the contractor to an equitable adjustment.

The owner of some schools in Indiana delayed issuing a notice to proceed to the contractor. As a result of that delay, the contractor was required to perform work in the winter and incurred delay damages of $155,207. The owner appealed from a judgment in favor of the contractor, contending that because the contractor submitted its claim for delay damages eight months after completion of the work, the contractor could not recover in view of the additional cost clause in the contract. That clause provided: "If the Contractor wishes to make a claim for an increase in the Contract Sum, he shall give the Architect written notice thereof within twenty days after the occurrence of the event giving rise to such claim. . . . No such claim shall be valid unless so made . . . ."

The decision contains the following discussion:

> Owner in essence is arguing that delay damages fall within the scope of additional costs. . . .
>
> Additional costs arise when construction is begun in a timely fashion and extra work is necessitated by changed specifications or unforeseen conditions. . . . [S]uch costs can be contemplated by the parties as they occur frequently in construction situations. On the other hand, delay damages are extra costs that arise solely as a result of delay by the owner, contractor, or subcontractor. Party-caused delays are not contemplated by the parties, and the damages resulting therefrom should not be treated as additional costs. Delays are breaches of implied or express contractual provisions, whereas additional costs do not stem from any such breach. Therefore, in the absence of express language in the construction contract equating the treatment of claims for additional costs and delay damages, they will not be treated the same.[46]

Some delays in construction work are the result of suspensions of all or part of the work by the owner under a *suspension of work clause*. These clauses give the owner the right to halt the contractor's work, an act which would otherwise be a breach of contract, and provide a contractual procedure for the payment of any increased costs. Some suspension of work clauses give the contractor rights in addition to the right to be paid its increased costs. Such a clause in one city's standard form contract provides, in part:

> The Department will, . . . pay the Contractor for each calendar day during which the entire work shall have been suspended, as provided in this Article, the sum set forth hereinafter. Said sum is hereby mutually agreed upon as fixed and liquidated damages in full settlement of all costs and expenses, losses, and damages which may accrue to the Contractor from any such entire suspension. . . . In the event that the

---

[45] GSBCA No. 6812, 84-2 B.C.A. (CCH) ¶ 17,275 (1984).

[46] Osolo School Bldgs. v. Thorleif Larsen & Son, 473 N.E.2d 643, 645–46 (Ind. Ct. App. 1985).

entire work shall be suspended by order of the Chief Engineer, as herein provided, and shall remain so suspended for a period of ninety consecutive days, through no fault of the Contractor, and notice to resume work shall not have been served upon the Contractor as hereinbefore provided, the Contractor may, at his option, by written notice to the Board, terminate the contract . . . .

§ 3.8 Nonpayment

Since payment of the contract price is probably the most important of the owner's few obligations under construction contracts, it is not surprising that most nonpayments by owners are breaches of contract. The nonpayments arise in many different contexts.

A Minnesota case[47] arose out of an owner's refusal to make the final payment to a house builder of $7,600, out of a total contract price of $67,000, until all defects were repaired and the builder gave up its lien rights. The builder refused to release its lien rights and did not make the repairs. The court found that the owner had breached the contract with the builder, and awarded the builder damages in the full amount of the final payment, less the amount saved by not repairing the defects.

In another Midwestern case,[48] an owner who was dissatisfied with the quality and timeliness of a house builder's work notified the builder that he wanted to change the weekly payments to a single payment after the owner was satisfied that all the work was completed. When the builder objected, the owner said that no more payments would be made until the house was completely finished. The builder walked off the job and sued the owner. The Indiana Court of Appeals held that the owner had repudiated the contract by refusing to make further payments and that the builder was justified in stopping performance and treating the contract as terminated.

By adopting an arbitrary performance evaluation system, the government determined certain work in one case to be unsatisfactory and deducted some $4,035 from the first monthly progress payment of about $11,095 due to a refuse removal and cleaning contractor.[49] The contractor then abandoned the job. The Board of Contract Appeals held that the substantial underpayment was a breach of contract justifying the contractor's abandonment of the work.

Progress or installment payments are the usual method of paying a contractor under construction contracts. The failure to make progress payments is usually not such a material breach as will authorize a contractor to abandon the work and sue for damages. However, a substantial failure by the owner to make progress payments, or an outright refusal to pay, entitles the contractor to rescind and recover the reasonable value of the labor and materials furnished.[50]

---

[47] Zobel & Dahl Constr. v. Crotty, 356 N.W.2d 42 (Minn. 1984).

[48] Burras v. Canal Constr. & Design Co., 470 N.E.2d 1362 (Ind. Ct. App. 1984).

[49] Building Maintenance Co., Eng. BCA No. 4115, 83-2 B.C.A. (CCH) ¶ 16,629 (1983).

[50] Integrated, Inc. v. Alec Fergusson Elec. Contractor, 250 Cal. App. 2d 287, 58 Cal. Rptr. 503 (1967).

CONTRACT CLAUSES WHICH DETER RECOVERY

§ 3.9 Exculpatory Clauses

Section 3.6 discussed the exculpatory clause in *Ell-Dorer Contracting Co. v. State.*[51] That type of specific disclaimer of responsibility for test results is rather common, particularly in public contracts.[52] However, such exculpatory clauses do not relieve the owner of responsibility for positive representations in the contract. In *E.H. Morrill Co. v. State,*[53] § 1A-12 of the contract provided:

> The site is situated on a terminal moraine. The soil is composed of granite boulders, cobbles, pebbles, and granite sand. Boulders which may be encountered in the site grading and other excavation work on the site vary in size from one foot to four feet in diameter. The dispersion of boulders varies from approximately six feet to twelve feet in all directions, including the vertical.

The Court said:

> Section 1A-12 did not purport merely to present the results of the state's own tests and investigations . . . but flatly asserts that the bidders could expect to confront only specified site conditions. It is clearly a "positive and material representation as to a condition presumably within the knowledge of the government . . ." (*Hollerbach v. United States* (1914) 233 U.S. 165, 169, 34 S.Ct. 553, 554, 58 L.Ed. 898). . .
>
> The responsibility of a governmental agency for positive representations it is deemed to have made through defective plans and specifications "is not overcome by the general clauses requiring the contractor to examine the site, to check up the plans, and to assume responsibility for the work . . ." (*United States v. Spearin,* 248 U.S. 132, 137, 39 S.Ct. 59, 61, 63 L.Ed. 166). Accordingly, the language in section 4 requiring the bidder to "satisfy himself as to the character . . . of surface and subsurface materials or obstacles to be encountered" cannot be relied upon to overcome those representations as to materials and obstacles which the state positively affirms in section 1A-12 not to exist, and plaintiff was entitled to rely and act thereon.[54]

In *Stenerson v. City of Kalispell,*[55] the lump-sum contract for rough grading of a golf course provided that the calculations on the "Rough Grading Plan" were

---

[51] 484 A.2d 356 (N.J. Super. Ct. App. Div. 1984).

[52] *See, e.g.,* Wunderlich v. State, 65 Cal. 2d 777, 423 P.2d 545 (1967).

[53] 65 Cal. 2d 787, 789–80, 423 P.2d 551, 553 (1967).

[54] *Id.* 792-93, 423 P.2d at 553–54.

[55] 629 P.2d 773 (Mont. 1981).

''approximate'' and that the contractors ''shall make [their] own determination as to the amount of topsoil and grading work to be done before submitting a bid.''[56] The contractors sued to recover the cost of moving an extra 27,477 cubic yards of earth which had not been included in the rough grading plan calculations. The owner argued that by reason of the language quoted above, the contractors had assumed the risk of extra quantities of earth. Nevertheless, in affirming the judgment awarding the contractors the total cost of the overrun, the Supreme Court of Montana said that contractors have the right to rely on plans and specifications furnished to them in bidding on and contracting a job. Further, the court stated:

> We affirmed the district judge in that decision but noted: ''We are not here holding that such exculpatory clauses may not be enforced in other situations, that detrimental reliance may be assumed in all cases, or that parties to such contracts are bound to exercise anything less than reasonable and prudent judgment. In other words we look to 'justifiable reliance.' ''[57]

§ 3.10 No Damage for Delay Clauses

A typical no damage for delay clause will be similar to the following:

> The contractor agrees to make no claim for damages for delay in the performance of this contract occasioned by any act or omission of the [owner] or any of its representatives and agrees that any such claim shall be fully compensated for by an extension of time to complete performance of the work as provided herein.[58]

The majority rule in the United States is that courts will generally enforce such a clause unless the delay is one (1) not contemplated by the parties; (2) amounting to an abandonment of the contract; (3) caused by bad faith of the owner; or (4) amounting to active interference by the owner.[59]

Some jurisdictions do not recognize the active interference exception,[60] or they construe the no damage for delay clause broadly (which favors the owner). Courts

---

[56] *Id.* 774.

[57] *Id.* 776.

[58] Kalisch-Jarcho, Inc. v. City of N.Y., 58 N.Y.2d 377, 488 N.E.2d 413, 461 N.Y.S.2d 746 (1983).

[59] Blake Constr. Co. v. C.J. Coakley Co., Inc., 431 A.2d 569, 579 (D.C. 1981) (''. . . delays . . . resulted from conduct amounting to active interference, largely due to Blake's improper work sequencing. Specifically, we refer to the trial court's findings of Blake's failure, after receiving notice from Coakley, to take effective steps to prevent further installation of pipes, ducts and electrical conduits contrary to the subcontract's terms; to provide heat for tarp-enclosed spaces; to assure that Coakley would have a reasonably clear work area; and to assure that fireproofing in place would not be damaged by other subcontractors or the elements.'').

[60] Kalisch-Jarcho, Inc. v. City of N.Y., 58 N.Y.2d 377, 448 N.E.2d 413, 461 N.Y.S.2d 746 (1983); Chicago College of Osteopathic Medicine v. George A. Fuller Co., 719 F.2d 1335 (7th Cir. 1983).

in other jurisdictions construe the clause narrowly,[61] thus favoring the contractor by limiting the types of delay to which the clause will apply.

REMEDIES

§ 3.11 Damages

A contractor's damage remedy will vary according to the situation giving rise to the claim. There are two main types of claims for which recovery may be sought: (1) claims for damages where the owner's breach prevents the contractor from completing the project; and (2) claims for extra costs incurred by the contractor because of extra work or delays.

§ 3.12 —Prevention of Completion

If the owner's breach prevents the contractor from completing work on the project, the contractor usually has a damage remedy in the amount of the profit the contractor could have expected to make had it been allowed to complete the work. This expected profit is usually measured as the contract price less the actual or estimated cost to the contractor of completing the work.[62] Some decisions provide that the contractor may additionally recover amounts already spent for material and labor, but must deduct prepayments or progress payments made by the owner.[63] In addition, a contractor may normally recover incidental and consequential damages arising from the owner's breach.[64]

[61] Giammetta Assocs., Inc. v. J.J. White, Inc., 573 F. Supp. 112 (E.D. Pa. 1983); U.S. Indus., Inc. v. Blake Constr. Co., 671 F.2d 539 (D.C. Cir. 1982). Neither of the clauses mentioned delays caused by suppliers, although the clause in *Blake* referred to delays due to the furnishing of materials. Held: the clauses do not apply to supplier-caused delays.

[62] Della Ratta, Inc. v. American Better Community Developers, Inc., 380 A.2d 627, 639 (Md. Ct. Spec. App. 1977); Holman v. Sorenson, 556 P.2d 499, 500 (Utah 1976): Williams v. Kerns, 265 S.E.2d 605, 606 (Ga. Ct. App. 1980).

[63] Williams v. Kerns, 265 S.E.2d 605, 606, 609 (Ga. Ct. App. 1980); First Atl. Bldg. Corp. v. Neubauer Constr. Co., 352 So. 2d 103, 104–05 (Fla. Dist. Ct. App. 1977) (contractor allowed to recover lost profit plus reasonable cost of labor and materials incurred in good faith in partial performance).

[64] *Consequential damages* include losses resulting from the breaching party's defective performance. *See* Restatement (2d) of the Law of Contracts § 347(b) and comment c (1981). In Gunderson's, Inc. v. James H. Tull, 678 P.2d 1061, 1064–65 (Colo. Ct. App. 1984), for example, the court awarded as consequential damages the cost of equipment leases where the equipment remained idle due to owner's breach. *Incidental damages* include costs incurred in a reasonable effort to mitigate losses. Restatement (2d) of the Law of Contracts § 347(b) and comment c (1981). In Spang Indus., Inc., Fort Pitt Bridge Div. v. Aetna Cas. & Sur. Co., 512 F.2d 365, 367, 370

Damage awards for lost profits plus consequential and incidental damages give expression to the general principle of contract law that damages normally fulfill a party's expectation interest. In awarding damages, the court tries to place the complaining party in the position it could have expected to be in had the contract been carried out.[65] A contractor's damage award is further subject to the general requirement, originating in the celebrated case of *Hadley v. Baxendale,*[66] that recovery for breach of contract includes only *foreseeable* damages.[67] Damages may be foreseeable either because they are likely to follow from the breach in the ordinary course of events, or because they are likely to result from special circumstances which the breaching party had reason to know of at the time the contract was formed.[68]

In construction cases, courts are likely to have little difficulty finding lost profits to have been a foreseeable element of damages following from an owner's breach. In *Della Ratta, Inc. v. American Better Community Developers, Inc.*, for example, the court found that the contractor's lost profits were foreseeable because the owner should have known that the contractor entered into the building contract to make a profit.[69]

The contractor may have more difficulty proving the foreseeability of consequential damages. To establish such damages, the contractor must prove that they follow from particular circumstances of which the owner had reason to know at the time the contract was formed. In *Traylor v. Henkels & McCoy, Inc.*, the court denied the contractor's claim for damages arising from the liquidation of the contractor's business, because the contractor failed to show that the owner had been aware that the contractor was depending on the owner's payments to stay in business.[70]

Another principle of contract law which may affect a contractor's recovery of consequential damages is the general requirement of *mitigation:* consequential damages which a claimant could have avoided are not recoverable. In *Gundersons, Inc. v. James H. Tull,*[71] for example, a contractor was denied recovery of the cost of retaining its superintendent after the owner terminated the project, since

---

(2d Cir. 1975), for example, the court upheld an allowance of damages covering extra costs incurred when contractor, in a good-faith effort to mitigate damages, embarked on a crash program to pour concrete before the onset of winter.

[65] Restatement (2d) of the Law of Contracts §§ 344, 347 (1981); Williams v. Kerns, 265 S.E.2d 605, 606 (Ga. Ct. App. 1980); Gunderson's, Inc. v. James H. Tull, 678 P.2d 1061, 1064 (Colo. Ct. App. 1984).

[66] 9 Ex. Ch. 341, 156 Eng. Rep. 145 (1854).

[67] Della Ratta, Inc. v. American Better Community Developers, Inc., 380 A.2d 627, 639 (Md. Ct. Spec. App. 1977); Ely v. Bottini, 179 Cal. App. 2d 287, 293–94, 3 Cal. Rptr. 756, 760 (1960). In *Ely*, a contractor sued a subcontractor rather than an owner. Where cited cases involve suits other than between contractors and owners, the principles they illustrate are of general application.

[68] Restatement (2d) of the Law of Contracts § 351(2) (1981).

[69] 380 A.2d 627, 639 (Md. Ct. Spec. App. 1977).

[70] 585 P.2d 970, 972–73 (Idaho 1979).

[71] 678 P.2d 1061, 1064–65 (Colo. Ct. App. 1984).

the contractor could have avoided this cost by temporarily laying off the superintendent. The court held, however, that the cost of maintaining long-term equipment leases was unavoidable and therefore recoverable.

All actual damages recovered by a contractor are subject to a final general contract law principle requiring that damages be certain. Unlike the foreseeability requirement, the certainty requirement may prevent recovery of lost profits as well as consequential damages. In *Della Ratta,* while the court held that lost profits were foreseeable, it denied the contractor's claim for lost profits because the contractor had failed to prove them with sufficient certainty.[72] While a contractor need prove the amount of its damages only with reasonable certainty, and not with absolute certainty, the court ruled that the contractor still must lay some foundation in the evidence enabling the factfinder to make a fair and reasonable estimate of the amount of damage.[73] The court found that the contractor had not proved lost profits beyond mere speculation.[74]

§ 3.13 —Extra Work

A different measure of damages determines the amount of a contractor's recovery for extra work. A contractor who incurs extra expense as a result of changes ordered after a project has begun may generally recover those extra costs as damages. The contractor's recovery may be based on a provision in the contract providing for equitable adjustments, or, where the extra work was not covered by a specific contract clause, on a quantum meruit theory, allowing the contractor to obtain the reasonable value of services and material benefits conferred on the owner.[75] Some courts may under certain circumstances allow quantum meruit recovery for extra costs even where an express contract exists.[76] Under either an equitable adjustments clause or a quantum meruit theory, recoverable extra costs include the costs of materials, labor, inefficiency due to owner's breach or changes, subcontractors, overtime, and additional equipment.[77]

---

[72] Della Ratta, Inc. v. American Better Community Developers, Inc., 380 A.2d 627, 639 (Md. Ct. Spec. App. 1977).

[73] *Id.* 641.

[74] *Id.*

[75] Affholder, Inc. v. Southern Rock, Inc., 736 F.2d 1007, 1011, 1014 (5th Cir. 1984) (recovery under specific contract provision); Black Lake Pipe Line Co. v. Union Constr. Co., Inc., 538 S.W.2d 80, 86 (Tex. 1976) (quantum meruit).

[76] *E.g.,* Southern Bell Tel. & Tel. Co. v. Acme Elec. Contractors, Inc., 418 So. 2d 1187 (Fla. 1983).

[77] *See, e.g.,* Affholder, Inc. v. Southern Rock, Inc., 736 F.2d 1007, 1014 (5th Cir. 1984) (cost of materials); *In re* King Enters., 678 F.2d 73, 77 (8th Cir. 1982) (cost of materials, labor costs); Black Lake Pipe Line Co. v. Union Constr. Co., Inc., 538 S.W.2d 80, 84 (Tex. 1976) (inefficiency costs); Union Bldg. Corp. v. J&J Bldg. & Maintenance Contractors, Inc., 578 S.W.2d 519, 520 (Tex. Civ. App. 1979) (subcontractors' extra costs); Natkin & Co. v. George A. Fuller Co., 347 F. Supp. 17, 35 (W.D. Mo. 1972) (overtime costs and additional equipment costs); General Ins. Co. of Am. v. Hercules Constr. Co., 385 F.2d 13, 21 (8th Cir. 1967) (overtime costs; suit by contractor against subcontractor's surety).

In contractors' suits against private owners, as in suits under government contracts, the acceptable methods of proving the extra costs resulting from a breach are a matter of some controversy. Most courts would prefer the contractor to give evidence showing the actual costs caused by each change or breach.[78] Such precision in proving causation and damages may be difficult, however, and contractors may resort to the *total cost method* to prove damages. Under the total cost method, the extra costs resulting from a change or breach are computed by deducting the bid estimate from the actual cost experienced by the contractor.[79]

The inadequacies of the total cost theory have long been recognized by the courts. The method may produce an inaccurate figure for the extra costs caused by a change or breach either if the bid was unrealistically low, or if factors other than the change or breach in question contributed to the added costs.[80] Some courts, however, may allow a contractor to prove extra costs using the total cost method under certain restricted circumstances which minimize its potential inaccuracies. The Utah Supreme Court, for example, has listed four conditions under which the total cost method is acceptable for determining damages for extra work: (1) if the nature of the particular losses makes it impossible or impracticable to determine the losses accurately; (2) if the bid or estimate was realistic; (3) if the actual costs incurred were reasonable; and (4) if the plaintiff was not itself responsible for the added costs.[81]

Principal elements of damages recoverable by a contractor for owner-caused delay include extended overhead and equipment rental costs and escalation costs.[82] Overhead and equipment rental costs are recoverable as delay damages only if incurred for an extended period of time. Extended overhead costs may be of two types: home-office overhead, including general and administrative expenses at the central office, such as rents and salaries; and field or job site overhead, including on-site expenses such as telephones, field payroll, electricity, and rents.[83]

Further controversy surrounds the method of calculating the home-office overhead[84] expenses attributable to a contract. In government contract cases, the

---

[78] *See, e.g.*, Highland Constr. Co. v. Union Pac. R.R. Co., 683 P.2d 1042, 1046-47 (Utah 1984).

[79] *Id.* 1046.

[80] *E.g.*, John F. Harkins Co. v. School Dist. of Philadelphia, 460 A.2d 260, 265 (Pa. Super. Ct. 1983) (suit by contractor against public entity).

[81] Highland Constr. Co. v. Union Pac. R.R. Co., 683 P.2d 1042, 1047 (Utah 1984).

[82] Guy James Constr. Co. v. Trinity Indus., Inc., 644 F.2d 525, 532–33 (5th Cir. 1981) (extended overhead; contractor sued supplier); Arcon Constr. Co. v. South Dakota Cement Plant, 349 N.W.2d 407, 413–15 (S.D. 1984) (extended equipment rental; contractor sued supplier); Higgins v. City of Fillmore, 639 P.2d 192, 194 (Utah 1981) (escalation; contractor sued public owner).

[83] *E.g.*, District Concrete Co., Inc. v. Bernstein Concrete Corp., 418 A.2d 1030, 1038 (D.C. 1980) (field overhead).

[84] Some courts require the contractor to prove that the extended home-office overhead costs claimed were additional to the overhead expenses the contractor would have experienced without the delay; i.e., that but for the delay other jobs would have been obtained to absorb the overhead. *See, e.g.*, Guy James Constr. Co. v. Trinity Indus., Inc., 644 F.2d 525, 532–33 (5th Cir. 1981). Other courts reject this strict requirement of proof and allow recovery of home-office expenses if they are merely allocated to the company's various jobs on some fair basis. *See, e.g.*, General Ins. Co. of Am. v. Hercules Constr. Co., 385 F.2d 13, 23 (8th Cir. 1967).

Board of Contract Appeals developed the so-called *Eichleay formula* for determining home-office overhead costs. Under this formula, the amount of overhead allocable to a particular job is calculated as the ratio of billings for the particular job to the total billings by the company for that period.[85] The acceptability of the *Eichleay* formula is somewhat uncertain at present, since several later decisions have rejected its use.[86] The Federal Circuit, however, has recently upheld the use of the formula,[87] and a contractor may be able to persuade a court to accept the formula in computing damages against a private owner.[88]

§ 3.14 Rescission and Restitution

In some suits equitable remedies rather than damages may be available to the contractor against the owner. The contractor may be able to rescind the contract and restore the precontract conditions through restitution of any benefits conferred by one side on the other. If the contract as drafted fails to reflect the agreement reached by the parties, the contractor may be able to reform the contract to bring it into conformance with the agreement.

Under general principles of contract law, rescission is available to a party if the contract is based on a mutual mistake of fact. The mistake must be material or concern a basic assumption underlying the contract. The *Restatement (Second) of Contracts* accordingly recognizes that rescission is available to the adversely affected party if the mutual mistake goes to a basic assumption on which the contract was made and has a material effect on the agreed exchange of performances.[89]

In construction cases, a number of courts have stated a four-part test for rescission based on mutual mistake. First, the mistake must go to a material feature of the contract. Second, the mistake must be of grave consequence so that enforcement of the contract would be unconscionable. Third, the mistake must have occurred despite the exercise of reasonable care. Finally, it must be possible to place the other party in the status quo ante (that is, to restore it to its precontract condition).[90]

In *John Burns Construction Co.*, the court granted rescission of the contract where subsoil conditions were concededly a material feature of the contract to construct a sewer pipeline, and where the problems encountered with subsoil conditions were beyond the contemplation of the parties so that it would have been unconscionable to enforce a predetermined price schedule for extra work.[91] The court

[85] Eichleay Corp., 60-2 B.C.A. (CCH) ¶ 2688 (A.S.B.C.A. 1960).

[86] *E.g.*, Berley Indus., Inc. v. City of N.Y., 45 N.Y. 2d 683, 385 N.E.2d 281, 412 N.Y.S.2d 589 (1978).

[87] Capital Elec. Co. v. United States, 729 F.2d 743, 747 (Fed. Cir. 1984).

[88] *E.g.*, PDM Plumbing & Heating, Inc. v. Findlan, 431 N.E.2d 594, 595 (Mass. App. Ct. 1982) (subcontractor sued both contractor and owner).

[89] Restatement (2d) of the Law of Contracts § 1521 (1981).

[90] *E.g.*, John Burns Constr. Co. v. Interlake, Inc., 433 N.E.2d 1126 (Ill. App. Ct. 1982).

[91] *Id.* 1130.

further found that the contractor had exercised reasonable care.[92] The court held, however, that restoration of the status quo ante was not a prerequisite for rescission where restoration has been made impossible through no fault of the party seeking rescission.[93]

When a court allows rescission of a contract, any benefits conferred by either party on the other generally must be restored to the party which possessed them before the contract was made.[94] Where a contractor has performed some work under the rescinded contract, restitution of benefits conferred takes the form of the recovery of the reasonable value of services rendered and material furnished.[95]

Under certain circumstances, rescission may be available to a contractor for unilateral mistakes in bidding. When the contractor discovers its bid mistake and seeks to withdraw its bid before acceptance by the owner, some courts have allowed rescission under essentially the same four conditions enumerated by the *John Burns* court for mutual mistake.[96] After a bid has been accepted and a contract formed, however, courts may be more reluctant to allow withdrawal of the bid and rescission of the contract due to unilateral mistake.[97] Rescission is more readily allowed if the owner knew or should have known of the bid mistake.[98]

A contractor permitted to withdraw its bid generally will not be required to forfeit its bid bond or deposit. The bond or deposit will be restored to the surety or contractor.[99]

In addition to mutual and unilateral mistake, material breach by the owner may be grounds for rescission of a contract by a contractor. The most commonly litigated material breach is the owner's failure to make progress payments.[1] In *Brady Brick*

---

[92] *Id.* 1131.

[93] *Id.*

[94] Restatement (2d) of the Law of Contracts § 376 (1981).

[95] Carnicle v. Swann, 314 N.W.2d 311, 313 (S.D. 1982).

[96] *E.g.*, Smith & Lowe Constr. Co. v. Herrera, 79 N.M. 239, 442 P.2d 197, 198 (1968). In the *Smith* case, the court permitted withdrawal of the bid where (1) the mistake was of such grave consequence that to enforce the contract would have been unconscionable; (2) the mistake concerned a material and fundamental feature of the contract; (3) the bidder acted in good faith and its mistake did not arise from the violation of a legal duty or from gross negligence; (4) the bidder was reasonably prompt in giving notice of the error; and (5) the offeree's status had not changed or the offeree had been restored to the status quo so that it suffered no serious hardship or prejudice other than the loss of the bargain. *See also* City of Baltimore v. De Luca-Davis Constr. Co., 124 A.2d 557, 562 (Md. 1956).

[97] *See* P. Bruner, *Mistakes in Bids,* 4 Construction Briefings 5 (Sept. 1978), in Construction Briefings Collection 43, 47 (1982) (stating that after a bid has been accepted, rescission will be allowed for unilateral mistake only if contractee knew or should have known of the error). Some courts, however, allow rescission after acceptance of the bid on basically the same grounds as before acceptance. *See, e.g.*, John J. Calnan Co. v. Talsma Builders, Inc., 367 N.E.2d 695, 698 (Ill. 1977).

[98] City of Baltimore v. De Luca-Davis Constr. Co., 124 A.2d 557, 562 (Md. 1956), citing Black, Rescission and Cancellation § 130 (2d ed. 1929).

[99] *See, e.g.*, M.F. Kemper Constr. Co. v. City of Los Angeles, 37 Cal. 2d 696, 705, 235 P.2d 7, 12 (1951); City of Baltimore v. De Luca-Davis Constr. Co., 124 A.2d 557, 567 (Md. 1956).

[1] See **§ 3.8.**

& Supply Co. v. Lotito,[2] the contract provided for payments to be made to the contractor as the job progressed. The owner refused the contractor's second request for payment and the trial court held that this was a material breach justifying rescission.[3] The court of appeals upheld the trial court's award of rescission and restitution of the value of labor and materials already furnished under the contract, finding that the trial court's verdict was not manifestly against the weight of the evidence.[4]

§ 3.15 Reformation

The remedy of reformation is generally available only where a contract as drafted fails to reflect the agreement actually reached by the parties because of mutual mistake as to the contents or effect of a writing. Under such circumstances, the instrument may be reformed so as to reflect the agreement accurately.[5]

In construction contracts, a contractor may seek reformation based on claims that the scope of the work agreed to by the parties is not accurately expressed in the contract document.[6] Alternatively, the contractor may dispute whether the bid price contained in the document accurately reflects the agreed price.[7] In the *Paterson* case, for example, both the contractor and the owner suspected a mistake in the bid at the time the bids were opened, but upon checking the computations, the contractor found no error.[8] After the bid was accepted and the contract was formed, the contractor discovered the error. The trial court granted reformation of the contract, but the court of appeals reversed on the grounds that the defendant-owner at no time agreed to any price other than the price contained in the contract document.[9] Since the error was not a mutual mistake about the contents of the instrument, reformation was unavailable.[10]

§ 3.16 Termination

A termination clause allows the owner to terminate a contract either by reason of a default by the contractor or at the convenience of the owner. If a contractor

---

[2] 356 N.E.2d 1126, 1130–31 (Ill. App. Ct. 1976).

[3] *Id.*

[4] *Id. See also* Integrated, Inc. v. Alec Fergusson Elec. Contractor, 250 Cal. App. 2d 287, 295–97, 58 Cal. Rptr. 503, 509–10 (1967) (substantial failure to make progress payments constitutes substantial breach justifying rescission).

[5] Restatement (2d) of the Law of Contracts § 155 (1981).

[6] Ed Sparks & Sons v. Joe Campbell Constr. Co., 578 P.2d 681, 682–84 (Idaho 1978).

[7] Paterson v. Board of Trustees of Montecito Union School Dist., 157 Cal. App. 2d 811, 321 P.2d 825 (1958).

[8] *Id.* 814, 321 P.2d at 826.

[9] *Id.* 814–15, 321 P.2d at 828.

[10] *Id.* 817, 321 P.2d at 828. The court also found that defendant, a public body, was not estopped from refusing to reform. 157 Cal. App. 2d at 819, 321 P.2d at 829.

believes that it has been improperly or wrongfully terminated by the owner, the contractor may have a suit against the owner for breach of contract. In *Allen Engineering Corp. v. Lattimore,*[11] a cement contractor sued a builder for breach after the builder terminated the contract. The contract's termination clause allowed the builder to terminate if the cement contractor caused undue delay or violated any condition in the contract.[12] The builder alleged that the cement contractor had violated the terms of the contract by not obtaining the required worker's compensation insurance, but the court of appeals upheld the trial court's finding that no violation had occurred.[13] The court affirmed the lower court's verdict for the cement contractor for breach of contract.[14]

Termination for convenience clauses allow owners to terminate a contract without cause and therefore can give rise to no such breach of contract actions.[15] These clauses are most often found in government contracts, or subcontracts under government contracts, but may appear in private contracts as well.[16] A contractor terminated for convenience may still, however, have an action to recover costs incurred plus profits already earned.[17]

§ 3.17 Liens

The laws of every state provide a remedy in the form of a mechanic's or materialman's lien for anyone who has contributed labor or materials to a construction project, including a contractor.[18] Since the mechanic's lien is entirely a creature of state statute, a contractor seeking to foreclose such a lien must look to the particular details of the statute in force in the jurisdiction in which the project is

---

[11] 201 A.2d 13, 13–15 (Md. Ct. Spec. App. 1964).

[12] *Id.* 14.

[13] *Id.* 15.

[14] *Id.* 16. *See also, e.g.*, Pathman Elec. Co. v. Hi-Way Elec. Co., 382 N.E.2d 453, 455–56, 458 (Ill. App. Ct. 1978) (subcontractor alleged contractor's termination for cause was improper because contractor failed to comply with notice requirements, but court upheld verdict for contractor because subcontractor had raised the issue for the first time on appeal).

[15] *See, e.g.*, Nolan Bros., Inc. v. United States, 405 F.2d 1250, 1253 (Ct. Cl. 1969) (termination for convenience clause gave the fullest discretion to the contracting officer to terminate work in best interests of the government).

[16] *E.g., id.;* United States Propellers v. Zenith Plastics Co., 187 Cal. App. 2d 780, 10 Cal. Rptr. 137 (1960) (government contracts or subcontracts thereunder); Arc Elec. Constr. Co. v. George A. Fuller Co., 299 N.Y.S.2d 129, 131 (N.Y. 1969) (private contract; clause allowed intermediate contractor to terminate subcontract at any time prior to completion).

[17] *See* Nolan Bros., Inc. v. United States, 405 F.2d 1250, 1253 (Ct. Cl. 1969) (government contract).

[18] *See, e.g.*, Cal. Civ. Code § 3110 (West 1974); *see also* Lifschitz & Little, *Mechanics' Liens: Basic Principles and Guidelines,* 5 Construction Briefings 1, 1 (Sept. 1982). Since stop notices are similar to mechanic's liens, this section does not discuss them directly. Those interested in differences in the details between these two remedies should consult their state's statutes. For a discussion of one state's stop notice remedy, *see* J. Acret, Attorney's Guide to California Construction Contracts and Disputes 151–52 (1976).

located.[19] Different states' mechanic's lien statutes vary in many specific respects, including, for example, the date from which the mechanic's lien exists[20] and the notice requirements imposed on the claimant under the lien.[21]

The amount the contractor may recover under a mechanic's lien generally depends on the nature of the contract and claim. Under a fixed-price contract, the contractor may recover the contract price less amounts already paid by the owner or credits due the owner.[22] Where the contract is not a fixed-price contract, or where recovery is sought for extra work, the amount of the contractor's recovery will be the reasonable value of labor and materials.[23] In a cost-plus contract, the contractor's recovery may include profit and overhead if specified in the contract.[24]

Unless the mechanic's lien statute provides that the statutory lien is the exclusive remedy, an equitable lien may in some instances be available if the claimant fails to perfect the mechanic's lien. A case in point is *Peninsular Supply Co. v. C.B. Day Realty.*[25] In this case, a materialman not in privity with the owner failed to comply with the 45-day notice requirement to perfect a mechanic's lien.[26] The court nevertheless held that the materialman had an equitable lien claim, since the mechanic's lien statute explicitly provided that the statutory lien was not the exclusive remedy.[27]

---

[19] *See, e.g.*, Wingler v. Niblack, 374 N.E.2d 252, 253 (Ill. App. Ct. 1978).

[20] In most states, the lien is deemed to exist from a date prior to the filing of the lien claim, i.e., the lien "relates back" to a period predating the lien claim. *See, e.g.*, Hodgins v. Marquette Iron Mining Co., 503 F. Supp. 88, 90 (D. Mich. 1980) (under Michigan law, liens attach at time of commencement of work). In some states, however, the lien does *not* relate back to a time before filing of the claim. *See, e.g.*, Mervin L. Blades & Son, Inc. v. Lighthouse Sound Marina & Country Club, 377 A.2d 523, 526 (Md. Ct. Spec. App. 1977) (under Maryland law, lien attaches only after claimant prevails on lien claim in court).

[21] *See, e.g.*, Peninsular Supply Co. v. C.B. Day Realty, 423 So. 2d 500, 501 (Fla. Dist. Ct. App. 1982) (notice to owner required within 45 days of first furnishing materials); William Moors, Inc. v. Pine Lake Shopping Center, Inc., 253 N.W.2d 658, 659 (Mich. Ct. App. 1977) (notice required within 90 days of first furnishing material or supplying labor).

[22] *See, e.g.*, Knoell Constr. Co. v. Hanson, 287 N.W.2d 435, 436, 438 (Neb. 1980).

[23] *See, e.g.*, Wingler v. Niblack, 374 N.E.2d 252, 253–54 (Ill. Ct. App. 1978) (extra work); Maxwell v. Anderson, 593 P.2d 29, 30, 32–33 (Mont. 1979) (where "estimated cost ceiling" in contract was not a firm limit, contractor could recover the value of labor and materials).

[24] *See, e.g.*, Sloane v. Malcolm Price, Inc., 339 A.2d 43, 45 n.1 (D.C. 1975). *See also generally* Lifschitz & Little, *Mechanics' Liens: Basic Principles and Guidelines*, 5 Construction Briefings 1, 3 (Sept. 1982).

[25] 423 So. 2d 500 (Fla. Dist. Ct. App. 1982).

[26] *Id.* 501–02.

[27] *Id.*

CHAPTER 4

SUING A GOVERNMENT: SPECIAL CONSIDERATIONS

Robert D. Wallick

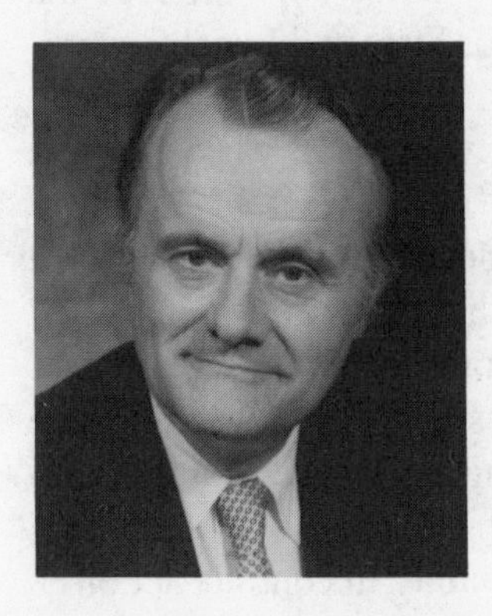

Robert D. Wallick is licensed to practice law in the District of Columbia, Maryland and before various federal courts and agencies and has been a partner at Steptoe & Johnson, Washington, D.C., since 1963. His principal practice has been government contract law, acquisition issues and related litigation. He holds a B.S. Degree in Electrical Engineering and a B.S. Degree in Business Administration from Lehigh University, Bethlehem, Pennsylvania, and received an L.L.B. Degree from George Washington University in 1956. Mr. Wallick is a past chairman of the American Bar Association's Section of Public Contract Law, president elect of the Federal Circuit Bar Association, a member of the Council of the U.S. Court of Claims, a past chairman of a national subcommittee of the ABA's Model Procurement Code Committee, a past president of the National Assistance Management Association and an active participant on various bar associations and other committees. He lectures and writes frequently on government contract issues.

§ 4.1 Introduction

Governments and governmental bodies—federal, state, and local—account for much of today's construction work. Because of inherent governmental complexities and their special legal status, contracting with them is different from contracting with private owners. Special pitfalls and hurdles await the aggrieved contractor who contemplates a claim against a governmental body.

§ 4.2 Sovereign Immunity and Its Waiver

The doctrine of sovereign immunity, which holds that a government may not be sued without its consent, is ancient but not universally hallowed. Part 9 of the American Bar Association's Model Procurement Code for State and Local Governments (ABA Model Code) calls for waiver of sovereign immunity for contracts. The federal and many state governments have enacted waivers in various forms and degrees.

Nevertheless, considerable vitality remains in the old doctrine. A few states have not waived immunity on construction contracts. Where there is a waiver, it is often limited to special tribunals and coupled with rigid temporal, procedural, and definitional strictures. The immunity doctrine and prejudices in favor of the doctrine may weigh heavily in judicial interpretations of unclear legislative grants of authority or relief.

In 1976, the State of Maryland waived sovereign immunity in contract actions.[1] Original jurisdiction over contract controversies with the state and its agencies was ceded to a special three-member Board of Contract Appeals.[2] In a case which arose before that encouraging development, *Charles E. Brohawn & Brothers, Inc. v. Board of Trustees,*[3] the Maryland Court of Appeals reaffirmed that the doctrine applied generally in breach of contract actions against a state agency,[4] explaining that

---

[1] Law of May 4, 1976, ch. 450, 1976 Md. Laws 1180 (codified at Md. State Gov't Code Ann. §§ 12-201 to 12-204; Md. Ann. Code art. 23A, § 1A (Supp. 1984); art. 25, § 1A (Supp. 1984); art. 25B, § 13A (Supp. 1984).

[2] Md. Ann. Code art. 21, § 7-202 (Supp. 1984).

[3] 304 A.2d 819, 820 (Md. 1973).

[4] Since *Brohawn* was a state court action, the Eleventh Amendment, which deprives the federal courts of jurisdiction, was not, as such, applicable. As academic propositions, there should be no significant distinction between the state's rights under the Eleventh Amendment and the state common law right of sovereign immunity. *See* Jones v. Scofield Bros., 73 F. Supp. 395, 396 (D. Md. 1947).

> a litigant is precluded from asserting an otherwise meritorious cause of action against this sovereign State or one of its agencies which has inherited its sovereign attributes, unless expressly waived by statute or by a necessary inference from such a legislative enactment. . . . And in the absence of statutory authorization, neither counsel for the State nor any of its agencies may, "either by affirmative action or by failure to plead the defense, waive the defense of governmental immunity. . . ." *Bd. of Education v. Alcrymat Corp.*, 258 Md. 508, 266 A.2d 349 (1970).

That is a good summary of this ancient doctrine—a doctrine which is viewed by some as a taxpayer's finger in the dike and by others as unfair, unwise, and unnecessary.

§ 4.3 Waivers by the Federal Government

The federal government's sovereign immunity in contract actions has been waived by three statutes—the Contract Disputes Act of 1978,[5] the Federal Courts Improvement Act of 1982,[6] and the Administrative Procedure Act[7]—and by the progeny of a landmark decision, *Scanwell Laboratories, Inc. v. Shaffer*.[8] The federal waivers are broad, extending to pre-award and post-award controversies; but, as explained in §§ **4.4** and **4.5**, the waivers are not unlimited.

§ 4.4 –Waiver for Contract Controversies

The Contract Disputes Act[9] governs controversies which arise after a procurement contract is formed. It applies to any express or implied contract for procurement of property (other than real property in being), services, or construction, alteration, repair, or maintenance of real property.[10] Jurisdiction under the Act is initiated as follows:

> All claims by a contractor against the government relating to a contract shall be in writing and shall be submitted to the contracting officer for a decision. All claims by the government against a contractor relating to a contract shall be the subject of a decision by the contracting officer.[11]

---

[5] 41 U.S.C. § 601 *et seq.* (1982).

[6] 28 U.S.C. §§ 1295, 1491 (1982).

[7] 5 U.S.C. § 702 (1982).

[8] 424 F.2d 859 (D.C. Cir. 1970).

[9] 41 U.S.C. § 601 *et seq.* (1982).

[10] 41 U.S.C. § 602(a) (1982).

[11] *Id.* § 605(a). In addition to its jurisdiction under the Contract Disputes Act over appeals of contracting officer decisions, the United States Claims Court has jurisdiction under 28 U.S.C. § 1491(a)(1) (1982):

> . . . to render judgment upon any claim against the United States founded either upon the Constitution, or any Act of Congress or any regulation of an executive department,

The contractor may appeal the contracting officer's decision either to an agency board of contract appeals (within 90 days of receipt of the decision) or to the United States Claims Court (within 12 months of the decision). The time limitations are jurisdictional.[12]

In either tribunal, the contractor is entitled, in effect, to a de novo hearing and determination on its claim.[13] Both forums have subpoena power and both allow discovery. Decisions by the agency boards and their pre-Contract Disputes Act progenitors and by the United States Claims Court are reported by private legal publishing services.[14]

The agency boards and the Claims Court are generally well respected and viewed by the private and government bars as professional and independent. Board membership is governed by 41 U.S.C. § 607(b)(1). The United States Claims Court is an Article I (Executive Department) court. Its 16 judges are appointed by the President, with approval by the Senate, to renewable 15-year terms.

The procedures followed by the boards and the Claims Court differ considerably. These differences may be of significance for a particular case in terms of time and cost of litigation. For example, pursuant to statutory requirements, the agency boards provide accelerated procedures for claims of less than $10,000 and expedited procedures for those of less than $50,000.[15] On the other hand, reimbursement of fees and costs under the Equal Access to Justice Act[16] may be available for small contractors (as defined in the Act) when the government litigates an unreasonable position before the Claims Court,[17] but not when it does so before an agency board.[18]

Judgments by the boards and the United States Claims Court under the Contract Disputes Act are payable "promptly" and ab initio from funds appropriated for the payment of judgments.[19] Appeals of both board and Claims Court decisions

---

or upon any express or implied contract with the United States, or for liquidated or unliquidated damages in cases not sounding in tort.

This parallels a provision in the now-superseded Tucker Act, formerly 28 U.S.C. § 1491 (1970 & Supp. V 1975). Prior to March 1, 1979, the effective date of the Contract Disputes Act, waiver of federal sovereign immunity over post-award procurement contract controversies had been under the counterpart general grant of jurisdiction to the United States Court of Claims under the Tucker Act. 28 U.S.C. § 1491(a)(1) continues to have vitality for actions not covered by the Contract Disputes Act, that is, other than procurement contract actions. *See, e.g.*, Chevron Chem. Corp. v. United States, 5 Cl. Ct. 807 (1984), where a partial summary judgment was issued for breach of a settlement and disposition agreement covering the disposal of silvex products.

[12] 41 U.S.C. §§ 606, 609(a) (1982).

[13] *See id.* § 605(b) (1982).

[14] Agency board decisions are reported by Commerce Clearing House in *Board of Contract Appeals Decisions* (B.C.A.); Claims Court decisions appear in West Publishing Company's *Claims Court Reporter*.

[15] 41 U.S.C. §§ 605(c)(1), 608 (1982).

[16] 28 U.S.C. § 2412 (1982).

[17] *See* Broad Ave. Laundry & Tailoring v. United States, 693 F.2d 1387 (Fed. Cir. 1982).

[18] *See* Fidelity Constr. Co. v. United States, 700 F.2d 1379 (Fed. Cir. 1983), *cert. denied*, 104 U.S. 97 (1984).

[19] 41 U.S.C. § 612 (1982).

rendered since October 1, 1982, are to the United States Court of Appeals for the Federal Circuit.[20] The standard of review of agency board decisions is prescribed by statute to be as follows:

> [T]he decision of the agency board on any question of law shall not be final or conclusive, but the decision on any question of fact shall be final and conclusive and shall not be set aside unless the decision is fraudulent, or arbitrary, or capricious, or so grossly erroneous as to imply bad faith, or if such evidence is not supported by substantial evidence.[21]

The standard of review of Claims Court decisions is the *clearly erroneous* standard applicable generally to federal circuit court reviews of lower court decisions.[22]

There are significant limitations upon the waiver of immunity granted by the Contracts Disputes Act. It extends only to the types of procurement contracts described in 41 U.S.C. § 602. Claims over $50,000 must have been certified.[23] Times for filing are jurisdictional. Interest does not start to accumulate until the claim has been received by the contracting officer, and, for claims over $50,000, interest does not start until after receipt of a certified claim.[24]

Moreover, equitable and declaratory relief[25] and punitive damages against the government appear to be outside the reach of the Contract Disputes Act's jurisdictional grant. Under the predecessor Tucker Act,[26] relief granted by the United States Court of Claims was limited to compensatory monetary damages;[27] the agency boards exercised a similarly limited jurisdiction. The Contract Disputes Act was grafted onto that existing process. Procedures under the Act are triggered by the filing of a claim (or certified claim). This (and other language in the Act) suggests that Congress intended to limit relief to compensatory money damages. Nothing in the Act suggests more.[28]

The Tucker Act's grant of jurisdiction over federal contract disputes was held to be exclusive and adequate.[29] Procurement contractors could not seek equitable

---

[20] 28 U.S.C. § 1295(a)(3), (10) (1982).

[21] 41 U.S.C. § 609(b) (1982).

[22] *See* Alger v. United States, 741 F.2d 391, 393 (Fed. Cir. 1984); *see also* Universal Minerals, Inc. v. C.A. Hughes & Co., 669 F.2d 98 (3d Cir. 1981).

[23] 41 U.S.C. § 605(c)(2) (1982). *See* W.M. Schlosser Co. v. United States, 705 F.2d 1336 (Fed. Cir. 1983).

[24] *See* Fidelity Constr. Co. v. United States, 700 F.2d 1379 (Fed. Cir. 1983), *cert. denied*, 104 U.S. 97 (1984).

[25] *See* Willow Beach Resort, Inc. v. United States, 5 Cl. Ct. 241, 243 (1984).

[26] See note 12 *supra* for a discussion of the former Tucker Act.

[27] *See* United States v. King, 395 U.S. 1 (1968).

[28] *See* Fidelity Constr. Co. v. United States, 700 F.2d 1379, 1383 (Fed. Cir. 1983), *cert. denied*, 104 U.S. 97 (1984), where the Federal Circuit, in concluding that interest on a claim over $50,000 did not accrue until the claim was certified, reasoned that "[n]ot only may the consent [to be sued] not be implied, but even a seemingly explicit consent will not be effective if the language used appears too sweeping and contrary to the overall statutory scheme as judicially deduced."

[29] International Eng'g Co. v. Richardson, 512 F.2d 573, 578 (D.C. Cir. 1975), *cert. denied*, 423 U.S. 1048 (1976); Warner v. Cox, 487 F.2d 1301 (5th Cir. 1974). In *Warner*, the court noted

or declaratory relief in United States district courts under the Administrative Procedures Act.[30]

§ 4.5 —Waiver for Bid Protests

In 1970, the United States Court of Appeals for the District of Columbia Circuit first established that an aggrieved bidder for a federal contract had "standing to sue," i.e., had a right to judicial review.[31] Thirty years earlier, in *Perkins v. Lukens Steel Co.*,[32] the Supreme Court had held to the contrary. In *Scanwell*, the District of Columbia Circuit based its conclusion that bidders had standing to sue upon § 10(a) of the Administrative Procedure Act,[33] which was enacted in 1946, subsequent to the *Lukens Steel* decision.

When confronted with *Scanwell* situations, the other circuit courts split. Thus, in *GF Business Equipment, Inc. v. TVA*,[34] the Sixth Circuit affirmed a district court decision that an aggrieved bidder on a direct federal procurement contract did not have standing to sue because the bidder was not within the protected "zone of interest."[35] However, many courts followed *Scanwell*.[36]

In *M. Steinthal & Co. v. Seamans*,[37] the D.C. Circuit set out what has become the standard for judicial review of procurement solicitation and award decisions where *Scanwell* is followed, concluding that:

> (1) courts should not overturn any procurement determination unless the aggrieved bidder demonstrates that there was no rational basis for the agency's decision; and

---

that "the APA does not provide for review under such circumstances. Specifically exempted from review is agency action for which there is some '. . . other adequate remedy in a court.' 5 U.S.C. & 704. Suit under the Tucker Act in the Court of Claims has been held such an adequate remedy." 487 F.2d at 1304.

[30] 5 U.S.C. § 702 (1982).

[31] Scanwell Laboratories, Inc. v. Shaffer, 424 F.2d 859 (D.C. Cir. 1970). We note, but do not here discuss, the fact that the competition in Contracting Act, 40 U.S.C. § 759(h) (Supp. 1985) provided a new forum for protests by aggrieved bidders for ADP contracts: the General Services Administration's Board of Contract Appeals.

[32] 310 U.S. 113 (1940).

[33] 5 U.S.C. § 702 (1982).

[34] 430 F. Supp. 699 (E.D. Tenn. 1975), *aff'd mem.*, 556 F.2d 581 (6th Cir. 1977).

[35] *See also* Cincinnati Elecs. Corp. v. Kleppe, 509 F.2d 1080 (6th Cir. 1975).

[36] *See, e.g.*, B.K. Instrument, Inc. v. United States, 715 F.2d 713 (2d Cir. 1983); Merriam v. Kunzig, 475 F.2d 1233 (3d Cir.), *cert. denied*, 414 U.S. 911 (1973); William F. Wilke, Inc. v. Department of the Army, 485 F.2d 180 (4th Cir. 1973); Hayes Int'l Corp. v. McLucas, 509 F.2d 247 (5th Cir.), *cert. denied*, 423 U.S. 864 (1975); Armstrong & Armstrong, Inc. v. United States, 514 F.2d 402 (9th Cir. 1975); Airco, Inc. v. Energy Research & Dev. Admin., 528 F.2d 1294 (7th Cir. 1975); *but see* People's Gas, Light & Coke Co. v. United States Postal Serv., 658 F.2d 1182 (7th Cir. 1981).

[37] 455 F.2d 1289 (D.C. Cir. 1971).

> (2) even in instances where such a determination is made, there is room for sound judicial discretion, in the presence of overriding public interest considerations, to refuse to entertain declaratory or injunctive actions in a pre-procurement context.[38]

Under the *Steinthal* standard of review, aggrieved bidders on federal procurements have had little success on the merits. Nonetheless, *Scanwell* and *Steinthal* opened up the possibility of judicial review to an aggrieved bidder with a strong case. The possibility of injunctive relief to maintain the status quo pending a General Accounting Office decision also existed, even if prospects for ultimately prevailing judicially against a well-reasoned administrative decision appeared unlikely.

The Court Reform Act of 1982[39] complicated this area. The jurisdictional grant to the United States Claims Court under that Act includes the following:

> (3) To afford complete relief on any contract claim brought before the contract is awarded, the court shall have exclusive jurisdiction to grant declaratory judgments and such equitable and extraordinary relief as it deems proper, including but not limited to injunctive relief. In exercising this jurisdiction, the court shall give due regard to the interests of national defense and national security.[40]

The government has argued that, as a result of the Court Reform Act, the district courts have no *Scanwell* jurisdiction (i.e., the Claims Court jurisdiction is exclusive) and that the Claims Court's jurisdiction is limited to legal actions initiated prior to award. Many in the contracting community have argued that, based on language in the legislative history, Congress intended to give concurrent jurisdiction to the district courts and the Claims Court, regardless of whether suit is brought before or after award.

The reach of the Claims Court's jurisdiction in this regard under 28 U.S.C. § 1491(a)(3) was resolved in *United States v. John C. Grimberg Co.*[41] Grimberg and Schlosser were disappointed bidders under two government procurements. Plaintiffs sought injunctive and declaratory relief from the Claims Court under the new 28 U.S.C. § 1491. Before award, the plaintiffs had filed protests with the contracting officer concerning the procurements. The latter had acknowledged the protests and had advised that "[t]he bids are being evaluated and you will be advised of our decision before award is made."[42] However, the contracting officer made awards to another a few days later, without advising Grimberg and Schlosser. The disappointed bidders petitioned the Claims Court on October 4, 1982, shortly after those awards. Sitting en banc, a divided Federal Circuit affirmed a Claims Court decision which concluded that the Claims Court lacked jurisdiction because suit had been initiated after award, and which transferred the case to the district court.

---

[38] *Id.* 1301.

[39] 28 U.S.C. §§ 1295, 1491 (1982).

[40] *Id.* § 1491(a)(3).

[41] 702 F.2d 1362 (Fed. Cir. 1983).

[42] *Id.* 1364.

The division within the Federal Circuit in *Grimberg* arose from the unfortunate use of the phrase "any contract claim brought before the contract is awarded" in describing the subject matter of the equitable jurisdiction granted. This does not fit well against many of the classic bid protest situations, such as post-award controversies, restrictive specifications, and exclusion from a solicitation. The precise reach of 28 U.S.C. § 1491(a)(3) in light of *Grimberg* is still being resolved judicially.

The extent of *Scanwell* jurisdiction in the United States district courts after the Court Reform Act remains unresolved. Not all circuits follow *Scanwell*. In those that do, jurisdiction generally is recognized where the suits are initiated after award under the solicitation.[43] One decision went further and found pre-award jurisdiction concurrent with that of the Claims Court.[44]

§ 4.6 Waiver by States and State Agencies

A contractor wishing to sue a state finds itself faced with the same sovereign immunity problems as the contractor attempting to sue the federal government, but without the benefit of express waivers of immunity. In November 1983, the Transportation Research Board, National Research Council, published a report entitled "Construction Contract Claims: Causes and Methods of Settlement." Its appendices A and (1) provide a timely overview of claims procedures and judicial review in each of the 50 states.

For highway construction contracts, three states[45] have no entry under "judicial remedies and procedures." Others have entries which suggest little or no assurance to an aggrieved contractor that its claim will be resolved impartially. For example, in West Virginia "Court of Claims suits to recover contract claims . . . [c]ourt awards must be approved by Governor and State Auditor." In Wisconsin, "[s]uit against the state on a claim denied by State Claims Board may be filed with special permission of the Legislature."[46] However, the board "is composed of representatives of [the] Governor, Attprney [sic] General, Department of Administration and chairpersons of Senate and Assembly Committees on Finance."[47]

---

[43] *See, e.g.*, B.K. Instrument, Inc. v. United States, 715 F.2d 713 (2d Cir. 1983); Merriam v. Kunzig, 475 F.2d 1233 (3d Cir.), *cert. denied*, 414 U.S. 911 (1973); William F. Wilke, Inc. v. Department of the Army, 485 F.2d 180 (4th Cir. 1973); Hayes Int'l Corp. v. McLucas, 509 F.2d 247 (5th Cir.), *cert. denied*, 423 U.S. 864 (1975); Armstrong & Armstrong, Inc. v. United States, 514 F.2d 402 (9th Cir. 1975); Airco, Inc. v. Energy Research & Dev. Admin., 528 F.2d 1294 (7th Cir. 1975); *but see* People's Gas, Light & Coke Co. v. United States Postal Serv., 658 F.2d 1182 (7th Cir. 1981).

[44] Coco Bros., Inc. v. Pierce, 741 F.2d 675 (3d Cir. 1984).

[45] Alabama, Arkansas, and Maine.

[46] Transportation Research Board, National Research Council, Construction Contract Claims: Causes and Methods of Settlement (Nov. 1983).

[47] *Id. See also* entries in the report's Appendix A for Texas (permission by legislature required), Maryland (judicial review only of legal issues), and Georgia ("jurisdictional requirements regarding claims that have been denied by department are uncertain").

Many other states are listed as providing special claims boards and arbitration panels. The report is unclear as to the degree of finality accorded to determinations by these bodies, and, of course, the report is silent on their qualifications or impartiality. A contractor doing business with a state or one of its agencies would be well advised to review in advance whether sovereign immunity has been effectively waived.

If the contracting party is a local government or an entity created by the state or local government, the doctrine of sovereign immunity may not apply. In *Charles E. Brohawn*,[48] the Maryland Court of Appeals phrased the issue as whether the agency is one which "inherited" the state's sovereign attributes.[49] A good example of this is *Dormitory Authority of State of New York v. Span Electric Corp.*,[50] in which the Court of Appeals of New York considered the obligation of the Authority to arbitrate a dispute pursuant to an arbitration provision in a contract. The authority sought a stay of that arbitration on the grounds of sovereign immunity but was rebuffed in a lower court. On appeal, the court held that the authority was obligated to arbitrate the dispute. The principal basis for sustaining the refusal to stay arbitration was the conclusion that the authority was not the alter ego of the state and could not, therefore, avail itself of the defense of sovereign immunity. The court stated:

> A reading of section 1678 discloses that the Authority has been given power: to sue and be sued; to have its own seal; to make its own by-laws; to appoint officers, agents and employees as well as fix their compensation; to acquire real property in the name of the State; to acquire personal property for its own corporate purposes . . . ; to enter contracts and execute instruments necessary for its purposes, subject only to the State Budget Director's approval as to cost; to fix and collect rentals for the use of the dormitories; and to borrow money and issue negotiable bonds. In addition, section 1678 bestows upon the Authority a broad grant of power to do all things necessary and convenient to carry out its purposes. . . .
>
> Considering and weighing all the above powers, functions, and obligations, it is clear that this Authority, enjoying a separate existence, transacting its own business, hiring and compensating its own personnel, is not identical with the State.[51]

Appendix A of the Transportation Research Board survey indicates that California, Delaware, and Rhode Island provide some form of arbitration as a final step in the procedures for resolving highway construction contract claims. Three other states[52] are listed as permitting some form of arbitration as an alternative to judicial review. An unresolved question remains: are agreements to arbitrate by their governmental agencies enforceable? It was decided early on that *federal* officials

---

[48] Charles E. Brohawn & Bros., Inc. v. Board of Trustees, 304 A.2d 819 (Md. 1973).

[49] The court there concluded that the contracting agency, the Chesapeake College, was such a state agency and therefore was immune from suit for breach of contract.

[50] 18 N.Y.2d 114, 218 N.E.2d 693 (1966).

[51] 218 N.E.2d at 695.

[52] Florida, Iowa, and North Dakota.

have no authority to commit the federal government to arbitration.[53] However, *Dormitory Authority* and other decisions have concluded that such agreements may be enforceable against *state* agencies.[54]

§ 4.7 Funding and Fiscal Restrictions

Governmental bodies seeking to contract must operate within their enabling and limiting fiscal rules— constitutional, legislative, charter, regulatory, and funding. These restrictions, grafted as they are onto vestiges of sovereign immunity plus judicial concern for the taxpayer's purse, and implemented often by exculpatory, excusatory, and limiting contract language, can have a considerable effect. Whether or not at fault, a contractor may find itself in considerable financial jeopardy if fiscal rules are not satisfied.

§ 4.8 —Federal Fiscal Restrictions

Fiscal restrictions imposed by Congress, and implemented by the General Accounting Office (GAO), may cause cancellation of a solicitation, impede a settlement, furnish a defense to a claim, delay the award, or cause a termination. These restrictions arise under the Anti-Deficiency Act and related federal fiscal statutes. They limit the federal government's ability to obligate itself, under contract, in excess of statutory authority and available appropriations. In enacting these statutes, Congress intended "to prevent executive officers from involving the Government in expenditures or liabilities beyond those contemplated or authorized by the law-making power."[55]

The statutes establish several relevant fiscal principles. The first, embodied in the Adequacy of Appropriation Act,[56] prohibits an agency from making a

---

[53] *See* United States v. Ames, 24 F. Cas. 784, 789 (C.C.D. Mass. 1845) (No. 14,441); Braucher, *Arbitration Under Government Contracts*, 17 Law & Contemp. Probs. 473 (1952). *But see* United States v. Farragut, 89 U.S. (22 Wall.) 406 (1875) (where pending case is referred by order of court to arbitrators, final award made by arbitral tribunal could be adopted as the decree of the court).

[54] *See* Dormitory Auth. of State of N.Y. v. Span Elec. Corp., 18 N.Y.2d 114, 218 N.E.2d 693 (1966); Commonwealth v. Eastern Paving Co., 288 Pa. 571, 136 A. 853 (1927); City of Auburn v. Nash, 34 A.D.2d 345, 312 N.Y.S.2d 700, 705 (1970); Pytko v. State of Conn., 255 A.2d 640 (1969); J.S. Watkins v. Department of Highways, 290 S.W.2d 28 (Ky. 1956).

[55] 21 Op. Att'y Gen. 244, 248 (1895). *See* 42 Comp. Gen. 272, 275 (1962). *See generally* Hopkins & Nutt, *The Anti-Deficiency Act (Revised Statutes 3679): And Funding Federal Contracts: An Analysis*, 80 Mil. L. Rev. 51 (1978); Efros, *Statutory Restrictions on Funding of Government Contracts*, 10 Pub. Cont. L.J. 254 (1978).

[56] 41 U.S.C. § 11 (1982).

commitment which existing appropriations are inadequate to fulfill[57] "unless otherwise authorized by law." If a contract violates this prohibition, it is void ab initio.[58]

The second rule, the Anti-Deficiency Act,[59] enjoins agency officials from creating obligations "under any appropriation or fund in excess of the amount available therein" or "in advance of an appropriation made for such purpose." This second rule builds on the first.[60] If there is enough money in an appropriation at the time a contract is made, there is no violation of 41 U.S.C. § 11, and the contract is valid and enforceable. If the applicable appropriation is exhausted sometime after the contract is executed, 31 U.S.C. § 665(a) is activated and forbids the involved agency to make payments otherwise due to the contractor or to create any other additional monetary obligations.[61]

A third rule, established by 31 U.S.C. § 628 (1982), provides that appropriations can only be used for the purpose for which they were appropriated. Under a *necessary expenses* concept developed by the GAO, this purpose-restriction rule may be extended to permit expenses to be charged against an appropriation for a particular object so long as the expenses are necessary to the proper execution of the object.[62] However, the necessary expenses concept cannot be used to justify the transfer of funds appropriated for a specific purpose to another appropriation account for which the funds would not have been available originally.[63] Similarly, the availability of an appropriation for a specific object precludes the use of a general appropriation to fund the object, even if the specific appropriation becomes exhausted.[64]

The last two rules are complementary. The first principle, known as the *one-year rule*, states that:

> Except as otherwise provided by law, all balances of appropriations contained in the annual appropriations bill and made specifically for the service of any fiscal year shall only be applied to the payment of expenses properly incurred during that year, or to the fulfillment of contracts properly made within that year.[65]

To enable this fiscal year limitation to work in practice, the GAO has evolved the *bona fide needs* rule.[66] This rule allows the government to obligate a fiscal

---

[57] *See* Comp. Gen. 272 (1962); Efros, *Statutory Restrictions on Funding of Government Contracts*, 10 Pub. Cont. L.J. 254, 262 (1978).

[58] Efros, *Statutory Restrictions on Funding of Government Contracts*, 10 Pub. Cont. L.J. 254, 262 (1978).

[59] 31 U.S.C. § 665(a) (1982).

[60] *See* 42 Comp. Gen. 272, 275 (1962).

[61] *See* Efros, *Statutory Restrictions on Funding Government Contracts*, 10 Pub. Cont. L.J. 254, 262 (1978).

[62] *See id.* 260-61.

[63] *See id.* 261-62. *See generally* Hopkins & Nutt, *The Anti-Deficiency Act (Revised Statutes 3679) And Funding Federal Contracts: An Analysis*, 80 Mil. L. Rev. 51, 105-15 (1978).

[64] *See, e.g.*, 36 Comp. Gen. 526 (1957); 19 Comp. Gen. 892 (1940).

[65] 31 U.S.C. § 712(a) (1982).

[66] *See* 78-2 Cont. Procurement Dec. ¶ 380 (1978); 56 Comp. Gen. 142 (1976); 55 Comp. Gen.

year's appropriation for goods or services to be paid for in a succeeding fiscal year, if the obligation is for a purpose which satisfies a genuine agency need arising during the year in which the obligation is incurred.

The second principle, embodied in the Advance Payment Statute,[67] furthers the same purposes as the one-year rule, in a more practical context, by placing explicit limitations on the government's ability to pay money out of the treasury to fulfill contractual obligations.[68] This rule prohibits the advancement of public money unless the payment is authorized by the "appropriation concerned or other law." The rule also forbids payment in excess of the value of the services rendered or goods delivered prior to the payment.

These funding limitation principles appear deceptively simple. They establish the present availability and amount of appropriations as the fundamental controls on the government's power to contract. An agency or contracting officer is explicitly prohibited from creating an obligation in the absence or in advance of an appropriation legally available therefor, allocating a current appropriation to liquidate overobligations of prior years, or using an appropriation to pay for bona fide needs of future years.

However, by incorporating the phrase "unless [otherwise] authorized by law," these statutory provisions permit deviations from their apparently strict fiscal accountability, if there is statutory support for the departure. Thus, the ultimate requirement is in fact that there be an adequate appropriation or statutory authority for any deviation from these strict fiscal limitations.

Moreover, in their application to specific situations, these rules prove to be complicated and controversial. Must an agency allow prospectively for changes and other potential charges against an appropriation? Will a contractor be reimbursed for costs incurred or effort expended before award and obligation? When does an authorizing statute qualify as a grant of authority to exceed appropriations? What activities fall within the legislative purpose for an appropriation? To what extent may appropriated funds be transferred?

In deciding such questions, there are basic philosophical and interpretive differences between the GAO, which is charged by Congress with the responsibility of enforcing the statutes, and executive agencies, which must exercise the government's contractual power in conformity with them. Often such differences have been resolved in particular cases by practical accommodations. However, more than once, divergent opinions on interpretation have resulted in the GAO's nullification of an agency's apparent contractual obligation.[69]

---

768 (1976); 42 Comp. Gen. 272 (1962); 36 Comp. Gen. 683 (1957); Efros, *Statutory Restrictions on Funding of Government Contracts*, 10 Pub. Cont. L.J. 254, 267-69 (1978).

[67] 31 U.S.C. § 529 (1982).

[68] *See* 57 Comp. Gen. 89 (1977).

[69] Efros, *Statutory Restrictions on Funding of Government Contracts*, 10 Pub. Cont. L.J. 254, 267-69 (1978). *See, e.g.*, Leiter v. United States, 271 U.S. 204 (1925); 56 Comp. Gen. 142 (1976); 42 Comp. Gen. 272 (1962); 36 Comp. Gen. 683 (1957).

§ 4.9 —Fiscal Restrictions on States and Other Governmental Bodies

Fiscal restrictions on contracting by the states and other nonfederal governmental bodies have their roots in the doctrine of sovereign immunity and in questions of authority. An example of the former is in *Charles E. Brohawn & Brothers v. Board of Trustees*,[70] where the Maryland Court of Appeals aptly explained:

> [I]t is established that neither in contract nor tort can a suit be maintained against a governmental agency, first, where specific legislative authority has not been given, second, even though such authority is given, if there are no funds available for the satisfaction of the judgment, or no power reposed in the agency for the raising of funds necessary to satisfy a recovery against it.[71]

The availability of authorized funds is critical to the contractor seeking to enforce contract rights against a state or its agencies. For example, in *Fisher & Carozza Brothers Co. v. Mackall*,[72] the court of appeals considered a case involving a contractor's breach of contract action against the State Roads Commission. The State Roads Commission claimed that it was immune from suit after the repeal of an authorizing statute. The court of appeals dismissed the action for damages, not because the statute had been repealed, but because there was no fund from which the plaintiff could have been paid.

On the other hand, a court may infer authority for an expenditure from the fact that funds are available. For example, in *State ex rel. Lane v. Dashiell*,[73] the state brought a declaratory judgment action to determine whether it was liable to the defendant contractor for additional costs incurred by reason of delays while the contractor was constructing a hospital for the state. The legislature had appropriated an amount not to exceed $250,000 to satisfy this defendant's claim for additional compensation if it were determined there was a legal obligation to pay. The legislature thereby sanctioned the suit and the court of appeals determined that the state was liable. In another case,[74] the court of appeals considered a declaratory judgment suit claiming that the State Roads Commission should bear the cost of removal, relocation, and construction of gas and electric utility facilities necessitated by some construction work. Under the common law, the costs to the utility company would have been an expense without a remedy (damnum absque injuria) because the action of the state did not amount to a taking. A statute, however, provided that all private property damaged or destroyed in the course of activities thereby authorized would be restored or repaired, or adequate compensation would be paid. The court of appeals held that the statute *did* change the common law. The court emphasized that, upon examining the legislative scheme, it became clear that the

---

[70] 304 A.2d 819 (Md. 1973).

[71] *Id*. 820.

[72] 114 A. 580 (Md. 1921).

[73] 75 A.2d 348 (Md. 1950).

[74] Baltimore Gas & Elec. Co. v. State Roads Comm'n, 134 A.2d 312 (Md. 1957).

cost of the project was to be paid out of bond revenues and that the users of the project were to bear all costs ultimately by the payment of tolls.[75]

In *Comptroller of Treasury v. Paintz*,[76] the court of appeals held that a plaintiff who had contracted with the University of Maryland for the printing of a magazine could recover the amount owed under the contract if there were funds available which had not been allocated for other purposes. Similarly, the nonavailability of funds was determinative in *University of Maryland v. Maas*,[77] a suit brought by a contractor who claimed damages from breach of contract in connection with a construction contract for buildings at the University of Maryland.

§ 4.10 Authority to Act: Federal Contract Actions

The general rule is that ordinary principles of estoppel and apparent authority, which might under similar circumstances bind a private party, cannot be used to bind the government. There must be actual authority. This principle was best articulated by the United States Supreme Court in the historic case of *Federal Crop Insurance Corp. v. Merrill*,[78] in which the Court explained:

> Whatever the form in which the Government functions, anyone entering into an arrangement with the Government takes the risk of having accurately ascertained that he who purports to act for the Government stays within the bounds of his authority. The scope of this authority may be explicitly defined by Congress or be limited by delegated legislation, properly exercised through the rule-making power. And this is so even though, as here, the agent himself may have been unaware of the limitations upon his authority.[79]

The obvious question is: Who may bind the government? Clearly, the warranted contracting officer (CO) may do so, but what about other government representatives with whom the contractor must deal on a more regular basis? Do they have authority, and if so, to what extent?

By definition, the term *contracting officer* includes "certain authorized representatives of the contracting officer acting within the limits of their authority."[80] Courts and boards finding authority in representatives other than the contracting officer have hinged their decisions on this basic definition. These cases primarily focus on the duties, responsibilities, and relationships of the representatives involved. The greater the duties and responsibilities of a representative, the wider

---

[75] *Id. See also* Masson v. Reindollar, 69 A.2d 482 (Md. 1949), in which the court assumed that the cost overruns on the first Chesapeake Bay bridge would be financed by additional bonds.

[76] 297 A.2d 289 (Md.1972).

[77] 197 A. 123 (Md. 1938).

[78] 332 U.S. 380 (1947).

[79] *Id*. 384.

[80] F.A.R. § 2.100.

his or her discretion becomes and the more likely that the officer will be deemed to have acted within his or her authority.

For example, very few decisions involving inspectors have found an inspector to have the authority to bind the government, since inspectors are ordinarily no more than the government's observers. They have very little CO contact. On the other hand, cases finding a field representative to be the authorized representative of the CO and thus capable of binding the government, are numerous. This result is not surprising when one considers the broader duties and correspondingly broader scope of authority inherent in the field representative's function. This reasoning is best explained by the Board of Contract Appeals in *Industrial Research Associates*, in which the board stated:

> [A] contracting officer can and sometimes does delegate his authority to technical representatives. And in these circumstances, the Government is bound by the directives given by the representative. . . . Similarly, an engineer or some other kind of technical advisor whose function is the giving of technical guidance to a contractor may also order a change in work. Thus, it seems clear to us that what is at issue here are the actual functions of the Government's engineer. . . .[81]

Increased responsibility translates into increased authority. This concept is illustrated by *Lillards*,[82] a 1961 ASBCA case involving a construction contract. In *Lillards*, the government's installation engineer ordered the contractor to provide manhole covers not required by the contract. The government argued that the engineer lacked authority to do so and consequently the government was not bound by the changes that he directed. Holding for the contractor, the board found that the engineer did in fact have authority. The board stated: ''He was in *direct control of the job*, the covers were obviously needed . . . it is inconceivable that he lacked authority to do what he did.[83]

Inspectors and field representatives are not the only representatives who may, under appropriate circumstances, bind the government. Several court and board decisions have found procurement personnel authorized to act on behalf of the CO and thus capable of binding the government. As with the field representative, and to a lesser extent the inspector, actual job function controls (i.e., circumstances, not job title, dictate the outcome). For example, in *Morrison-Knudsen Co.*,[84] a 1972 ASBCA decision, the board determined that a contracting officer's assistant who gave guidance at a prebid conference was, in this limited situation, the CO's authorized representative and had authority equivalent to the CO.

In another ASBCA decision, *Wickham Contracting Co.*,[85] an estimator was found to have authority. The controversy in *Wickham* centered on an error in one of

---

[81] 68-1 B.C.A. (CCH) ¶ 7069, at 32,685 (1968).

[82] ASBCA No. 6630, 61-1 B.C.A. (CCH) ¶ 3053 (1961).

[83] 61-1 B.C.A. (CCH) at 15,801 (emphasis added).

[84] ASBCA No. 16483, 72-2 B.C.A (CCH) ¶ 9733 (1972).

[85] ASBCA No. 19069, 75-1 B.C.A. (CCH) ¶ 11,248 (1975).

the drawings in the invitation for bids for an underground electrical cable replacement. Apparently, the government's estimator knew of the error but failed to report it. The ASBCA stated, ''[W]e find that the Government's estimator was the authorized representative of the contracting officer for the purposes of preparing the Government's estimate of the cost of the work, and for the purpose of initiating drawings corrections. . . .''[86]

§ 4.11 –Ratification by Contracting Officer

In cases involving contract language clearly restricting the authority of the government's representatives, courts and boards have often circumvented the authorized representative issue by finding authority in the related concepts of contracting officer (CO) ratification, acquiescence, and imputation of knowledge.[87] A typical decision binding the government on the basis of imputed knowledge is *United States Federal Engineering and Manufacturing*.[88] In *Federal Engineering* the CO was not informed that the project manager had approved certain additions correcting defective specifications. The board nevertheless allowed the contractor to be compensated for the extra work, holding:

> The fact that the contracting officer did not have actual knowledge of the additions to be made to the device does not insulate the Government from the consequences that actual knowledge would impose. His various representatives are his eyes and ears (if not his voice) and their knowledge is treated for all intents and purposes as his.[89]

The closer the working relationship of the representative is with the CO, the more likely the representative's knowledge will be presumed to be the knowledge of the CO.[90] Consequently, inspector decisions rarely find the type of imputed knowledge found in *United States Federal Engineering and Manufacturing*.

Decisions based on contracting officer acquiescence hold that where the CO was or should have been aware of the changes proposed by its representative, the CO will be held to have approved them. A typical example is found in *W. Southward Jones*,[91] a case in which the contractor, the government technical personnel, and the contracting officer were all operating on the same military base. In finding that the CO had acquiesced to changes ordered by technical personnel, the ASBCA stated:

---

[86] 75-1 B.C.A. (CCH) at 53,576.

[87] Tepfer, *Authority and the Contracting Officer's Representatives*, 24 Air Force L. Rev. 1, 3 (1984).

[88] ASBCA No. 19909, 75-2 B.C.A. (CCH) ¶ 11,578 (1975).

[89] 75-2 B.C.A. (CCH) at 55,298-55,299.

[90] Tepfer, *Authority and the Contracting Officer's Representatives*, 24 Air Force L. Rev. 1, 3 (1984).

[91] ASBCA No. 6321, 61-2 B.C.A. (CCH) ¶ 3182 (1961).

> Performance proceeded almost from the outset under the eyes of assigned supervisory personnel in a manner patently at variance with the contract drawings. . . . The digging of the conduit in question continued in plain view along the taxiway at the Base every day for almost a month. Base technical personnel clearly knew what was going on and as clearly intended that the original drawings should not be followed. . . . Conceding therefore that the contracting officer was the only one empowered to authorize changes in the contract and that a special clause was included to emphasize the limitations of authority of the technical personnel, we must hold under the particular circumstances of this case that he had timely notice of changes, if not actually, then certainly constructively.[92]

Whether contractor recovery in a particular case is based upon a finding that the representative was the authorized representative of the contracting officer or, alternatively, that there was CO ratification, acquiescence, or constructive knowledge, thus depends on the circumstances of each case.

§ 4.12 Authority for Contract Actions by States and Other Governmental Bodies

Court and board decisions involving the authority for state and local actions are couched in terms of the liability of a state or local governmental body on contracts entered into by unauthorized representatives. The decisions finding liability rely on theories of implied contract, ratification, and estoppel. Presumably, courts and boards would apply the same theories in determining a government's liability for unauthorized changes.

Courts finding liability based on ratification hold that ratification or affirmance can create an implied promise of payment if the unauthorized contract is one which the state or local government may lawfully make.[93] Other cases have reached the same result by utilizing the concept of estoppel. These cases hold that a municipality or other governmental body may be estopped from asserting that its agents lacked contracting authority where the contract was one which the government could lawfully make and the government has accepted the contract's benefits. While this estoppel theory has been applied to local governments, its applicability (if any) to states is unclear. As the New York Supreme Court explained in *DiGiamo v. City of New York:*[94] "[T]here can be no estoppel against the state which will impede the exercise of sovereign powers for the protection of the people, [although] in a proper case the state may be estopped, the same as an individual, by representations and conduct *provided its officers and agents were acting with authority.*"

---

[92] 61-2 B.C.A. (CCH) at 16,498.

[93] *See, e.g.*, Township of Ridley v. Haulaway Trash Removal, Inc., 68 Pa. Commw. 176, 448 A.2d 654 (1982) (partial payment tantamount to ratification of unauthorized contract). *See generally* 154 A.L.R. 373 (1945); 33 A.L.R.3d 1164 (1970).

[94] 397 N.Y.2d 632, 638 (1977), quoting 21 N.Y. Jur. § 77.

The cases that find liability based on quasi-contract/quantum meruit theories hold that the state or other political subdivision must pay the reasonable value of the services rendered by the contractor notwithstanding the fact that the services were rendered pursuant to an unauthorized contract. For example, in *Ridley v. Pipe Maintenance Services, Inc.*,[95] the Pennsylvania Commonwealth Court allowed contractor recovery for the cost of maintenance and repair work performed pursuant to an unauthorized contract. The court reasoned that when a municipality has voluntarily accepted and retained the benefits of a contract which (1) it had the power to make, but (2) was defective in its making, and (3) thus was invalid, the party who has conferred those benefits may recover compensation therefor—not on the invalid contract itself, but in quantum meruit.

An equal, if not greater, number of decisions, however, have denied contractor recovery in cases involving contracts entered into by unauthorized representatives. These cases deny that a governmental body must pay for the benefits derived from an unauthorized contract.[96]

---

[95] 477 A.2d 610 (Pa. Commw. Ct. 1984).

[96] *See, e.g.*, Dour v. Village of Port Jefferson, 89 Misc. 2d 197, 390 N.Y.S.2d 965 (Sup. Ct. 1976); Seif v. Long Beach, 286 N.Y. 382, 36 N.E.2d 630 (1941); Sorenson v. City of N.Y., 202 F.2d 857 (2d Cir. 1953), *cert. denied*, 47 U.S. 951 (1951).

CHAPTER 5

SUING THE CONSTRUCTION MANAGER

John B. Tieder, Jr. and Julian F. Hoffar

John B. Tieder, Jr., is a senior partner in the Washington, D.C. law firm of Watt, Tieder, Killian & Hoffar. He specializes in the representation of owners, architects, engineers, contractors, and subcontractors in construction matters. Mr. Tieder received his A.B. degree from the Johns Hopkins University in 1968 and his J.D. from the American University Washington College of Law in 1971. Mr. Tieder is the author and co-author of a variety of publications on construction matters, and is currently a visiting lecturer in government contracts law at Marshall-Wythe School of Law, College of William and Mary.

Julian F. Hoffar, senior partner in the Washington, D.C. law firm of Watt, Tieder, Killian & Hoffar, specializes in national and international contract law. Mr. Hoffar earned his undergraduate degree from Wittenberg University and is an Honor Graduate of the National Law Center, George Washington University, where he was a member of the Law Review. Mr. Hoffar has written articles on construction contracting, and has lectured before several industry groups on the subject of construction law.

INTRODUCTION

INTRODUCTION

§ 5.1 Role of Construction Manager

As more owners rely on fast-track[1] and/or multi-prime construction or simply wish to withdraw from the active management of their construction projects, construction management has evolved as a separate field of construction practice. The *construction manager* (CM) is employed by the owner to perform some or all of the services traditionally provided by a design architect, a supervisory engineer, a quality control inspector, or a general contractor. This can include all or portions of the design, preparation of bidding documents, review and evaluation of bids, inspection, scheduling, and coordination. Usually the CM does not perform any actual construction work, although it is not uncommon for the CM to provide and maintain basic project services such as temporary power and sanitary facilities. Typically, the CM does not have any contractual relationship with the actual prime contractors, although this is not always the case.[2]

The concept of the construction manager is fairly new. Thus, to date, there are not a large number of reported cases dealing specifically with the liability of the CM to the contractors on the project. Since, however, the CM's duties encompass

---

[1] See § 5.2.

[2] On some projects denominated as construction management projects, the CM does have direct contracts with the several prime contractors. *See, e.g.*, United States v. Johnson Controls, Inc., 2 Fed. Procurement Dec. ¶ 15 (Fed. Cir. 1983).

many of the traditional duties of an architect/engineer and a general contractor, it is possible to rely on cases discussing these parties in order to determine the CM's liability to the contractor.

§ 5.2 Construction Manager Defined

The term *construction management* has two similar but distinct meanings. First, it is a method of construction, and second, it is a separate discipline or profession. As a method of construction, the term *construction management* refers to a system whereby each of the construction trades on a particular project have a direct contract with the owner, and the work is administratively controlled by a separate entity which performs little or no direct construction work—the construction manager (CM).[3] It differs from a traditional single-prime project in that most contractors which typically performed as subcontractors now perform as primes. Although it is possible to have a construction management project with very few or even one contractor, the typical construction management project will have several primes; it can be as many as 50 or more.[4]

Although it is not an inherent part of construction management, many projects constructed by this method have some element of *fast-tracking* or *phased* construction. This simply means that certain portions of the project are designed and let for construction while other portions are still being designed.[5] Although this process can reduce overall project time,[6] since the design is not complete, it is virtually impossible to award a single overall prime contract. Thus, individual contracts are awarded as various portions of the design are completed.

The concept of construction management, almost of necessity, had to give rise to the discipline of a construction manager. Although the role of the CM is fairly clear (i.e., to administer the construction for the owner), the status of the CM is a subject of considerable controversy. Is the CM a professional, and thus susceptible to the liability of other professionals (e.g., architects and engineers), or is it a general contractor?[7] On most projects, the CM enters into a contract with the owner for administration of all or some portion of the overall construction project.

---

[3] Conner, *Contracting for Construction Management Services,* 46 Law & Contemp. Probs. 5 (Winter 1983).

[4] Tieder & Cox, *Construction Management & the Specialty Trade (Prime) Contractors,* 46 Law & Contemp. Probs. 39 (Winter 1983).

[5] *Id.* 44.

[6] *Overall project time* means the time between the start of design by the architect/engineer and the completion of the project to a stage when it is available for use by the owner.

[7] This issue is discussed in detail in § **5.6.**

FORMS OF CONSTRUCTION MANAGEMENT PRACTICE

§ 5.3 Agency Construction Management

The agency form of construction management practice is similar to the owner-architect relationship, in that the construction manager (CM) is an agent of the owner managing the project for a fixed or cost-plus fee. An example of an agency construction management contract is American Institute of Architects Document B801, "Standard Form of Agreement Between Owner and Construction Manager," which provides as follows:

> The Construction Manager covenants with the Owner to further the interests of the Owner by furnishing the Construction Manager's skill and judgment in cooperation with, and in reliance upon, the services of an architect. The Construction Manager agrees to furnish business administration and management services and to perform in an expeditious and economical manner consistent with the interests of the Owner.[8]

Agency construction management is typically performed by architects or engineers who are familiar with the rights and responsibilities owed by a design professional to an owner. The owner's cost savings, if any, come from the CM's success in reducing the cost and time of completion of the building. In cases where the CM assumes both design and management roles (often called A/CM), the construction management services can be viewed as an extension of the architect's duties, since there is no independent design review function. In situations where the CM is also acting as a general contractor (by letting contracts, performing some of the work, or guaranteeing a maximum price), the identity of interests necessary to support the "relationship of trust and confidence" of an agency usually does not exist because the CM-contractor has a financial interest in all or part of the work.

§ 5.4 Guaranteed Maximum Price Construction Management

Guaranteed maximum price (GMP) construction management retains the most important characteristic of a general contractor-owner relationship: it offers cost security to the owner. In this arrangement, the construction manager (CM) assumes control over the performance of the work and guarantees a price ceiling, with all

[8] American Institute of Architects, AIA Document B801, Standard Form of Agreement Between Owner and Construction Manager, Art. 1 (1980 edition).

overruns deducted from the CM's fee. In some cases, there is also a cost-cutting incentive which nets the CM a percentage of the owner's savings if the job is completed under budget. An example of the GMP format is an Associated General Contractors form which provides that "[c]osts in excess of the Guaranteed Maximum Price" are included in the CM's fee.[9]

The GMP format creates a potential conflict of interest between the CM and the owner by establishing financial incentives for the CM to keep construction costs low. While this is one owner goal, another is reducing long-term operating costs, which can often be achieved at a slight increase in construction costs by using higher quality material. The incentive to cut costs may adversely influence the CM's decision whether to install more durable material or implement other changes that may benefit the owner in the long run.

§ 5.5 Extended Service Construction Management

Construction management services, while nominally limited to activities related to construction and design, have been extended by some practitioners to include a complete line of ancillary services. For example, some construction management contracts provide for standard design and construction services, as well as claim resolution, tenant solicitation and space planning, and occupancy scheduling. An extended service agreement is similar to the design-build relationship in that the owner delegates all but the most general duties concerning the project to the construction manager (CM).

Guaranteed maximum price construction management is sometimes combined with design services to create a professional design-build package. In this format, the architect/CM/contractor offers an integrated building concept to the owner, relieving the owner of all supervisory and coordination responsibilities. The disadvantage of this system is that the owner often has no independent representative on the project, since both the architect/engineer and the CM are allied with the contractor.

STANDARD OF CARE

§ 5.6 Professional or Contractor

Unlike the traditional positions of design architect or supervisory engineer, the role of the construction manager (CM) has not yet been clearly defined by statute, case law, or industry practice. Thus, to identify the standard of care owed by the

---

[9] Associated General Contractors of America, AGC Document No. 500, Standard Form of Agreement Between Owner and Construction Manager (Guaranteed Maximum Price Option) Art. 7.3.5 (1980 edition).

CM to the contractor, it must first be determined if the CM is functioning as a professional or as a general contractor. This distinction is important because it establishes the basic standard of care to which the CM will be held.

An architect/engineer or other engineering professional owes a duty of ordinary skill and competence of members of that profession.[10] This standard applies to both the design and administration of the project.[11] If that duty is not met, the engineering professional will be subject to liability for negligence. In the absence of a specific contractual agreement to the contrary, a professional is not required to provide a perfect design or to perform other engineering services to perfection, but is only liable for a failure to exercise reasonable care and skill in the design or supervision of the work.[12] The standard by which the professional is judged is variable, depending on the nature and location of the work.[13] A professional in charge of a large, complex industrial plant will be held to a higher degree of care than the professional in charge of a small office building. Likewise, a professional who holds itself out as a specialist will be held to a higher degree of care than a general practitioner. Finally, a professional who regularly works in a large metropolitan area will be subject to closer scrutiny than a small, rural practitioner.[14]

Proving that the standard was not met in a particular case invariably requires expert testimony. A general contractor's duty to third parties is one of ordinary and reasonable care.[15] It should be noted, however, that some jurisdictions hold that the duty of care owed by a contractor, or any other party possessing special knowledge or skill, is commensurate with the circumstances.[16] In these jurisdictions,

---

[10] *See, e.g.*, Gravely v. Providence Partners, 549 F.2d 958 (4th Cir. 1977); Allied Properties v. John A. Blume & Assocs., Eng'rs, 25 Cal. App. 3d 848, 102 Cal. Rptr. 259 (1972); La Rossa v. Scientific Design Co., 402 F.2d 937 (3d Cir. 1968); Borman's Inc. v. Lake State Dev. Co., 60 Mich. App. 175, 230 N.W.2d 363 (1975); Queensbury Union v. Jim Walter Corp., 91 Misc. 2d 804, 398 N.Y.S.2d 832 (1977); Ryan v. Morgan Spear Assocs., Inc., 546 S.W.2d 678 (Tex. Civ. App. 1977). Some jurisdictions apply a stricter standard and hold that the architect/engineer is deemed to have warranted the adequacy of its plans and specifications. *See, e.g.*, Prier v. Refrigeration Eng'g Co., 442 P.2d 621 (Wash. 1968); Broyles v. Brown Eng'g Co., 151 So. 2d 767 (Ala. 1963); Bloomsburg Mills, Inc. v. Sordoni Constr. Co., 164 A.2d 201 (Pa. 1960); Hill v. Polar Pantries, 64 S.E.2d 885 (S.C. 1951).

[11] Normoyle-Berg Assocs. v. Village of Deer Creek, 39 Ill. App. 3d 744, 350 N.E.2d 559 (1976).

[12] American Institute of Architects, AIA Document B801, Standard Form of Agreement Between Owner and Construction Manager, Art. 1 (1980 edition).

[13] Mounds View v. Walijarvi, 263 N.W.2d 420, 424 (Minn. 1978).

[14] Bondumin-McClure, Inc. v. Snow, 183 Cal. App. 2d 58, 6 Cal. Rptr. 52 (1960); *but see* Carroll-Boone Water Dist. v. M.&P. Equip. Co., 280 Ark. 560, 661 S.W.2d 560 (1984), in which the court held that the "same locality" rule only applied to medical professionals, not architect/engineers.

[15] Green Constr. Co. v. William Form Eng'g Corp., 506 F. Supp. 173, 177 (W.D. Mich. 1980); Hawthorne v. Kober Constr. Co., 640 P.2d 467, 470 (Mont. 1982).

[16] *See* La Vine v. Clear Creek Skiing Corp., 557 F.2d 730 (10th Cir. 1977), which discusses the standard of reasonable prudence as it is affected by the special skills of the actor charged with negligence. The court stated that "a person having special knowledge must exercise a quantum of care which is commensurate with the circumstances, one of which is his or her special skill and training." *Id.* 734.

a CM, even if regarded as a general contractor, will be held to a greater standard than ordinary care. The standard of care will depend on such matters as the complexity of the project, the skill and expertise needed to effect the work, and similar surrounding circumstances.

Thus, the determination of whether a CM is acting as a professional or a general contractor establishes the standard of care owed to third parties. **Sections 5.7-5.9** set forth several factors which courts examine in making this determination: licensing requirements, bond and insurance requirements, and bidding requirements for public projects.

§ 5.7 Licensing Requirements

All states require professionals such as architects and engineers to obtain licenses. Twenty-six states require contractors to obtain some type of license or registration.[17] Although no states have specific licensing requirements for construction managers (CMs), in approximately 28 states the definition of architectual and engineering practices includes at least some of the functions of a CM.[18] At least one state, Illinois, specifically includes construction management in its definition of architect/engineering services.[19] Several other states define *contracting* in such a manner as to include many of the the services provided by a CM.[20] At least three states[21] specifically include construction management within their definitions of *contracting*.[22] Tennessee, while not using the specific term, makes it clear that a CM is a contractor within the meaning of the state licensing statute.[23] In those states where a CM must be licensed as either an architect or a contractor, the standard of care should be the same as for any other licensee of that category.

A CM who fails to obtain a required license is subject to a variety of criminal and civil sanctions. For example, the Utah licensing statute provides that anyone "acting in the capacity of contractor . . . without a license as herein provided shall be guilty of a misdemeanor."[24] In addition, a CM who fails to obtain the proper license may be prohibited from instituting a lawsuit against the owner for

[17] Associated General Contractors of America, *Legal Trends—Does the CM Need a License?*, 1982 CM Bulletin.

[18] L.D. Phillips, The Legal Status of the Construction Management Project Delivery System in the United States (unpublished paper, Michigan Technological Univ., May 1984).

[19] *Id.*

[20] *Id.*

[21] Nevada, New Mexico, and Alabama.

[22] L.D. Phillips, The Legal Status of the Construction Management Project Delivery System in the United States (unpublished paper, Michigan Technological Univ., May 1984).

[23] *Id.*

[24] Utah Code Ann. §§ 58-23-3(3), 58-23-18 (1953). *See* Meridian Corp. v. McGlynn/Garmaker Co., 567 P.2d 1110 (Utah 1977).

collection of a fee or even from asserting a counterclaim or setoff if sued by another party.[25]

§ 5.8 Bond and Insurance Considerations

Contractors are routinely required to obtain bonds guaranteeing performance of the contract and proper payment to all subcontractors and suppliers.[26] Architects and engineers are typically exempt from bonding requirements. The Oregon Court of Appeals explained the rationale behind this exception: ''[S]ince performance of a contract involving esthetic judgment cannot be measured objectively, it would not be reasonable to require an architect to post a performance bond.''[27] In that case, a county contracted with an architect, a space-use planner, and a construction manager (CM) during the design and fast-track construction of a courthouse annex. The contracts were challenged as violating state procurement laws requiring bids and bonds of public contractors. The court held that all three contracts were properly let under the professional services exemption of the bid and bond laws.

The design professional's closest counterpart to the contractor's performance bond is its errors and omissions (E & O) or malpractice insurance. These policies typically indemnify the owner for damages caused by the negligence of the design professional, such as drafting errors in the plans and specifications or failure to make required inspections during construction. A typical E & O policy covers ''liability arising out of any negligent act, error, mistake, or omission in rendering or failing to render professional services.''[28] The requirement of E & O insurance in a CM contract, as is the case in the American Institute of Architects Form B801,[29] indicates that the CM must meet the design professional's standard of care. Thus, the type of bond or insurance coverage maintained by a CM should assist in determining the classification of a CM as a design professional or general contractor.

---

[25] *Id.*; Cochran v. Ozark Country Club, Inc., 339 So. 2d 1023 (Ala. 1976); Bird v. Pan W. Corp., 261 Ark. 56, 546 S.W.2d 417 (1977); Gorrell v. Fowler, 248 Ga. 801, 286 S.E.2d 13 (1982); United Stage Equip., Inc. v. Charles Carter & Co., 342 So. 2d 1153 (La. Ct. App. 1977).

[26] C. Foster, *Challenges for Suretyship: Bonding the CM and Design-Build/Fast Track Construction* 10 (1982; unpublished manuscript presented at Duke Univ. Symposium on Construction Management and Design/Build ''Fast Track'' Construction, Jan. 21–22, 1983), scheduled for publication in *Law & Contemporary Problems*.

[27] Mongiovi v. Doerner, 24 Or. App. 639, 546 P.2d 1110, 1113 (1976).

[28] C. Foster, *Challenges for Suretyship: Bonding the CM and Design-Build/Fast Track Construction* 17 (1982; unpublished manuscript presented at Duke Univ. Symposium on Construction Management and Design/Build ''Fast Track'' Construction, Jan. 21–22, 1983).

[29] American Institute of Architects, AIA Document B801, Standard Form of Agreement Between Owner and Construction Manager, Art. 10 (1973 edition).

§ 5.9 Bidding Requirements

On public contracts, the issue of the construction manager's (CM's) classification and the resultant standard of care can at least be narrowed, if not resolved, by reference to the procurement codes and building requirements of the particular public body soliciting the contract. These codes frequently limit the types of services for which a public body can contract or, at least, categorize the CM as either a design professional or a contractor.

Most public agencies are required to procure construction services by public competitive bidding. They are also allowed, however, to procure professional services by negotiation. Thus, the question of whether a CM is performing as a design professional or as a contractor can be resolved, in part, by the method under which the services are procured.

Approximately eight states have specific statutes providing for the procurement of construction management services and define the procedure by which these services will be procured.[30] In most states, however, the method of procurement of construction management services is not specifically set forth. The courts have, therefore, taken a pragmatic view in determining whether a contract should be competitively bid as a construction contract or negotiated as a design professional contract. In California, for example, the Supreme Court held that a construction management contract guaranteeing the maximum contract price was "too closely akin to traditional lump sum general construction contracting to be held exempt from the statutory competitive bidding requirements."[31] The court reasoned that the purpose behind the statutory exemption was that professional service contracts cannot be evaluated by objective criteria, but that the same rationale does not apply to a guaranteed maximum price contract, which can be judged on the basis of price. The Indiana Court of Appeals reached a different conclusion in the case of an agency construction management contract, holding that the contract at issue was for a professional service. The court held that the contract did not involve any duties or responsibilities which would not otherwise be required of a licensed engineer, and stated, "Significantly, [the contract] did not guarantee that the total cost of each addition would be at or below the total value of the bids. A decision that implies exemption of all CM contracts from public bidding was reached in Oregon in 1976."[32] The Oregon Court of Appeals held that a construction management contract was exempt, stating that "when making a contract involving professional or esthetic judgment, the legislature could not have intended lowest price to be the ultimate determining factor."[33]

---

[30] L.D. Phillips, The Legal Status of the Construction Management Project Delivery System in the United States (unpublished paper, Michigan Technological Univ., May 1984).

[31] City of Inglewood-Los Angeles Conty Civic Center Auth. v. Superior Court, 7 Cal. 3d 861, 500 P.2d 601, 103 Cal. Rptr. 689 (1972).

[32] Attlin Constr., Inc. v. Muncie Community Schools, 413 N.E.2d 281 (1980).

[33] Mongiovi v. Doerner, 24 Or. App. 639, 546 P.2d 1110, 1113 (1976).

DUTY OWED BY CONSTRUCTION MANAGER

§ 5.10 Generally

Although the standard of care owed by a construction manager (CM) is determined first by its classification as a professional or a general contractor and then by the degree of care owed by that category depending on the nature and locale of the project,[34] the duty owed by the CM depends primarily on the CM's specific undertaking on a project.[35] A CM cannot be held liable to a third party for its failure to perform tasks which it has no obligation to perform. For the most part, the CM's obligations are set forth in its contract with the owner.[36] These contractual obligations, however, can be either extended or limited by statutes or regulations applicable to a project and by the tasks undertaken *in fact* by the CM regardless of its contractual obligation to perform such work. Each of these matters, which determines the duty owed by the CM, is discussed below.

§ 5.11 Contract Obligations

Even though the contractor does not typically enter into a contract with the construction manager (CM), and is not a party to the owner/CM contract, the scope of the duty owed by the CM to the contractor is defined by the owner/CM contract. In *Bates & Rogers Construction Corp. v. North Shore Sanitary District,*[37] the court held:

> [A] duty runs from the architect or engineer who engaged in a tortious action which hinders and damages the contractor. The scope of that duty, although based upon tort rather than contract, is nevertheless defined by the architect and engineer's contract with the owner.[38]

Virtually all jurisdictions which have addressed the issue have taken a similar approach.[39] Thus, before initiating an action against a CM, the contractor must know what contractual obligations were owed by the CM to the owner. For example, a contractor cannot sue the CM for a failure to schedule the work properly unless

---

[34] See § **5.6**.

[35] Plan-Tec v. Wiggins, 443 N.E.2d 1212 (Ind. 1983).

[36] Krieger v. S.E. Griner Co., Inc., 282 Md. 50, 832 A.2d 1069 (1978).

[37] 92 Ill. App. 3d 90, 414 N.E.2d 1274 (1980).

[38] 414 N.E.2d at 1280.

[39] Krieger v. S.E. Griner Co., Inc., 282 Md. 50, 832 A.2d 1069 (1978); Vanasik v. Hirsch & Stevens, 65 Wis. 2d 1, 221 N.W.2d 815 (1974); Craig v. Everett M. Brooks Co., 351 Mass. 497, 222 N.E.2d 752 (1967).

the CM's contract with the owner required it to prepare, monitor, and maintain the construction schedule.

§ 5.12 Procurement Codes

Even though the owner/construction manager (CM) contract is the primary source for the determination of the scope of the CM's duty, a secondary source is the statutes, regulations, and standards applicable to a given contract. The procurement codes of various public agencies frequently define the types of services on which public funds can be expended. In New York, for example, the state is prohibited from delegating its duty to coordinate construction projects.[40] This legislative prohibition against the delegation of coordination responsibilities has been interpreted so strictly that a contract clause requiring a prime contractor to submit a progress schedule was held invalid.[41] Thus, a construction management contract to coordinate the work of several prime contractors would be illegal, and the CM could not be held responsible for its failure to perform a duty which, by statute, it would not be permitted to perform.

The federal government is also prohibited from delegating certain duties to a CM:

> Notwithstanding any other provision of this section the Administrator shall be responsible for all construction authorized by this chapter, including the interpretation of construction contracts, the approval of materials and workmanship supplied pursuant to a construction contract, approval of changes in the construction contract, certification of vouchers for payment due the contractor, and final settlement of the contract.[42]

A CM could not be held liable for its failure to perform functions which are by statute the sole duty of the owner.

§ 5.13 Building Codes and Industry Standards

Virtually all construction, public or private, must be designed and erected in accordance with various codes and standards. These include building and safety codes mandated by federal, state, and local governments and industry standards. Compliance with these codes is either required by contract or mandated by law,

---

[40] N.Y. Gen. Mun. Law § 101 (McKinney 1977). *See also* B. Gladstone & D. Hartzog, Allocation of Risk in the Construction Industry: The Nonprofessional Owner and His Construction Manager 8 (1982; unpublished manuscript presented at the Duke Univ. Symposium on Construction Management and Design/Build "Fast Track" Construction, Jan. 21–22, 1983), scheduled for publication in *Law & Contemporary Problems*.

[41] General Bldg. Contractors v. County of Oneida, 54 Misc. 2d 260, 282 N.Y.S.2d 385 (1967).

[42] 40 U.S.C. § 609(c) (1976).

regardless of contract. All construction professionals are charged with knowledge of and a duty to comply with these codes and standards.

Design professionals and contractors must comply with applicable building, safety, and fire codes. A construction manager (CM) who is serving the function of either a design professional or a contractor must also comply with these codes. Even the CM serving in the most limited role as overall project coordinator may be held responsible to a contractor for a failure of the designer or another contractor to comply with an applicable code. For example, in a recent Illinois case, an architect/engineer was held liable to an owner for damages resulting from a subcontractor's failure to comply with the local electrical code in the design of telephone cable encasement.[43]

Most construction contracts require work to be "of good quality" and "performed in a good and workmanlike manner." These phrases are given meaning by industry standards, which are often codified and endorsed by industry councils or associations. The methods of calculating minimum steel dimensions prescribed by the American Institute of Steel Construction (AISC), concrete reinforcing formulas specified by the American Concrete Institute (ACI), and ductile iron pipe connections recommended by the American Water Works Association (AWWA) are only a small sample of the industry standards available to design professionals and contractors. The CM should be held responsible for either complying with these standards or assuring that others comply with them.

§ 5.14 Construction Manager De Facto Rule

Even though the owner/construction manager (CM) contract and the applicable statutes, regulations, and standards define the scope of the CM's duties, a CM can be liable to third parties for tasks it undertakes *in fact*. Some jurisdictions will not hold a CM or similarly situated party liable for an action unless there is a clear contractual obligation to perform such an action.[44] Other jurisdictions, however, have taken a more expansive view.[45] In *Plan-tec v. Wiggins*,[46] the court held:

> A duty of care may also arise where one party assumes such a duty, either gratuitously or voluntarily [citations omitted]. The assumption of such a duty creates a special relationship between the parties and a corresponding duty to act in the manner of a reasonably prudent person [citations omitted]. Failure to act in a reasonable manner will give rise to an action for negligence.[47]

---

[43] Himmel Corp. v. Stade, 52 Ill. App. 3d 294, 367 N.E.2d 411 (1977).

[44] Wheeler & Lewis v. Slifer, 195 Colo. 291, 577 P.2d 1092 (1978).

[45] Swarthaut v. Beard, 33 Mich. App. 395, 90 N.W.2d 373 (1971), *rev'd on other grounds*, 388 Mich. 637, 202 N.W.2d 300 (1972); Miller v. DeWitt, 37 Ill. 2d 273, 226 N.E.2d 630 (1967).

[46] 443 N.E.2d 1212 (Ind. 1983).

[47] *Id.* 1219.

It is important to note that the majority of the cases dealing with the liability of the CM for de facto duties are in the area of safety on the work site.[48] There is, however, nothing in the reasoning of these cases which would make them inapplicable to a contractor's claim against the CM. Thus, a contractor should look to the actual role or course of conduct of the CM on a given project, as well as those set forth in the contract, to determine the scope of the CM's duties.

§ 5.15 Parties to Whom Duty Is Owed

The majority of the cases which have addressed the issue of construction manager (CM) or architect/engineer liability to third-party contractors concern prime contractors. Since the CM's liability to the prime contractor in most cases is based in tort, and not on the existence of a contract agreement between the CM and the contractor, the duty can be owed to subcontractors and suppliers as well as primes. At least one court has found that a CM can be liable to a subcontractor in negligence.[49] Likewise, a supervisory architect/engineer has been found liable to a supplier for interference with contract.[50]

THEORIES OF RECOVERY

§ 5.16 Negligence

There are four potential theories of recovery in an action by a contractor against a construction manager (CM): 1) negligence; 2) third-party beneficiary; 3) interference with contract; and 4) implied warranty. Of these four theories, negligence is the most common; however, each is discussed below.

To establish a cause of action in negligence against a CM, the contractor must prove three things: 1) that the contractor was owed a duty by the CM; 2) that the CM failed to carry out that duty in accordance with the applicable standard of care; and 3) that the contractor was damaged as a result of the CM's breach of its duty.[51]

As set forth in § **5.11**, the issue of whether a particular duty was owed to the contractor by the CM is established primarily by the contract between the CM

---

[48] Caldwell v. Bechtel, 631 F.2d 989 (D.C. Cir. 1980).

[49] Case Prestressing Corp. v. Chicago College of Osteopathic Medicine, 455 N.E.2d 811 (Ill. App. Ct. 1983).

[50] Waldinger Corp. v. Ashbrook-Simon-Hartley, Inc., 564 F. Supp. 970 (C.D. Ill. 1983).

[51] *See, e.g.*, Morrison v. MacNamara, 407 A.2d 555, 560 (D.C. 1979); Sides v. Richard Mach. Works, Inc., 406 F.2d 445 (4th Cir. 1969).

and the owner, as well as the obligations imposed by any applicable statutes, regulations, or standards and the tasks which the CM has in fact undertaken on a project. There is no requirement that there be a direct contractual relationship or privity of contract between the contractor and the CM.[52]

As set forth in § **5.6**, the standard of care is determined by whether the CM is regarded as a professional or a general contractor on a particular project. The issue of whether the CM breached the applicable standard of care is determined by comparing the CM's performance on the project in dispute to the standard of performance applicable to that CM.[53] This comparison typically requires expert testimony.[54]

In order to establish that it was damaged by the CM, the contractor must show that it incurred additional costs as a direct result of the CM's negligence. Otherwise stated, the negligence must be the *proximate cause* of the contractor's damage.[55]

§ 5.17 Third-Party Beneficiary

A contractor can recover from the construction manager (CM) under a third-party beneficiary theory only if it can establish that it was an *intended*, as compared to an *incidental*, beneficiary of the owner/CM contract. This means that the contractor must prove that the owner and the CM specifically intended that the contractor be benefited by the CM's performance of its contract for the owner. In view of the fact that owner/CM contracts rarely identify third-party beneficiaries, the proof of this intent is difficult. In the absence of any contract language clearly designating the contractor as a beneficiary of the owner/CM contract, the contractor will most likely not be deemed a third-party beneficiary. In *Gateway Erectors v. Lutheran General Hospital*,[56] an Illinois court allowed an action in tort against a CM while specifically denying an alternative count on a third-party beneficiary theory holding: ''Morse/Diesel [Project Manager] was the agent of a disclosed principal and did not agree to become personally liable under the contract.

---

[52] Arizona seems to have been the last to abandon the ''privity'' rule. *See* Donnelly Constr. Co. v. Oberg/Hunt/Gilleland, 667 P.2d 1298 (Ariz. 1984). *See also* Conforti & Eisele, Inc. v. John C. Morris Assocs., 175 N.J. Super. 341, 418 A.2d 1290 (1981); A.R. Moyer, Inc. v. Graham, 295 So. 2d 397 (Fla. 1973).

[53] *E.g.*,Gateway Erectors v. Lutheran Gen. Hosp., 102 Ill. App. 3d 300, 430 N.E.2d 20 (1981); James McKinney & Son, Inc. v. Lake Placid 1980 Olympic Games, Inc., 461 N.Y.S.2d 483 (Sup. Ct. App. Div. 1983).

[54] *See, e.g.*, Alfieri v. Butler, 38 A.D.2d 672, 327 N.Y.S.2d 108 (1971); Dillon Properties, Inc. v. Minmar Builders, Inc., 257 Md. 274, 262 A.2d 740 (1970).

[55] Morrison v. MacNamara, 407 A.2d 555 (D.C. 1979); Sides v. Richard Mach. Works, Inc., 406 F.2d 445 (4th Cir. 1969).

[56] 102 Ill. App. 3d 300, 430 N.E.2d 20 (1981).

Consequently, it is not personally liable on a contract theory.''[57] Likewise, most courts have denied a third-party beneficiary approach in traditional architect/engineer and owner contracts or owner/prime contracts.[58]

In at least one case, however, a third-party beneficiary action has been allowed. In *Broadway Maintenance Corp. v. Rutgers*,[59] a contractor sued the owner for its alleged failure to schedule and coordinate properly the work on a project. The court denied the action and held that, pursuant to the terms of the owner's contract with its construction manager, the CM[60] had specifically agreed to schedule and coordinate the work for the benefit of the contractors. In addition, each of the contractors' prime contracts with the owner stated that the CM would be responsible for the scheduling and coordination. Absent such explicit language making the CM responsible to other contractors for scheduling and coordination, however, it is unlikely that a contractor can succeed on a third-party beneficiary cause of action.

§ 5.18 Interference with Contract

A contractor can also sue the construction manager (CM) for interference with contract. The only case in which an interference with contract action specifically against a CM has been attempted to date is *John E. Green Plumbing & Heating Co., Inc. v. Turner Construction Co.*[61] Although the contractor was not able to prove its case, the cause of action was allowed.

In order to establish a cause of action for interference with the contract, the contractor must establish five elements:

1. The existence of a valid contract
2. The defendant's knowledge of the contract
3. Malicious interference by the defendant
4. A causal relationship between the defendant's interference and the breach for nonperformance of the contract
5. Resulting damages.[62]

---

[57] 430 N.E.2d at 23.

[58] Valley Landscape Co., Inc. v. Rollader, 218 Va. 257, 237 S.E.2d 120 (1977); A.R. Moyer, Inc. v. Graham, 285 So. 2d 397 (Fla. 1977); Buchman Plumbing Co. v. Regents of Univ. of Minn., 298 Minn. 328, 215 N.W.2d 479 (1974).

[59] 434 A.2d 1125 (N.J. 1981).

[60] In the *Broadway Maintenance* case, the CM was also the general contractor on the project.

[61] 500 F. Supp. 910 (E.D. Mich. 1980).

[62] Downey v. United Weatherproofing, Inc., 253 S.W.2d 976 (Mo. 1953).

Obviously, the most difficult element of this cause of action is to prove that the interference was *malicious*. This was the element on which the *John E. Green Plumbing & Heating* case foundered.

Similar causes of action have been allowed against architect/engineers. For example, in *Waldinger Corp. v. Ashbrook-Simon-Hartley, Inc.*,[63] a supplier of equipment was able to recover from an engineer on the grounds that the engineer unreasonably failed to approve its shop drawings. The court held that this was an intentional interference with the supplier's contract with its contractor. In a similar vein, a contractor successfully sued an architect/engineer for interference with a potentially beneficial financial relationship; the architect/engineer interfered with the contractor's bid to the owner.[64]

§ 5.19 Implied Warranty

It may be possible to have a cause of action against a construction manager (CM) on the theory of implied warranty. In *Taramac Development Co., Inc. v. Delamater Fround & Associates, P.A.*,[65] the Supreme Court of Kansas held:

> It can be said that certain professionals, such as doctors and lawyers, are not subject to such an implied warranty. However, an architect and an engineer stand in much different posture as to insuring a given result than does a doctor or a lawyer. The work performed by architects and engineers is an exact science: that performed by doctors and lawyers is not. A person who contracts with an architect or engineer for a building of a certain size and elevation has a right to expect an exact result.[66]

A few other jurisdictions have also taken an *implied warranty* approach as relates to an architect/engineer.[67] Although to date there have been no similar causes of action by a contractor against a CM, there is no reason why such a theory could not be maintained if there was in fact a warranty, express or implied, running from the CM to the owner. Since the scope of the CM's duty to the owner is defined by the owner/CM contract,[68] a breach of warranty action could be allowed under either a negligence or a third-party beneficiary theory.[69]

---

[63] 564 F.Supp. 970 (C.D. Ill. 1983).

[64] Joba Constr. Co. v. Burns & Roe, Inc., 329 N.W.2d 760 (Mich. Ct. App. 1982).

[65] 675 P.2d 361 (Kan. 1984).

[66] *Id.* 364.

[67] *See, e.g.*, Broyles v. Brown Eng'g Co., 151 So. 2d 767 (Ala. 1963); Bloomburg Mills, Inc. v. Sordoni Constr. Co., 164 A.2d 201 (Pa. 1960).

[68] See § **5.11**.

[69] At least one court, however, has refused to allow a warranty action by a contractor because of the lack of privity between the contractor and the CM. James McKinney & Son, Inc. v. Lake Placid 1980 Olympic Games, Inc., 461 N.Y.S.2d 483 (Sup. Ct. App. Div. 1983).

BASES FOR LIABILITY

§ 5.20 Generally

As set forth in **§ 5.11**, the grounds upon which a contractor can sue the construction manager (CM) are determined by the scope of the duties owed by the CM to the owner. Thus, the breach of any obligation owed by the CM to the owner can form the basis of a contractor's action. Since there is still limited case law specifically referring to a CM by name, in several situations analogies must be made to architect and/or engineer cases. The various bases for liability are set forth in the general chronological order of a construction project.

§ 5.21 Interference with Bids

Construction managers are frequently charged with reviewing and evaluating the bids of the various potential contractors on a project and recommending to the owner which bid to accept. In the case of *Joba Construction Co. v. Burns & Roe, Inc.*,[70] the plaintiff was a bidder on a contract for an agency of the City of Detroit. The defendant was a consulting engineer who advised the city not to accept the plaintiff's bid. Plaintiff sued the defendant on the grounds that the defendant did not have adequate justification for recommending to the owner that its bid be rejected. The court agreed and awarded the contractor its probable lost profits for the contract plus punitive damages on the theory that the defendant had interfered with a beneficial economic relationship. Thus, a contractor can sue a construction manager or similarly situated party if it improperly interferes with the contractor's bid to the owner.

§ 5.22 Design Defects

The construction manager (CM) is not generally responsible for the overall design of a construction project. It may, however, participate in design reviews and make recommendations to the owner for changes or improvements. Although there have been no cases to date where a contractor has sued a CM for design defects, it is clear that such a cause of action would be allowed. This conclusion is based on two lines of cases which have already fully developed: 1) contractor claims

---

[70] 500 F. Supp. 910 (E.D. Mich. 1980).

against design architect/engineers;[71] and 2) owner claims against CMs for their participation in the design process.[72]

A contractor is clearly entitled to recover from an architect if design defects cause the contractor to incur increased performance costs. Thus, where a contractor incurred additional costs to overcome an allegedly defective concrete design, it was entitled to seek recovery of those costs from the architect who designed the project.[73] In a similar vein, an owner has been able to recover from its CM where the CM contributed to a defective design.[74] Thus, there seems to be no reason why a contractor cannot recover from a CM for a defective design.

§ 5.23 Delayed Site Access

One of the characteristics of a construction manager-directed project is the large number of prime contractors involved.[75] One of the construction manager's (CM's) standard duties is to schedule the access of the several prime contractors to their portions of the overall work. If the CM fails to provide adequate or timely access, it can be liable to the contractor adversely affected. In one case, the CM and the owner were both found liable to a prime contractor for failure to provide adequate site access because they allowed other contractors on the site to drain water into the plaintiff's work area.[76] On another project, it was held that a subcontractor could maintain an action against the CM because the CM directed the subcontractor to commence its work before there was adequate site access.[77]

§ 5.24 Administration of Drawings

One of the construction manager's (CM's) duties on most projects is to review and approve the shop drawings submitted by the contractor. This activity can result in CM liability to the contractor in two situations: 1) when the CM rejects drawings that should have been approved; and 2) when the CM delays the approval of drawings. In the case of *Waldinger Corp. v. Ashbrook-Simon-Hartley, Inc.*,[78]

---

[71] *See, e.g.*, Conforti & Eisele, Inc. v. John C. Morris Assocs., 175 N.J. Super. 341, 418 A.2d 1290 (1981); Detweiler Bros., Inc. v. John Graham Co., 412 F. Supp. 416 (E.D. Wash. 1976); A.R. Moyer, Inc. v. Graham, 285 So. 2d 397 (Fla. 1973).

[72] R. Cushman, Construction Litigation: Representing the Owner 105 (John Wiley & Sons 1984).

[73] United States v. Rogers & Rogers, 161 F. Supp. 132 (S.D. Cal. 1958).

[74] First Nat'l Bank of Akron v. William F. Cann, 503 F. Supp. 419 (W.D. Ohio 1980).

[75] See § **5.2**.

[76] R.S. Noonan, Inc. v. Morrison-Knudsen Co., Inc., 522 F. Supp. 1186 (E.D. La. 1981).

[77] Gateway Erectors v. Lutheran Gen. Hosp., 102 Ill. App. 3d 300, 430 N.E.2d 20 (1981).

[78] 564 F. Supp. 970 (C.D. Ill. 1983).

a contractor and its supplier sued the owner's project engineer for its failure to approve certain items of equipment proffered by the supplier. As a result of this failure, the equipment had to be obtained from another source. The court, applying Illinois law, found the engineer liable under both negligence and interference with contract theories, because there was no justification for the failure to approve the drawings.[79]

With regard to delays and approval of shop drawings, most owner/prime contracts set forth a time within which the shop drawings will be reviewed and returned by the CM. If the CM fails to return the drawings within the stated time, there is a possible cause of action for negligence.[80] It is unlikely, however, that occasional short delays in the approval of shop drawings will provide the basis for a cause of action. It is only when there have been considerable delays on several occasions that the requisite standard of care could be deemed to have been breached.

§ 5.25 Failure to Provide Timely Information

The construction manager (CM) is typically required to provide the contractors on a project with the information necessary to carry out the work. This includes such items as clarification of drawings, schedules of other contractors, advance notice of proposed changes to the work, and similar items. In one Illinois case, the contractor sued the project engineer for its failure to provide timely responses to requests for instructions and complete details to the plans.[81] The court held that "the supervising engineer owes a duty to a general contractor to avoid negligently causing extra expenses for the contractor in the completion of the construction project."[82] In another case, the court allowed an action in negligence for the CM's failure to approve change orders in a timely fashion, although the CM eventually prevailed on other grounds.[83]

§ 5.26 Failure to Provide Construction Support Services

The construction manager (CM) is frequently required to provide basic construction support services, e.g., heat, light, power, surveying sanitary facilities, and related items, to the several contractors on a project.The CM's failure to provide

---

[79] *Id.* 980–81.

[80] John E. Green Plumbing & Heating Co., Inc. v. Turner Constr. Co., 500 F. Supp. 910 (E.D. Mich. 1980).

[81] Normoyle-Berg Assocs. v. Village of Deer Creek, 39 Ill. App. 3d 744, 350 N.E.2d 559 (1976).

[82] 350 N.E.2d at 561.

[83] John E. Green Plumbing & Heating Co., Inc. v. Turner Constr. Co., 742 F.2d 965 (6th Cir. 1984).

these items can be the basis of liability. For example, in an Illinois case, a contractor was allowed to bring an action both for the CM's failure to properly survey and stake the project and for its failure to provide electrical power.[84] In another case, the failure of the CM to provide heat to the project during the winter season was the basis for a cause of action by one of the project contractors.[85] The contractor's action can be based on both the inadequacy of the services provided and any delay or interruption in the service.

§ 5.27 Failure to Schedule and Coordinate

On most projects, the construction manager's (CM's) primary duty is to schedule and coordinate the work of the various contractors. For example, AIA Document B801, "Standard Form of Agreement Between the Owner and the Construction Manager," provides as follows:

> **1.1.3** Provide for the Architect's and the Owner's review and acceptance, and periodically update, a Project Schedule that coordinates and integrates the Construction Manager's services, the Architect's services and the Owner's responsibilities with anticipated construction schedules.
>
> **1.1.5.3** Develop a Project Construction Schedule providing for all major elements such as phasing of construction and times of commencement and completion required of each separate Contractor. Provide the Project Construction Schedule for each set of Bidding Documents.
>
> **1.1.5.4** Investigate and recommend a schedule for the Owner's purchase of materials and equipment requiring long lead time procurement, and coordinate the schedule with the early preparation of portions of the Contract Documents by the Architect. Expedite and coordinate delivery of these purchases.
>
> **1.2.2** Provide administrative, management and related services as required to coordinate Work of the Contractors with each other and with the activities and responsibilities of the Construction Manager, the Owner and the Architect to complete the Project in accordance with the Owner's objectives for cost, time and quality. Provide sufficient organization, personnel and management to carry out the requirements of this Agreement.
>
> **1.2.11** Record the progress of the Project. Submit written progress reports to the Owner and the Architect including information on each Contractor and each Contractor's Work, as well as the entire Project, showing percentages of

---

[84] Bates & Rogers Constr. Corp. v. North Shore Sanitary Dist., 92 Ill. App. 3d 90, 414 N.E.2d 1274 (1981).

[85] John E. Green Plumbing & Heating Co., Inc. v. Turner Constr. Co., 742 F.2d 965 (6th Cir. 1984).

> completion and the number and amounts of Change Orders. Keep a daily log containing a record of weather, Contractors' Work on the site, number of workers, Work accomplished, problems encountered, and other similar relevant data as the Owner may require. Make the log available to the Owner and the Architect.[86]

This obligation to the owner is mirrored in American Institute of Architects Document A201/CM, "Construction Management Edition General Conditions of the Contract for Construction" for use by the owner and its contractors. It provides as follows:

> **2.3.7** The Construction Manager will schedule and coordinate the Work of all contractors on the Project including their use of the site. The Construction Manager will keep the Contractor informed of the Project Construction Schedule to enable the Contractor to plan and perform the Work properly.[87]

Similar provisions appear in most owner/CM and owner/contractor contracts.[88]

Under such provisions, the CM is required to create and implement a reasonable construction schedule. Such a schedule must accommodate the site access, concurrent usage, material and equipment deliveries, and labor critical paths of the owner and all the project contractors. In addition, the CM must monitor that schedule throughout the course of the project, assure that all parties meet their portions of the schedule, take corrective action when a party is behind schedule, and modify the schedule if necessary. The failure of the CM to perform any of these activities can result in liability on the part of the CM. In Illinois, the court held that an engineer fulfilling the role of the CM was liable for improper supervision:

> A supervising engineer owes a duty to a general contractor to avoid negligently causing extra expenses for the contractor in the completion of a construction project. A supervising engineer must be held to know that a general contractor will be involved in a project and will be directly affected by the conduct of the engineer. This relationship, supervising engineer and general contractor, gives rise to a duty of care on the part of each party to the other. Such a duty exists even in the absence of a direct contractual relationship. We do not express an opinion as to the duty owed by the supervising engineer to parties connected with the construction project other than the general contractor.[89]

---

[86] American Institute of Architects, AIA Document B801, Standard Form of Agreement Between Owner and Construction Manager, Art. 1 (1980 edition).

[87] American Institute of Architects, AIA Document A201/CM, Construction Management Edition, General Conditions of the Contract for Construction, Art. 2.3.7 (1980 edition).

[88] *E.g.*, Associated General Contractors of America, AGC Document No. 510, Standard Form of Agreement Between Owner and Construction Manager (Owner Awards All Trade Contracts), Arts. 2.1.1, 2.2.1.5 (1979 edition).

[89] Normoyle-Berg Assocs. v. Village of Deer Creek, 39 Ill. App. 3d 744, 350 N.E.2d 559, 561 (1976).

It should be noted that although the above case only referred to the CM's duty to a prime contractor, a subsequent Illinois case extended that duty to include subcontractors.[90]

Another case which addresses the issue of a CM's liability for delay was *Casson Construction Inc.*[91] In that case, a CM was hired by the government to coordinate 21 separate contractors. One contractor's late completion forced acceleration of a follow-on contractor, for which the government sued the late contractor. The GSA Board of Contract Appeals disallowed the GSA's claim, stating that the delay was actually caused by the improper coordination of the CM.

CONSTRUCTION MANAGER DEFENSES

§ 5.28 Owner/Construction Manager Contract

Since the construction manager's (CM's) duties to the contractor are defined in large part by the owner/CM contract,[92] the CM may be able to rely on other provisions of the same contract to defend against a contractor's claims. If the owner/CM contract does not require the CM to perform a certain function, or provides that the CM is not responsible for a specific function, then the CM should not be liable to a contractor for a failure to perform that function. If, however, a CM actually does undertake a function, it may well be liable for a failure properly to perform that function, regardless of the provisions of its contract with the owner.[93]

In some cases the owner/CM contract may state that the contractor cannot rely on provisions of that contract. Many contracts contain provisions similar to the following:

> Nothing contained herein shall be deemed to create any contractual relationship between the [CM] and the Architect or any of the Contractors, Subcontractors or material suppliers on the Project; nor shall anything contained in this Agreement be deemed to give any third party any claim or right of action against the Owner or the [CM] which does not otherwise exist without regard to this Agreement.[94]

Such language has been held to defeat a contractor's action based on a third-party beneficiary theory, but not a negligence action.[95]

---

[90] Case Prestressing Corp. v. Chicago College of Osteopathic Medicine, 455 N.E.2d 811 (Ill. App. Ct. 1983).

[91] GSBCA Nos. 4884 *et al.*, 83-1 B.C.A. (CCH) ¶ 16,523 (1983).

[92] See § **5.11.**

[93] See § **5.14.**

[94] American Institute of Architects, AIA Document B801, Standard Form of Agreement Between Owner and Construction Manager ¶ 13.2 (1980 edition).

[95] Gateway Erectors v. Lutheran Gen. Hosp., 102 Ill. App. 3d 300, 430 N.E.2d 20 (1981).

§ 5.29 Owner/Prime Contract

On many construction management projects, the owner/prime contract specifically refers to the functions that will be performed by the construction manager (CM). For example, AIA Document A201/CM provides as follows:

> **2.3.1** The Architect and the Construction Manager will provide administration of the Contract as hereinafter described.
>
> **2.3.2** The Architect and the Construction Manager will be the Owner's representatives during construction and until final payment to all contractors is due. The Architect and the Construction Manager will advise and consult with the Owner. All instructions to the Contractor shall be forwarded through the Construction Manager. The Architect and the Construction Manager will have authority to act on behalf of the Owner only to the extent provided in the Contract Documents, unless otherwise modified by written instrument in accordance with Subparagraph 2.3.22.[96]

The courts have held that, since the contractor can rely on such provisions in bringing an action against a CM, the CM can rely on other provisions of the owner/prime contract in defending against claims by a contractor. One court has held:

> We believe it would be manifestly inequitable to permit [the contractor] to both claim that [the CM] is liable to it for its failure to perform the contractual duties described in the [Owner-Contractor] agreement and at the same time deny that [the CM] is a party to that agreement in order to avoid arbitration of claims clearly within the ambit of the arbitration clause. ''In short, the plaintiff cannot have it both ways. It cannot rely on the contract when it works to its advantage and repudiate it when it works to [its] disadvantage.''[97]

Owner/prime contracts contain a variety of provisions which may be used by a CM in defending against contractor claims. The most common of such provisions are discussed below.

§ 5.30 —Duty to Review Drawings and Prior Work

Many contracts attempt to shift the burden of reviewing the drawings to the contractor. A typical such provision provides as follows:

> The Contractor shall carefully study and compare the Contract Documents and shall at once report to the Architect and the Construction Manager any error, inconsistency

---

[96] American Institute of Architects, AIA Document A201/CM, Construction Management Edition, General Conditions of the Contract for Construction, Arts. 2.3.1, 2.3.2 (1980 edition).

[97] Hughes Masonry v. Greater Clark County School Bldg. Corp., 659 F.2d 836 (7th Cir. 1981) (citations omitted).

> or omission that may be discovered. The Contractor shall not be liable to the Owner, the Architect or the Construction Manager for any damage resulting from any such errors, inconsistencies or omissions in the Contract Documents. The Contractor shall perform no portion of the Work at any time without Contract Documents or, where required, approved Shop Drawings, Product Data or Samples for such portion of the Work.[98]

Although this language does not shift the total responsibility for the plans and specifications to the contractors, it does attempt to shift to them at least a portion of the duty to verify the accuracy of the plans and specifications.

An even more difficult provision from a contractor's viewpoint is one which makes it responsible for the conformance to plans and specifications of a prior contractor's work. A typical such clause reads as follows:

> If any part of the Contractor's Work depends for proper execution or results upon the work of the Owner or any separate contractor, the Contractor shall, prior to proceeding with the Work, promptly report to the Construction Manager any apparent discrepancies or defects in such other work that render it unsuitable for such proper execution and results. Failure of the Contractor so to report shall constitute an acceptance of the Owner's or separate contractor's work as fit and proper to receive the Work, except as to defects which may subsequently become apparent in such work by others.[99]

Although the effectiveness of such clauses has not been ruled on in the context of contractor/CM litigation, similar provisions have been held to bar claims by a contractor against an owner.[1] At the very least, it seems that such provisions shift at least a portion of the obligation to discover design defects or prior nonconforming work to the contractor, and its failure to discover such items could at least constitute contributory negligence.[2]

§ 5.31 —No Damages for Delay Clauses

A *no damages for delay* clause typically provides that, in the event of a delay, a contractor will only be entitled to a time extension. Alternatively, it will provide that the contractor cannot recover damages for delay or interference. A typical clause reads as follows:

> Should the Contractor be delayed in the commencement, prosecution or completion of the Work by the act, omission, neglect or default of the Owner, Construction

---

[98] American Institute of Architects, AIA Document A201/CM, Construction Management Edition, General Conditions of the Contract for Construction, Art. 4.2.1 (1980 edition).

[99] American Institute of Architects, AIA Document A201/CM, Construction Management Edition, General Conditions of the Contract for Construction, Art. 6.2.2 (1980 edition).

[1] Granite Constr. Co., IBCA No. 1-1442-3-81, 82-2 B.C.A. (CCH) ¶ 15,975 (1982).

[2] See § **5.40.**

> Manager, or of anyone employed by them, or of any other Contractor or Subcontractor on the Project, or by any other cause beyond the Contractor's control, none of which are due to any fault, neglect, act or omission on its part, then the Contractor shall be entitled to an extension of time only, such extension to be for a period equivalent to the time lost by reason of any and all of the aforesaid causes, as determined by the Construction Manager. The Contractor shall not be entitled to any such time, however, unless a claim therefor is presented in writing to the Construction Manager within twenty-four (24) hours of the commencement of such claimed delay. Such extension of time, as determined by the Construction Manager, shall release and discharge the Owner and Construction Manager from any and all claims of whatever character by the Contractor on account of said delay.[3]

The above clause even purports to absolve the owner and construction manager (CM) of liability for their own negligence.

As a general rule, courts will uphold no damages for delay clauses. They will, however, strictly construe such clauses and only apply them where the allocation of risk is clear and unambiguous.[4] Moreover, any ambiguities in such clauses will be resolved against the author of the contract. Thus, in the recent case of *John E. Green Plumbing & Heating Co., Inc. v. Turner Construction Company,*[5] the court (applying Michigan law) held that a no damages for delay clause would not bar a contractor's claim against a CM for damages for extra labor due to interference because the clause referred only to ''delay damages.''[6]

Notwithstanding the fact that most courts will give effect to no damages for delay clauses, four judicial exceptions to the application of the clause have developed. These were summarized in *Peter Kiewit Sons' Co. v. Iowa Southern Utilities Co.*:[7]

1. When the delay was not of a kind contemplated by the parties
2. When the length of the delay was unreasonable
3. When there was evidence of bad faith
4. When the delay was caused by the active interference of the owner.[8]

These exceptions are applied differently in the various jurisdictions.

§ 5.32 —Not of a Kind Contemplated by the Parties

In many jurisdictions delay damages can be awarded, despite a no damages for delay clause, when the delay was not of a kind in the contemplation of the parties.

---

[3] This provision is a composite from several private contracts.

[4] City of Seattle v. Dyad Constr., Inc., 565 P.2d 423 (Wash. Ct. App. 1972).

[5] John E. Green Plumbing & Heating Co., Inc. v. Turner Constr. Co., 500 F. Supp. 910 (E.D. Mich. 1980).

[6] *Id.*

[7] 355 F. Supp. 376 (S.D. Iowa 1973).

[8] *Id.* 397.

In *Franklin Contracting Co. v. State,*[9] a New Jersey court allowed recovery for delay caused by lack of access to the worksite. The contract specified that the contractor would become familiar with the means of access and would assume the costs of providing access. When Franklin requested proof of the availability of a right-of-way, the state's representatives gave assurances of access. Because there was no such right-of-way, construction was delayed for six months. The court held that this situation was not within the contemplation of the parties at the time of contracting:

> In a case where inquiry of the Director has disclosed that the status of one or more rights-of-way is such that they have not been obtained, it would be clear that in such circumstances this [no damages] clause would be operative in foreclosing the contractor from making any claim for delay. But where,as here, inquiry as to the status brought forth a representation that there was no problem with respect to same, it can hardly be claimed that this clause was intended to render the State immune from liability or that the parties contemplated delay because the State did not have a valid right-of-way.[10]

The same result has been reached by other courts.[11]

Some state courts, however, view a no damages for delay clause as extending to unforeseeable causes. The Texas Court of Appeals explained this position in *City of Houston v. R.F. Ball Construction Co., Inc.*:[12]

> The clause does not limit its application to those delays and hindrances that were foreseen by the parties when they entered into the contract. Instead, it embraces all delays and hindrances which may occur during the course of the work, foreseen and unforeseen. Indeed, it is the unforeseen events which occasion the broad language of the clause since foreseeable ones could be readily provided for by specific language.[13]

Similar results were reached in Washington[14] and Utah.[15]

§ 5.33 —Unreasonable Period of Delay

An unreasonable or excessive delay on the part of the owner may also override a no damages for delay clause, especially when the delay is close to an abandonment

---

[9] 365 A.2d 952 (N.J. Super. Ct. 1976).

[10] *Id.* 956.

[11] Buckley & Co., Inc. v. State, 356 A.2d 56 (N.J. Super. Ct. 1975); McGuire & Hester v. City & County of San Francisco, 247 P.2d 934 (Cal. Dist. Ct. App. 1952).

[12] 570 S.W.2d 75 (Tex. Civ. App. 1978).

[13] *Id.* 78.

[14] Nelse Mortensen v. Group Health, Inc., 566 P.2d 560 (Wash. Ct. App. 1977).

[15] Western Eng'rs, Inc. v. State Road Comm'n, 20 Utah 2d 294, 437 P.2d 216 (1968).

of the project. In a leading New York case,[16] damages for a three-year suspension of performance were awarded despite a limitation of liability clause:

> The contract in question, Clause J, does not admit of a construction that the board of education could by the exercise of arbitrary power extend the time of realtor's work and delay the same for 3 years or longer. The clause must be construed as inapplicable to delays so great or so unreasonable that they may fairly be deemed equivalent to an abandonment of the contract and thus interpreted was not a bar to realtor's claim.[17]

A delay of eight months in the performance of a 350-day contract was considered "sufficiently unreasonable to strike at the heart of the contract and justify the contractor in abandoning it," notwithstanding a no damages for delay clause.[18] In another New York case, although the court found that the state had actively interfered with construction, it indicated that the no damages for delay clause would probably be inapplicable to a 21-month delay.[19] However, in a more recent case, the New York Court of Appeals ruled that the no damages for delay clause was intended to reach a range of unreasonable delays and that active interference, absent bad faith or deliberate intent, would not make such a clause unenforceable. A strongly worded dissent would have upheld the right of the contractor to recover upon a finding of active interference for an extraordinarily long delay—28 months.[20]

One court has held that extensive delay without actions showing actual abandonment of the contract is not enough to make the no damages for delay clause unenforceable.[21] Other courts have found the specific delay in question not excessive.[22]

§ 5.34 —Bad Faith or Fraud

If an owner's delay is found to be a result of bad faith, fraud, concealment, or willful misrepresentation, a no damages for delay clause can be set aside.[23] Because every contract implies fair dealing between the parties, bad faith or tortious intent can withdraw the benefit of a bargain from the party abusing it.[24] In *Housing Authority v. Hubbell,*[25] bad faith was found when an owner caused delays willfully

---

[16] People v. Craig City, 133 N.E. 419 (N.Y. 1921).

[17] *Id.* 426.

[18] Brady v. Board of Educ., 222 A.D. 504, 506, 226 N.Y.S. 707, 709 (1926).

[19] American Bridge Co., Inc. v. State, 245 A.D. 535, 283 N.Y.S. 577 (1935).

[20] Kalisch-Jarcho, Inc. v. City of N.Y., 58 N.Y.2d 377, 448 N.E.2d 413, 461 N.Y.S.2d 746 (1983).

[21] F.D. Rich Co. v. Wilmington Hous. Auth., 392 F.2d 841 (3d Cir. 1968).

[22] Endres Plumbing Corp. v. State, 95 N.Y.S.2d 574 (Ct. Cl. 1950) (six-month delay); E.C. Ernst, Inc. v. Manhattan Constr. Co., 551 F.2d 1026 (5th Cir. 1977) (two-month delay); Dickerson Co., Inc. v. Iowa State Dep't of Transp., 300 N.W.2d 112 (Iowa 1981) (two-year delay).

[23] *See* Annot., 74 A.L.R.3d 187, 215 n.38 (1976).

[24] Psaty & Fuhrman, Inc. v. Housing Auth., 76 R.I. 87, 68 A.2d 32, 36 (1949).

[25] 325 S.W.2d 880 (Tex. Civ. App. 1959).

by "unreasoning action." In that case, the delays were caused by the arbitrary and capricious requirements of the owner's architect, and recovery was allowed despite the no damages for delay cause.

Extreme negligence can also supersede a damages disclaimer. In *Ozark Dam Constructors v. United States,*[26] the government delayed delivery of cement to a construction site because the cement plant was closed by a strike. Although the government was aware of the likelihood of a strike, it failed to warn bidders or to procure cement through other channels.

> The possible consequences were so serious, and the action necessary to prevent those consequences was so slight, that the neglect was almost willful. It showed a complete lack of consideration for the interest of the Plaintiffs. If the Plaintiffs really included in their bid an amount to cover the contingency of such inconsiderate conduct on the part of the Government's representatives, the Government was buying and the public was paying for things that were worth less than nothing.[27]

Similarly, a city's failure to obtain a right-of-way was termed a "knowing delay of the public authority, which transcends lethargy or bureaucratic bungling," and therefore a no damages for delay clause was held inapplicable.[28]

§ 5.35 —Active Interference

A no damages for delay clause will not protect the owner from liability for delay caused by *active interference.* The standard for active interference was defined in *Peter Kiewit Sons' Co. v. Iowa Southern Utilities Co.*:[29]

> [s]ome affirmative, willful act, in bad faith, to unreasonably interfere with plaintiff's compliance with the terms of the construction contract. . . . The use of the term "active" to modify interference . . . is significant and clearly implies that more than a simple mistake, error in judgment, lack of total effort is needed for plaintiff to prove the "active interference" necessary to render unenforceable an otherwise clear and unambiguous "no damages" clause.[30]

Although there have been numerous cases on the issue of active interference, one of the most recent has specifically addressed the owner's responsibility on a multi-prime project. In *United States Steel Corp. v. Missouri Pacific Railroad Co.,*[31] the court, applying Arkansas law, decided that an owner's directive to a follow-on prime to commence performance, even though a prior prime had not completed preceding work, constituted active interference. Specifically, the owner directed

---

[26] 127 F. Supp. 187 (Ct. Cl. 1955).

[27] *Id.* 191.

[28] Southern Gulf Utils., Inc. v. Boca Ciega Sanitary Dist., 238 So. 2d 458 (Fla. Dist. Ct. App. 1970).

[29] 355 F. Supp. 376 (S.D. Iowa 1973).

[30] *Id.* 399.

[31] 668 F.2d 435 (8th Cir.), *cert. denied*, 459 U.S. 836 (1982).

a steel erector on a bridge project to commence performance before the preceding concrete piers had been completed. In another recent case, *Blake Construction Co. v. C.J. Coakley Co., Inc.*,[32] with significant potential for application in the construction manager context, the court found that a prime had actively interfered with one of its subcontractors by failing to schedule and sequence properly the work of other subcontractors. Thus, a subcontractor was allowed to recover delay damages in spite of a comprehensive no damages for delay clause.

Delays which are in the contemplation of the parties are not, of course, interference.[33] Likewise, mere negligence will not override a no damages for delay clause.[34] In a recent case, the New York Court of Appeals ruled that the active interference exception would not apply unless such interference was "actuated by bad faith and deliberate intent."[35] This is, however, a dramatic departure from prior New York decisions.

§ 5.36 —Statutory Exceptions

At least one state, Washington, has declared no damages for delay clauses, at least in certain situations, unenforceable. The statute provides:

> Any clause in a construction contract, as defined in RCW 4.24.370, which purports to waive, release, or extinguish the rights of a contractor, subcontractor, or supplier to damages or an equitable adjustment arising out of unreasonable delay in performance which delay is caused by the acts or omissions of the contractee or persons acting for the contractee is against public policy and is void and unenforceable.[36]

Thus, owners and construction managers cannot protect themselves from delays which they cause. A similar statute was enacted in California for public contracts.[37]

§ 5.37 —Delegation of Scheduling Function to Contractors

Many owner/prime contracts attempt to delegate the duty to schedule and coordinate the overall project work to the several prime contractors, regardless of whether there is a construction manager (CM) on the project. In a similar vein, some owner/prime contracts attempt to shift the liability for delay directly to the individual contractor who caused the delay.

---

[32] 431 A.2d 569 (D.C. 1981).

[33] Gherardi v. Board of Educ., 147 A.2d 535 (N.J. Super. Ct. 1958).

[34] Cunningham Bros., Inc. v. City of Waterloo, 117 N.W.2d 46 (Iowa 1962).

[35] Kalisch-Jarcho, Inc. v. City of N.Y., 58 N.Y.2d 377, 380, 448 N.E.2d 413, 461 N.Y.S.2d 746 (1983).

[36] Wash. Rev. Code Ann. § 4.34.360 (1984).

[37] Act of Mar. 18, 1984, ch. 42, § 7102, 1984 Cal. Laws Reg. Sess. (to be codified at Cal. Pub. Contracts Code § 7102).

One of the earliest cases to address directly the obligations of the various parties on a CM project was *Pierce Associates.*[38] This was a United States General Services Administration (GSA) contract, which contained a provision that the contractors were to coordinate with each other and that, if one prime was delayed by another, it should seek recovery from that other contractor.[39]

One of the contractors was delayed approximately five months and brought its claim directly against the GSA. The GSA defended on the grounds that the contract limited the contractor's recovery to actions against other contractors. The GSA Board of Contract Appeals denied the GSA's defense and held that the GSA contracting officer still had an affirmative duty to coordinate the work.[40] A similar result was reached in *Jacobson & Co.*,[41] where the GSA Board of Contract Appeals held:

> The Government can decide on whatever method of construction and contracting it chooses, but, in doing so, it also assumes the responsibilities inherent in its choice. With phased construction, it is obvious that the contracting officer has the obligation and duty to demand that the various contractors cooperate with the construction schedule and not interfere with the work of any other contractor.[42]

Although these cases deal with the owner's obligations, they would be equally applicable to the CM.

In a series of cases arising out of the construction of the University of New Jersey Medical Facilities at Rutgers, a different conclusion was reached. On that project, the owner entered into six separate prime contracts, but designated one of the contractors as responsible for overall scheduling and coordination. Each of the other contractors was directed to comply with the directives of this one contractor.[43]

Disregarding this language, at least two of the primes elected to file suit directly against the owner. In both cases, the court ruled that since the duty of coordination was clearly delegated to a party other than the owner, the owner was not liable for delays and lack of coordination.[44] In reaching this overall decision, the trial court seemed to rely heavily on its determination that each contractor was a third-party beneficiary of each of the other contractor contracts and had a right to sue one another. It also appears that one of the contractors was fulfilling the role of a CM.

The difference between the GSBCA's decision in *Pierce Associates* and the New Jersey decisions in *Dobson* and *Broadway Maintenance* seems to be that in the former, no one party was clearly responsible for scheduling and coordination, while in the latter two cases, the duty was clearly assigned to a prime contractor who

---

[38] Pierce Assocs., Inc., GSBCA No. 4163, 77-2 B.C.A. (CCH) ¶ 12,746 (1977).

[39] 77-2 B.C.A. (CCH) at 61,939.

[40] *Id.* at 61,943.

[41] GSBCA No. 5605, 80-2 B.C.A. (CCH) ¶ 14,521 (1980).

[42] 80-2 B.C.A. (CCH) at 71,565.

[43] Broadway Maintenance Corp. v. Rutgers, 434 A.2d 1125, 1127–28 (N.J. 1981).

[44] Edwin J. Dobson, Jr., Inc. v. Rutgers State Univ., 157 N.J. Super. 357, 384 A.2d 1121 (1978).

was fulfilling the role of a CM. Thus, it seems that a general delegation of the duty to schedule and coordinate to several contractors will not relieve the CM of responsibility.

§ 5.38 –Forum Selection

Many owner/prime contracts designate a particular forum for resolving disputes between them. In at least one case, a court has held that a contractor with a negligence claim against the construction manager (CM) was required to submit the matter to arbitration, because its contract with the owner required that all owner/contractor disputes be arbitrated.[45] The court ruled that it would be "manifestly inequitable" to permit the prime to bring a negligence suit against the CM based on the CM's duties as defined in the owner/CM contract and not require the contractor to be bound by other terms of that contract.[46]

§ 5.39 Owner's Agent

It may be possible for a construction manager (CM) to argue that since it is performing as an agent of an owner, it is not independently liable for its negligence. The general test for the existence of an agency relationship is whether the owner has the right to control and direct the work, not only as to the result of the accomplished work, but also as to the manner and means by which that result is accomplished.[47] The CM who is the owner's agent is obligated to obey the owner's reasonable orders and carry out all duties in the owner's interest.[48] An independent contractor, on the other hand, is not required to comply with these duties; it is only responsible for carrying out the terms of its contract with the owner.

To date, there have been no cases where a CM avoided liability to a contractor because it asserted that it was acting as an agent of the owner. If a CM is functioning as the owner's agent, however, it may be able to obtain indemnification from the owner for any damages sought by a contractor.[49] Likewise, it may be able to avail itself of all defenses available to the owner.[50]

§ 5.40 Contributory and Comparative Negligence

Under general principles of tort law (i.e., negligence), a plaintiff who is negligent in a particular situation cannot recover in negligence from another party. This is

[45] Hughes Masonry v. Greater Clark County School Bldg. Corp., 659 F.2d 836 (7th Cir. 1981).

[46] *Id.* 838-39.

[47] S. Williston, A Treatise on the Law of Contracts § 1012A (3d ed. 1967).

[48] *Id.* § 1012B.

[49] Restatement (2d) of the Law of Agency § 439 (1958).

[50] *Id.* § 334.

called the doctrine of *contributory negligence* and is applied in many jurisdictions.[51] In other jurisdictions, a concept of *comparative negligence* has evolved. This means that the courts will compare the negligence of the plaintiff and the defendant and apportion damages based on the degree of negligence of each party.[52] For example, if the court finds that a plaintiff and a defendant are each 50 percent responsible for a situation, the plaintiff's damages could be reduced by 50 percent.

Although neither the doctrines of contributory nor comparative negligence have yet been applied in the construction manager (CM) field, there is absolutely no reason why they would not be applicable. Thus, a contractor who is suing a CM for delays to a project could lose its case under the doctrine of contributory negligence, or have its damages reduced under the doctrine of comparative negligence, if the CM could establish that the contractor contributed to or caused part of the delay.

§ 5.41 No Recovery for Economic Loss

Although there is little doubt that a contractor can sue a construction manager (CM) for negligence, there is considerable controversy as to whether the CM will be liable for the contractor's economic loss. Although the term *economic loss* is not precisely defined, it generally means any money damages other than compensation for direct damage to a person or property.[53]

Some jurisdictions hold that a design professional is liable only to its client, i.e., the owner or the other party with whom it has a contract, for economic loss. For example, in *Local Joint Executive Board of Las Vegas v. Stern*,[54] the court held as follows:

> The well established common-law rule is that absent privity of contract or an injury to person or property, a plaintiff may not recover in negligence for economic loss. . . . The primary purpose of the rule is to shield a defendant from unlimited liability for all of the economic consequences of a negligent act, particularly in a commercial or professional setting, and thus to keep the risk of liability reasonably calculable.[55]

Other jurisdictions, however, have held that design professionals are liable for economic loss. A recent Illinois case held as follows:

> The architect maintains that *Moorman* represents a philosophical statement about the limits of tort and contract law and accordingly urges that it be read expansively to preclude Rosos' recovery in the instant case. He contends that an architect's

---

[51] Prager v. City of N.Y. Hous. Auth., 112 Misc. 2d 1034, 447 N.Y.S.2d 1013 (1982).

[52] Sprayregen v. American Airlines, Inc., 570 F. Supp. 16 (D.C.N.Y. 1983).

[53] Steckmar Nat'l Realty & Inv. Corp., Ltd. v. J.I. Case Co., 99 Misc. 2d 212, 415 N.Y.S.2d 946 (1979).

[54] 651 P.2d 637 (Nev. 1982).

[55] *Id.* 638.

> contractual duties encompass not only those expressed in the contract, but many arising *ex lege* from the relationship created by the contract [citations omitted]. As a result, Hansen concludes Rosos must be limited to its remedy for loss on its contract claim. We disagree.
>
> Among the reasons architects have been found answerable in malpractice actions is because they hold themselves out and offer services to the public as experts in their line of endeavor. Those who employ them perceive their skills and abilities to rise above the levels possessed by ordinary laymen. Such persons have the right to expect that architects, as other professionals, possess a standard minimum of special knowledge and ability, will exercise that degree of care and skill as may be reasonable under the circumstances and, when they fail to do so, that they will be subject to damage actions for professional negligence, as are other professionals [citations omitted]. The broad reading of *Moorman* urged by Hansen in the instant case would, simply by inference, effectively eliminate and stand squarely in conflict with the body of law defining the scope of an architect's liability for professional negligence [citations omitted].[56]

Many other jurisdictions have followed the lead of Illinois and allowed for the recovery of economic loss.[57] Other jurisdictions, notably New York, do not allow economic loss in negligence actions, absent property damage or personal injury.[58]

§ 5.42 Statutes of Limitations

All legal actions are governed by statutes of limitations, which establish time limits within which a cause of action may be brought. These statutes vary from state to state, and generally apply different time limits to contract and negligence actions.[59] As a general rule, the statutes of limitations for negligence actions are shorter than those for written contracts; they range from two to five years from the date of accrual of the cause of action. However, they vary widely from state to state.[60] Some states also have special statutes applying to any improvements to real property.[61] A contractor should be sure to verify the statute of limitations in its jurisdiction.

---

[56] Rosos Litho Supply Corp. v. Richard T. Hanson, 123 Ill. App. 3d 290, 462 N.E.2d 566, 571 (1984).

[57] Conforti & Eisele, Inc. v. John C. Morris Assocs., 175 N.J. Super. 341, 418 A.2d 1290 (1981); A.R. Moyer, Inc. v. Graham, 285 So. 2d 397 (Fla. 1973); Detweiler Bros., Inc. v. John Graham & Co., 412 F. Supp. 416 (E.D. Wash. 1976).

[58] *See, e.g.*, John R. Dudley Constr., Inc. v. Drott Mfg. Co., 66 A.D.2d 368, 412 N.Y.S.2d 512 (1979); Steckmar Nat'l Realty & Inv. Corp., Ltd. v. J.I. Case Co., 99 Misc. 2d 212, 415 N.Y.S.2d 946 (1979).

[59] R. Cushman, Construction Litigation: Representing the Owner 116 (John Wiley & Sons 1984).

[60] *Id.*

[61] A. Kornblut, Statutory Trends: Statutes of Limitations and Repose and Anti-Indemnification Statutes (paper for ABA Forum Comm. on the Constr. Indus., Oct. 19–20, 1984).

§ 5.43 —Accrual

A statute of limitations for negligence starts to run when the tort occurs.[62] The determination of when the tort occurs requires an evaluation of all the facts and circumstances of the particular case, but some generalizations are possible. For actions arising during the course of construction, the statute of limitations usually does not begin to run until substantial completion of the construction.[63] Similarly, actions originating from defective design usually do not accrue until the design is completed.[64]

For those jurisdictions which have special statutes of limitations for improvements to real property, the statute typically starts to run from the date the project was substantially complete.[65] *Substantial completion* is that stage of completion when the project is ready for its intended purpose.[66]

§ 5.44 —Tolling

There are several exceptions to the rule that the statute of limitations begins to run when the tort occurs. These circumstances are said to *toll* or delay the start of the running of the statutory time limit.

The most significant exception that tolls the limitation is the discovery rule, which holds that certain causes of action do not accrue until the plaintiff knew or reasonably should have known about the harm. The discovery rule has been widely applied to tort claims but is not generally accepted for breach of contract actions.[67] As a general precept, however, the discovery rule would not apply to contractor claims against the construction manager. The contractor should know of its harm by the time of project completion.

Evasive or deceptive tactics by the defendant to frustrate or mislead a plaintiff can also toll the statute. For example, a contractor was not allowed to plead the statute of limitations when it avoided the plaintiff's attempt to sue it by lying about its residence.[68]

---

[62] R. Cushman, Construction Litigation: Representing the Owner 116-17 (John Wiley & Sons 1984).

[63] *See, e.g.*, President & Directors v. Madden, 660 F.2d 91 (4th Cir. 1981).

[64] *See, e.g.*, McCloskey & Co. v. Wright, 363 F. Supp. 223 (E.D. Va. 1973).

[65] A. Kornblut, Statutory Trends: Statutes of Limitations and Repose and Anti-Indemnification Statutes (paper for ABA Forum Comm. on the Constr. Indus., Oct. 19–20, 1984).

[66] Liptak v. Diane Apartments, Inc., 109 Cal. App. 3d 762, 167 Cal. Rptr. 440 (1980).

[67] *See* Board of Directors v. Regency Tower Venture, 2 Hawaii App. 506, 635 P.2d 244 (1981).

[68] Kessler v. Peck, 266 Ala. 669, 98 So. 2d 606 (1957).

PRACTICE CONSIDERATIONS IN LITIGATION DECISIONS

§ 5.45 Issues for Evaluation

Not every claim should be litigated. Ideally, all issues should be amicably settled by a candid and open negotiation between the parties. Unfortunately, informal dispute resolution is frequently unsuccessful, so litigation must be considered by the aggrieved party. This part of the chapter highlights several issues that must be considered in the contractor's litigation evaluation.

§ 5.46 Cost

Although it varies from locale to locale, any judicial system generally has a large backlog of cases waiting to be tried, with waiting lists of over five years in some jurisdictions. Attorneys' fees and the cost of experts to testify and prepare claims are so high that even the most rudimentary claim can cost $20,000 to $30,000. Extremely complicated construction cases frequently cost over a million dollars to present.

Both parties are affected by these costs, but it is the plaintiff who is most injured because of being deprived of the use of the claimed compensation for the pretrial period. Contractors who have unquestionably been damaged by the negligence of a construction manager may decide that settlement at a fraction of the total claim is worth as much as full recovery after a long litigation process.

§ 5.47 Evaluating Recovery Potential

The essential factors in an evaluation of recovery potential are the probability of success on each claim, the amount of provable damages, and the defendant's capacity to satisfy the judgment. The value of each claim must be candidly evaluated in light of legal theories and trends, precedents, potential witnesses and evidence, attorneys' litigation skills, type of trier of fact (judge, jury, or arbitrator), character of the parties, and other intangibles unrelated to the merit of the claim. The claimed damages must be carefully reviewed, documented, and judged against the evidence the defendant is likely to produce. Finally, the construction manager's ability to satisfy a judgment five or ten years in the future (appeals can last as long as trials) must be considered.

§ 5.48 Rule of Reason

The recovery potential must be balanced against the costs of litigation and, unless there is a nonmonetary benefit to be gained from a lawsuit,[69] the recovery must exceed the cost before the contractor should consider litigation. Of course, there are intermediate goals that can be achieved by pursuing a lawsuit to a certain point and then negotiating. Often a defendant is unwilling to talk about settlement until forced to confront the possibility of defending a costly lawsuit. Therefore, the rule of reason permits litigants to evaluate each proposed stage of the judicial proceedings from the same standpoint.

§ 5.49 Arbitration versus Litigation

Since actions against a construction manager (CM) are not usually based on a contract but on a tort theory, litigation will typically take place in court.[70] If there is a diversity of citizenship (i.e., the contractor and the CM are legal citizens of different states or countries), the litigation can be conducted in a United States district court.[71] If, on the other hand, both parties are citizens of the same state, litigation must take place in state court.

It may be possible for the contractor to obtain the CM's agreement to resolve a dispute by arbitration. This private method of adjudication has several advantages over litigation. First, there is no backlog of cases or shortage of arbitrators; hence, the five-year wait for a trial is reduced to the amount of time required by both parties to prepare for arbitration. Second, the arbitration panel is usually selected from a pool of experienced lawyers, contractors, architects, and owners who should be better able to grasp the technical issues than a harried trial judge. Finally, arbitration decisions are final and will not be set aside on appeal, absent fraud or egregious error.

Arbitration's strengths are also its weaknesses. The arbitrators are not as impartial as a judge would be. They are members of the industry and can be expected to hold strong opinions on issues that affect the parties. As they are also selected and paid by the parties, arbitrators tend to render compromise decisions which leave something for everyone. This factor alone tends to weaken strong cases and strengthen weak ones. The absence of a lengthy trial docket creates no incentive to speed up the proceedings, so a hearing can be greatly extended by an opponent. The finality of the decision increases the risk of being saddled with a bad judgment, since the right of appeal is virtually nonexistent.

---

[69] R. Cushman, Construction Litigation: Representing the Owner 114 (John Wiley & Sons 1984).

[70] But see § **5.48.**

[71] 28 U.S.C. § 1332.

§ 5.50 Third-Party Practice

An action against a construction manager (CM) alone is a two-party lawsuit. In many cases, the CM will elect to crossclaim against the party it deems ultimately responsible: the owner, architect/engineer, or other contractors. These parties are referred to as the *third parties*.[72]

Basically, a crossclaim is an entirely new lawsuit against a third party. It is permitted where the factual issues of the original claim and the crossclaim are so closely intermingled that a joint resolution of the cases is practical. For example, if a contractor sues the CM for failure to provide site access, the CM could crossclaim against another contractor who did not vacate that site in a timely manner.

---

[72] 28 U.S.C. § 7(a).

CHAPTER 6

SUING THE DESIGN PROFESSIONAL

Kenneth I. Levin and Robert F. Cushman

Kenneth I. Levin is a partner in the national law firm of Pepper, Hamilton & Scheetz, Philadelphia, Pennsylvania. He received his undergraduate degree from Cornell University and graduated magna cum laude from Villanova Law School, where he was an editor of the Law Review. Mr. Levin is a coauthor of *Construction Contracts 1985: Rights and Responsibilities of the General Contractor, Sub-Contractor and Material Supplier* (Practising Law Institute 1985). Mr. Levin has also authored papers for and spoken before bar association and industry groups on construction litigation. In 1983, he presented a paper on design professionals' liabilities to a special joint seminar on construction failures of the American Bar Association and the American Society of Civil Engineers. Mr. Levin is a member of the American Bar Association's Section of Litigation, its Construction Litigation Committee, the Section of Torts and Insurance Practice, its Fidelity and Surety Committee, and the Forum Committee on the Construction Industry. He specializes in construction and surety litigation, representing primarily contractors, subcontractors, suppliers, sureties, and owners.

Robert F. Cushman is a partner in the national law firm of Pepper, Hamilton & Scheetz and a recognized specialist and lecturer on all phases of real estate and construction law. He serves as legal counsel to numerous trade associations, construction, development, and bonding companies.

Mr. Cushman is the editor and co-author of *Doing Business In America, The Dow Jones Business Insurance Handbook, The Businessman's Guide to Construction and High Tech Real Estate*, published by Dow Jones-Irwin; *The McGraw-Hill Construction Business Handbook, The Construction Industry Formbook*, McGraw-Hill; *Avoiding Liability in Architecture, Design and Construction, Construction Litigation: Representing the Owner, The John Wiley Handbook on Managing Real Estate in the 1980s, Handling Property and Casualty Claims* and *Handling Fidelity and Surety Claims*, published by John Wiley & Sons.

Mr. Cushman, who is a member of the Bar of the Commonwealth of Pennsylvania and who is admitted to practice before the Supreme Court of the United States and the United States Claims Court, has served as Executive Vice President and General Counsel to the Construction Industry Foundation, as well as Regional Chairman of the Public Contract Law Section of the American Bar Association. He is a member of the International Association of Insurance Counsel.

---

§ 6.1 Historical and Theoretical Bases for Contractor's Right of Action

The recognition of the contractor's right of action against the design professional, though no longer a novelty, is still a development of relatively recent origin. Moreover, while it appears that the right has been sanctioned by the majority of jurisdictions which have considered the issue of its existence in recent years, acceptance has been neither universal nor unqualified.

The 1958 decision in *United States ex rel. Los Angeles Testing Laboratory v. Rogers & Rogers*[1] marked the beginning of a new era in the law of architectural malpractice. Before *Rogers,* the defense of privity had effectively insulated the design professional from malpractice liability for economic losses to all members of the contracting group except its client, the owner. In *Rogers,* a prime contractor sued a project architect. The architect had, after reviewing tests performed on concrete supplied by the plaintiff's subcontractor, authorized the incorporation of the concrete in the building under construction. The owner-architect agreement required the architect to review the test reports. Upon later inspection, a state agency refused to pass the structural members in which the concrete had been incorporated, and the architect stopped the work while corrective measures were taken. The prime contractor sued the architect to recover costs incurred in

---

[1] 161 F. Supp. 132 (S.D. Cal. 1958).

compensating for the defects and resulting from the ensuing delay, on the theory that the architect had been negligent in interpreting the test reports.

The architect moved for summary judgment, asserting that because he was not in contractual privity with the contractor, he owed no duty of care to the contractor (as opposed to the owner) in interpreting the test reports. However, the court held that the architect owed a duty to the contractor to supervise the project with due care under the circumstances alleged, notwithstanding that his sole contractual relationship was with the owner. Regarding the question of whether the absence of privity should bar a right of action, the court said:

> The determination whether in a specific case the defendant will be held liable to a third person not in privity is a matter of policy and involves the balancing of various factors, among which are the extent to which the transaction was intended to affect the plaintiff, the foreseeability of harm to him, the degree of certainty that the plaintiff suffered injury, the closeness of the connection between the defendant's conduct and the injury suffered, the moral blame attached to the defendant's conduct, and the policy of preventing future harm.[2]

Weighing these factors, the court concluded that a right of action should exist, stating:

> Considerations of reason and policy impel the conclusion that the position and authority of a supervising architect are such that he ought to labor under a duty to the prime contractor to supervise the project with due care under the circumstances, even though his sole contractual relationship is with the owner, here the United States. Altogether too much control over the contractor necessarily rests in the hands of the supervising architect for him not to be placed under a duty imposed by law to perform without negligence his functions as they affect the contractor. The power of the architect to stop the work alone is tantamount to a power of economic life or death over the contractor. It is only just that such authority, exercised in such a relationship, carry commensurate legal responsibility.[3]

---

[2] *Id.* 135 (citations omitted).

[3] *Id.* 135–36. The court also held that the contractor's claim against the architect was actionable on an alternative theory of negligent misrepresentation. The court stated:

> Additionally, it should be noted that insofar as the counterclaim alleges that the architect negligently interpreted and construed reports of tests on the concrete and then authorized the incorporation of the bents into the building, the contractor in effect asserts a claim of negligent misrepresentation by the architect, with reasonably foreseeable reliance thereon by the contractor to the latter's detriment. This is so because authorization to incorporate the bents into the building, given by the architect, who is admittedly responsible for overseeing the work of the testing laboratories and for supervising construction to assure conformity with specifications, must be taken to imply a representation, first, that the architect has inspected the bents and reviewed the tests performed on the concrete of which they were made, and second, that both the bents and the concrete therein conform to specifications.

The *Rogers* decision has been criticized, not so much because it rejected the defense of privity as an absolute bar to a contractor's right of action for professional negligence, but because of the way in which it stated the test for liability. One commentator observed that the characterization of the design professional as an all-powerful "master builder" with independent authority to stop the work is no longer universally accurate. The language of the standard form owner-professional agreements has in recent years been modified to divest the design professional of the independent authority to stop the work. Further, the design professional's supervisory, quality control, administrative, and coordination functions are now often shared with a separate construction manager. Therefore, the commentator argued, it is inappropriate to impose a liability upon the professional simply on the basis of assumptions regarding his or her status and powers. Rather, whether a duty is owed by a professional to a contractor should depend upon (1) the scope of the professional's obligation under the owner-professional agreement (or otherwise assumed by the professional apart from that agreement) and (2) whether the contractor's reliance upon the professional's careful performance of such obligations was reasonably foreseeable.[4] In fact, it appears that considerations of scope of duty and foreseeability of reliance have been implicitly, if not explicitly, operative in the cases which followed *Rogers*. Thus, it seems appropriate that such considerations should be the primary focus of inquiry.

§ 6.2 —Economic Loss Limitation

In *Rogers'* wake, numerous decisions recognized the existence of a contractor's right of action in negligence against the design professional.[5] However, the cases

> The later cases indicate that the California courts would impose liability for such a negligent representation. . . .

Id. 136 (citations omitted).

[4] *See* Note, 92 Harv. L. Rev. 1076, 1084–85 (1979).

[5] *See, e.g.*, E.C. Ernst v. Manhattan Constr. Co., 551 F.2d 1026 (5th Cir. 1977) (applying Alabama law); Donnelly Constr. Co. v. Oberg/Hunt/Gilleland, 677 P.2d 1298 (Ariz. 1984); A.R. Moyer, Inc. v. Graham, 285 So. 2d 397 (Fla. 1973); Gateway Erectors v. Lutheran Gen. Hosp., 102 Ill. App. 3d 300, 430 N.E.2d 20 (1981) (actually involving construction manager); Bates & Rogers Constr. Corp. v. North Shore Sanitary Dist., 92 Ill. App. 3d 90, 414 N.E.2d 1274 (1980); W.H. Lyman Constr. Co. v. Village of Guerney, 403 N.E.2d 1325 (Ill. App. Ct. 1980); Normoyle-Berg Assocs. v. Village of Deer Creek, 39 Ill. App. 3d 744, 350 N.E.2d 559 (1976); Peter Kiewit Sons' Co. v. Iowa S. Utils. Co., 355 F. Supp. 376 (S.D. Iowa 1973); Gurtler, Hebert & Co., Inc. v. Weyland Mach. Shop, Inc., 450 So. 2d 660 (La. Ct. App. 1981); City of Columbus v. Clark-Dietz & Assocs.-Eng'rs, Inc., 550 F. Supp. 610 (N.D. Miss. 1982); Owen v. Dodd, 431 F. Supp. 1239 (N.D. Miss. 1977); Lane v. Geiger-Berger Assocs., P.C., 608 F.2d 1148 (8th Cir. 1979) (applying Missouri law); Conforti & Eisele, Inc. v. John C. Morris Assocs., 175 N.J. Super. 341, 418 A.2d 1290 (Law Div. 1981); Shoffner Indus. v. W.B. Lloyd Constr. Co., 257 S.E.2d 50 (N.C. Ct. App. 1979); Davidson & Jones, Inc. v. County of New Hanover, 255 S.E.2d 580 (N.C. Ct. App. 1979); Detweiler Bros., Inc. v. John Graham & Co., 412 F. Supp. 416 (E.D.

have not been unanimous in their recognition of the contractor's right of action,[6] nor have all those which refused to permit a right of action turned upon the bar of privity. Some questioned a contractor's ability to recover in tort for purely economic losses.[7]

The rationale for the economic loss limitation in the context of the contractor-professional relationship has not been explored to any substantial extent in the cases which relied on it. One of the principal arguments presented by design professionals for such a limitation is that it is consistent with a general rule applied in the law of torts that recovery may not be had in actions for negligence where a plaintiff's losses are entirely economic in character. The professionals also argue that the allocation of liabilities among members of the contracting group should be purely a matter of contract.

As against arguments for an economic loss limitation, a contractor could be expected to present the following counterarguments. To be sure, courts often recognize the general tort-law rule which holds that recovery may not be had for purely economic losses in an action based solely upon negligence.[8] The rationale for requiring physical injury to property or person as a predicate to recovery for economic losses is that a *bright-line* rule is necessary to protect against an otherwise virtually limitless scope of liability for negligence in terms of dollar magnitude and number of potential plaintiffs.[9] For example, as a matter of policy, it would be inappropriate for an individual who negligently caused an automobile collision during the morning rush hour to be held liable for lost wages to all commuters who were delayed on their way to work. As a matter of policy, the law has drawn lines short of the broader limits of foreseeability in such circumstances.

However, a contractor might argue that the rationale for the economic loss limitation should be inoperative where it is possible to confine the right of action to

---

Wash. 1976). *See also* Coac, Inc. v. Kennedy Eng'rs, 136 Cal. Rptr. 890 (Ct. App. 1977) (holding contractor to be third-party beneficiary of engineer's duty under owner-engineer agreement); John E. Green Plumbing & Heating Co., Inc. v. Turner Constr. Co., 742 F.2d 965 (6th Cir. 1984) (applying Michigan law; holding that construction manager owes duty to contractor on third-party beneficiary theory); James McKinney & Son, Inc. v. Lake Placid 1980 Olympic Games, Inc., 461 N.Y.S.2d 483 (Sup. Ct. App. Div. 1983) (holding that responsibilities of project/construction manager were such as to establish duty in tort to subcontractor).

[6] *See, e.g.*, Bryant Elec. Co., Inc. v. City of Fredericksburg, 762 F.2d 1192 (4th Cir. 1985) (applying Virginia law); Peyronnin Constr. Co., Inc. v. Weiss, 208 N.E.2d 489 (Ind. Ct. App. 1965); Hogan v. Postin, 695 P.2d 1042 (Wyo. 1985); Bernard Johnson, Inc. v. Continental Constructors, Inc., 630 S.W.2d 365 (Tex. Civ. App. 1982).

[7] *See, e.g.*, Bernard Johnson, Inc. v. Continental Constructors, Inc., 630 S.W.2d 365 (Tex. Civ. App. 1982); Bryant Elec. Co., Inc. v. City of Fredericksburg, 762 F.2d 1192 (4th Cir. 1985); Waldinger Corp. v. Ashbrook-Simon-Hartley, Inc., 564 F. Supp. 970 (C.D. Ill. 1983).

[8] *See, e.g.*, Restatement (2d) of the Law of Torts § 766C (providing, among other things, that as a general rule there is no liability in tort for negligence which interferes with another's performance of his or her contract or makes the performance more expensive or burdensome).

[9] *See, e.g.*, State of Louisiana *ex rel.* William J. Guste, Jr. v. M/V Testbank, 752 F.2d 1019 (5th Cir. 1985), *appeal pending*.

a narrow and readily identifiable group, whose reliance upon the performance of a duty with care is reasonably foreseeable. For example, the *Restatement (Second) of Torts* notes that the general bar to an action for negligent interference with contract does not apply where the plaintiff stands in a special relationship to the tortfeasor:

> e. *Duty of care to prevent pecuniary loss.* Outside the scope of this Section are certain cases in which the actor renders a service or has some other contractual relation ship in which he owes a duty to use reasonable care to avoid a risk of pecuniary loss to the person with whom he is directly dealing, *and that same duty is held to extend to another person whom he knows to be pecuniarily affected by the service rendered.* Most of the cases coming within this category are covered by § 552, involving information negligently supplied for the guidance of others. Other cases, involving other services than the supplying of information, may not fall within the exact provisions of § 552 but are covered by the general principle underlying it. . . .[10]

Similarly, the section of the *Restatement* dealing with the scope of liability for negligent misrepresentation could support an argument that members of the contracting group are a sufficiently confined group as to invest in them a right to be free of economic loss resulting from a design professional's malpractice. That section provides:

> 9. The City of A is about to ask for bids for work on a sewer tunnel. It hires B Company, a firm of engineers, to make boring tests and provide a report showing the rock and soil conditions to be encountered. It notifies B Company that the report will be made available to bidders as a basis for their bids and that it is expected to be used by the successful bidder in doing the work. Without knowing the identity of any of the contractors bidding on the work, B Company negligently prepares and delivers to the City an inaccurate report, containing false and misleading information. On the basis of the report C makes a successful bid, and also on the basis of the report D, a subcontractor, contracts with C to do a part of the work. By reason of the inaccuracy of the report, C and D suffer pecuniary loss in performing their contracts. B Company is subject to liability to B [sic C] and to D.[11]

In short, it appears that there are arguments available to a contractor against the application of an economic loss limitation in assessing liabilities in tort among members of the contracting group. It remains to be seen how the courts will resolve this issue; however, the majority of courts which have followed *Rogers* to date have done so without regard to any economic loss limitation. This is not, however, to say that the imposition of such a limitation is inappropriate where claims are brought by members of the public outside the contracting group who do not stand in any special relationship to the professional.[12]

---

[10] Restatement (2d) of the Law of Torts § 766C, Comment e, at 26 (emphasis added).

[11] *Id.* § 552, Illustration No. 9, at 135.

[12] *Cf.* Moore v. Pavex, Inc., No. 2843 S 1984, slip op. (Ct. of Common Pleas of Dauphin County, Pa., Oct. 4, 1985) (owner and contractors on construction project could not be held liable to class

§ 6.3 Negligence as the Basis for Design Professional's Liability

Before proceeding to a review of the circumstances in which the scope and nature of the design professional's obligations to contractors have been considered, it is appropriate briefly to consider the legal standard by which a design professional's liability is measured: the *negligence* standard.

In the absence of express contractual warranties or stipulations imposing a different standard of care, proof of negligence has traditionally been required as a basis for the imposition of liability in damages upon design professionals. The basic rule has been stated as follows:

> Plaintiff was an engineer and was employed as such. In performing the work which he undertook, it was his duty to exercise such care, skill and diligence as men engaged in that profession ordinarily exercise under like circumstances. He was not an insurer that the contractors would perform their work properly in all respects; but it was his duty to exercise reasonable care to see that they did so.[13]

While the doctrines of implied warranty and strict liability (liability without fault) are widely applied as a basis for imposing liability upon manufacturers of defective products, the courts, for the most part, reject these doctrines in assessing the liability of design professionals.[14] The rationale for refusing to impose strict liability upon design professionals has been articulated as follows:

---

consisting of entire business community of City of Harrisburg for purely economic losses resulting from puncture of water main).

The courts have divided on whether design professionals are liable for economic losses claimed to have been suffered by tenants or subsequent purchasers of facilities which they have designed. Cooper v. Jevne, 129 Cal. Rptr. 724 (Ct. App. 1976) (purchasers of condominium units may sue for economic losses); San Francisco Real Estate Investors v. J.A. Jones Constr. Co., 524 F. Supp. 768 (S.D. Ohio 1981), *aff'd,* 703 F.2d 976 (6th Cir. 1983) (subsequent purchaser of structure may not sue architect).

In A.E. Inv. Corp. v. Link Builders, Inc., 214 N.W.2d 764 (Wis. 1974), the court held that a tenant, the operator of a supermarket, could sue an architect for damages consisting of loss of profits, loss of fixtures, equipment, and merchandise, and loss of reputation and good will, where it claimed that the architect's design negligence had resulted in floor subsidence which made the leased structure untenantable. *Id.* 766. The court, however, ruled that the architect would, at trial, be permitted to present evidence on whether public policy considerations should in the particular case preclude an award of damages for such economic losses. *Id.* 770.

[13] Pastorelli v. Associated Eng'rs, Inc., 176 F. Supp. 159, 166 (D.R.I. 1959), quoting Cowles v. City of Minneapolis, 151 N.W. 184, 185 (Minn. 1915).

[14] *See* Del Mar Beach Club Owners Ass'n, Inc. v. Imperial Contracting Co., Inc., 176 Cal. Rptr. 886 (Ct. App. 1981); Castaldo v. Pittsburgh-Des Moines Steel Co., 376 A.2d 88 (Del. 1977); Borman's Inc. v. Lake State Dev. Co., 60 Mich. App. 175, 230 N.W.2d 363 (1975); City of Mounds View v. Walijarvi, 263 N.W.2d 420 (Minn. 1978); Board of Trustees of Union College v. Kennerly, Slomanson & Smith, 400 A.2d 850 (N.J. Super. Ct. Law Div. 1979); Sears, Roebuck & Co. v. Enco Assocs., Inc., 372 N.E.2d 555 (N.Y. 1977). *Contra* Federal Mogul Corp. v. Universal Constr. Co., 376 So. 2d 716 (Ala. Civ. App. 1976) (dicta); Broyles v. Brown Eng'g

> Architects, doctors, engineers, attorneys, and others deal in somewhat inexact sciences and are continually called upon to exercise their skilled judgment in order to anticipate and provide for random factors which are incapable of precise measurement. The indeterminate nature of these factors makes it impossible for professional service people to gauge them with complete accuracy in every instance. Thus, doctors cannot promise that every operation will be successful; a lawyer can never be certain that a contract he drafts is without latent ambiguity; and an architect cannot be certain that a structural design will interact with natural forces as anticipated. Because of the inescapable possibility of error which inheres in these services, the law has traditionally required, not perfect results, but rather the exercise of that skill and judgment which can be reasonably expected from similarly situated professionals. . . .
>
> [Further], while it is undoubtedly fair to impose strict liability on manufacturers who have had ample opportunity to test their products for defects before marketing them, the same cannot be said of architects. Normally, an architect has but a single chance to create a design for a client which will produce a defect free structure. Accordingly, we do not think it just that architects should be forced to bear the same burden of liability for their products as that which has been imposed on manufacturers generally.[15]

While a professional may by contract agree to a more rigorous standard of care than the negligence standard, it has been held that the benefit of such higher standard will run solely to those in privity and not to contractors claiming in tort.[16] Further, as a general rule, proof of professional negligence must be established by expert testimony. The standard of care must be determined from the testimony of experts, unless the conduct involved is within the common knowledge of laymen.[17]

§ 6.4 Design Professional's Liability to Contractor

Against this background, it is appropriate to consider in more detail representative cases which have considered the design professional's liabilities and duties to contractors. The courts have evaluated the liabilities of design professionals to contractors for errors or omissions in the following areas:

Co., 151 So. 2d 767 (Ala. 1963) (limited to provision of routine services); Bloomsburg Mills v. Sordoni Constr. Co., 164 A.2d 201 (Pa. 1960) (warranty of fitness). *See also* Doundoulakis v. Town of Hempstead, 368 N.E.2d 24 (N.Y. 1977) (holding that strict liability could apply where design was for a project which in itself constituted an abnormally dangerous activity as defined in Restatement (2d) of the Law of Torts § 520).

[15] City of Mounds View v. Walijarvi, 263 N.W.2d 420, 424–25 (Minn. 1978).

[16] *See* Peter Kiewit Sons' Co. v. Iowa S. Utils. Co., 355 F. Supp. 376, 394 (S.D. Iowa 1973).

[17] *See* Allied Properties v. John A. Blume & Assocs., Eng'rs, 25 Cal. App. 3d 848, 102 Cal. Rptr. 259 (1972); Paxton v. County of Alameda, 259 P.2d 934 (Cal. Ct. App. 1953). *See generally* Annot., 3 A.L.R.4th 1023 (1981).

1. Design
2. Supervision and inspection
3. Administration (including review of submittals, consideration of substitutes, and reaction to unanticipated conditions)
4. Scheduling and coordination.

The cases demonstrate that, in assessing such liabilities, the courts have been attentive to the scope of the design professional's responsibilities as set forth in the owner-professional agreement or as otherwise established by the professional's conduct during the project, to the foreseeability of the contractor's reliance on the professional's performance of such duties, and to evidence of the professional's compliance or noncompliance with prevailing standards of care.

§ 6.5 —Defective Design

City of Columbus v. Clark-Dietz & Associates-Engineers, Inc.[18] arose out of the construction of a wastewater treatment plant, and, more particularly, the failure of a protective levee surrounding the construction site at a time when the project was nearly complete. As a result of the failure, the site was inundated, and there was extensive damage and delay in completion of the facility. The owner brought suit against the contractor and the engineer for damages resulting from the failure, and the contractor both counterclaimed against the city and crossclaimed against the engineer to recover for extra construction expenses resulting from negligent design. The court concluded that the failure of the levee was due to negligent design on the part of the engineer, that the contractor had a right of action in tort against the engineer for negligence, and that the owner and the engineer were therefore jointly and severally liable to the contractor for the damages resulting from the failure.

In *Columbus,* the parties agreed that the levee failures occurred where pipes encased by concrete seepage collars penetrated a slurry wall. They also agreed that the failure occurred as a result of water seepage and consequent "piping" through the wall at the level of the pipes, which eventually permitted the water to surge through and cause the wall to collapse. The contractor claimed that the failure was due to professional negligence and defective design. The engineer claimed that the failure was a result of defective construction.

The original design called for pipes to be placed through the levee with a slurry wall then to be placed around the pipes. When the necessary pipes were unavailable at the outset of the slurry wall construction, the engineer decided that the slurry wall construction could proceed without the pipes in place and that cuts could be made in the wall for later installation of pipe. The engineer's representatives determined that seepage collars would be adequate to restore the cuts in the wall.

[18] 550 F. Supp. 610 (N.D. Miss. 1982).

During the course of installation of the pipes and collars, a disagreement arose as to the method of installation of the collars. The contractor maintained that the collars should be encased with clay to seal the concrete, thus eliminating direct contact between the slurry and the collars themselves. The engineer, however, insisted that the collars be keyed directly into the slurry. Based on expert testimony presented by the contractor, the court ultimately concluded that the vastly differing deformability characteristics of the concrete collar and the cement-bentonite slurry in the wall were the primary cause of the seepage and piping which led to the failure of the levee. The court also concluded that the engineer's conduct with respect to the placement of the collars constituted professional negligence in that: (1) the engineering personnel who made the design decisions did not have sufficient prior experience with cement-bentonite slurry to provide them with a basis for evaluating the sufficiency of the collar design; (2) they did not consult a report available to them, from which the compression strength of the slurry could have been computed; (3) they did not run tests to compute the strength of the slurry; and (4) they made no effort to determine the suitability of the collar design either through consultation with slurry experts or with specialty engineers in-house at their own firm. The court therefore concluded that the engineer was liable to the contractor for the losses resulting from the levee failure.[19]

The *Columbus* case establishes that the professional's liability to the contractor for defective design requires not only proof of a design deficiency, but also proof that the deficiency was a result of professional negligence.

§ 6.6 —Negligence in Supervision and Inspection

Also at issue in the *Columbus*[20] case were the relative responsibilities of the contractor and the engineer for deficiencies in other areas of the levee that had not failed, but which nevertheless required correction. Although these deficiencies were found, in part, to be a result of deficiencies in construction work, as opposed to design, the contractor argued that the engineer should bear full responsibility because of its supervisory responsibilities on the project. Given that the owner-architect agreement, however, did not impose upon the engineer a duty of continuous and exhaustive inspection, the court held that the engineer owed no duty of supervision to the contractor other than to exercise reasonable care when it provided instructions and test results at the job site. The court's analysis of a design

---

[19] Closely related to the concept of liability for negligent design is liability for negligent misrepresentation. For example, in Davidson & Jones, Inc. v. County of New Hanover, 255 S.E.2d 580 (N.C. Ct. App. 1979), the court held that a contractor could maintain an action against the owner's soils consultant for negligent misrepresentation of subsurface conditions. This right in tort was held to exist notwithstanding provisions of an exculpatory character contained in the owner-contractor agreement. Similarly, in Lane v. Geiger-Berger Assocs., P.C., 608 F.2d 1148 (8th Cir. 1979), the court held that a subcontractor's reliance on the professional's representation as to the character of fill which would be permitted on a project would provide a basis for a claim by the subcontractor against the professional.

[20] City of Columbus v. Clark-Dietz & Assocs.-Eng'rs, Inc., 550 F. Supp. 610 (N.D. Miss. 1982).

professional's supervisory responsibilities to the contractor is highly instructive and bears repetition at length. The court stated:

> Basic also asserts that despite our finding that its poor workmanship and use of improper soil materials in the levee was partly responsible for the necessity of a second slurry wall, Clark-Dietz should bear full responsibility. Basic contends that Clark-Dietz, by supervising construction, performing soil tests and approving work on a lift by lift basis, accepted *all* work as complying with plans and specifications. To determine the extent of Clark-Dietz' duty to Basic for supervision, we must first examine the pertinent contract provisions. Sections of the construction contract provide:
>
> 5. AUTHORITY AND RESPONSIBILITY OF THE ENGINEER/ARCHITECT
>
> All work shall be done under the general supervision of the Engineer/Architect. The Engineer/Architect shall decide any and all questions which may arise as to the quality and acceptability of material furnished, work performed, rate of progress of work, interpretation of Drawings and Specifications, and all questions as to the acceptable fulfillment of the Contract on the part of the Contractor.
>
> **ARTICLE 7. THE ENGINEER'S STATUS:**
>
> The Engineer shall have general supervision and direction of the Work. He is the Agent of the Owner only to the extent provided in the Contract Documents. He has authority to stop the Work whenever such stoppage may be necessary to insure proper execution of the Contract.
>
> **ARTICLE 11. INSPECTION OF WORK:**
>
> At any time during the course of construction of this project when, in the opinion of the Engineer, provisions of the Plans, Specifications or Contract Conditions are being violated by the Contractor or his employees, the Engineer shall have the right and authority to order all construction to cease or materials to be removed, until arrangements satisfactory to the Engineer are made by the Contractor for resumption of the Work in compliance with the provisions of the Contract. It shall not be construed as a waiver of defects if the Engineer shall not order the Work stopped or more material removed, as the case may be.
>
> Other contractual provisions deal with the approval of compaction equipment, testing soil in the borrow pits, and the owner's right to perform soil density tests. These provisions give Clark-Dietz considerable authority to evaluate Basic's work and bring construction to a halt should it deem such action necessary. *Unquestionably, Basic cannot be held responsible for work done as Clark-Dietz explicitly required and/or approved. Oral directions of [the engineer's representatives], at variance with the written plans and specifications, such as the clay key between the slurry wall and embankment, were actually field modifications of the plans and specifications; and by following them, Basic constructed the project in accordance with the plans and specifications.* . . . Furthermore, once Clark-Dietz undertook to perform soil tests and supervise Basic's work, the law imposed a duty on Clark-Dietz to discharge

such activities with reasonable care. . . . Since Clark-Dietz exercised extensive control over the project, *Basic had the right to rely on specific approval of its work and testing results provided by Clark-Dietz.* . . .

Any duty of Clark-Dietz to Basic for supervising construction in the first instance is dependent upon the obligation owed by Clark-Dietz to the City. . . . This makes relevant various provisions of the agreement for professional services made by the City with Clark-Dietz:

> Paragraph 1.6.2 [The engineer will] [m]ake periodic visits to the site to observe the progress and quality of the executed work and to determine in general if the work is proceeding in accordance with the contract documents; he will not be required to make exhaustive or continuous on-site inspections to check the quality or quantity of the work; . . . but he will not be responsible for the contractor's failure to perform the construction work in accordance with the contract documents . . .
>
> Paragraph 1.6.7 The engineer shall not be responsible for the acts or omissions of the contractor, any sub-contractor or any of the contractor's or sub-contractor's agents or employees or any other persons performing any of the work under the construction contract.
>
> Paragraph 6.6 Nothing herein shall be construed as creating any personal liability on the part of any officer or agent of any public body which may be a party hereto, nor shall it be construed as giving any rights or benefits hereunder to anyone other than the owner and the engineer.

These paragraphs unambiguously limit Clark-Dietz' duty for supervising construction to an obligation to observe the general progress of the work, and not to make continuous and exhaustive inspections. We hold that Clark-Dietz performed this contractual duty by generally overseeing construction and conducting soil tests with reasonable care. . . .

The language in the Clark-Dietz—City contract clearly does not create a requirement for Clark-Dietz to inspect and verify every step of Basic's work. In the absence of an active undertaking to guarantee the contractor's work, courts have ordinarily held that similar language absolves the architect of any liability for the contractor's poor workmanship. . . . Thus, we hold that Clark-Dietz owed no duty of supervision to Basic other than to exercise reasonable care when it provided instructions and test results at the job site. Having found that Clark-Dietz was not negligent in these respects, Basic must bear responsibility for the unacceptable soil material found in the embankment except for sand seams directly above the slurry wall, and is liable to the City for damages resulting therefrom.

As discussed earlier, the presence of sand seams immediately above the slurry wall was a proximate result of Clark-Dietz' defective design for which Basic is not liable.[21]

While *Columbus* absolves the professional of a duty of continuous inspection, it does reaffirm the professional's responsibilities for specific approvals, as did the *Rogers* case discussed in § **6.1**.[22] Similarly, in *Shoffner Industries v. W.B. Lloyd*

---

[21] *Id.* 626–27.

[22] In connection with the holding in *Columbus,* it should be noted that even though an architect has not contracted to provide continuous on-site inspection, he or she still owes a duty of care in

Construction Co.,[23] the court held that a contractor had a right of action against a design professional for negligence in approving the work of its supplier. The case arose out of the collapse of a roof structure, which was the result of defective trusses supplied by the contractor's supplier. The contractor claimed that the architect had inspected and approved the trusses before they were incorporated in the roof structure, and that the contractor had relied on that approval. The allegations were held adequate to state a cause of action against the professional.

§ 6.7 —Negligence or Intentional Interference in Contract Administration

E.C. Ernst, Inc. v. Manhattan Construction Co.[24] deals primarily with the issue of the design professional's liability to the contractor for negligence in the performance of its obligations of project administration. In *Ernst,* an electrical subcontractor on a hospital construction project sued to recover damages for delays from the owner, the general contractor, the architect, and one of its suppliers. Following a lengthy bench trial, the district court denied the subcontractor's claims for damages. The subcontractor appealed.

On appeal, the court of appeals concluded that a no damage for delay clause in the subcontract barred Ernst's claims against the general contractor. The court also concluded that the subcontractor could not recover directly from the owner, because it was not a third-party beneficiary of the owner-general contractor agreement. Nevertheless, rejecting defenses of privity and arbitral immunity, the court concluded that the subcontractor did have a viable right of action in tort against the architect.

Ernst's claims against the architect were predicated upon claims that the architect had been negligent in (1) refusing to approve a particular bed light fixture application and (2) with respect to an emergency generator system, in drawing faulty specifications and failing to act promptly on submittals. The court of appeals held that the architect's procrastination and inconsistency in acting on generator submittals constituted negligence as a matter of law for which the subcontractor could recover damages. It also remanded to the district court, for further consideration, the issues of whether the architect had been negligent in its actions upon the bed

---

making such periodic observations or inspections as he or she does make. *See* Dickerson Constr. Co., Inc. v. Process Eng'g Co., Inc., 431 So. 2d 646 (Miss. 1977). Further, in making periodic inspections or observations, it has been held that an architect has a duty to specifically inspect those aspects of the construction likely to create safety hazards if installation is improper. *See* Pastorelli v. Associated Eng'rs, Inc., 176 F. Supp. 159 (D.R.I. 1959). Accordingly, even though the design professional's obligation might be contractually limited to an obligation to make periodic observations, a failure to spot obvious deficiencies in critical aspects of the work (e.g., the work of other prime contractors) might subject the professional to liability to the contractor if a failure should subsequently occur.

[23] 257 S.E.2d 50 (N.C. Ct. App. 1979).

[24] 551 F.2d 1026 (5th Cir.), *cert. denied,* 434 U.S. 1067 (1978).

light submittals and in drafting the plans and specifications for the emergency generator system.

The pattern of action with respect to the emergency generator submittals, which the court concluded amounted to professional negligence as a matter of law, was marked by an initial approval ("providing that the requirements of the plans and specifications [would be] met"), followed by a series of criticisms based initially upon claims of foreign manufacture, and later upon claims that the submitted generator was not a standard product and failed to meet horsepower requirements. The architect initially urged resubmittals, then suggested that it would accept the generator subject to tests, and then refused to permit tests and ultimately disapproved the system. During the process, which spanned more than a year, the professional consistently failed to list its reasons for disapprovals in a manner which might have resolved the impasse. Reviewing the history of this process, the court concluded:

> [The architect's] activities regarding its arbitral responsibilities on the emergency generator submittals constituted negligence as a matter of law. They represent a pattern of procrastination which, in view of the interdependence of effort so vital on a construction project, falls below the standard of care required in this situation for professional architects.[25]

Significant, also, was the court of appeals' conclusion that the professional could not avail itself of the traditional defense of arbitral immunity (based upon its status under the owner-contractor agreement as the resolver of disputes between owner and contractor and as the interpreter of the contract documents). The court stated that the architect's pattern of delay and of tentative, incongruous, and inconsistent action deprived it of the benefit of this defense:

> In his role as interpreter of the contract and as private decision maker, the arbitrator has a duty, express or implied, to make reasonably expeditious decisions. Where his action, or inaction, can fairly be characterized as delay or failure to decide rather than timely decisionmaking (good or bad), he loses his claim to immunity because he loses his resemblance to a judge. He has simply defaulted on a contractual duty to both parties.
>
> We are mindful of the problems of characterization that may attend the distinction between delay in deciding and bad judgment in the decision itself. The idea of a misfeasance-nonfeasance dichotomy has been subject to question. *See* W. Prosser, *Torts* § 56, at 339-40 (4th ed. 1971). But with respect to the generator submittals here, the pattern seem obvious. The tentative, incongruous, and often contradictory nature of McCauley's actions constitutes a default to all parties in the contractual sense that we have described.[26]

Another case, *Waldinger Corp. v. Ashbrook-Simon-Hartley, Inc.*,[27] establishes that a design professional's exclusionary enforcement of equipment specifications can

---

[25] 551 F.2d at 1032.

[26] *Id.* 1033–34.

[27] 564 F. Supp. 970 (C.D. Ill. 1983).

give rise to liability for intentional interference with contract relations. *Waldinger* arose out of an engineer's refusal to permit the substitution of sludge dewatering equipment on a wastewater treatment project. Notwithstanding EPA regulations prohibiting restrictive specifications, the engineer drafted the specifications for the equipment around the product of a particular manufacturer, and also interpreted the specifications so as to exclude the use of equipment of a competing manufacturer. The engineer's conduct was held to constitute intentional interference with the contractor's contract rights.

Among the factors which the court found significant in support of the conclusion that the engineer's refusal to permit the substitution was intentional and wrongful were the following. The engineer: (1) insisted on literal compliance with certain design characteristics, without scientific foundation or empirical basis; (2) asked the contractor to establish that the substitute equipment had capabilities which had not been established with respect to the specified product; and (3) insisted upon the receipt of performance data from the substitute supplier which was not available from the specified supplier. The court concluded that such conduct amounted to intentional interference with contract relations, stating:

> [I]t is manifest that Dietz patterned the sludge dewatering equipment specifications on equipment produced by the Carter Manufacturing Company. It is also clear that Dietz insisted on the use of Carter equipment and conformity to Carter design purposefully and in deliberate disregard of [EPA regulations] which required open and free competition.
>
> There was no rational basis for Dietz' insistence on the Carter mechanical subsystems. There was no showing that the Carter mechanical subsystems were reasonable and necessary for the purpose intended for the sludge dewatering equipment. . . .
>
> Dietz took Carter's performance claims at face value and made no independent study of them. In fact, the specifications were patterned on the Carter design without a test of that company's 15-31 machine by anyone. . . . *In sum, Dietz arbitrarily insisted on literal conformity to the Carter mechanical subsystems and did so without any rational or scientific engineering basis.*
>
> That Dietz' conduct was deliberate and intentional is borne out by the testimony of Sakolosky, Cantello and Nevers. What those three individuals or Dietz or the Sanitary District stood to gain from insistence on use of the Carter equipment is not at all clear from the evidence. The most that can be said is that early in its design work, the Dietz team formed the opinion that the Carter equipment was superior, although the team provided no basis for that conclusion. It is also apparent that Dietz designed the structure housing the sludge dewatering equipment on the basis of the Carter specifications and, when a competitor of Carter became the successful bidder, structural difficulties arose in trying to fit an alien machine into the space designed and the team may have been trying to avoid those difficulties.[28]

Other cases have also considered conduct of design professionals in their administrative functions which might give rise to liability to a contractor on tort theories.

---

[28] *Id.* 980 (emphasis added). *See also* Craviolini v. Scholer & Fuller, 357 P.2d 611 (Ariz. 1961), imposing liability on a professional for intentional interference with contract.

For example, in *Detweiler Brothers, Inc. v. John Graham & Co.*,[29] the court held that an engineer could incur liability to a contractor for reversal of a decision upon which the contractor had relied to its detriment. In *Detweiler,* allegations that an engineer had first approved a mechanical subcontractor's submittal to substitute grooved piping for welded or threaded pipe, and then later ordered the contractor to stop work and replace the grooved pipe with welded pipe, stated a cause of action in negligence.

Similarly, in *W.H. Lyman Construction Co. v. Village of Guerney,*[30] the court held that, although an engineer had no liability to a contractor for an alleged failure to disclose a subsurface water condition, the engineer could be liable in tort to the contractor for delay in devising a design solution to the problem. In *Gurtler, Hebert & Co., Inc. v. Weyland Machine Shop, Inc.*,[31] the court held that allegations that an architect had delayed in the approval of shop drawing submittals was adequate to support a claim in tort against the architect.

§ 6.8 —Scheduling and Coordination

Peter Kiewit Sons' Co. v. Iowa Southern Utilities Co.[32] involved a contractor's claim against an engineer for alleged negligence in scheduling and coordinating the construction of a power plant project. While the contractor's claim was ultimately denied, the decision is significant because it holds that an engineer with scheduling and coordination functions does owe a duty of care to a contractor in the performance of those functions, and because it illustrates the manner in which the performance of those functions will be evaluated.

The work for the construction of the project was let on a multiple-prime, modified fast-track basis. Both the contract for engineering services and the various prime contracts contained provisions requiring completion by May 1, 1968. Completion by that date was essential because the utility-owner's contracts for outside power were to expire then. The engineer's contract with the owner required the engineer to coordinate and expedite the work on the project, and the various prime contracts also contained provisions pertaining to the engineer's authority in coordination and scheduling.

The claim in *Kiewit* was brought by the prime contractor for general construction. Delays in precedent work under the structural steel fabrication and erection contract and the boiler fabrication and erection contract (which were determined to have occurred essentially without the fault of either contractor) considerably hindered the general construction contractor in its ability to proceed with its work in a productive fashion. In particular, the delays affected the contractor's sequencing and scheduling of concrete slab pours. The contractor's work was further adversely affected when efforts to bring the project back on schedule after the initial

---

[29] 412 F. Supp. 416 (E.D. Wash. 1976).

[30] 403 N.E.2d 1325 (Ill. App. Ct. 1980).

[31] 405 So. 2d 660 (La. Ct. App. 1981).

[32] 355 F. Supp. 376 (S.D. Iowa 1973).

delays led to congestion and further disruption of its activities. The general contractor sought to recover damages resulting from the disruption and loss of productivity from the engineer on a theory of negligent scheduling and coordination.

Initially, the court held that, despite provisions in the owner-contractor agreement which gave the engineer broad discretion in the planning and management of the project, the engineer's discretion was tempered by a duty of care owed to the contractor. The court stated:

> The Engineer's discretionary powers under the contract between Kiewit and Iowa Southern, however, do have limitations. This discretionary power is subject to the implied limitations of reasonableness and the duty to exercise care commensurate with the standards of his profession.
>
> We think an architect whose contractual duties include supervision of a construction project has the duty to supervise the project with reasonable diligence and care. An architect is not a guarantor or an insurer but as a member of a learned and skilled profession he is under the duty to exercise the ordinary, reasonable technical skill, ability and competence that is required of an architect in a similar situation. . . .[33]

Ultimately, however, the court concluded that the engineer did not breach the duty of care which it owed to the contractor, stating:

> There is no evidence that Black & Veatch failed to provide adequate and competent personnel to administer the contract documents on the job site. . . . [T]he evidence is that Black & Veatch handled the site supervision very efficiently, by coping with potentially disastrous setbacks and yet completing the job substantially on time. In fact, the Engineer's efficiency in this regard is one of the matters about which Kiewit most bitterly complains. Black & Veatch certainly was not negligent in this respect. . . . There is no evidence that Black & Veatch administered the contract in any manner other than that accepted by the trade. In the first place, Kiewit put on no evidence but its own self-serving declarations as to how a supervising engineer should function. In the larger respect, however, the Court cannot fathom how Black & Veatch could have operated any differently than it did, considering the circumstances, and still have performed its obligations to the Owner. . . .
>
> Under the circumstances, the Engineer adhered, and caused the work to proceed, as closely to [the original schedule] as was possible. As the Court has concluded . . . , [that schedule] was not absolutely binding on the Owner. Deviations were anticipated by the parties, depending upon the circumstances. All contractors were bound by [that schedule], or schedules closely approximating it. The circumstances on the site caused the Engineer to order substantial deviations, in some instances, from [that schedule]. The Engineer's actions in this regard, however, were reasonable and were within the contemplation of the contracts of all parties involved. . . .
>
> Kiewit alleges that the Engineer required Plaintiff to work under costly and inefficient circumstances not contemplated by the contract documents. Kiewit did have to work under less than ideal circumstances, but all other contractors also suffered from the same problem. The Court can find nothing in the contract that suggests that the Engineer had to give preference to Kiewit, with respect to working conditions. Rather it was the Engineer's duty to treat Kiewit equitably, considering its

[33] *Id.* 394 (citations omitted).

> similar duty to all other contractors on the site. The Engineer undeniably treated Kiewit the same as all other contractors, giving Kiewit preference when the totality of the job would be advanced by such treatment, and holding Kiewit back when circumstances dictated that other contractors should be given priority in their work. Kiewit was an experienced contractor; it knew of the possible difficulties and delays which could beset a job such as the instant one. This job had more than its share of difficulties, but the contracts clearly allowed Black & Veatch to do what it did. Black & Veatch would have breached its contract with the Owner, had it done differently. Kiewit was the victim of those circumstances that sometimes construction contractors experience. The fault here was not with the Engineer.[34]

The factors identified in the findings of fact which appeared to lend support to the conclusion that the engineer had not breached its duty of care to the contractor were:

1. When it appeared that the project might be delayed by default in the preparation of shop drawings by the structural steel contractor, the engineer actively worked to assist the contractor in the preparation of shop drawings, and also assigned additional staff to the review of such drawings, thus avoiding a delay in that regard
2. When delays did unavoidably occur in the steel erection work due to delays in material deliveries, the engineer required the erection contractor to run longer day shifts at premium time in order to speed up the work. Other alternatives, such as night shift work, were considered, but rejected because of concerns about safety and the lack of sufficient qualified ironworkers to man such a shift
3. The engineer relaxed specifications and sequencing requirements affecting the general construction contractor's work in order to assist it in maintaining productivity
4. The engineer actively worked to alleviate safety hazards (created by other contractors) to the general contractor's performance and also to mitigate damage by other contractors to the general contractor's work
5. When it became apparent that the project was falling behind, the engineer adopted a monthly CPM update process (not called for in the original contract documents), and often made daily field scheduling adjustments to accommodate events as they occurred on the project
6. The engineer provided adequate and competent staffing on site, and the project was never delayed for want of such staffing
7. There was no hindrance of the general construction contractor's work due to defects in plans and specifications
8. The general construction contractor conceded that there had been no favoritism in scheduling decisions by the engineer, and it appeared that

---

[34] *Id.* 395–96.

> the engineer had considered the economic impact upon various contractors and the overall goal of the May 1, 1968 completion date as prime considerations in scheduling decisions which it made.

Therefore, although the *Kiewit* case denied a recovery to the contractor, the case also stands for the proposition that an engineer with scheduling and coordination responsibilities can be held liable to a contractor on negligence theory if the engineer: (1) fails promptly to address, acknowledge, or respond to delays occurring on the project; (2) fails to consider available alternatives for bringing the project on schedule; (3) fails properly to update or adjust schedules; (4) shows favoritism in its scheduling or coordination decisions; or (5) provides inadequate or incompetent staffing to perform the scheduling and coordination functions.

§ 6.9 Conclusion

As the foregoing discussion demonstrates, in most jurisdictions which have considered the issue, the contractor's right of action in tort against the design professional is no longer barred by the doctrine of privity. The existence of the right of action can be a valuable tool in the contractor's effort to obtain compensation for losses incurred in contract performance. As in *Ernst,*[35] the right may present the contractor's only hope of recovery, given the existence of exculpatory provisions or other bars to recovery on a contractual basis. Moreover, in the case of owner insolvency, the design professional and its insurer may be the only available sources of reimbursement for a contractor's losses.

However, the prosecution of a claim against the design professional is not necessarily an easy matter. First, in those jurisdictions which have not yet decided the issue, the contractor must be prepared to confront arguments regarding the absence of privity or the existence of an economic loss limitation. (In others, which have decided the issue adversely, pursuit of a claim may be futile.) Second, the contractor will, in pleading and at trial, be required to establish both the existence of an undertaking of the professional with respect to the act or omission in issue and the foreseeability of the contractor's reliance on that undertaking. Third, the contractor will be required to establish, often by expert testimony, the professional's negligence. This requires identification of the standard of care applicable to the activity in question, the professional's deviation from that standard, and the causal relationship between the deviation and the contractor's loss.

Thus, for example, in the case of a design error, prosecution of a claim against the professional may be more difficult than prosecution of a claim against the owner. This is because the owner warrants the adequacy of the plans and specifications which it provides to a contractor (without regard to negligence in preparation); to establish the professional's liability, it is generally necessary to establish not

---

[35] E.C. Ernst, Inc. v. Manhattan Constr. Co., 551 F.2d 1026 (5th Cir.), *cert. denied,* 434 U.S. 1067 (1977).

only a design deficiency, but also that the deficiency was a result of professional negligence.

Finally, the contractor may have to surmount affirmative defenses raised by the professional, such as the defense of *arbitral immunity* (where that defense is recognized). These considerations mean that the decision to prosecute a claim against a design professional requires competent and careful advance consideration from the business, legal, and technical standpoints.

CHAPTER 7

SUING THE CONSTRUCTION LENDER

Richard E. Alexander and David S. Barlow

Richard E. Alexander has been in private practice for approximately 14 years and is a partner in the Portland, Oregon firm of Stoel, Rives, Boley, Fraser & Wyse. He represents prime contractors, subcontractors, suppliers, architects, and engineers, as well as owners in the construction and design field. His practice includes litigation, arbitration, and claims work, as well as the negotiation and finalization of agreements in the construction and design area. He is a member of Associated General Contractors of America, Inc. and the Public Contracts and Litigation Sections of the American Bar Association. He is an arbitrator appointed by the American Arbitration Association for construction disputes. He has lectured in the area of construction and design law and is a co-author of the 1983 and 1984 editions of *Oregon Construction Law*.

David S. Barlow is an associate with Stoel, Rives, Boley, Fraser & Wyse. He works primarily in the labor and construction fields. Mr. Barlow received a B.A. from Harvard University in 1975, a Masters from the London School of Economics in 1977, and a J.D. from the University of Washington in 1983.

§ 7.1 Introduction

Actions against construction lenders are a relatively uncommon remedy for contractors, subcontractors, and materialmen. The existence of more accessible defendants, statutory remedies, built-in limitations in the legal theories available for such suits, and judicial assumptions about who should bear the risk in a construction project all account for this slow development.

Although claims against a lender are generally not the preferred remedy for an unpaid contractor, subcontractor, or materialman, a contractor will often encounter situations where it is required to make a claim against a construction lender. For example, a lender may be the only remaining solvent party, or the contractor, because of its contract, its own error, or the nature of a particular jurisdiction's lien laws, may have no other potential source of recovery.

This chapter explores the theories available to a contractor, subcontractor, or materialman[1] seeking to recover from a construction lender. It reviews the available means for the contractor to bring an action against a lender, the requirements and limitations of the various theoretical approaches, and how these theories have fared in contractor-construction lender litigation to date. **Sections 7.2-7.5** discuss potential statutory remedies. Where available, these may well be the contractor's best remedy. **Sections 7.7-7.11** explore means to establish and enforce contractual rights against the lender and discuss possible arguments for grouping the lender with the owner for privity purposes and third-party beneficiary theories. The remaining sections of this chapter discuss estoppel theories, equitable remedies, and possible tort theories.

STATUTORY REMEDIES

§ 7.2 Importance of Statutory Remedies

A contractor considering making a claim against a construction lender should initially review the particular jurisdiction's available statutory remedies. Statutory schemes may give the contractor various remedies against a lender. They also may dictate what theories are available for the contractor attempting to sue a lender.[2]

§ 7.3 Specific Statutory Remedies—Mechanics' Lien Statutes

A contractor should consider the relative priority between the lender and the contractor pursuant to the applicable mechanics' lien law. Though most jurisdictions will treat the construction lender with a properly recorded security interest as superior to any lien filed by a contractor, this is not the case in all jurisdictions.[3] Where a contractor enjoys priority over a lender, resort to more exotic remedies may not be necessary. Where the secured lender enjoys priority over the contractor or subcontractor, the need for alternative approaches will be greater.

---

[1] The term *contractor* is used here to describe prime contractors, subcontractors, and materialmen, unless otherwise noted.

[2] California, for instance, has statutorily abolished the equitable lien. *See, e.g.*, Cal. Civ. Code § 3264 (West 1974); Pankow Constr. Co. v. Advance Mortgage Corp., 618 F.2d 611, 616-17 (9th Cir. 1980).

[3] *E.g.*, Oregon: Or. Rev. Stat. § 87.025(1), (2), (6) (1983).

§ 7.4 —Stop-Notice Provisions

Some states provide a statutory mechanism to inform a lender that a subcontractor or materialman has not been paid on a project. Once notice of nonpayment is given, the lender either is barred from making further payments or bears the risk of proceeding.[4] The statutes protect only subcontractors and materialmen.

Each statute has its idiosyncrasies. As with lien remedies, the stop-notice statutes generally have short time periods in which to file. The subcontractor or materialman also must meet various statutory requirements. Therefore, these statutes should be checked promptly, and their specific requirements should be carefully followed. Where such a statutory remedy exists, though, it provides a useful mechanism to freeze assets sufficient to pay for labor and materials provided on a project.

§ 7.5 —Requirements that Loan Agreement Be Recorded

Other states require the construction lender to record the loan agreement to protect its statutory lien priority.[5] Such public recording statutes can be a useful tool for the contractor, since public recording allows the contractor to review the loan agreement before deciding to proceed. The recording statutes can also be used as a weapon in litigation with lenders. A lender can also lose its priority for knowingly filing a contract that misrepresents the amount of loan funds available or by modifying the agreement without recording the change.[6] Again, the recording statutes generally protect only subcontractors and materialmen.

CONTRACT THEORIES

§ 7.6 Introduction

Where statutory remedies are not available or not practical, a contractor should next consider any rights it might have under the construction loan agreement.

---

[4] *E.g.*, Cal. Civ. Code §§ 3158-3162 (West 1974); Wash. Rev. Code § 60.04.210 (1961 & Supp 1985). Other states having stop-notice statutes include Alabama, California, Indiana, Mississippi, New Jersey, North Carolina, Texas, and Washington. Sweet, Legal Aspects of Architecture and Engineering 471-72 (2d ed. 1977); Reitz, *Construction Lenders' Liability to Contractors, Subcontractors, and Materialmen,* 130 U. Pa. L. Rev. 416, 433 (1981); Note, *Mechanics' Liens: The Stop Notice Comes to Washington*, 49 Wash. L. Rev. 689 (1974).

[5] *E.g.*, N.Y. Lien Law §§ 13, 22 (McKinney 1966).

[6] *See, e.g.*, Nanuet Nat'l Bank v. Eckerson Terrance, Inc., 47 N.Y.2d 243, 391 N.E.2d 983, 417 N.Y.S.2d 901 (1979).

If a contractor is fortunate, it can negotiate with the lender and provide itself with protection. Normally, though, a lender has no express contractual relationship with a project's contractors, because only the borrower has dealt directly with the lender. Therefore, the contractor seeking to enforce rights pursuant to the construction agreement must establish that it is in privity or has a right to enforce the agreement's terms.

In asserting contractual rights, the contractor should consider several approaches. The contractor may argue that the formal contractual separation between the owner and lender should be ignored, or that it is a third-party beneficiary of the agreement between the lender and borrower. The contractor can also argue that estoppel requires the lender to perform certain obligations or not to take certain actions.

§ 7.7 Treating Lender and Borrower as a Single Entity

One approach is to argue that the lender and contractor in fact should be treated as being in privity despite the parties' formal contractual separation. Essentially, the contractor argues that the lender's involvement in the project extends beyond simply supplying funds, and that its role in the project justifies treating the lender and borrower as a unit. If the contractor has privity with the borrower, treating the lender and borrower as one creates the possibility of direct contractual claims against the lender.

This view has not been widely adopted. Most courts view the lender only as a supplier of capital and not as an active participant in the construction project. Cases suggest that this would not be so if the lender's activities on a project clearly exceed those of a normal lender, but do little to define what those activities might be.[7]

One area in which courts have been willing to treat the owner and lender as a unit has been in litigation involving projects financed by the government through the National Housing Act.[8] On these projects, otherwise assetless nonprofit entities contract with the contractors to build housing. The owner receives its financing completely from a lender. The lenders are willing to make these loans because they are insured by the government; when a project fails, the government steps in for the lender. In cases arising from these projects, courts have found that prime contractors can assert rights against the government.[9] The courts have recognized that the government's role in financing and controlling the project makes it an active participant in the project and thus subject to suit by the contractor.

---

[7] *E.g.*, Wierzbicki v. Alaska Mut. Sav. Bank, 630 P.2d 998 (Alaska 1981); Mortgage Assocs., Inc. v. Monona Shores, Inc., 177 N.W.2d 340, 349-50 (Wis. 1970); Connor v. Great W. Sav. & Loan Ass'n, 69 Cal. 2d 850, 477 P.2d 609, 73 Cal. Rptr. 369 (1969).

[8] 12 U.S.C. § 1715Z-1 *et seq.* (1982).

[9] S.S. Silberblatt, Inc. v. East Harlem Pilot Block, 608 F.2d 28 (2d Cir. 1979); Trans-Bay Eng'rs & Builders, Inc. v. Hills, 551 F.2d 370, 381-82 (D.C. Cir. 1976).

Similar arguments could probably be made in the private sector where the lender, by the exercise of control, sharing of profits, and other direct participation in the project, becomes in effect a joint venturer.[10] One commentator suggests that the lender and borrower could be compressed into a single entity for privity purposes using partnership, limited partnership, or joint venture theories.[11]

These theories to date have not met with much success.[12] They do offer a potential basis for providing the privity needed for a contractually based claim against the lender. One commentator summarizes the possibilities as follows:

> There may be sound basis in many projects to characterize the construction lender as part of the developer's team rather than as an independent entity merely supplying secured credit. This may be the most principled path toward establishing the normative standards for construction lenders liability, and would meld well with the growing body of contract law that is extending the bounds of traditional privity concepts.[13]

§ 7.8 Third-Party Beneficiary Status

A more tested theory for asserting contractual rights is to argue that the contractor is a third-party beneficiary of agreements between the lender and borrower. As such, the contractor can enforce obligations created in the construction loan agreement.

The basic limitation on any third-party beneficiary argument is the need to show that the lender and borrower intended to make the contractor a third-party beneficiary of their agreement. Under the *Restatement (Second) of Contracts* § 302, two basic ways exist to make a party a third-party beneficiary.[14] The third party can establish

---

[10] *See* Reitz, *Construction Lenders' Liability to Contractors, Subcontractors and Materialmen,* 130 U. Pa. L. Rev. 416, 453-59 (1981), for an excellent discussion on the idea of treating the lender and borrower as a single unit.

[11] *Id.* 457.

[12] *See,* for instance, the California Supreme Court's rejection of a joint venture argument involving an owner and lender in Connor v. Great W. Sav. & Loan Ass'n, 69 Cal. 2d 850, 477 P.2d 609, 73 Cal. Rptr. 369, 375 (1969). As with other theories discussed in this chapter, it is often difficult to meet the legal requirements to establish that a joint venture exists.

[13] Reitz, *Construction Lenders' Liability to Contractors, Subcontractors, and Materialmen,* 130 U. Pa. L. Rev. 416, 459 (1981).

[14] Restatement (2d) of the Law of Contracts § 302 (1981):

§ 302. Intended and Incidental Beneficiaries.

(1) Unless otherwise agreed between promisor and promisee, a beneficiary of a promise is an intended beneficiary if recognition of a right to performance in the beneficiary is appropriate to effectuate the intention of the parties and either

(a) the performance of the promise will satisfy an obligation of the promisee to pay money to the beneficiary; or

(b) the circumstances indicate that the promisee intends to give the beneficiary the benefit of the promised performance.

(2) An incidental beneficiary is a beneficiary who is not an intended beneficiary.

that the promisor's performance under the contract discharges a duty otherwise owed the third party by the promisee. Second, the third party can show an intent to benefit. Other courts continue to analyze third-party beneficiary issues using the *donee, creditor,* and *incidental beneficiary* language of the *First Restatement of Contracts.*[15] Under either approach, the court must also determine that the beneficiary's enforcement of the agreement would be appropriate. If the contractor cannot satisfy these tests, it has no right as a third-party beneficiary to enforce the lender's obligations in its contract with the borrower.[16]

What is sufficient to establish the necessary intent to create a third-party beneficiary varies among different jurisdictions. Some courts require that the intent to benefit third parties affirmatively appear in the contract.[17] Other courts vary on whether both parties must manifest an intent to benefit, or only the promisee.[18] A fairly standard formulation of the necessary intent is set forth below:

> Though the third party beneficiary need not be named in the contract, the terms of the contract must express directly and clearly, an intent to benefit the specific party or an identifiable class of which the party asserting rights as a third party beneficiary is a member. Inasmuch as people usually contract and stipulate for themselves and not for third persons, a strong presumption arises that such was their intention, and the implication to overcome that presumption must be so strong as to amount to an express declaration.[19]

§ 7.9 —Establishing Third-Party Beneficiary Status

In attempting to establish rights as a third-party beneficiary, contractors have relied on various factors, including the terms of the construction loan agreement, industry custom, and the contractor's direct dealings with the parties to the loan agreement.

Construction loan agreements often include an express disclaimer of any intent to create third-party rights. These clauses may announce in general terms that no third party is entitled to rely on the agreement's provisions, or may specifically state that the lender owes no obligation and has no relationship with the project's

---

[15] Restatement of the Law of Contracts § 133 (1932).

[16] Gee v. Eberle, 420 A.2d 1050, 1056 n.4 (Pa. Super. Ct. 1980); Knight Constr. Co. v. Barnett Mortgage Trust, 572 S.W.2d 381, 382-83 (Tex. Civ. App. 1978), *writ refused n.r.e.*

[17] *Compare* Knight Constr. Co. v. Barnett Mortgage Trust, 572 S.W.2d 381, 382-83 (Tex. Civ. App. 1978), *writ refused n.r.e., with In re* Gebco Inv. Corp., 641 F.2d 143 (3d Cir. 1981).

[18] Khabbaz v. Swartz, 319 N.W.2d 279, 285 n.9 (Iowa 1982); Gee v. Eberle, 420 A.2d 1050, 1056 (Pa. Super. Ct. 1980).

[19] Laclede Inv. Corp. v. Kaiser, 596 S.W.2d 36, 42 (Mo. Ct. App. 1980).

contractors. The presence of such a clause will normally negate any attempt to establish third-party beneficiary rights.[20]

Some courts have indicated that such an express disclaimer would prevent a third-party beneficiary claim even if other evidence suggested such an intent.[21] In *Silverdale Hotel v. Lomas & Nettleton Co.*,[22] a Washington appellate court held that the trial court's finding that the lender promised to pay the contractor through an escrow account in the face of such a clause violated the parol evidence rule. The court found that the oral agreement was inconsistent with a contract term denying lender liability to any contractor or materialman.[23] A contractor considering suing on a third-party beneficiary theory should review the construction loan agreement for any disclaimer of intent to benefit third parties.

In other situations, contractors have argued that the construction loan agreement's provisions show an intent to make the contractor a third-party beneficiary. Pennsylvania courts in particular have been active in determining if a construction loan agreement creates third-party rights for a contractor.[24]

The Pennsylvania cases show, though, that often the contract language alone will not be determinative. The parties' conduct during a project is also an important factor in determining the contracting parties' intent. For instance, in *Clardy v. Barco Construction Co.*,[25] and *Demharter v. First Federal Savings and Loan Association of Pittsburgh*,[26] the Pennsylvania court construed identical payout

---

[20] Gee v. Eberle, 420 A.2d 1050, 1055 n.4 (Pa. Super. Ct. 1980); Pioneer Plumbing Supply Co. v. Southwest Sav. & Loan Ass'n, 428 P.2d 115, 122 (Ariz. 1967). *But see* Twin City Constr. Co. v. ITT Indus. Credit Co., 358 N.W.2d 716 (Minn. Ct. App. 1984). In *Twin City,* the contractor performed approximately $600,000 worth of work on a project. When the borrower failed to pay, the contractor filed liens. The borrower then sought an additional loan to rescue the project. As a condition of the new loan, the lender required the contractor to release its lien, subordinate its interest, and waive any lien rights. The agreement between the lender and borrower for the new loan expressly stated that no third party could rely on it. Subsequently, the lender did not make a final payment to the contractor.

The Minnesota court found that the contractor could establish third-party beneficiary status by either showing that the lender assumed the duty owed by the promisee or by showing an intent to benefit. The court found that the ''duty owed'' test had been met, stating that the lender assumed the duty to pay for the work when it committed itself to making the loan and making progress payments. It found an intent to benefit the contractor from the documents as a whole. The court particularly emphasized the lender's requirement that the contractor forego its statutory remedies in exchange for the lender's additional loan for the project.

[21] Knight Constr. Co. v. Barnett Mortgage Trust, 572 S.W.2d 381, 383 (Tex. Civ. App. 1978), *writ refused.*

[22] 36 Wash. App. 762, 677 P.2d 773, 779 (1984).

[23] The contractor, though, was able to enforce the agreement on a promissory estoppel theory.

[24] For a detailed analysis of the Pennsylvania cases discussed in this section, *see* Reitz, *Construction Lenders' Liability to Contractors, Subcontractors, and Materialmen,* 130 U. Pa. L. Rev. 416, 424-26 (1981); *see also In re* Gebco Inv. Corp., 641 F.2d 143 (3d Cir. 1981); R.P. Russo Contractors v. C.J. Pettinato Realty, 482 A.2d 1086, 1091-92 (Pa. Super. Ct. 1984).

[25] 208 A.2d 793, 795-96 (Pa. Super. Ct. 1965).

[26] 194 A.2d 214 (Pa. 1963).

provisions in two construction loan agreements. In *Clardy*, the court found that a subcontractor suing on the prime's rights was a third-party beneficiary.[27] In *Demharter*, the court had held that no third-party beneficiary status existed for subcontractors and materialmen.

The Pennsylvania Superior Court distinguished the cases based on the *Demharter* lender's ability to choose to pay either the subcontractors or the contractor-owner. In *Clardy,* the Superior Court found that other portions of the contract obligated the lender to pay the contractor directly. The court emphasized that the contract provided that the contractor would be paid in full before the owner was entitled to any return of the money, that the contractor was named in the contract, and that the contractor was a signatory to it. The Superior Court also emphasized the parties' conduct, pointing out that payments under the loan agreement had been made directly to the contractor.[28]

Contractors have also tried unsuccessfully to establish themselves as third-party beneficiaries based on industry practice. In one New Jersey case, subcontractors sought to rely in part on a lender's policy manual to establish how the lender should have conducted its affairs.[29] The subcontractors argued—unsuccessfully—that a manual suggesting that a lender make payment only after obtaining affidavits from the contractor establishing that it paid subcontractors and materialmen and a contract clause requiring such affidavits created third-party rights.

Plaintiffs in some of the cases noted in **§ 7.24**, discussing negligent disbursement of loan funds, have also attempted to use industry custom to establish lender liability. Though plaintiffs have not succeeded with this argument, industry custom and practice is a potential source of an implied obligation to pay.

Contractors have also sued as third-party beneficiaries when the contractor led an owner/borrower to a lender and the lender did not make the loan or stopped payment,[30] but these arguments receive short shrift. Courts have noted the lack

---

[27] ". . . it is mutually agreed between the parties hereto that the Association *shall:*

. . .

> 6. Pay out of the funds to contractors, subcontractors or materialmen, as the Association may elect, for work performed, services rendered, and materials furnished in and about the construction of the building. Such payments shall be made at such times and in whatever amounts the Association may deem expedient, and shall be made according to requisitions approved by a building inspector designated by the Association, which requisition shall be in such form and shall contain such information as the Association may require; it being the intention of the parties hereto that the Association shall be free to make the payments heretofore mentioned in such manner that the Association's security shall at all times be protected."

Clardy v. Barco Constr. Co., 208 A.2d 793, 795 n.1 (Pa. Super. Ct. 1965) (emphasis in original).

[28] *Id.* 795-96. *See also* Kreimer v. Second Fed. Sav. & Loan Ass'n of Pittsburgh, 176 A.2d 132, 133-34 (Pa. Super. Ct. 1961); *In re* Gebco Inv. Corp., 641 F.2d 143 (3d Cir. 1981).

[29] First Nat'l State Bank of N.J. v. Carlyle House, Inc., 246 A.2d 22, 31-34 (N.J. Super. Ct. 1968).

[30] Winnebago Homes, Inc. v. Sheldon, 139 N.W.2d 606, 609 (Wis. 1966); Burns v. Washington Sav., 171 So. 2d 322 (Miss. 1965); Stephens v. Great S. Sav. & Loan Ass'n, 421 S.W.2d 332, 335-37 (Mo. Ct. App. 1967).

of intent to benefit the contractor and the failure of the owner/borrower to perform obligations that were conditions precedent to the lender's duty to pay.

In other cases, plaintiffs seek to impose liability on facts similar to those used to support equitable estoppel arguments.[31] In these cases, the contractor contends that the lender somehow induced it to provide additional services or material. Even when the court finds the facts sufficient to support an equitable estoppel argument, it also normally finds the contractor unable to show any intent by the contracting party to make it a third-party beneficiary.

§ 7.10 –Enforcing Third-Party Rights

Even if the contractor establishes third-party beneficiary status, it must still establish its right to be paid. The party alleging third-party status has no greater rights under the contract than those of the party through which it claims.[32] Failure of the owner to provide an assignment of rents or evidence of clear title, or to perform other conditions precedent to the lender's obligation to pay, can prevent a contractor's enforcement of contract rights.[33] Similarly, the contractor's own material failure of performance can prevent it from asserting third-party beneficiary rights.[34]

§ 7.11 –Summary

Establishing a third-party beneficiary status is a way for the contractor to enforce the terms of the loan agreement. However, the requirements of this approach are difficult to meet, as a contractor must establish both its third-party status and the fact that the lender's performance is due. Though courts have shown flexibility when the equities favor the contractor, the criteria discussed in **§§ 7.8-7.10** limit the widespread use of a third-party beneficiary theory.

RIGHTS BASED ON ESTOPPEL

§ 7.12 Introduction

Another approach to suing a construction lender rests on the concept of estoppel. Contractors may argue that a lender is estopped from either not paying the

---

[31] Apex Siding & Roofing Co. v. First Fed. Sav. & Loan Ass'n, 301 P.2d 352, 354-55 (Okla. 1956); Winnebago Homes, Inc. v. Sheldon, 139 N.W.2d 606, 609 (Wis. 1966); Silverdale Hotel v. Lomas & Nettleton Co., 36 Wash. App. 762, 677 P.2d 773 (1984).

[32] Stephens v. Great S. Sav. & Loan Ass'n, 421 S.W.2d 332, 337 (Mo. Ct. App. 1967).

[33] *Id.* 335-37; L.B. Herbst Corp. v. Northern Ill. Corp., 241 N.E.2d 125 (Ill. App. Ct. 1968).

[34] R.P. Russo Contractors v. C.J. Pettinato Realty, 482 A.2d 1086, 1091-92 (Pa. Super. Ct. 1984).

contractor for work or from asserting a lien priority because of representations or promises made by the lender on which the contractor relied to its detriment.[35]

§ 7.13 Estoppel Theories

Different forms of estoppel exist. Plaintiffs in cases involving construction lenders have argued theories of promissory estoppel, equitable estoppel, and even quasi-estoppel. Under the *Restatement (Second) of Contracts,* promissory estoppel requires showing

> [a] promise which the promisor should reasonably expect to induce action or forbearance on the part of the promisee or a third person and which does induce such action or forbearance is binding if injustice can be avoided only by enforcement of the promise. The remedy granted for breach may be limited as justice requires.[36]

Equitable estoppel requires assertion of a position by conduct or word, reasonable reliance, and prejudice.[37] It differs from *promissory estoppel* in requiring only a representation, and not a promise, to the relying party. *Quasi-estoppel* does not require reliance but looks for a showing that facts and circumstances make the assertion of a contrary or inconsistent position unconscionable. Though reliance is not a necessary element, courts will consider whether the party asserting the inconsistent position has gained an advantage or produced some disadvantage to the other party.[38]

§ 7.14 Elements of Successful Estoppel Claims

Regardless of the precise estoppel theory articulated, reported cases involving contractor suits against lenders focus on two factors: first, that a promise or representation was made to the contractor; and second, that the contractor reasonably relied on the representation to its detriment.

§ 7.15 —Actionable Representations

As a rule, the contractor must show more than that the lender promised to make funds available to the borrower. California cases, for a period prior to the statutory

---

[35] Strouss v. Simmons, 657 P.2d 1004 (Hawaii 1982); Fretz Constr. Co. v. Southern Nat'l Bank, 626 S.W.2d 478 (Tex. 1982).

[36] Restatement (2d) of the Law of Contracts § 90(1) (1981).

[37] *E.g.,* Alaska Statebank v. Kirschbaum, 662 P.2d 939, 942 (Alaska 1983).

[38] *Id.* 943.

death of the equitable lien in that state, found that knowledge of the existence of the loan fund was a sufficient basis to establish the reliance necessary for an equitable lien.[39] Other courts, however, require a showing that the lender made some type of separate representation or enforceable promise to the contractor beyond indicating the existence of a loan.[40]

In *Apex Siding & Roof Co. v. First Federal Savings and Loan Association,*[41] a court found that estoppel applied when a contractor specifically inquired about whether a loan was made and relied on the lender's affirmative reply.[42] Similarly, in *Silverdale Hotel v. Lomas & Nettleton Co.*,[43] the court found that a lender's statement to a contractor that the borrower was not in default and that it would pay according to the loan agreement established promissory estoppel. Representations that money would be set aside to pay for work on a project also have been found sufficient evidence to support a promissory estoppel argument.[44]

In *Gee v. Eberle,*[45] the court held that lender representations that sufficient financing existed and that the owners were ''strong'' and ''everything seems okay'' were sufficient evidence to support a reliance argument. Other jurisdictions, however, have found similar representations insufficient to establish the basis for an equitable lien.[46]

Courts have found that lenders have no duty to inform contractors that a borrower is in default or faces financial difficulties.[47] Accordingly, claimed reliance on a lender's failure to warn will normally not result in estoppel.

In short, the cases require showing that the lender made independent representations or promises to the contractors. Neither silence nor representations that a loan was made are normally enough to estop the lender. Generally, independent assurances that adequate funding exists, that funds will be set aside, or that the borrower is financially sound are required.

---

[39] *E.g.*, Doud Lumber Co. v. Guaranty Sav. & Loan Ass'n, 254 Cal. App. 2d 585, 60 Cal. Rptr. 94, 96-97 (1967).

[40] Strouss v. Simmons, 657 P.2d 1004, 1012-13 (Hawaii 1982); Gee v. Eberle, 420 A.2d 1050, 1063-64 (Pa. Super. Ct. 1980); Silverdale Hotel v. Lomas & Nettleton Co., 36 Wash. App. 762, 677 P.2d 773 (1984); Pioneer Plumbing Supply Co. v. Southwest Sav. & Loan Ass'n, 428 P.2d 115, 122 (Ariz. 1967).

[41] Apex Siding & Roofing Co. v. First Fed. Sav. & Loan Ass'n, 301 P.2d 352 (Okla. 1956).

[42] *See also* H.O. Bragg Roofing, Inc. v. First Fed. Sav. & Loan Ass'n, 226 Cal. App. 2d 24, 37 Cal. Rptr. 775 (1964).

[43] 36 Wash. App. 762, 677 P.2d 773, 779 (1984).

[44] Fretz Constr. Co. v. Southern Nat'l Bank, 626 S.W.2d 478, 482-83 (Tex. 1982).

[45] 420 A.2d 1050, 1062-64 (Pa. Super. Ct. 1980).

[46] *E.g.*, Indiana Mortgage & Realty Inv. v. Peacock Constr., 348 So. 2d 59 (Fla. Dist. Ct. App.), *cert. denied,* 353 So. 2d 677 (Fla. Dist. Ct. App. 1977).

[47] *E.g.*, L.B. Herbst Corp. v. Northern Ill. Corp., 241 N.E.2d 125, 126 (Ill. App. Ct. 1968); Mortgage Assocs., Inc. v. Monona Shores, Inc., 177 N.W.2d 340, 349 (Wis. 1970).

§ 7.16 —Reliance

The second essential element of an estoppel claim is reliance. Contractors usually establish their reliance by showing either that they continued work based on the lender's representation or promise or that they forewent use of their statutory rights.[48] Estoppel theory, however, could be expanded to fit other types of detrimental reliance.

Furthermore, the reliance must be reasonable. An argument that a contractor relied on representations that a loan would be made, for instance, was found unreasonable in the face of clauses specifically conditioning how the loan would be paid.[49]

§ 7.17 Summary

Promissory and equitable estoppel are helpful theories when the contractor has had independent contact with the lender. Unlike a third-party beneficiary claim, the contractor does not need to show that any contracting parties intended to benefit it. Demonstrated reliance on the lender's representations or promises is all that is needed. Reliance is usually based on continuing work or foregoing statutory remedies, although the contractor must show that its reliance was reasonable.

UNJUST ENRICHMENT THEORIES

§ 7.18 Generally

The difficulties contractors face in establishing privity or their right to enforce a loan agreement on other grounds lead to the use of theories not based on privity. These include various equitable remedies and torts.

Unjust enrichment involves the basic principle that allowing a person to retain the money or property of another is not equitable. Some courts proceed under the general rubric of unjust enrichment without specifying the exact nature of the equitable remedy involved,[50] while others attempt to fit the contractor-lender relationship into a traditional equitable remedy, such as the equitable lien or constructive trust.[51]

---

[48] Strouss v. Simmons 657 P.2d 1004, 1012-14 (Hawaii 1982); First Nat'l State Bank of N.J. v. Carlyle House, Inc., 246 A.2d 22 (N.J. Super. Ct. 1968); H.O. Bragg Roofing, Inc. v. First Fed. Sav. & Loan Ass'n, 226 Cal. App. 2d 24, 37 Cal. Rptr. 775 (1964).

[49] Strouss v. Simmons, 657 P.2d 1004, 1012-14 (Hawaii 1982).

[50] S.S. Silberblatt, Inc. v. East Harlem Pilot Block, 608 F.2d 28, 38 (2d Cir. 1979); Twin City Constr. Co. v. ITT Indus. Credit Co., 358 N.W.2d 716, 719 (Minn. Ct. App. 1984).

[51] Gee v. Eberle, 420 A.2d 1050 (Pa. Super. Ct. 1980); Ralph C. Sutro Co. v. Paramount Plastering Inc., 216 Cal. App. 2d 433, 31 Cal. Rptr. 174 (1963).

Both the *equitable lien* and *constructive trust* theories are based on property concepts. There is no requirement of a contractual relationship. The equitable lien essentially grants a person a security interest in specific property. The constructive trust theory finds that a person holding identifiable property inequitably is the trustee to the person actually entitled to possession. The two theories overlap; hence, **§§ 7.19-7.23** discuss unjust enrichment in generic terms.

§ 7.19 Defining Unjust Enrichment in Construction

Defining unjust enrichment has been a problem in construction contractor-construction lender litigation, and courts vary considerably in how they approach, define, and measure the enrichment. They wrestle with issues such as when a lender has been unjustly enriched, how the amount of unjust enrichment is measured, and what conditions precedent, if any, must be satisfied before a claim of unjust enrichment exists.

§ 7.20 —When Does Unjust Enrichment Occur?

Courts differ on when unjust enrichment can occur between a construction lender and contractors. Florida, whose courts are active in the equitable lien area, specifically limits unjust enrichment to situations where the construction project has been completed. Several Florida cases hold that when completion of a project leaves a lender with both the finished improvement as security and the undistributed construction loan funds, the retention of both the improvement created by the contractor and the loan funds meant to pay for the improvement is inequitable. In those cases, Florida courts have granted the contractor an equitable lien on the undisbursed loan funds.[52] On uncompleted projects, however, the Florida courts indicate that the value of the labor and material provided will likely exceed the structure's market value, and thus they will not allow an equitable lien.[53]

Other courts have not drawn such a sharp distinction between completed and uncompleted projects, indicating that unjust enrichment should be determined on a case-by-case basis. Unjust enrichment is available if a contractor can show that its work in some manner enhanced the lender's return from its security. Some courts have found no need for the claiming contractor to have finished its work on the project if it has not materially breached the contract.[54]

---

[52] Blosam Constr., Inc. v. Republic Mortgage Inv., 353 So. 2d 1225 (Fla. Dist. Ct. App. 1977), *cert. denied,* 359 So. 2d 1218 (Fla. Dist. Ct. App. 1978); J.G. Plumbing Serv., Inc. v. Coastal Mortgage Co., 329 So. 2d 393 (Fla. Dist. Ct. App.), *cert. denied,* 339 So. 2d 1169 (Fla. Dist. Ct. App. 1976).

[53] J.G. Plumbing Serv., Inc. v. Coastal Mortgage Co., 329 So. 2d 393, 395 (Fla. Dist. Ct. App.), *cert. denied,* 339 So. 2d 1169 (Fla. Dist. Ct. App. 1976).

[54] S.S. Silberblatt, Inc. v. East Harlem Pilot Block, 608 F.2d 28, 37 (2d Cir. 1979); Gee v. Eberle, 420 A.2d 1050, 1057 n.5, 1059 (Pa. Super. Ct. 1980); Miller v. Mountain View Sav. & Loan

§ 7.21 —Measuring the Amount of Unjust Enrichment

Cases are not explicit about the proper measure of a lender's unjust enrichment. Generally, courts require evidence of undistributed construction loan funds and evidence that the claimant increased the value of the improvement for the lender.[55] In those cases, the courts have still had difficulty evaluating the extent to which a particular party has been enriched,[56] since the value of the labor and materials provided will not necessarily equal the value of the uncompleted improvement. In trying to value the contractor's improvement, courts have at times looked to the resale value of the improvement involved.[57]

If the lender has paid out all the money in the fund or has paid for the specific work in question, courts reason that the lender may retain the work as part of its security interest in the improvement, because it has paid the intended amount for the work. In such cases, courts find that no unjust enrichment exists.[58]

§ 7.22 Potential Limitations on Use of Unjust Enrichment Theories

In *Gee v. Eberle*,[59] the Pennsylvania Superior Court, in attempting to identify a lender's liability to subcontractors under an equitable lien theory, analogized the situation to those in the more numerous cases discussing owner-subcontractor relationships. The court noted that the weight of authority was against finding such liability when the subcontractor lacked privity with the owner, and found that past decisions generally advanced three reasons therefor: (1) the subcontractor must show wrongdoing by the lender; (2) the subcontractor must exhaust its statutory or contractual remedies; and (3) a direct contractual relationship must exist. The *Gee* court suggested that none of these reasons should prevent recovery in either the owner-subcontractor or contractor-construction lender situation.

With regard to intent, the court indicated that it is the result of unjust enrichment, not the wrongful intent, that should be the court's concern.[60] Other courts,

---

Ass'n, 238 Cal. App. 2d 644, 48 Cal. Rptr. 278, 292 (1965) (prior to statutory end of equitable lien doctrine in California).

[55] Trans-Bay Eng'rs & Builders, Inc. v. Hills, 551 F.2d 370, 382-83 (D.C. Cir. 1976); Gee v. Eberle, 420 A.2d 1050, 1061 n.9 (Pa. Super. Ct. 1980).

[56] Reitz, *Construction Lenders' Liability to Contractors, Subcontractors, and Materialmen*, 130 U. Pa. L. Rev. 416, 445-46 (1981); *cf.* Miller v. Mountain View Sav. & Loan Ass'n, 238 Cal. App. 2d 644, 48 Cal. Rptr. 278, 288-89 (1965).

[57] Gee v. Eberle, 420 A.2d 1050 (Pa. Super. Ct. 1980); Miller v. Mountain View Sav. & Loan Ass'n, 238 Cal. App. 2d 644, 48 Cal. Rptr. 278 (1965).

[58] Gee v. Eberle, 420 A.2d 1050 (Pa. Super. Ct. 1980).

[59] Gee v. Eberle, 420 A.2d 1050 (Pa. Super. Ct. 1980).

[60] *Id.* 1059.

however, appear to look for wrongful conduct by a lender before allowing a contractor to use an equitable remedy against a lender.[61]

As to requiring the contractor to exhaust statutory remedies, the *Gee* court stated that the mechanics' lien would generally be the preferred remedy. Depriving the subcontractor of the advantage associated with lien laws was penalty enough. The court indicated further that denying relief for failure to utilize other remedies when unjust enrichment existed was inconsistent with equitable principles. This would not be applicable, though, where the statutory remedies were intended to stop use of equitable remedies.[62]

Finally, the court rejected a requirement imposed by some other courts that unjust enrichment required a contractual relationship between the parties. The court argued that the essence of an unjust enrichment is that no direct contractual remedy exists.[63]

In *Gee v. Eberle* the court anticipated and rejected many arguments that a contractor may face in bringing an unjust enrichment argument against a lender. Particularly in jurisdictions where the law is unsettled or hostile, the *Gee* court's decision may serve as a helpful guide to a contractor trying to sue a lender on unjust enrichment theories.

§ 7.23 Summary

When undistributed loan proceeds exist for a completed project, many courts recognize the right of the contractor to impose an equitable lien on the undisbursed fund. In other factual settings, such as an incomplete project or where loan funds have been disbursed, the availability of an equitable lien is much more disputed. It is an area where creative reliance on decisions involving HUD-financed projects and analogies with suits involving other relationships in the construction process should be used.

TORT THEORIES

§ 7.24 Negligence

Contractors have also tried various tort theories to reach a construction lender. To date, these theories have not met with significant success. However, negligence appears to be an area where development is possible. Arguments could be made

---

[61] *Cf.* Superior Glass Co. v. First Bristol County Nat'l Bank, 8 Mass. App. 356, 394 N.E.2d 972 (1979), *aff'd,* 406 N.E.2d 672 (1980).

[62] Gee v. Eberle, 420 A.2d 1050, 1060 (Pa. Super. Ct. 1980); Cal. Civ. Code § 3264 (West 1974); Pankow Constr. Co. v. Advance Mortgage Corp., 618 F.2d 611, 616 (9th Cir. 1980).

[63] Gee v. Eberle, 420 A.2d 1050, 1060 (Pa. Super. Ct. 1980).

that a lender was negligent in making a loan, in administering construction payments, or in disbursing loan proceeds. Another potential area for attack is the lender's failure to observe accepted banking practices, such as failing to require signed releases or necessary bonds.

At least one contractor was successful in a suit against a lender on a negligent disbursement of loan funds theory. In *Cook v. Citizens Savings & Loan Association,*[64] the contractor requested payment from the lender after completing work. The lender paid money directly to the owner without receiving any account of its use, which resulted in the contractor's not being paid. The court concluded that the lender owed the contractor a duty of reasonable diligence to see that the funds were actually used for paying the construction costs. Nevertheless, subsequent courts have read *Cook* narrowly, tying it to the particulars of Mississippi law.[65]

In California, a surety successfully sued a lender for negligent disbursement in *Commercial Standard Insurance Co. v. Bank of America.*[66] The suit arose when the owner sued the contractor and the contractor's surety. The surety alleged that the bank had made disbursements exceeding the value of the work done. The court specifically rejected the bank's contention that it owed no duty to the surety. It stated that the question of duty was "question begging" and that it was reasonably foreseeable that the surety would be injured by the bank's failure to exercise care.[67]

Negligence theories, though, usually meet defeat on the issue of duty. While courts have at times recognized that lenders owe a duty to other parties involved in the construction project, they generally do not extend this duty to include contractors.[68] It is an area in which a contractor will want to analogize to other relationships with a lender on a construction project.

Additionally, a contractor will need to establish what the duty is. Industry custom, internal policy manuals, and other documents can be helpful in this regard.[69]

§ 7.25 Misrepresentation

Another possible tort theory is *negligent misrepresentation.* The *Restatement (Second) of Tort* recognizes that liability can exist for information negligently supplied for the guidance of others:

---

[64] 346 So. 2d 370 (Miss. 1977).

[65] *E.g.*, Rockhill v. United States, 418 A.2d 197, 202 (Md. Ct. App. 1980).

[66] 57 Cal. App. 3d 241, 129 Cal. Rptr. 91, 95 (1976).

[67] 129 Cal. Rptr. at 94-95.

[68] Rockhill v. United States, 418 A.2d 197, 199-205 (Md. Ct. App. 1980); Daniels v. Big Horn Fed. Sav. & Loan Ass'n, 604 P.2d 1046, 1048-50 (Wyo. 1980); Superior Glass Co. v. First Bristol County Nat'l Bank, 8 Mass. App. 356, 294 N.E.2d 972, 975-76 (1979), *aff'd,* 406 N.E.2d 672 (1980); Bescor, Inc. v. Chicago Title & Trust Co., 446 N.E.2d 1209 (Ill. App. Ct. 1983).

[69] Daniels v. Big Horn Fed. Sav. & Loan Ass'n, 604 P.2d 1046 (Wyo. 1980); Commercial Standard Ins. Co. v. Bank of Am., 57 Cal. App. 3d 241, 129 Cal. Rptr. 91 (1976); First Nat'l State Bank of N.J. v. Carlyle House, Inc., 246 A.2d 22 (N.J. Super. Ct. 1968).

> (1) One who, in the course of his business, profession or employment, or in any other transaction in which he has a pecuniary interest, supplies false information for the guidance of others in their business transactions, is subject to liability for pecuniary loss caused to them by their justifiable reliance upon the information, if he fails to exercise reasonable care or competence in obtaining or communicating the information.[70]

The *Restatement* limits this liability to cases where the loss is suffered by a person for whose benefit or guidance the information was given and involves a transaction (or one substantially similar) that the maker intends to influence or knows that the recipient of the information intends to influence.[71]

In a Wisconsin case,[72] a trial court allowed a subcontractor to allege negligent misrepresentation when a lender informed the subcontractor that $46,000 was available for construction when in fact $20,000 of that amount had been withdrawn. The appellate court affirmed on another theory and did not address the negligent misrepresentation issue. Other courts find, however, that mistaken representations about the amount of undisbursed loan funds are not an ''affirmative deception'' similar to fraud and misrepresentation that would justify equitable relief.[73]

A negligent misrepresentation theory may be available in other factual situations. A lender's representation, for instance, that an adequate bond existed or that the owner had unencumbered title could be actionable, in the right factual setting.

§ 7.26 Other Tort Theories

One's facts and imagination are the principal limitations on the development of other tort theories. One possibility is *intentional interference with contract.*[74]

In *Lincor Contractors, Ltd. v. Hyskell,*[75] a contractor successfully sued a lender on an intentional interference with contract theory. The lender insisted that the contractor be removed from a project, and told the owner that no more loan proceeds would be made available until the owner removed the contractor. The Washington court found that the lender had no right equal or superior to the contractor's right to perform that justified this interference, and awarded the contractor its anticipated profit.

---

[70] Restatement (2d) of Torts § 552 (1977).

[71] *Id.* § 522(2).

[72] *E.g.*, Klein-Dickert Oshkosh, Inc. v. Frontier Mortgage Corp., 287 N.W.2d 742 (Wis. 1980).

[73] Indiana Mortgage & Realty Inv. v. Peacock Constr., 348 So. 2d 59, 61 (Fla. Dist. Ct. App.), *cert. denied,* 353 So. 2d 677 (Fla. Dist. Ct. App. 1977); J.G. Plumbing Serv., Inc. v. Coastal Mortgage Co., 329 So. 2d 393 (Fla. Dist. Ct. App.), *cert. denied,* 339 So.2d 1169 (Fla. Dist. Ct. App. 1976).

[74] Koppers Co., Inc. v. Garling & Langlois, 594 F.2d 1094 (6th Cir. 1979); Pankow Constr. Co. v. Advance Mortgage Corp., 618 F.2d 611, 617 (9th Cir. 1980).

[75] 692 P.2d 903, 908 (Wash. Ct. App. 1984).

A similar claim may exist where a lender in bad faith diverts loan funds to the detriment of contractors and subcontractors. A principal issue in these intentional interference with contract uses will be whether the lender's own interests justified its interference with the contractor.[76]

Another possible theory is *outrageous conduct.* The contours of this tort vary between jurisdictions, but the hallmark of the tort is that the conduct in question is beyond the bounds tolerated by society.[77] When no other remedy seems available, outrageous conduct may be a possible approach. While there do not appear to be any cases on point, a jury issue could exist if a lender used excessive or illegal means to protect its interests, to the expense of contractors.

§ 7.27 Conclusion

This chapter has explored statutory and common law remedies available to contractors seeking to recover from a construction lender. Statutory remedies include mechanics' liens, stop notices, and recording statutes. These are generally the preferred remedies, particularly for subcontractors and materialmen.

Next, the contractor may attempt to assert contractual rights. The key issues are establishing the necessary privity or an intent to benefit the contractor. If the necessary elements for enforcement of contract rights are not present, estoppel or equitable arguments should be explored.

A final potential remedy is tort liability. Courts have considered negligence, misrepresentation, and intentional interference with contract theories. To date, these theories generally have been unsuccessful, but hold possibilities for future development.

---

[76] *Id.*

[77] *See generally* Turman v. Central Billing Bureau, 279 Or. 443, 568 P.2d 1382 (1977); W. Prosser, Prosser on Torts 57-58 (4th ed. 1971).

CHAPTER 8

SUING THE SUBCONTRACTOR AND MATERIAL SUPPLIER

Robert J. Hoffman and David L. Simmons

Robert J. Hoffman received his Juris Doctor degree, summa cum laude, from Indiana University— Indianapolis Law School in 1974 and his Bachelor's Degree from Indiana University in 1968, majoring in political science and history. He is a partner in the Indianapolis, Indiana law firm of Lowe Gray Steele & Hoffman and concentrates his legal practice in the area of construction and public works law. He has frequently lectured on this topic before various groups and construction organizations, and is the author of numerous publications and articles on the subject.

Mr. Hoffman maintains membership in the Indianapolis and Indiana State Bar Associations and is also active in the American Bar Association, Public Contracts Section and Forum Committee on the Construction Industry.

David L. Simmons received his Juris Doctor degree from Indiana University School of Law at Indianapolis in 1980, his Master of Business Administration degree from Indiana University in 1977, and his Bachelor's degree from Indiana University in 1975, majoring in political science and English. He is associated with the Indiana law firm of Lowe Gray Steele & Hoffman and concentrates his legal practice on construction and commercial litigation. Mr. Simmons maintains memberships in the Indianapolis and Indiana State Bar Associations, as well as the American Bar Association, Public Contracts Section and Forum Committee on the Construction Industry.

§ 8.1 Introduction

At an early time in history when construction became discernable as a semi-distinct trade or enterprise, all the talent, materials, and manpower involved in the building process were furnished and controlled by the master builder. The economic prospects of the master builder's success on a project (and in some cases his ability to keep his head) depended only upon his ability to manage and control his own talent and resources, not upon the performance and quality of work furnished by people who were not ultimately subject to his domination. The master builder proved to be the ancestor of the modern general contractor, one who contracts directly with an owner to perform all work and furnish all material and equipment necessary to transform a design into a finished product ready for use and occupancy.

The modern contractor's links to those remote origins have been all but severed. No longer does a contractor perform all the work with its own forces. Instead, it has become more of a broker or middleman who commits to an owner to construct a project; however, to fulfill these commitments, it must retain and rely upon a host of subcontractors (those who are to perform some portion of *work* on the project) and material suppliers (those who are primarily expected to *sell* and furnish items of material and equipment to be used in construction). Thus, to a significant extent, the modern contractor's contractual *reliance* upon lower tiers has replaced the master builder's physical or economic *control* over his forces. The contractor's ability to fulfill its promise to the owner is, from the bidding stage all the way through project completion and subsequent warranty periods, subject to and governed by the qualitative and timely performance by these lower-tier elements. Today's prime contractor is much like the conductor of a symphony orchestra; it is the one upon whom the owner, sitting in the audience, relies for a coordinated, well-rehearsed presentation of the advertised masterpiece; in order to meet its obligation to give an acceptable performance, it must in turn rely upon the ability of each musician to play his or her respective instrument properly and in harmony with the others. Given this uncertain and risk-filled environment, it is not at all unexpected that disputes involving subcontractors and material suppliers have become so prevalent.

A contractor faces a continuous risk of liability to an owner due to the failure of its subcontractors and suppliers to perform as they must if it is to meet the demands placed on it by its own contract with the owner. Disputes involving subcontractors and suppliers can arise during any phase of construction and, in some instances, long after the project is completed. Subcontractor-contractor controversies occur in the same types of factual settings and normally are governed by the same legal principles that exist or apply as between the owner and contractor. Hence, a claim or suit by an owner against a prime contractor will almost certainly result in the same type of claim, in the same lawsuit or independently, by the prime contractor against one or more subcontractors; this represents the contractor's attempt, as middleman, to pass all its liabilities on to the ultimately responsible lower-tier party. A claim by the contractor against a subcontractor may exist on its own in cases where the contractor incurs its own distinct damages as a result of the subcontractor's failure to perform. The outcome of the contractor's claim, whether independent of or corollary to the owner's claim, will frequently depend upon the contractor's ability to shift or allocate liabilities and duties to its subcontractors and suppliers, at least to the same extent that risks have been allocated to it by the terms of its prime contract with the owner.

This chapter describes general principles governing a contractor's relationship with its subcontractors and material suppliers, and describes some of the more prevalent remedies, theories of recovery, and (most importantly) protective subcontract terminology, which will enhance the contractor's most fundamental objective: *assuring that the risk of loss due to the conduct of a lower-tier party is ultimately borne by that party and not by the contractor who engaged him or her.*

Contractor claims against subcontractors and material suppliers can develop from a multitude of factual settings and may involve literally hundreds of different expressions found in the written subcontract, purchase order, or one or more of the contract documents which form the integrated contract between the contractor and owner and which may (and should) be incorporated into the subcontract by reference. It is impossible to discuss each such factual or contractually derived dispute within the confines of one chapter. Instead, this chapter focuses on some of the more common dispute categories, together with general principles and contract language which enhance a contractor's chances of either winning disputes with subcontractors or avoiding them altogether.

For purposes of brevity, the term *subcontractor* is used throughout this chapter without necessarily being accompanied by the term *material supplier;* however, in most cases the subject of discussion applies equally to both types of lower-tier participants on a project. Note, though, that a dispute involving a material supplier (as opposed to a subcontractor) may be subject to additional, and in some situations conflicting, rules and standards as provided by the Uniform Commercial Code, specifically Article 2 dealing with the sale of goods, which has been enacted in every state except Louisiana.

§ 8.2 Bidding and Contract Formation

The construction bidding process, viewed from the contractor's perspective, is one involving numerous parties; it is at this stage that the contractor must, in order to submit its own bid commitment to an owner, assemble, through a sub-bidding process, a subcontractor and material supplier team to perform the job in the event that its bid is accepted by the owner.

Traditional concepts of contract law govern the bidding process, as well as the subsequent stages of construction. A contract between two or more parties creates a legally enforceable relationship defining the obligations and reciprocal rights of the contracting parties. In a traditional sense, contracts for construction are formed by the making of an offer (bid) by one party and the acceptance (award) of that offer by the other party, thereby resulting in a *meeting of the minds,* which serves as a system of reciprocal promises and commitments which is legally enforceable.

Before submitting its bid to the owner, the contractor must know to a high level of certainty what its costs will be to perform the work. To accomplish this, the contractor solicits and receives bid proposals from subcontractors for certain assigned portions of the work (or certain items of material or equipment). The contractor should assure that a subcontractor's bid proposal is in fact a *definite offer to perform* which will become an enforceable contract if the bid is accepted, as opposed to a naked price quotation, which courts have frequently viewed as mere preliminary negotiations.[1]

Without contrary language appearing in the bid or invitation to bid, or some enforceable type of reliance by the contractor upon the subcontractor's bid, the subcontractor may, under traditional contract formation rules, withdraw its bid at any time before communication of acceptance by the contractor. The amount of the subcontractor's proposal and the scope of work and other conditions to which the subcontractor commits is an indispensable ingredient in the contractor's formulation of its bid and bid price to the owner. Therefore, a contractor who receives and acts upon subcontractor quotations and proposals is in a predicament: it has no need or desire to accept the proposal until it knows whether or not its own bid is accepted by the owner. This time lapse between submission and owner acceptance (or nonacceptance) creates the risk, under traditional contract law principles, that the subcontractor will withdraw its bid in the meantime. In an effort to deal with this risk and avoid disputes when a subcontractor withdraws its bid in such a manner, contractors frequently require subcontractor offers to be *firm* or irrevocable for a definite period of time after submittal. A *firm offer* may be stated on the face of the bid or may be required by the bid solicitation.

Contractors themselves usually make the same type of firm offer to owners; likewise, the contractor should extract this same commitment from candidate subcontractors and suppliers so that all subcontractors will be bound on their bids for as long a period as the contractor is bound on its bid to the owner. A firm, irrevocable offer has been found by courts to be a commitment which is contractual in nature and legally enforceable as such. A subcontractor who fails to perform or honor such a bid may be liable to the contractor for resulting damages incurred by the contractor.[2]

---

[1] *See, e.g.,* Mitchell v. Siqueiros, 99 Idaho 396, 582 P.2d 1074 (1978).

[2] Illinois Valley Asphalt, Inc. v. J.F. Edwards Constr. Co., 413 N.E.2d 209 (Ill. 1980).

The Uniform Commercial Code (UCC) provides additional protection for the contractor under certain circumstances. The UCC provides that a written bid which assures that it will be held open is not revocable during the time stated. If no specific period of time is stated, the bid cannot be withdrawn for a "reasonable period" of time, not to exceed three months.[3] It is important to note, however, that the UCC applies only to the sale of goods and movable equipment, and to contracts where the sale of goods is the dominant transaction.[4] Application of UCC provisions pertaining to sale of goods is therefore useful in the material supplier context, but will rarely control in a dispute involving a subcontractor.

A bidding contractor will not always be able to extract binding irrevocable proposals from subcontractors. Nevertheless, the contractor must still rely upon such proposals in pricing its bid to the owner, thus giving rise to the following dispute scenario: The contractor uses a subcontractor's bid price in formulating a bid to the owner, following which the owner awards the contract to that contractor, thereby locking in the contractor to its bid price. Before the contractor can formally accept the subcontractor's proposal, the subcontractor withdraws its bid and refuses to perform. Courts in numerous jurisdictions have, with almost complete uniformity, recognized a contractor's peculiar vulnerability in this situation and have afforded contractors relief under the theory of *promissory estoppel.* This doctrine essentially permits a court to transform the subcontractor's bid into a firm, irrevocable offer even though the bid itself does not express such irrevocability, based upon the *justifiable reliance* on the part of the contractor. To avoid the inequities which would otherwise be imposed upon a contractor in such a situation, courts have held that a subcontractor's bid is binding upon the subcontractor if the subcontractor knew, or had reason to know, that its bid would be justifiably relied upon by the contractor in preparing and submitting its bid to the owner, and that the contractor would be damaged if the subcontractor withdrew its bid before acceptance.[5]

The UCC contains a potential trap for a contractor who receives an oral proposal from a supplier involving the sale of goods or equipment. The statute of frauds section of the UCC provides that, subject to defined exceptions, a party cannot enforce a contract for the sale of goods for $500 or more unless the contract is in writing and signed by the party against whom enforcement is sought.[6] The question arises as to whether application of the doctrine of promissory estoppel may remove a case from the ambit of the statute of frauds, thus enabling the contractor-buyer to enforce an oral agreement based on its detrimental reliance.

Court decisions are not uniform on this issue. Some decisions have held that the statute of frauds must be enforced despite the existence of justifiable reliance

---

[3] U.C.C. § 2-205.

[4] U.C.C. Art. 2.

[5] C.E. Frazier Constr. Co. v. Campbell Roofing & Metal Works, Inc., 373 So. 2d 1036 (Miss. 1979); Montgomery Indus. Int'l, Inc. v. Thomas Constr. Co., 620 F.2d 91 (5th Cir. 1980); Harry Harris, Inc. v. Quality Constr. Co. of Benton, Ky., Inc., 593 S.W.2d 872 (Ky. 1979); James King & Son, Inc. v. DeSantis Constr. No. 2 Corp., 413 N.Y.S.2d 78 (Sup. Ct. 1977).

[6] U.C.C. § 2-201.

which in other cases could bring the promissory estoppel doctrine into play.[7] Other courts recognize promissory estoppel as a device to avoid the UCC statute of frauds provisions and have held that oral bids for the sale of goods will be enforceable where there was justifiable reliance.[8]

§ 8.3 —Bid Mistakes

Bidding mistakes are another source of subcontract disputes during the bidding process. Subcontractors, like the contractors who engage them, can make mistakes in preparing bids, often because the hurried conditions under which they must submit their bids. If the mistake is significant, the bidder may attempt to withdraw its proposal or refuse to go forward with performance, thus giving rise to a suit by the contractor. Courts have fashioned various theories for resolving bid mistake situations, depending on the size of the mistake and the reasonableness of the contractor's reliance.

As a general principle, a bid mistake will relieve the bidder of its obligation to perform where the bid is at such a variance from the other bids submitted that the contractor should reasonably have concluded or suspected that a mistake had been made.[9] In such a situation, given the obviousness of the error, there has not been the meeting of the minds necessary for the formation of a contract. Contractors therefore should review subcontractor bids to ascertain their reasonableness before assuming that the subcontractor will be bound. If there is any question about the amount of a subcontractor's bid, the contractor should verify the amount before submitting its own bid.

Subcontractors have been allowed to withdraw their bids and escape performance obligation in some circumstances where the contractor purported to accept the bid but added terms which were not made known to the subcontractor prior to submission of the subcontractor's bid.[10] This may occur in the common situation where "acceptance" takes the form of a contractor submitting its subcontract agreement to the subcontractor for signature when the agreement contains terms which were never part of the proposal. To avoid this problem, the contractor's invitation to bid should expressly state that "the subcontractor shall perform its work according to the terms and provisions of all relevant contract documents, including plans and specifications, which are available for inspection at [the contractor's office]."

---

[7] Anderson Constr. Co. Inc. v. Lyon Metal Prods., Inc., 370 So. 2d 935 (Miss. 1979).

[8] Edward Joy Co. v. Noise Control Prods., Inc., 443 N.Y.S.2d 361 (N.Y. 1981).

[9] Robert Gordon, Inc. v. Ingersoll-Rand Co., 117 F.2d 654 (7th Cir. 1941); Southeastern Sales & Serv. Co. v. T.T. Watson, Inc., 172 So. 2d 239 (Fla. Dist. Ct. App. 1965).

[10] C.H. Leavell & Co. v. Grafe & Assocs., Inc., 414 P.2d 873 (Idaho 1966).

§ 8.4 The All-Important Flow-Down Clause

A prime contractor, especially one whose contract results from the bidding process as opposed to negotiation, has very little bargaining leverage with the owner, and must essentially take the contract as he or she finds it. The assorted contract documents which combine to form the contract with the owner can be expected to represent an attempt by the owner to shift a wide range of risks to the prime contractor and away from the owner. When construction disputes develop which involve a claim by the contractor against its subcontractors, it is in the contractor's interests to assure that no gaps exist between the obligations expressed in the subcontract and those in the primary contract with the owner. A well-drafted subcontract should therefore always contain a clause which places the contractor on exactly the same footing against the subcontractor as the owner occupies against the contractor.

The so-called *flow-down* or *conduit* clause is the primary protective language in the subcontract which accomplishes this; it does so by an express statement that the subcontractor is to be obligated to the contractor to the same extent as the contractor, under the provisions of the contract documents, is obligated to the owner. Such a clause should contain a proviso permitting conflicting provisions within the subcontract instrument itself to take precedence over other contract obligations which are incorporated by reference.

If drafted effectively, a flow-down clause goes a long way toward assuring the contractor that it may step into the shoes of the owner and thereby employ all the risk-shifting terminology in the prime contract against a nonperforming subcontractor or material supplier, so that it can measure the subcontractor's liability and its own recovery of damages by its obligations to the owner.

§ 8.5 Description of Work and Materials

A flow-down clause is not helpful in defining the work or materials to be furnished to a project by a subcontractor or supplier, since the prime contract obviously includes all work, whereas the subcontract or material supply purchase order is intended to cover only a portion of the whole. A vast number of disputes between contractors and their subcontractors arise because a subcontractor has refused to perform some item of work because "it wasn't in my price," thereby forcing the contractor to have the work performed with its own crew, or through others, with resulting additional expense which is sought to be charged back against the subcontractor. The contractor's ability to recover these additional expenses, usually as damages arising from the subcontractor's breach of contract, rests upon the contractor's ability to point to a specific description of the scope of the subcontractor's work contained in the subcontract agreement. It is the description in the agreement, *not* whether or not the subcontractor had an item of work included in its bid price, that governs.

At a minimum, the subcontract should refer to the scope of work in terms of referencing the section of the prime contract specifications which is to be performed by the subcontractor, so that there will be no doubt that the description of the quality and quantity of work is exactly the same as that which defines the contractor's obligation to the owner. However, this may not always be enough. Frequently, supplementary terminology is needed to broaden the subcontractor's obligations beyond those set forth in a specific section of the specifications, to encompass additional work (found in other sections or contract documents) normally to be performed by the trade to which the subcontractor belongs.

The contractor must assume that good draftsmanship alone will not solve all disputes with subcontractors concerning scope of work. When such disputes develop, the contractor must have some means of assuring that the disputed work will be performed without unduly interfering with scheduled job progress and coordination. Therefore, the subcontract should contain a clause which obligates the subcontractor to perform such work upon the unilateral directive of the contractor, or face termination of the subcontract. Such a clause should be coupled with language imposing time limits within which the subcontractor must make or reserve a claim for additional compensation, in the event that the subcontractor prevails and the disputed work is found not to be in the original scope. The contractor thus receives a measure of assurance that work will continue at an uninterrupted pace without the contractor having to retain a replacement subcontractor or incur expense to perform the work with its own resources.

Virtually every prime contract contains a provision whereby the owner may unilaterally make changes, additions, deletions, or other modifications in the scope of work by written change order. In the subcontract agreement, the contractor must, by appropriate language, reserve the same right to make unilateral changes in the work, and obligate the subcontractor to perform that work as changed. Typically, a subcontract deals with the subject of changed and extra work by providing that the subcontractor shall have no right to claim or receive extra compensation for such work unless it is authorized and directed in writing by the contractor. Some subcontracts add terminology which limits the subcontractor's recovery to the amount which is collected by the contractor from the owner for that work.

§ 8.6 Coordinated and Timely Performance

Delay and disruption on a job site caused by the actions of a subcontractor can have devastating results to a prime contractor on at least three fronts. First, disruptions and delays significantly expose the contractor to liability for economic damages suffered by the owner. The prime contract frequently contains a liquidated damage provision which imposes hundreds or even thousands of dollars in damages against the contractor for each day that project completion is delayed. Secondly, the contractor may be subjected to delay damage claims asserted by other subcontractors or material suppliers. Thirdly, the contractor can expect to incur losses and expenses within its own organization because of the delay.

The subcontract should expressly confirm the subcontractor's liability for such loss and damages incurred by the contractor in all three areas. A subcontract agreement which merely requires a subcontractor to complete its work by a certain date is not sufficient for this purpose. Additional contract language should be included which obligates the subcontractor to coordinate and schedule its work in cooperation with the contractor and with *other subcontractors*. In the drafting and administration of the subcontract, the contractor should explicitly tie the subcontractor to the construction schedule which governs the interrelated performance of work *by all trades*.

If a dispute or lawsuit develops concerning a delay of significant proportions, the contractor and its attorney can expect that everyone affected by the delay, including subcontractors, the contractor, the owner, and others, will point to everyone else as the culprit. From its own point of view, the contractor should assume that in litigation which focuses upon project delays, it will be forced to assume a defensive posture as much as an offensive one. It has become common for contractors to attempt to insulate themselves from delay damage claims asserted by subcontractors by employing *no damage for delay* clauses in their subcontract forms. Such clauses in substance state that the subcontractor agrees that its sole remedy in the event of delay will be an extension of time and that it shall have no right to recover delay damages as against the contractor. Some contractors adopt a softer form of the clause which permits a subcontractor to recover delay-related damages, but only to the extent that the contractor recovers and collects such damages from the owner or some other party. Such clauses have received a mixed reception in the courts. Some jurisdictions construe no damage for delay clauses strictly against the contractor, and do not apply them in cases where the contractor has caused the delay by actively interfering with the subcontractor's work[11] or in cases where the delay was brought about by circumstances totally beyond the contemplation of the parties.[12] It is the opinion of the author that judicial willingness to find ways not to enforce such clauses has reached a high-water mark, leaving perhaps a more visible line of authority which upholds such clauses in favor of the contractor.[13]

§ 8.7 Correcting Defective Work

The General Conditions or other contract documents forming the contractor's agreement with the owner typically require the contractor to correct or replace any work which is rejected or found by the owner's architect or engineer to be defective or otherwise not in accordance with the specifications. Arguably, a well-drafted

---

[11] Blake Constr. Co. v. C.J. Coakley Co., Inc., 431 A.2d 569 (D.C. 1981). See also §§ **3.10, 5.31-5.35.**

[12] Phoenix Contractors, Inc. v. General Motors Corp., 135 Mich. App. 787, 355 N.W.2d 673 (1984). See also § **5.32.**

[13] Owen Constr. Co. v. Iowa State Dep't of Transp., 274 N.W.2d 304 (Iowa 1979).

flow-down clause[14] will impose a similar obligation upon a subcontractor who performed or placed the rejected work. However, it is perhaps more prudent for a contractor to include a clause within the subcontract agreement which specifically imposes the obligation upon the subcontractor to replace or correct any work which, in the opinion of the architect *or contractor,* is deficient or not in accordance with the applicable contract documents. The subcontract should further provide a means for the contractor to notify the subcontractor of such deficiencies, and provide for deadlines for the subcontractor to take appropriate corrective action.

The contractor who is confronted with work which is rejected by the owner's architect should immediately take steps to notify the subcontractor of such rejection, in exact accordance with the procedure outlined in the subcontract agreement. Then, if the subcontractor fails to perform corrective action as and when necessary, the subcontract should make clear that the contractor has the right to declare the subcontractor in default and to terminate the subcontract accordingly.[15] Whether or not the subcontract is terminated, the subcontract should expressly give the contractor the right to have the corrective work performed by others and charge the expense resulting from such remedial work against the subcontractor.

Apart from such recommended express contractual language, a contractor's case against a material supplier may be enhanced by important implied warranties. The Uniform Commercial Code provides for an important implied warranty that the materials and equipment furnished will be "merchantable," meaning that they will "pass without objection in the trade under the contract description" and will be "of fair average quality as required by the contract."[16] Under the Code, a material supplier impliedly warrants that the goods will be fit for the purchaser's (contractor's) particular purpose, if the supplier has reason to know the purpose to which the goods will be used at the time they are sold and that the purchaser is relying upon the supplier's superior skill and judgment.[17] These implied warranties can provide an important foundation for recovery of damages by a contractor against a defaulting subcontractor or supplier, but, in appropriate circumstances, they may be contractually disclaimed. The contractor will no doubt avoid such disclaimer language in its own subcontract or purchase order forms; however, if any other document is included in the contract with the subcontractor or supplier, the contractor should scrutinize it carefully in order to uncover disclaimer language which could eliminate the contractor's implied warranty remedies.

§ 8.8 Indemnification

Occasionally a contractor will find itself subjected to claims for bodily injury, death, or loss or damage to property arising out of a subcontractor's performance of work on a project. In such cases, it is essential for a contractor to maximize its ability

---

[14] See § **8.4.**

[15] See § **8.9.**

[16] U.C.C. § 2-314.

[17] U.C.C. § 2.315.

to transfer its own claim exposure to the subcontractor. To accomplish this, contractors regularly employ some form of indemnification clause in their subcontract forms, under which the subcontractor agrees to indemnify the contractor and hold the contractor harmless against such claims, demands, liabilities, losses, etc., including attorneys' fees which may be incurred by the contractor in defending the claim. Such clauses fall within three general categories, according to the scope of protection offered: (1) narrow coverage, subjecting the contractor to liability only for claims which result from its own acts or omissions, and not the negligence of others, (2) intermediate coverage, subjecting the subcontractor to liability to the extent it is caused in whole or in part by any negligent act of the subcontractor, and (3) broad indemnity, which purports to extend protection to the contractor to include claims brought about by the contractor's own negligence, and without negligence on the part of the subcontractor.

A subcontractor can be expected to be held within the grip of the first two categories of clauses. The enforceability of the third type, the broad-form clause, however, is far more questionable. Many states have enacted so-called *anti-indemnity* statutes which severely limit or prohibit the enforcement of a contractual clause which indemnifies one against one's own negligence.[18] In other states, similar nonenforceability has become the rule by judicial decree. Therefore, in drafting a subcontract indemnity clause, a contractor would be well advised to consult the law of the jurisdiction where the work will be performed in order to determine the permissible outer boundaries of an indemnity clause. Contractors which regularly engage in work in numerous states would be well advised to employ the intermediate form of indemnity clause.

In some instances, contractors have the right to a form of implied indemnification against a subcontractor, even in the absence of express indemnification language in the subcontract instrument. *Implied indemnification* arises when a contractor is exposed to liability due to the subcontractor's negligence or breach of contract. The contractor who is held liable and incurs damages as a result of such negligence or breach may seek indemnification against the subcontractor for its losses.[19]

§ 8.9 Default and Termination

Occasionally, the trouble generated by a subcontractor's presence and involvement (or perhaps lack thereof) on a job reaches a magnitude which requires the contractor to procure a replacement subcontractor to complete the work of the subcontractor in default. The most obvious example of such a situation involves the subcontractor who, because of some dispute involving the work or because of sheer lack of economic resources to complete, walks off the job and abandons

---

[18] Cal. Civ. Code § 1782 (West); Ga. Code § 20-50A; Ill. Rev. Stat. ch. 29, § 61; N.Y. Gen. Oblig. Law § 5-324.

[19] Leaseway Warehouses, Inc. v. Carlton, 568 F. Supp. 1041 (N.D. Ill 1983); Maxfield v. Simmons, 449 N.E.2d 110 (Ill. 1983).

the project. When a dispute arises, subcontractors frequently see the threat of "closing down the job" as giving them a substantial amount of leverage. A subcontractor who contemplates such action often feels that it is a superior form of leverage for extracting concessions from the contractor, but fails to realize that abandoning the project will almost assuredly result in a court decision which renders it liable for all additional expenses incurred by the contractor to retain a replacement to complete the work.

If the subcontractor's abandonment is necessitated by its economic inability to perform, then the contractor unhappily has a hollow remedy against the subcontractor for damages which will no doubt be uncollectible. In these uncertain economic times, this prospect has led many contractors to require major subcontractors to furnish labor and material payment bonds which permit the contractor, in the event of default, to recover damages against the contract surety.

A more complex scenario exists when the subcontractor remains on the job and professes its willingness and ability to complete the agreement but, because of some situation or occurrence, the contractor considers the subcontractor to be in default under the contract, and desires to terminate the contract and procure a different subcontractor to finish the work.

Many contractors, faced with severe subcontractor-caused problems on a project, go through the process of terminating the defaulting subcontractor without fully appreciating the legal consequences which could flow from the termination. It is true that a defaulting subcontractor whose breach goes to the heart of its ability to perform its contract may be terminated, in which case the contractor may administer the default and termination clauses of the subcontract agreement and may recover all excess costs and related damages incurred in retaining a new subcontractor to finish the work. However, significant factual and legal issues may exist as to whether or not the problem or occurrence which motivated the contractor to take the action actually constituted a *material* breach of the subcontract, as opposed to a lesser form of breach which entitles the contractor only to recover damages, but not to terminate. In short, if the contractor is unable to establish that the subcontract permits the action which was taken, then the contractor may not recover its excess costs to complete through a replacement subcontractor, and may in fact find itself liable to the defaulting subcontractor for the profits which that subcontractor would have recovered had it been permitted to complete its agreement. A contractor who makes the wrong decision, or who has a subcontract agreement which inadequately treats the subject of termination, will be unable to pass these combined losses on to the owner.

In cases of subcontractor default, it is essential for the contractor to (1) maximize its chances of recovering completion-related damages, and (2) avoid potential liability to the subcontractor because the termination was wrongful. This is done by employing a default and termination clause in the subcontract which endorses the contractor's right to terminate the subcontractor. The default clause in the subcontract should explicitly define the types of occurrences which constitute an event of default warranting termination, such as (1) failure on the part of the subcontractor to pay its sub-subcontractors and suppliers, (2) failure to diligently prosecute the work as directed by the contractor or as required by the agreed project schedule, (3) failure to perform any term or condition set forth in the subcontract agreement, or other contract document which is incorporated by reference, (4) refusal to perform corrective work as directed, (5) violation of applicable laws and regulations, or (6) bankruptcy.

A well-drafted default and termination provision in a subcontract should also establish the exact procedure by which the contractor may terminate the subcontractor. The most common form of default/termination procedure involves an initial notice sent to the subcontractor by the contractor, describing the occurrence which the contractor has determined warrants termination and providing a period of time during which the subcontractor must cure the event of default and thereby retain its entitlement to continue on the job. The default clause should then provide for a second, follow-up notice to be sent to the subcontractor after the expiration of the grace period, notifying the subcontractor of the contractor's intent to terminate its right to proceed.

The default provision should also describe the remedies which are available to the contractor upon termination. Frequently, termination clauses provide that the contractor may take possession of the subcontractor's materials, tools, and equipment and finish the subcontractor's work by whatever means the contractor may determine to be appropriate. It is also wise to provide in the termination clause that the contractor shall be entitled to all monies expended and all other damages and extra expense (including overhead and attorneys' fees) which result from the subcontractor's default, termination, and completion of work, and that the subcontractor shall not be entitled to any additional payment under the contract until the contractor's damages are ascertained, i.e., after the subcontractor's work is completed by the replacement or by the contractor's own forces.

A contractor should consider using a separate, additional clause which permits termination of the subcontract "for convenience." Federal government contracts and many state and local public works contracts employ such clauses in the prime contract, whereby the owner may, in the absence of breach, nevertheless terminate the contract. Upon such termination, the contractor's recovery is limited to the work completed, without additional profit on work uncompleted. Such a clause, if employed by the contractor in its subcontract agreements, gives similar protection. Additionally, the contractor may limit its exposure to a defaulting subcontractor by providing that a termination for default, if determined to be incorrect, will constitute a termination for convenience, thereby limiting the subcontractor's recovery and preventing the subcontractor from recovering lost profits on unperformed work.

§ 8.10 Documenting the Contractor's Claim

As suggested in the preceding sections of this chapter, a variety of legal rules and theories exist which, when combined with a comprehensive, well-drafted subcontract agreement, provide a sound and formidable foundation to support a contractor's suit against a subcontractor. However, it is only a foundation: it still remains for the contractor to prove its case against the subcontractor. To accomplish this ultimate objective, it is imperative that the contractor's claim be supported with sufficient documentary and other evidence to establish its right to relief.

It is indeed paradoxical that many contractors go to great lengths to use subcontract forms which give them every possible legal and economic leverage and then, when it comes time to implement the "boiler plate" and prove their case against a subcontractor, utterly fail to furnish the factual documentation necessary to persuade a court or jury that the facts entitle them to the remedy being sought. A contractor who has been damaged by the activity of a subcontractor must ultimately convince the trier of fact of the relationship of the parties, their conduct which relates to the dispute, and the monetary damages resulting from the subcontractor's breach of its obligations. It is therefore critical for the contractor to be able to document and maintain sufficient proof to substantiate its position.

A more detailed discussion of the types and items of documentation which a contractor should maintain as a matter of course on a construction project is set forth in other chapters.[20] These suggested means of documentation apply with equal force to the contractor-subcontractor relationship. Regardless of the documentation which is *customarily* maintained on a project, when trouble signs develop to suggest even a remote likelihood of a claim being made by the contractor against a subcontractor (or anyone else for that matter), it is time for the contractor's office and front-line personnel to institute immediate steps to build a record which focuses upon the specific problem involved and which follow an issue through its development to the point of full and complete resolution.

In the area of subcontract relations and dispute handling, as with other areas of construction law, the legal and economic environment which now surrounds modern construction serves as a warning to contractors that times have changed and the construction industry must change accordingly. Handshake agreements, which served as the backbone and theme of construction in the past, have now lost their appeal. The use of a blank purchase order as a means of engaging a subcontractor to perform work on a project serves today as a potential shortcut to business failure. Reliance upon a telephone conversation rather than a written confirmation is no longer prudent.

Legal rules and theories will always be present to protect the contractor and enable it to formulate a legal claim for relief against a delinquent subcontractor. *Good draftsmanship* can arm the contractor with the contractual basis for administering and enforcing its expectations and reliance. *Effective documentation, recordkeeping,* and *evidence preservation* supply the third indispensable member of the triumvirate. Together, they furnish the contractor with a powerful tool which may be employed should it become necessary to assert a claim against a subcontractor, and to recover upon that claim so that the contractor may fulfill the legitimate objective of allowing liability to flow to the party at fault.

---

[20] See **chs. 13, 14.**

CHAPTER 9

THE CONTRACTOR'S INSURANCE COVERAGE UNDER ITS LIABILITY AND BUILDER'S RISK POLICIES

Barry L. Bunshoft and Richard L. Seabolt*

Barry L. Bunshoft is the managing partner of Hancock, Rothert & Bunshoft, San Francisco, California. Mr. Bunshoft received his undergraduate degree from the University of Massachusetts and his law degree from Harvard University. Mr. Bunshoft has extensive experience in litigation and arbitration involving international and domestic construction and insurance coverage disputes. He has represented contractors in proceedings under the International Chamber of Commerce Rules and in hearings before the United States—Iran International Claims Tribunal at The Hague.

Richard L. Seabolt is a partner in the San Francisco law firm of Hancock, Rothert & Bunshoft. He received his undergraduate degree from the University of Michigan and his law degree from the University of California, Hastings College of the Law. Mr. Seabolt devotes a major portion of his general litigation practice to construction litigation and insurance coverage litigation. Mr. Seabolt is a member of the American Bar Association Forum Committee on the Construction Industry and an affiliate member of The Associated General Contractors of America.

* The authors acknowledge the assistance of Ernest J. Beffel, Jr. and Scott Hammel, associates at Hancock, Rothert & Bunshoft, in preparing certain portions of this chapter.

§ 9.1 Introduction

The most productive time and energy a lawyer can expend to benefit a contractor client faced with a claim is to recognize that the claim may be covered by insurance, to identify all potentially applicable insurance policies, and to tender the case to the insurers. From the client's financial perspective, full insurance coverage is a result only slightly less favorable than a defense verdict.

A lawyer representing a contractor should review the client's insurance coverage as soon as a potentially insured claim is brought to the lawyer's attention. Defending a contractor without notifying all potentially involved insurers may jeopardize the client's insurance coverage.[1]

[1] Windt, Insurance Claims and Disputes: Representation of Insured and Insurers (Shepard's/McGraw-Hill 1982) includes a chapter that provides a thorough discussion of issues related to insurance

If there is some arguable basis for even partial coverage of the claims made against the client, most jurisdictions require the insurer to provide a defense as a separate obligation under a comprehensive general liability policy.[2] In some states, the existence of a coverage issue and a reservation of rights by the insurer may require the insurer to pay the attorney's fees of an independent defense attorney selected by the insured client.[3] Because of the high cost of defending most construction disputes, an insurer-provided defense can result in substantial savings for the client.[4]

While lawyers often are not in a position to advise their contractor clients on insurance matters until after a liability-producing event has occurred, they should encourage their clients to have a close working relationship with an insurance broker who is experienced with construction insurance issues. The broker can provide the contractor client with advice on its insurance needs, including compliance with contract insurance requirements.[5] The contractor client should supplement the insurance broker's advice by consulting a lawyer to review areas in which case law may affect the client's insurance needs. There is no substitute for preventative legal advice. As one author has stated:

> There is still, however, no substitute for familiarizing oneself with the details of a particular policy. This is most crucial in preventative law practice so that the client can be assured of receiving the protection he needs prior to a loss rather than searching the policy afterwards in an attempt to find coverage.[6]

---

notice requirements, including a breakdown of those jurisdictions in which (1) late notice without prejudice is a sufficient policy defense; (2) late notice is a defense if the insured can prove no prejudice to the insurer; and (3) late notice is not a defense unless the insurer can prove prejudice. *Id.* 8–11.

[2] St. Paul Fire & Marine Ins. Co. v. Sears, Roebuck & Co., 603 F.2d 780, 786 (9th Cir. 1979), following Gray v. Zurich Ins. Co., 65 Cal. 2d 263, 419 P.2d 168 (1966); *see also* Windt, Insurance Claims and Disputes: Representation of Insured and Insurers 104–107 (Shepard's/McGraw-Hill 1982).

In Central Mut. Ins. Co. v. Del Mar Beach Club Owners Ass'n, 123 Cal. App. 3d 916, 930, 176 Cal. Rptr. 895 (1981), the court indicated that the insurers of various construction entities could not use a declaratory relief action to terminate their duty to defend, since the court found that the applicability of the various exclusions had to await the outcome of the underlying litigation against the insureds.

[3] Prashker v. United States Guar. Co., 1 N.Y.2d 584, 136 N.E.2d 871, 154 N.Y.S.2d 910 (1956); Executive Aviation, Inc. v. National Ins. Underwriters, 16 Cal. App. 3d 799, 809, 94 Cal. Rptr. 347 (1971); San Diego Navy Fed. Credit Union v. Cumis Ins. Soc'y, Inc., 162 Cal. App. 3d 358, 208 Cal. Rptr. 494 (1984); *see generally* Windt, Insurance Claims and Disputes: Representation of Insured and Insurers 139–142 (Shepard's/McGraw-Hill 1982); Berg, *Losing Control of the Defense—The Insured's Right to Select His Own Counsel*, For The Defense (July 1984); Berg, *After Cumis: Regaining Control of the Defense*, For The Defense (Aug. 1985).

[4] A lawyer may wish to consult with the client contractor's insurance broker before taking an extremely aggressive position on coverage or defense obligations. A poor claims history could have an adverse impact on a client's future premiums or even insurability in a tight insurance market.

[5] *See generally* Derk, Insurance for Contractors 3–5 (Fred S. James & Co. 1981).

[6] Henderson, *Insurance Protection for Products Liability and Completed Operations—What Every Lawyer Should Know,* 50 Neb. L. Rev. 415, 446 (1971).

The moment of truth is when a claim arises. If the contractor client failed to purchase an appropriate insurance policy or the insurer denies coverage in a gray situation, it is essential that the contractor's lawyer be familiar with the most significant coverage issues facing construction contractors.

§ 9.2 Contra Proferentem and the Reasonable Expectations Doctrine

All insurance coverage disputes result from the conflict between the conscious intent of the insurance underwriter to leave certain risks uninsured and the reasonable (or unreasonable) expectation of the insured that the particular claim is covered. Historically, the insurance industry has had mixed success in drafting standard policy language which conforms to this intent to leave certain risks uninsured. Courts have applied the traditional contra proferentem rule that a contract should be interpreted against the drafter to find coverage whenever ambiguous policy language permits that result.[7] Recently, however, some courts have refused to apply this rule when the insured is sophisticated and has equal bargaining power with the insurer.[8]

Where the courts find ambiguity, they may apply the *reasonable expectations doctrine* to protect the objectively reasonable expectations of the group of policyholders in the position of the particular insured. In theory, this doctrine prevents insurers from enforcing provisions which are unclear, inconspicuous, misleading, or unfair.[9] In practice, however, some courts, have pushed this doctrine so far as to abandon the fundamental contract principle that both parties are bound by the *unambiguous* terms of the insurance contract. These courts require the insurer to meet an almost impossible burden of proving that the insured was aware of and understood the effect of the particular provision or exclusion upon which the insurer relies.[10] What emerges is a sui generis body of insurance law which construes standard policy language to maximize the insured's recovery from the insurer in the particular case, at the expense of general certainty of interpretation.

The lawyer's ability to stretch a contractor's insurance coverage beyond the insurance underwriter's intent at the time the policy was drafted or issued depends on the particular jurisdiction's view of insurance policies. Some courts sympathetic

[7] *See generally* Windt, Insurance Claims and Disputes: Representation of Insured and Insurers 225–233 (Shepard's/McGraw-Hill 1982).

[8] Garcia v. Truck Ins. Exch., 35 Cal. 3d 426, 438, 682 P.2d 1100, 204 Cal. Rptr. 435 (1984); Ostrager & Ichel, *The Role of Bargaining Power Evidence in the Construction of the Business Insurance Policy: An Update,* 18 Forum 577–90 (Summer 1983).

[9] Windt, Insurance Claims and Disputes: Representation of Insured and Insurers 236 (Shepard's/McGraw-Hill 1982).

[10] Hionis v. Northern Mut. Ins. Co., 230 Pa. Super. 511, 327 A.2d 363, 365 (1974); *see generally* Windt, Insurance Claims and Disputes: Representation of Insured and Insurers 234 (Shepard's/McGraw-Hill 1982).

to the complex language of insurance policies have stated that "[i]t is widely recognized that language of insurance policies is often necessarily complicated and difficult for the layman to fully understand."[11] Other courts hostile to insurers have characterized some insurance policies as "an inexplicable riddle, a mere flood of darkness and profusion . . . printed in such small type, and in lines so long and crowded, that perusal of it was made physically difficult, painful and injurious."[12]

Such wide-ranging attitudes among various jurisdictions about insurance policies tend to promote forum shopping by lawyers familiar with these judicial attitudes. One recent California court, however, dismissed a coverage case brought by an East Coast insured on an East Coast claim against an East Coast insurer, stating that "California courts do not throw their doors wide open to forum shopping."[13]

The contra proferentem rule reflects the predisposition of various courts to find the language of insurance policies ambiguous. The reasonable expectations doctrine in practice can mean open season on insurers. The history of the development of the standard liability policy forms reflects the efforts of the insurance industry to tighten the policy language that courts have found ambiguous or unenforceable. To understand the current liability policy form, it is necessary to review the development of the form which resulted from this tug of the courts and pull of the insurance industry.

THE LIABILITY POLICY

§ 9.3 Introduction

The insurance industry first prepared a standard general liability policy in 1940, with revisions in 1943, 1955, 1966, and 1973.[14] The basic format of the *comprehensive general liability* (CGL) policy was set in the 1966 version. The 1973 revisions clarify and refine the 1966 policy language.[15]

---

[11] McGann v. Hobbs Lumber Co., 150 W. Va. 364, 145 S.E.2d 476, 481 (1965).

[12] Delancey v. Insurance Co., 52 N.H. 581, 587–88 (1873), quoted in Storms v. United States Fidelity & Guar. Co., 388 A.2d 578 (N.H. 1978).

[13] Appalachian Ins. Co. v. Superior Court (Union Carbide Corp.), 162 Cal. App. 3d 427, 208 Cal. Rptr. 627 (1984).

[14] Tinker, *Comprehensive General Liability Insurance—Perspective and Overview,* FIC Quarterly 221 (Spring 1975).

[15] The Manufacturers and Contractors coverage part of a general liability policy provides narrower coverage than a standard CGL policy in the areas of products hazard, completed operations, independent contractor liability at owned or rented premises of the insured, and structural alterations. General Liability Manual II.A.2 (The International Risk Management Institute 1983). In general, the focus here is on the standard CGL policy rather than the Manufacturers and Contractors coverage part. The completed operations hazard is discussed because of the substantial amount of litigation that surrounds that exclusion. See § **9.15**.

The Insurance Services Office (ISO) recently completed an entirely new form of liability policy scheduled for use after January 1, 1986, to be called the *Commercial General Liability* policy (1986 CGL policy). Although this chapter focuses primarily on the current CGL form, the 1986 CGL policy form is discussed where comparison is appropriate. Coverage issues under the pre-1986 form of CGL policy will continue to arise for some years. Courts no doubt will interpret the 1986 policy, as they did its predecessors, with consideration of the pre-1986 form of the CGL policy as historical background.

As the most difficult coverage issues faced by contractors under CGL policies involve property damage claims, not bodily injury claims, this chapter deals primarily with property damage issues. Most bodily injury claims either are covered by the contractor's CGL policy or are covered under other policies or endorsements (e.g., automobile, mobile equipment, aircraft, watercraft, workers' compensation, or completed operations). In contrast, many property damage claims against contractors are not covered by either the standard CGL policy or any other commonly available standard insurance policy.

The coverage provided by a CGL policy is very different from the coverage under a professional liability or errors and omissions policy, in that it is "not intended to indemnify the contractor . . . for direct damage resulting because the contractor furnished defective materials or workmanship."[16] A CGL policy thus provides much narrower coverage than a professional liability policy, since the CGL policy does not purport to provide coverage for all liability that can result from a contractor's errors and omissions.[17]

The ISO and its predecessors drafted the CGL policy with the intent of avoiding coverage for *business risks* which are within the effective control of the insured. Meeting contract requirements for quantity, quality, and timeliness of the work create business risks which can be reduced by competent business management. The CGL policy is not intended to serve as a warranty or performance bond.[18]

The significant negative risks of cost overruns, contractual penalties for failure to complete on time, and costs for correction of defective work are undertaken by the contractor for the positive expectation of a substantial profit. Insurance underwriters traditionally have been unwilling to insure these business risks.[19] If an

---

[16] Rafeiro v. American Employers' Ins. Co., 5 Cal. App. 3d 799, 808, 85 Cal. Rptr. 701 (1970).

[17] Many major construction contractors have insurance programs which provide professional liability coverage in addition to general liability coverage. Such professional liability insurance is particularly important to a contractor who performs design services or construction management services, since any liability for construction delays would not be covered by a CGL policy. Construction delay claims do not involve an *occurrence* or *property damage* as defined by those policies, and are excluded by the business risk exclusion. This chapter does not discuss the question of whether a contractor may be covered for damage to others caused by its delay in performance of the work under a policy that provides professional liability coverage.

[18] *See generally* Tinker, *Comprehensive General Liability Insurance—Perspective and Overview,* FIC Quarterly 217, 223–24 (Spring 1975).

[19] Sandridge, *Professional Liability Coverage for the CM,* in Liability in Construction Management 38–39 (ASCE 1983). The author also points out that a performance bond surety also does not insure against these risks, since the surety only acts as a guarantor to protect the owner and the surety retains the right to proceed against the contractor in the event of default.

insured could maximize its profits by performing defective work and passing the costs of correction onto the insurer, it would be a type of financial guaranty not contemplated by the relatively modest premium structure of the CGL policy. What the CGL policy does intend to cover are *fortuities*—unexpected and unintended occurrences.

§ 9.4 Insuring Agreement and Definitions

The insuring agreement of the pre-1986 comprehensive general liability policy obliges the insurer to indemnify and defend the insured. It states:

> The company will *pay on behalf of* the insured all sums which the insured shall become legally obligated to pay as damages because of
>
> Coverage A. bodily injury or
> Coverage B. *property damage*
>
> to which this insurance applies, caused by an *occurrence,* and the company shall have the right and duty to defend any suit against the insured seeking damages on account of such bodily injury or property damage. . . . (emphasis added).

The agreement to "pay on behalf of" distinguishes the liability policy from an agreement to idemnify against loss. The obligation is to pay third parties on behalf of the insured, not simply to reimburse the insured.[20]

§ 9.5 —Definition of Occurrence

Occurrence is defined in the definition portion of the comprehensive general liability (CGL) policy as follows:

> "Occurrence" means an accident, including continuous or repeated exposure to conditions, which results in bodily injury or property damage neither expected nor intended from the standpoint of the insured.

The "accident" and "neither expected nor intended" language emphasizes the essential nature of insurance: an insurer, in exchange for the payment of a premium, accepts the *risk* that the insured will become legally obligated to pay damage covered by the policy. If the element of risk has been eliminated so that the bodily injury or property damage is either expected or intended, there is no coverage.[21]

---

[20] This distinction is important, for instance, to a general contractor who has an insurance-covered claim against a bankrupt subcontractor or supplier. A liability insurer must "pay on behalf" of the insured all claims which are covered under the policy, whether or not the insured is bankrupt. 11 Couch on Insurance 2d §§ 44:250-44:252 (rev. ed. 1982).

[21] Bartholomew v. Appalachian Ins. Co., 655 F.2d 27, 29 (1st Cir. 1981) ("the carrier insures against a risk, not a certainty").

The trigger of coverage under the current CGL policy, which is written only on an occurrence basis, is "physical injury to or destruction of tangible property which occurs *during the policy period.*" The date of the negligent act causing the property damage generally is unimportant in determining whether there is coverage.[22] The date on which a claim was first made is also unimportant in determining whether there is coverage.[23]

The insurance industry adopted the term *occurrence* in the 1966 general liability policy revisions to replace the former term *accident.* These revisions were in recognition of a judicial trend eliminating the suddenness requirement that earlier courts have found in the term *accident.*[24] The 1973 revisions further clarified the term *occurrence* by adding the language "including continuous or repeated exposure to conditions."

The trigger of coverage where there is continuing damage is one of the most controversial issues in insurance law. Continuing damage issues can arise in many construction-related contexts: for example, roof leaks, foundation settlement cracks, pipeline leakage, spalling concrete, etc. Continuing damage may have the effect of broadening a contractor's coverage by permitting a stacking of policy limits so that, for example, two consecutive policies with $100,000 limits provide coverage totaling $200,000.[25] In *California Union Insurance Co. v. Landmark Insurance Co.*,[26] a California appellate court held that where there has been one occurrence involving continuous, progressive, and deteriorating damage, such as earth movement caused by leakage from a defectively constructed swimming pool, "the carrier in whose policy period the damage first becomes apparent remains on the risk until the damage is finally and totally complete." The court also held that the subsequent insurer was required to pay half of the damage that occurred during its policy period.[27]

When confronted with a continuing damage case, it is prudent for the contractor's lawyer to notify all insurers on the risk for all periods during which there

---

[22] Remmer v. Glens Falls Indem. Co., 140 Cal. App. 2d 84, 88, 295 P.2d 19 (1956); Home Mut. Fire Ins. Co. v. Hosfelt, 233 F. Supp. 360 (D. Conn. 1962). *See generally* Berg, *"Occurrence": An Elusive Description,* in 1982 DRI Monograph No. 2, "Occurrence" and Other Insurance Coverage Issues (Defense Research Institute 1982). *But see* Sylla v. United States Fidelity & Guar. Co., 54 Cal. App. 3d 895, 127 Cal. Rptr. 38 (1976).

[23] The new 1986 revised CGL form has an option to have coverage written on a *claims-made* basis. Coverage is triggered under a claims-made or discovery policy when a claim, whether formal or informal, is first made against the insured.

[24] *Compare* Arthur A. Johnson Corp. v. Indemnity Ins. of N. Am., 6 A.D.2d 97 (N.Y. App. Div. 1958), *aff'd*, 7 N.Y.2d 222, 164 N.E.2d 704 (1959) *with* Hauenstein v. St. Paul-Mercury Indem. Co., 242 Minn. 354, 65 N.W.2d 122 (1954) (cracked plaster held to fall within meaning of *accident*).

[25] Insurance Co. of N. Am. v. Forty-Eight Insulations, 633 F.2d 1212, 1226 n.28 (6th Cir. 1980).

[26] 145 Cal. App. 3d 462, 193 Cal. Rptr. 461 (1983).

[27] This later damage, even though resulting from the same damage-causing forces, was unexpected and unintended when the second insurer came on the risk, since the later damage had not yet manifested itself. 145 Cal. App. 3d at 477. The *Landmark* apportionment formula, in which earlier insurers have greater liability than later insurers, was rejected by the *Forty-Eight Insulations* court in favor of an apportionment based on the number of years each insurer was on the risk.

is evidence of actual damage. As a practical matter, as the facts frequently are unclear, it is wise to notify all insurers from the period the work was performed to the period the claim was made against the contractor. CGL insurers which insured a general contractor before the completion of construction probably would not accept even partial coverage, since damage occurring before completion usually would be excluded entirely by the care, custody, and control exclusion.[28] CGL insurers which came on the risk after the insured reasonably could expect that damage would continue (and, certainly, those which came on the risk after the insured received a demand or complaint) would not accept coverage, since the damage by that time had become expected from the standpoint of the insured.[29] Within these parameters, however, the determination of the periods in which continuing damage claims are covered seems limited only by the imagination of the lawyer and the mood of the judge.

The insurance industry's response to the battering it is taking in the courts over the trigger of coverage is to change from an occurrence to a *claims-made* basis. The proposed 1986 Insurance Services Office (ISO) revision provides, in the Insuring Agreement:

> . . . This insurance does not apply to ''bodily injury'' or ''property damage'' which occurs before the Retroactive Date, if any, shown in the Declarations. . . . This insurance applies to ''bodily injury'' and ''property damage'' only if a claim for damages because of the ''bodily injury'' or ''property damage'' is first made in writing against any insured during the policy period.

To hedge the bet, in case the claims-made form does not find commercial acceptance, the ISO also proposed an alternative 1986 revision written on an occurrence basis. If the market for liability insurance remains tight in 1986, however, most liability insurers probably will offer claims-made insurance only. The effect would be to resolve for the future the current controversy over trigger of coverage in favor of manifestation, discovery, and prompt notification of claims.

---

[28] See **§ 9.9**.

[29] Bartholomew v. Appalachian Ins. Co., 655 F.2d 27 (1st Cir. 1981) (insurer which came on the risk after the insured had been sued for providing a defective piece of equipment did not provide coverage); *see also* Keene Corp. v. Insurance Co. of N. Am., 667 F.2d 1034, 1044 (D.C. Cir. 1981). The California Supreme Court recently ordered that American Star Ins. Co. v. American Employers Ins. Co., 165 Cal. App. 3d 728, 210 Cal. Rptr. 836 (1985), a poorly reasoned California intermediate appellate court decision, not be published in the California Official Reports. The intermediate court had held that a particular liability insurer of a mechanical subcontractor was responsible for a pro rata portion of defense costs even though the insurer came on the risk after the cause of the defective leaky pipes was reasonably discoverable. The intermediate appellate court decision ignored the insurance contract requirement that the damage not be ''expected . . . from the standpoint of the insured.''

There is support for an argument that a contractor has full coverage for a continuing damage claim even if the contractor was not insured for some period of time during which the damage continued. Keene Corp. v. Insurance Co. of N. Am., 667 F.2d 1034, 1044, 1047-48 (D.C. Cir. 1981). *But see* Insurance Co. of N. Am. v. Forty-Eight Insulations, 633 F.2d 1212, 1224 (6th Cir. 1980) (insured bore pro rata share of liability risk for insured time period, although insurers were required to prorate entire cost of defense).

§ 9.6 —Definition of Property Damage

The term *property damage* is defined in the current comprehensive general liability (CGL) policy, in part, as "the *physical* injury to or destruction of *tangible* property." The significance of this definition of property damage is best understood by reviewing the evolution of the current definition in response to earlier court decisions.

CGL insurers historically have attempted to avoid coverage for pure warranty-type claims—defective products or work that do not cause other damage—by using narrowly defined terms in the insuring agreement and specific exclusions. Despite the effort of the insurance industry to avoid coverage for the cost of repair or replacement of defective products or work of the insured, the courts usually granted coverage where the defective product or work was a component integrated into the whole. In the leading case in this area, *Hauenstein v. Saint Paul-Mercury Indemnity Co.*,[30] the Minnesota Supreme Court held that a plaster distributor was covered under its liability policy for a warranty claim made because the plaster shrunk and cracked after it was used in the construction of a hospital. The court found that the claim satisfied the policy requirement of "injury to or destruction of property" since the presence of the defective plaster diminished the value of the building. The *Hauenstein* decision was widely followed.[31]

The 1966 revisions to the standard liability policy added the requirement that there be "injury or destruction to *tangible* property," probably on the theory that the diminution in value caused to a structure by incorporation of a defective product was an intangible loss.[32] The addition of the word "tangible" seems to have had little effect. The Nevada Supreme Court found that property damage existed under the 1966 standard language when the insured cement company incurred shoring expenses to strengthen a structure weakened by the insured's defective concrete.[33] The court relied on *Hauenstein* and its progeny without discussing the effect of the changed definition of property damage.

The 1973 revisions again attempted to reverse the trend begun by *Hauenstein* by changing the definition of property damage to the current definition: "*physical* injury to or destruction of tangible property." This latest change has achieved the desired result in at least some states. In *Wyoming Sawmills v. Transportation Insurance Co.*,[34] the Supreme Court of Oregon distinguished the *Hauenstein* line of cases as decided before "property damage" was defined to require "*physical* injury to . . . tangible property." The court stated: "the inclusion of this word ['physical'] negates any possibility that the policy was intended to include

---

[30] 242 Minn. 354, 65 N.W.2d 122 (1954).

[31] *See, e.g.*, Geddes & Smith, Inc. v. St. Paul-Mercury Indem. Co., 51 Cal. 2d 558, 344 P.2d 881 (1959); Dakota Block Co. v. Western Cas. & Sur. Co., 81 S.D. 213, 132 N.W.2d 826 (1965).

[32] Bowers, *General Liability Insurance Coverage for Defective Materials and Workmanship,* in B.C. Hart, Construction Litigation 181 (1979).

[33] United States Fidelity & Guar. v. Nevada Cement Co., 561 P.2d 1335 (Nev. 1977).

[34] 282 Or. 401, 578 P.2d 1253 (1978); *see also* St. Paul Fire & Marine Ins. Co. v. Coss, 80 Cal. App. 3d 888, 892–93, 145 Cal. Rptr. 836 (1978).

'consequential or intangible damage,' such as depreciation in value, within the term 'property damage.' ''[35] Recently, the Minnesota Supreme Court, the very court that decided *Havenstein*, rejected application of the *Havenstein* rationale to post-1973 policies and held that ''diminution in value'' did not satisfy the post-1973 definition of *property damage*.[36]

Other cases construing post-1973 policies, such as the California *Economy Lumber* case [37] have continued to follow *Hauenstein* and its progeny so that there is coverage for subcontractors who perform defective component work or for suppliers of defective components if the component work or the components reduce the value of the entire structure.

The 1986 revision to the CGL policy makes no substantive change in the definition of property damage. While the 1973 revisions and the *Wyoming Sawmills* case would seem to end the diminution in value trick, the *Economy Lumber* case shows that this issue may survive the 1986 revision.

§ 9.7 Exclusions

The standard comprehensive general liability (CGL) policy contains exclusions which narrow the scope of the insuring agreement. As the saying goes, ''What the insuring agreement giveth, the exclusions taketh away.'' There are several purposes for the exclusions:

1. To avoid duplication of coverage typically provided in other liability insurance policies (e.g., automobile, mobile equipment, watercraft, pollution, war risk, liquor liability, employer's liability, and worker's compensation)

2. To identify specific risks which the contractor can buy back or modify for an additional premium (e.g., contractually assumed liability; care, custody, and control; damage to the work performed on behalf of the insured; product recall; and explosion, collapse, or underground hazard)

3. To eliminate risks inherent in the contractor's business, which the contractor cannot buy back or modify for an additional premium (e.g., delay in or lack or performance of the contract, and failure of products or work performed to meet the level warranted).

Endorsements are available to provide coverage for many of the exclusions identified in item 2. The most important endorsement for a contractor is the broad

[35] Wyoming Sawmills v. Transportation Ins. Co., 282 Or. 401, 578 P.2d 1253, 1256 (1978).

[36] Federated Mut. Ins. Co. v. Concrete Units, 363 N.W.2d 751 (Minn. 1985).

[37] Economy Lumber Co. v. Insurance Co. of N. Am., 157 Cal. App. 3d 641, 203 Cal. Rptr. 135 (1984) (court quoted but did not discuss post-1973 property damage definition; court limited its discussion to whether there was an ''occurrence'' and whether various exclusions were applicable).

form property damage endorsement, which narrows the care, custody, and control and the damage to work exclusions. The version of this endorsement that includes completed operations is highly recommended for all contractors.[38]

This chapter does not deal with exclusions designed to avoid duplication of coverage available in other liability policies or with other limitations on coverage, such as the controversial issue of whether public policy prohibits coverage of punitive damages.[39] It does focus on the exclusions most troublesome for contractors.

§ 9.8 –Contractually Assumed Liability Exclusion

Exclusion (a), the *contractually assumed liability* exclusion, of the current comprehensive general liability (CGL) policy states:

> This insurance does not apply:
>
> (a) to liability assumed by the insured under any contract or agreement except an incidental contract; but this exclusion does not apply to a warranty of fitness or quality of the named insured's products or a warranty that work performed by or on behalf of the named insured will be done in a workmanlike manner . . . .

The purpose of the contractual liability exclusion is to confine coverage to the insured's tort liability. The CGL insurer, unlike a surety, does not underwrite the insured's contractual undertakings.[40] The exception for the defined term *incidental contract* rarely will effect coverage for the contractor.

The classic example of an excluded contractually assumed liability is an indemnity agreement that requires the insured to indemnify another for losses not caused by the insured's own negligence. The rationale is that the insurer has accepted a premium in exchange for indemnifying the insured from the consequences of the insured's *own* negligence. The premium, which is set at the beginning of the

---

[38] The Broad Form Property Damage (BFPD) Liability Coverage (Including Competed Operations) endorsement replaces and narrows exclusion (k), the care, custody, and control exclusion, and exclusion (o), the damage to work exclusion. During the course of construction, BFPD coverage generally narrows the application of these exclusions to the particular part of any property upon which operations are being performed by the insured or its subcontractors at the time of the property damage. After completion of operations, BFPD coverage generally narrows the application of exclusions (k) and (o) so that damage to work arising out of a subcontractor's work would be covered. *See* II General Liability Manual IV.B.18-.20 (The International Risk Management Institute 1983).

[39] This issue of coverage for punitive damages is discussed in Annot., *Insurance Coverage—Punitive Damages*, 17 A.L.R.4th 11 (1982); and Ghiardi & Kucher, *Insurance against Punitive Damages*, in 1982 DRI Monograph No. 2, ''Occurrence'' and Other Insurance Coverage Issues (Defense Research Institute 1982). A chart which provides a summary of the treatment of this issue by each state is contained in the ''Policy Terms'' section of Warren, McVeigh & Griffin, The Umbrella Book (Griffin Communications Inc. 1984).

[40] Couch on Insurance 2d § 44A:36 (1981); Haugan v. Home Indem. Ins. Co., 86 S.D. 406, 197 N.W.2d 18 (1972).

policy, is not intended to cover risks later undertaken by the insured through indemnity agreements.

Tort liability arising from a contractual relationship is not excluded by the contractual liability exclusion. For example, a claim for negligent breach of contract does not fall within the exclusion. The exclusion applies only when the liability is based *solely* upon a contract.[41]

Some indemnity and other contract provisions only obligate the contractor to be responsible for the contractor's own negligence. Such provisions do not add any obligation to the contractor's general tort obligation to perform work with reasonable care. If a contractor negligently causes damage which also qualifies as a breach of such a provision, the contractually assumed liability exclusion will not preclude coverage.

In *American Casualty Co. v. Timmons,*[42] the contractor expressly agreed in its contract to be responsible for all damages that resulted from his fault or negligence. The court found that the contractually assumed liability exclusion did not exclude coverage since the liability was imposed by law and since the contractor "would have been liable for negligence even in the absence of contractual liability." The court commented that the very purpose of the insurance policy was to protect against liability for such negligence.[43] The test is whether that particular contract enlarged the obligations that would be implied by law from the insured's contract.[44]

As the exception to the exclusion shows (i.e., "but this exclusion does not apply to a warranty of fitness or quality of the named insured's products or a warranty that work performed by or on behalf of the named insured will be done in a workmanlike manner"), the contractually assumed liability exclusion does not apply where the insured would be "legally obligated to pay" even absent any express contractual assumption of liability. It clarifies the fact that implied warranties or express warranties that simply reiterate warranties usually implied by law are not contractually assumed liabilities. While liability for damage to the insured's work or products is excluded under other exclusions, the exception to the contractually assumed liability exclusion means that coverage for damage to *other property* resulting from a breach of the described warranties is not barred by this exclusion.[45]

The insurance industry anticipated and attempted to avoid court decisions that would use the exception to the contractually assumed liability exclusion to undercut

---

[41] Couch on Insurance 2d §§ 44A:37–44A:38 (1981); Colonial Refrigerated Transp., Inc. v. Worsham, 705 F.2d 821, 827 (6th Cir. 1983).

[42] 352 F.2d 563 (6th Cir. 1968).

[43] *Id.* 568; *see also* United States Fidelity & Guar. Co. v. Virginia Eng'g Co., 213 F.2d 109 (4th Cir. 1954); Lumbermen's Mut. Cas. Co. v. Town of Pound Ridge, 362 F.2d 430 (2d Cir. 1966).

[44] 63 A.L.R.2d 1122 at 1123 (1959).

[45] Oliver B. Cannon & Son, Inc. v. Fidelity & Cas. Co., 484 F. Supp. 1375, 1384 n. 35 (D. Del. 1980) (["Exclusion (a) . . . or parts of products."]

> Exclusion (a) is obviously intended to exclude liability assumed under a contract but is not meant to exclude the types of liability subsumed under the category 'products liability.' This is the reason for the proviso to (a). This proviso is thus much broader than the exclusion[s] . . . which apply only to exclude the costs of replacing the insured's own defective products or parts of products).

the effectiveness of other exclusions. The 1966 revisions attempted to forestall court decisions finding a grant of coverage in an exception to an exclusion that was worded "except with respect to." The post-1966 language "but this exclusion does not apply to" was intended to make clear that the exception affected only the exclusion in which it was contained.

One well-respected commentator predicted in 1969 that "this unambiguous language avoids any inference that another exclusion may not apply."[46] Despite his prediction, this exception to the contractually assumed liability exclusion has spawned extensive litigation and a major split of opinion among various jurisdictions. The issue is whether the language is so ambiguous that, although couched as an exception to an exclusion, it creates coverage for the insured's warranty-type obligations associated with the insured's defective products or faulty workmanship.

Other exclusions in the CGL policy exclude coverage for "property damage to the named insured's products arising out of such products or any part of such products" or "property damage to work performed by or on behalf of the named insured arising out of the work or any portion thereof, or out of materials, parts or equipment furnished in connection therewith." A number of jurisdictions, however, have held that if the liability of an insured is premised on the warranties described in the exception to the contractually assumed liability exclusion, these later exclusions are not applicable. This view has been taken by courts in Arizona, New Hampshire, North Dakota, Maine, Missouri, and Colorado.[47]

In contrast, the majority of courts have held that the exclusion simply does not apply to claims based on the warranties described in the exception to the exclusion. They hold that such claims are subject to the other exclusions of the policy. This view has been taken by courts in South Dakota, the Eighth Circuit (applying Missouri law), California, New Jersey, Alaska, Tennessee, Florida, Indiana, Montana, New York, Virginia, West Virginia, Michigan, Hawaii, Illinois, Massachusetts, and Washington.[48] These courts generally base their decisions on the

---

[46] 3 Long, The Law of Liability Insurance App-41; App. B, § 10 (1969).

[47] Federal Ins. Co. v. P.A.T. Homes, Inc., 113 Ariz. 135, 139, 547 P.2d 1050 (1976); Commercial Union Assurance Cos. v. Gollan, 118 N.H. 744, 748, 394 A.2d 839 (1978) (3-2 decision); Aid Ins. Servs., Inc. v. Geiger, 294 N.W.2d 411, 414 (N.D. 1980); Baybutt Constr. Corp. v. Commercial Union Ins. Co., 455 A.2d 914, 921–22 (Me. 1983) (4-3 decision); McRaven v. F-Stop Photo Labs, Inc., 660 S.W.2d 459, 462 (Mo. Ct. App. 1983); Worsham Constr. Co. v. Reliance Ins. Co., 687 P.2d 988 (Colo. Ct. App. 1984).

[48] Haugan v. Home Indem. Ins. Co., 86 S.D. 406, 413, 197 N.W.2d 18 (1972); Biebel Bros. v. United States Fidelity & Guar. Co., 522 F.2d 1207, 1212 (8th Cir. 1975); St. Paul Fire & Marine Ins. Co. v. Coss, 80 Cal. App. 3d 888, 896, 145 Cal. Rptr. 836 (1978); Weedo v. Stone-E-Brick, Inc., 81 N.J. 233, 247, 405 A.2d 788 (1979) (6-1 decision); United States Fire Ins. Co. v. Colver, 600 P.2d 1, 3-4 (Alaska 1979); Vernon Williams & Son Constr., Inc. v. Continental Ins. Co., 591 S.W.2d 760, 763–65 (Tenn. 1979); LaMarche v. Shelby Mut. Ins. Co., 390 So. 2d 325, 326–27 (Fla. 1980) (disapproving Fontainebleau Hotel Corp. v. United Filigree Corp., 298 So. 2d 455 (Fla. Dist. Ct. App. 1974)): Indiana Ins. Co. v. DeZutti, 408 N.E.2d 1275, 1278–79 (Ind. 1980); Stillwater Condominium Ass'n v. American Home Assurance Co., 508 F. Supp. 1075, 1079 (D. Mont. 1981), *aff'd*, 688 F.2d 848 (9th Cir. 1982), *cert. denied*, 460 U.S. 1038 (1983); Zandri Constr. Co. v. Fireman's Ins. Co., 81 A.D.2d 106, 109 (N.Y. App. Div.), *aff'd*,

principle that exclusion clauses subtract from coverage rather than grant it[49] and that claims coming within an exception to an exclusion remain subject to all other policy exclusions.[50]

Contractual liability insurance can be obtained by paying an additional premium. This coverage is still subject to many other standard exclusions.[51]

The 1986 Insurance Services Office Commercial General Liability Coverage Form substantially changes standard coverage for contractually assumed liability. The new provision states:

> This insurance does not apply to:
>
> b. "Bodily injury" or "property damage" for which the insured is obligated to pay damages by reason of the assumption of liability in a contract or agreement. This exclusion does not apply to liability for damages:
>
> (1) Assumed in a contract or agreement that is an "insured contract"; or
> (2) That the insured would have in the absence of the contract or agreement.

Subpart (1) attempts to eliminate the exception to the exclusion problem, as suggested in a footnote to a recent court decision,[52] by defining *contractual liability* in the definitions section and removing the exception from the text of the exclusion. The definition of *insured contract* includes contracts pertaining to the insured's business under which the insured assumes the tort liability of another. This change has the effect of narrowing the applicability of the contractually assumed liability exclusion. Business-related contracts assuming tort liability of another is a much broader exception to the exclusion than the warranty exception used in the current CGL policy.

Subpart (2) specifically states that the exclusion does not apply to liability that the insured would have in the absence of the contract. This is consistent with judicial decisions interpreting the current CGL policy.

---

54 N.Y.2d 999, 430 N.E.2d 922, 446 N.Y.S.2d 45 (1981); Nationwide Mut. Ins. Co. v. Wenger, 222 Va. 263, 267, 278 S.E.2d 874 (1981); Helfeldt v. Robinson, 290 S.E.2d 896, 901 (W. Va. 1981); Fresard v. Michigan Millers Mut. Ins. Co., 414 Mich. 686, 327 N.W.2d 286 (1982), *aff'g but overruling as to this issue* 97 Mich. App. 584, 590, 296 N.W.2d 112 (1982); Sturla, Inc. v. Fireman's Fund Ins. Co., 684 P.2d 960, 964 (Hawaii 1984); Qualls v. Country Mut. Ins. Co., 123 Ill. App. 3d 831, 834, 462 N.E.2d 1288 (1984); Bond Bros., Inc. v. Robinson, 471 N.E.2d 1332 (Mass. 1984); Harrison Plumbing & Heating, Inc. v. New Hampshire Ins. Group, 37 Wash. App. 621, 681 P.2d 875 (1984). It is ironic that the Florida intermediate appellate court decision which first held that the exception granted coverage and which subsequently became a direct or indirect basis for all other minority view decisions has been overruled by the Florida Supreme Court.

[49] Weedo v. Stone-E-Brick, Inc., 81 N.J. 233, 405 A.2d 788, 795 (1979).

[50] St. Paul Fire & Marine Ins. Co. v. Coss, 80 Cal. App. 3d 888, 896, 145 Cal. Rptr. 836 (1978).

[51] II General Liability Manual V.A.2–V.A.5 (The International Risk Management Institute 1983).

[52] Oliver B. Cannon & Son, Inc. v. Fidelity & Cas. Co., 484 F. Supp. 1375, 1384 n.35 (D. Del. 1980).

§ 9.9 —Care, Custody, or Control Exclusion

Exclusion (k), the *care, custody, or control* exclusion, of the current comprehensive general liability (CGL) policy states:

> This insurance does not apply: . . .
>
> (k) to property damage to
> (1) property owned or occupied by or rented to the insured
> (2) property used by the insured, or
> (3) property in the care, custody or control of the insured or as to which the insured is for any purpose exercising physical control;
>
> but parts (2) and (3) of this exclusion do not apply with respect to liability under a written sidetrack agreement and part (3) of this exclusion does not apply with respect to property damage (other than to elevators) arising out of the use of an elevator at premises owned by, rented to or controlled by the named insured . . . .

This exclusion substantially narrows the coverage under the CGL policy for occurrences during the course of construction. It typically eliminates coverage for damage to any property owned, occupied, or rented by the insured or under the care, custody, or control of the insured.[53] It avoids potential overlap between coverage under the standard CGL policy and the builder's risk policy.[54]

The courts enforce this exclusion broadly against general contractors, but very narrowly against subcontractors.[55] By contract, general contractors assume care, custody, and control of an entire project; subcontractors do not assume that degree of responsibility and are not afforded that degree of control. The difference between their roles is crucial to the interpretation of the exclusion.

While general contractors are not covered for damage to their project during the course of construction as a result of this exclusion,[56] the care, custody or control exclusion can be narrowed by a broad form property damage endorsement

---

[53] *See generally* Annot., *Scope of Clause Excluding from Contractor's or Similar Liability Policy Damage to Property in Care, Custody, or Control of Insured,* 8 A.L.R.4th 563 (1979); 44A Couch on Insurance 2d § 44A:13 *et seq.* (1981); 2 Long, The Law of Liability Insurance § 10.17 (1984).

[54] Hart, *The Comprehensive General Liability Insurer: Claims By and Against,* in B.C. Hart, Construction Contract Claims 394, 399 (American Bar Ass'n 1978); Estrin Constr. Co., Inc. v. Aetna Cas. & Sur. Co., 612 S.W.2d 413, 425–27 (Mo. Ct. App. 1981); *see* North Am. Iron & Steel Co., Inc. v. Isaacson Steel Erectors, Inc., 36 A.D.2d 770, 321 N.Y.S.2d 254, 256 (1971), *aff'd*, 30 N.Y.2d 640, 331 N.Y.S.2d 667 (1972) (mem. op.) (general contractor bought builder's risk policy). See also **§§ 9.16–9.23**.

[55] This discussion treats all contractors either as general contractors, who have general responsibility for and control over an entire project, or as subcontractors, who have limited responsibility for an aspect of a project. For coverage purposes, a prime contractor on a project with multiple primes would be analogous to a subcontractor.

[56] Estrin Constr. Co., Inc. v. Aetna Cas. & Sur. Co., 612 S.W.2d 413, 427 (Mo. Ct. App. 1981); Bor-Son Bldg. Corp. v. Employees Commercial Union Ins. Co., 323 N.W.2d 58 (Minn. 1982); McCord, Condron & McDonald, Inc. v. Twin City Fire Ins. Co., 607 S.W.2d 956 (Tex. Civ.

for an additional premium.[57] In *Estrin Construction Co. v. Aetna Insurance Co.*,[58] the Missouri Court of Appeals enforced the exclusion and absolved the CGL insurer of any duty to defend a third-party complaint related to collapse of a wall in heavy winds. The court reviewed the terms of the exclusion and the sophistication of the corporate officer who regularly purchased CGL and builder's risk policies for the general contractor. It observed that the contractor did not obtain the available broad form property damage endorsement. It enforced the care, custody, or control exclusion because the contract imposed a duty on the general contractor to supervise the entire job and to protect the property during construction. The contract also granted the corresponding control over the property to the contractor. Most of the courts which have considered the application of this exclusion to general contractors have reached the same conclusion, based upon similar reasoning.[59] One court permitted a general contractor to escape the effect of the care, custody, or control exclusion under the rationale that the general contractor "regains custody and control over the property within the meaning of the exclusion clause [only] after the termination of the work by the subcontractor."[60] Despite the passage of nearly 30 years since this rationale was articulated by a commentator, only one decision in Pennsylvania has repeated it.[61]

Subcontractors, in contrast, regularly avoid this exclusion under a number of theories.[62] Courts enforce it against subcontractors only when the damaged property was under the supervision of the subcontractor and it was a necessary element

---

App. 1980); Maryland Cas. Co. v. Farmers Alliance Mut. Ins. Co., 566 P.2d 168 (Okla. Ct. App. 1977); L.L. Jarrell Constr. Co., Inc. v. Columbia Cas. Co., 130 F. Supp. 436 (S.D. Ala. 1955); *but see* Dubay v. Trans-America Ins. Co., 75 A.D.2d 312, 429 N.Y.S.2d 449, 452-53 (1980) (crane and operator hired by general contractor, but not excluded).

[57] Hart, *The Comprehensive General Liability Insurer: Claims By and Against,* in B.C. Hart, Construction Constract Claims 394, 399 (American Bar Ass'n 1978); *see* Estrin Constr. Co., Inc. v. Aetna Cas. & Sur. Co., 612 S.W.2d 413, 425 (Mo. Ct. App. 1981). Another method of extending coverage, particularly to subcontractors, is to obtain a wrap-up policy endorsed to provide coverage "as if separate policies had been issued to each insured." Continental Cas. Co. v. Gilborne Bldg. Co., 391 Mass. 143, 461 N.E.2d 209 (1984).

[58] 612 S.W.2d 413 (Mo. Ct. App. 1981).

[59] Bor-Son Bldg. Corp. v. Employees Commercial Union Ins. Co., 323 N.W.2d 58, n.9 (Minn. 1982) (contractor's control determined by reference to contract documents); McCord, Condron & McDonald, Inc. v. Twin City Fire Ins. Co., 607 S.W.2d 956 (Tex. Civ. App. 1980) (general contractor in possession until contract completed); L.L. Jarrell Constr. Co., Inc. v. Columbia Cas. Co., 130 F. Supp. 436, 437 (S.D. Ala. 1955) (quoting provisions of the contract specifications); *see* Hartford Accident & Indem. Co. v. Shelby Mut. Ins. Co., 208 So. 2d 465, 467–68 (Fla. Dist. Ct. App. 1968) (interpreting general conditions and specifications of the contract).

[60] Annot., *Scope of Clause Excluding from Contractor's or Similar Liability Policy Damage to Property in Care, Custody, or Control of Insured,* 62 A.L.R.2d 1242, 1254 (1958).

[61] Slate Constr. Co. v. Bituminous Cas. Corp., 228 Pa. Super. 1, 323 A.2d 141 (1974); *but see* Standard Venetian Blind Co. v. American Empire Ins. Co., 469 A.2d 563 (Pa. 1983) (overruling *Hionis*, which had required proof that the insured actually understood particular exclusions); *see also* Dubay v. Trans-America Ins. Co., 75 A.D.2d 312, 429 N.Y.S.2d 449, 452-53 (1980) (general contractor never gained exclusive control over crane and operator which it hired).

[62] *See generally* Elcar Mobil Home, Inc. v. D.K. Baxter, Inc., 66 N.J. Super. 478, 169 A.2d 509 (1961); 44A Couch on Insurance 2d §§ 44A:13, at 17 (1981).

of the subcontractor's work.[63] In one case, a New Jersey court applied the exclusion to a boiler explosion, where the insured was responsible for completely dismantling and refurbishing the boiler.[64] In another, a Connecticut court barred a tile installer from coverage for damage arising from improper use of acid to remove grout from the face of tiles.[65]

Subcontractors have successfully narrowed the application of this exclusion with a variety of arguments: injury to items which are "merely incidental" to the subcontractor's work is not excluded;[66] damage arising out of the subcontractor's access to its work area may not be excluded;[67] diminution in the overall value of realty which results from the subcontractor's work on a portion of the realty may not be excluded;[68] injury to items over which the subcontractor shares control with the general contractor is not excluded in some states;[69] injury to items to which the subcontractor is given "temporary or limited access" is not excluded in some states;[70] and injury to realty which is not in the overall control of the subcontractor

---

[63] *See, e.g.*, Herbison v. Employers Ins. Co. of Ala., 593 S.W.2d 923 (Tenn. Ct. App. 1979) (painter has care of sprinkler heads in ceiling which was painted); *see* Hartford Accident & Indem. Co. v. Shelby Mut. Ins. Co., 208 So. 2d 465, 469 (Fla. Dist. Ct. App. 1968) (subcontractor had joint control under contract).

[64] Condenser Serv. & Eng'g Co., Inc. v. American Mut. Liab. Ins. Co., 58 N.J. Super. 179, 155 A.2d 789 (1959); *see* H.E. Wiese, Inc. v. Western Stress, Inc., 407 So. 2d 464, 466 (La. Ct. App. 1981) (stress relief contractor had control of boiler tube sheet).

[65] Ceramic Tiles of Fairfield, Inc. v. Aetna Cas. & Sur. Co., 39 Conn. Supp. 477, 466 A.2d 348 (1983).

[66] Thomas W. Hooley & Sons v. Zurich Gen. Accident & Liab. Ins. Co., 235 La. 289, 103 So. 2d 449, 450–51 (1958) (ornamental urn at base of statue); Meiser v. Aetna Cas. & Sur. Co., 8 Wis. 2d 233, 98 N.W.2d 919, 923 (1959) (damage to windows by plastering subcontractor during clean-up was merely incidental); Royal Indem. Co. v. T.B. Smith, 121 Ga. App. 272, 173 S.E.2d 738, 739–40 (1970) (damage to boom in tank due to repositioning was merely incidental to sandblasting).

[67] McCreary Roofing Co., Inc. v. Northern Ins. Co. of N.Y. 275 A.2d 388 (Pa. Super. 1971) (roofing installation).

[68] Petrol Indus., Inc. v. Gearhart-Owen Indus., Inc., 424 So. 2d 1059, 1064 (La. Ct. App. 1982) (oil well).

[69] Home Indem. Co. v. Leo L. Davis, Inc., 79 Cal. App. 3d 863, 872, 145 Cal. Rptr. 158 (1978) (crane operator could not see object being lifted); McCreary Roofing Co., Inc. v. Northern Ins. Co. of N.Y., 275 A.2d 388, 389 (Pa. Super. Ct. 1971) (concurring opinion); General Mut. Ins. Co. v. Wright, 161 N.Y.S.2d 974, 976 (Sup. Ct. 1957) (steel erector did not have custody or control of steel supplied by the general contractor); *see* Bor-Son Bldg. Corp. v. Employees Commercial Union Ins. Co., 323 N.W.2d 58, n.9 (Minn. 1982) (mentioning, but not applying, exclusive care and control test); Dubay v. Trans-America Ins. Co., 75 A.D.2d 312, 429 N.Y.S.2d 449, 453 (1980) (general contractor did not have exclusive control); Applegreen v. Milbank Mut. Ins. Co., 268 N.W.2d 114, 118 (N.D. 1978) (duty to defend based on allegations in complaint).

[70] Krichner v. Hartford Accident & Indem. Co., 440 S.W.2d 751, 756–57 (Mo. Ct. App. 1969) (swimming pool repair contractor did not have access to pool during night when groundwater pressure raised bottom of pool); Bigelow-Liptak Corp. v. Continental Ins. Co., 417 F. Supp. 1276, 1286–87 (E.D. Mich. 1976) (boiler lining repair contractor); Boston Ins. Co. v. Grable, 352 F.2d 368 (5th Cir. 1965) (Georgia law) (subcontractor's workers were the only occupants of the house when a fire started, but did not have control); *see* Baldwin v. Auto-Owners Ins.

usually is not excluded.[71] These cases have produced substantial inconsistencies among courts of different states and even among courts of a single state at different times. As a result, situations which are practically indistinguishable have produced diametrically opposed results.[72]

The Insurance Services Office 1986 revision changes the language of the exclusion substantially. Exclusion (j) would exclude property damage to:

> (1) Property you own, rent or occupy; . . .
> (4) Personal property in your care, custody or control;
> (5) That particular part of real property on which you or any contractors or subcontractors working directly or indirectly on your behalf are performing operations, if the ''property damage'' arises out of these operations. . . .

This change specifically limits the care, custody, and control exclusion to personal property.

The purpose of the change generally is to incorporate the coverage of the current broad form property damage endorsement into the coverage of the standard policy. The new exclusion applies only to the particular work on which either the contractor or its subcontractor was performing operations.

Co., 5 Ill. App. 3d 300, 282 N.E.2d 204, 206 (1972) (tenant did not have custody of building during construction of improvements by landlord).

[71] Bigelow-Liptak Corp. v. Continental Ins. Co., 417 F. Supp. 1276, 1286–87 (E.D. Mich. 1976) (boiler lining repair contractor); Klapper v. Hanover Ins. Co., 39 Misc. 2d 215, 240 N.Y.S.2d 284 (1963) (window washer); Leiter Elec. Co. v. Bituminous Cas. Corp., 99 Ill. App. 2d 386, 241 N.E.2d 325 (1968) (welding contractor who attached cables linking pipe sections).

[72] 2 Long, The Law of Liability Insurance § 10.17, at 10-41 to 10-43 (1984); *see* Bigelow-Liptak Corp. v. Continental Ins. Co., 417 F. Supp. 1276, 1288 (E.D. Mich. 1976) (genuine split of authority); Estrin Constr. Co., Inc. v. Aetna Cas. & Sur. Co., 612 S.W.2d 413, 417–18 (Mo. Ct. App. 1981) (harmonizing local cases). *Compare* General Mut. Ins. Co. v. Wright, 161 N.Y.S.2d 974 (Sup. Ct. 1957) (steel erection contractor did not have control over work) *with* North Am. Iron & Steel Co., Inc. v. Isaacson Steel Erectors, Inc., 36 A.D.2d 770, 321 N.Y.S.2d 254, 256 (1971), *aff'd*, 30 N.Y.2d 640, 331 N.Y.S.2d 667 (1972) (mem. op.) (steel erection contractor had control of beams which collapsed). *Compare* McLouth Steel Corp. v Mesta Mach. Co., 214 F.2d 608 (3d Cir. 1954) (Illinois law) (machine installation contractor did not have control over machine it was installing) *with* International Derrick & Equip. Co. v. Buxbaum, 240 F.2d 536, 538–39 (3d Cir. 1956) (Pennsylvania law) (criticizing *McLouth*; antenna mast erector had control over the mast). *Compare* Glen Falls Ins. Co. v. Fields, 181 So. 2d 187, 189 (Fla. Dist. Ct. App. 1965) (crane operator did not have control over silo it was moving) *with* Phoenix of Hartford v. Holloway Corp., 268 So.2d 195, 199 (Fla. Dist. Ct. App. 1972) (crane operator had physical control over reactor head it was moving). *Compare* Hendrix Elec. Co., Inc. v. Casualty Reciprocal Exch., 297 So. 2d 470 (La. Ct. App. 1974) (electrical contractor had only temporary access to fuse box) *with* Visant Elec. Contractors v. Aetna Cas. & Sur. Co., 530 S.W.2d 76 (Tenn. 1975) (electrical contractor had control of switchboard in which it attempted to install circuit breakers). *Compare* Goldberry Operating Co., Inc. v. Cassity, Inc., 367 So. 2d 133, 135 (La. Ct. App. 1979) (oil well contractor hired to perforate lower section of well had control over entire well) *with* Petrol Indus., Inc. v. Gearhart-Owen Indus., Inc., 424 So. 2d 1059, 1964 (La. Ct. App. 1982) (oil well contractor hired to patch middle section of well did not have control over entire well; opinion did not cite *Goldberry*).

§ 9.10 —Premises Alienated Exclusion

Exclusion (1) of the standard comprehensive general liability (CGL) policy states:

> This insurance does not apply:
>
> (1) to property damage to premises alienated by the named insured arising out of such premises or any part thereof

The application of this exclusion in a construction setting generally is limited to owners and developers which have sold real property with structures and other improvements which cause property damage. Since the typical construction contractor does not own and later sell the real property on which the construction work is performed, this exclusion ordinarily does not affect the coverage provided to a contractor.[73] It would affect, however, contractor-developers which perform turnkey construction or build on speculation.

The Insurance Services Office 1986 revision joins premises alienated with care, custody, or control in exclusion (j). It would exclude property damage to:

> (2) Premises you sell, give away or abandon, if the "property damage" arises out of any part of these premises.

It would except from paragraph (2), however, contractor-developers.

§ 9.11 —Performance or Business Risk Exclusion

Exclusion (m), the *performance or business risk* exclusion, of the current standard comprehension general liability (CGL) policy states:

> This insurance does not apply: . . .
>
> (m) to loss of use of tangible property which has not been physically injured or destroyed resulting from
>
> a delay in or lack of performance by or on behalf of the named insured of any contract or agreement, or the failure of the named insured's products or work performed by or on behalf of the named insured to meet the level of performance, quality, fitness or durability warranted or represented by the named insured;
>
> but this exclusion does not apply to loss of use of other tangible property resulting from the sudden and accidental physical injury to or destruction of the named insured's products or work performed by or on behalf of the named insured

---

[73] The exclusion would affect a contractor which sold real property that was improved in a turnkey project. II General Liability Manual II.C.8 (The International Risk Management Institute 1983).

> after such products or work have been put to use by any person or organization other than an insured . . . .

Because the 1966 revisions broadened the definition of *property damage* to include loss of use of tangible property caused by an *occurrence* (which itself was redefined to eliminate the element of suddenness), it was necessary to retain a business risk exclusion to eliminate coverage for business risks involving loss of use, such as construction delay claims.[74]

The current exclusion only applies to limit coverage for claims for loss of use of tangible property not physically injured or destroyed; it does not limit coverage for third-party property damage claims arising from an occurrence. This narrowing of the scope of the exclusion occurred during the 1973 revisions, which also reduced the complexity of the exclusion and eliminated an ''active malfunctioning'' exception which had proved difficult to apply to particular fact situations.[75]

The exception to the exclusion is a narrow exception, since it reintroduces the requirement that the property damage be ''sudden and accidental.''[76] This eliminates loss of use coverage in fact situations which fall within the post-1966 expanded *occurrence* definition, which includes in the description of *accident* the concept of ''continuous and repeated exposure to conditions.''

The 1986 Insurance Services Office revision, while worded differently, makes no substantive changes in exclusion (m) as modified by the broad form endorsement.

§ 9.12 —Damage to Insured's Product or Work Exclusion

The standard comprehensive general liability (CGL) policy contains two similar exclusions, (n) and (o), which are intended to exclude coverage for damage to the insured's own products or damage to the insured's own work which arises out of such product or such work. The exclusions state:

> This insurance does not apply:
>
> (n) to property damage to the named insured's products arising out of such products or any part of such products;
>
> (o) to property damage to work performed by or on behalf of the named insured arising out of the work or any portion thereof, or out of materials, parts or equipment furnished in connection therewith . . . .

---

[74] Tinker, *Comprehensive General Liability Insurance—Perspective and Overview,* FIC Quarterly 217, 279 (Spring 1975).

[75] Hart, *The Comprehensive General Liability Insurer: Claims By and Against,* in B.C. Hart, Construction Contract Claims 404–05 (American Bar Ass'n 1978).

[76] *But see* Federated Mut. Ins. Co. v. Concrete Units, 363 N.W.2d 751 (Minn. 1985) (court found that supplier of defective concrete had coverage for lost use of slip-formed concrete grain elevator because lost use resulted from ''sudden and accidental physical injury'' to the concrete's reinforcing rods placed by the contractor and the slip-forms used by the contractor).

Contractors generally are concerned primarily with the insured's work exclusion, since contractors usually provide services, not products. However, the two exclusions are so similar that there is considerable crossover in court decisions which have interpreted these exclusions.

The work and product exclusions are the heart and soul of the intent of the insurance industry to avoid assuming the insured's business risks in the CGL policy. These exclusions are designed to exclude the insured's warranty-type obligations to repair or replace its own defective work or defective products.

The intent of these exclusions was illustrated in an early decision which stated:

> if a contractor builds a house and as a result of an improper mixture of the stucco, water is absorbed into the walls and the stucco cracks and falls off and a child is injured by the falling stucco, the injury to the child would not be excluded . . . but the replacement cost of the stucco would be excluded. Also, if the water absorbed into the walls should reach the interior walls and injure a valuable painting hanging there the damage to the painting would be recoverable under the policy while the damage to the walls would not.[77]

The pre-1966 standard work product exclusion language excluded injury to products or "work completed by or for the named insured, out of which the accident arises."[78] Because some courts held that this language only excluded coverage for the cost of repairing or replacing the component part that failed,[79] the 1966 revisions added the language "or any part of such products" and "or any portion thereof" to clarify the insurance industry's intent that even nondefective portions of the insured's work product were excluded if they were damaged by defective components of the insured's work product. Two recent cases concluded that courts which have interpreted the post-1966 policy language "have consistently held that this language unambiguously excludes coverage for the cost of repairing or replacing non-defective as well as defective components of the insured's work product."[80]

These exclusions have a much greater effect on a general contractor than on a subcontractor. The work exclusion excludes "property damage to work performed by *or on behalf of the* named insured." From the general contractor's standpoint, this excludes damage to all work performed at a construction project, since

---

[77] Liberty Bldg. Co. v. Royal Indem. Co., 177 Cal. App. 2d 583, 587, 2 Cal. Rptr. 329 (1960).

[78] Henderson, *Insurance Protection for Products Liability and Completed Operations—What Every Lawyer Should Know,* 50 Neb. L. Rev. 415, 441–442 (1971).

[79] *See, e.g.,* S.L. Rowland Constr. Co. v. St. Paul Fire & Marine Ins. Co., 72 Wash. 2d 682, 434 P.2d 725 (1967); Eichler Homes, Inc. v. Underwriters at Lloyd's, London, 238 Cal. App. 2d 532, 47 Cal. Rptr. 843 (1965); Blackfield v. Underwriters at Lloyd's, London, 245 Cal. App. 2d 271, 53 Cal. Rptr. 838 (1966); Owens Pac. Marine, Inc. v. Insurance Co. of N. Am., 12 Cal. App. 3d 661, 90 Cal. Rptr. 826 (1970); Pittsburgh Bridge & Iron Works v. Liberty Mut. Ins. Co., 444 F.2d 1286 (3d Cir. 1971).

[80] Todd Shipyards Corp. v. Turbine Serv., Inc., 674 F.2d 401, 422 (5th Cir. 1982); *see also* Western Employers Ins. Co. v. Arciero & Sons, 146 Cal. App. 3d 1027, 1031, 194 Cal. Rptr. 688 (1983).

subcontractors perform their work "on behalf of" the general contractor.[81] This is consistent with an intent to provide the general contractor with an incentive to exercise care in supervising the workmanship of its subcontractors and thus reduce the risk of covered losses involving personal injury and damage to property of others.[82]

In contrast, a subcontractor which performs defective work which damages the work of another subcontractor will have coverage for the damage to the other subcontractor's work, even though the insured subcontractor will not have coverage for damage to its own work. The general contractor can eliminate the "on behalf of" problem, however, by paying an additional premium for the broad form property damage endorsement.

The 1986 Insurance Services Office revision would broaden the standard coverage to provide substantially the same coverage as provided by the provisions of the current broad form property damage endorsement. It excludes property damage to:

> (5) That particular part of real property on which you and any contractors or subcontractors working directly or indirectly on your behalf are performing operations, if the "property damage" arises out of those operations; or
> (6) That particular part of any property that must be restored, repaired or replaced because "your work" was incorrectly performed on it.

The 1986 revision provides that paragraph (6) does not apply with respect to the products-completed operations hazard.

§ 9.13 —Sistership or Recall Exclusion

Exclusion (p), the *sistership or recall* exclusion of the current comprehensive general liability (CGL) policy states:

> This insurance does not apply: . . .
> (p) to damages claimed for the withdrawal, inspection, repair, replacement, or loss of use of the named insured's products or work completed by or for the named insured or of any property of which such products or work form a part, if such products, work or property are withdrawn from the market or from use because of any known or suspected defect or deficiency therein.

This exclusion bars coverage for damages associated with products or work withdrawn from the market due to a known or suspected deficiency. It is referred to as the *sistership* exclusion since it was first used in connection with the recall

---

[81] Of course, if the particular insurance policy contains the standard exception to the contractually assumed liability exclusion, and the policy is governed by the laws of a minority-view jurisdiction, a claim based on a warranty theory will be covered despite the work product exclusions.

[82] Western Employers Inc. Co. v. Arciero & Sons, 146 Cal. App. 3d 1027, 1031–32, 194 Cal. Rptr. 688 (1983).

of aircraft ("sisterships") similar to an aircraft which was known to have a defect.[83] This exclusion refers to costs of recalling other similar products or other similar work, not damage to the product itself or the work itself (which are excluded by other exclusions). Since this exclusion applies almost exclusively to product or material suppliers, a construction contractor's coverage is not significantly affected thereby.

The 1986 Insurance Services Office revision would make no substantive changes to this exclusion.

§ 9.14 —XCU Exclusion

The standard comprehensive general liability policy contains an exclusion that excludes property damage, but not bodily injury, claims for explosion, collapse, or underground property damage. Since the language defining these hazards is quite specific, the exclusion has created few legal issues. Contractors which engage in operations which involve these risks currently may buy back the applicable coverage by endorsement. This exclusion would disappear altogether in the proposed 1986 revision.

§ 9.15 —Completed Operations Hazard

While the current standard comprehensive general liability (CGL) policy does not contain a completed operations exclusion,[84] it does contain a definition of the *completed operations hazard,* as follows:

> "*completed operations hazard*" includes *bodily injury* and *property damage* arising out of operations or reliance upon a representation or warranty made at any time with respect thereto, but only if the *bodily injury* or *property damage* occurs after such operations have been completed or abandoned and occurs away from premises owned by or rented to the *named insured.* "Operations" include materials, parts or equipment furnished in connection therewith. Operations shall be deemed completed at the earliest of the following times:
>
> (1) when all operations to be performed by or on behalf of the *named insured* under the contract have been completed,
> (2) when all operations to be performed by or on behalf of the *named insured* at the site of the operations have been completed, or

[83] Yakima Cement Prods. Co. v. Great Am. Ins. Co., 590 P.2d 371, 374 (Wash. Ct. App. 1979), *rev'd on other grounds,* 608 P.2d 254 (Wash. 1980).

[84] Additional discussion of the completed operations exclusion can be found in Landis & Rahdert, *The Completed Operations Hazard,* 19 Forum 570–90 (Summer 1984); Annot., *Liability Insurance—"Completed Operations,"* 58 A.L.R.3d 12 (1974).

(3) when the portion of the work out of which the injury or damage arises has been put to its intended use by any person or organization other than another contractor or subcontractor engaged in performing operations for a principal as a part of the same project.

Operations which may require further service or maintenance work, or correction, repair or replacement because of any defect or deficiency, but which are otherwise complete, shall be deemed completed.

The *completed operations hazard* does not include *bodily injury* or *property damage* arising out of

(a) operations in connection with the transportation of property, unless the *bodily injury* or *property damage* arises out of a condition in or on a vehicle created by the loading or unloading thereof,

(b) the existence of tools, uninstalled equipment or abandoned or unused materials, or

(c) operations for which the classification stated in the policy or in the company's manual specifies "including completed operations" . . . .

The completed operations hazard is significant to contractors in several ways. First, the exclusion is contained in the Manufacturers and Contractors liability insurance form, which some insurers still require. It states: "This insurance does not apply: to bodily injury or property damage included within the completed operations hazard or the products hazard."

Second, the standard CGL policy includes completed operations in its list of hazards for rating purposes and in its list of hazards subject to aggregate limits of liability.

Third, the broad form property damage endorsement replaces exclusion (o) with an exclusion which states that with respect to the completed operations hazard, the insurance will not apply to "property damage to work performed by the named insured arising out of the work or any portion thereof or out of materials, parts or equipment furnished in connection therewith." This eliminates an exclusion for damage to work performed by a subcontractor on behalf of an insured contractor that occurred after the operation has been completed.

The completed operations hazard and the products hazard are somewhat parallel in the manner in which they affect the coverage provided by the standard CGL policy to providers of services and manufacturers, respectively. Since contractors generally are viewed as providing services, a contractor's concerns focus primarily on the completed operations hazard.[85]

Before 1966, the standard policy language treated the completed operations hazard as a subclassification of the products hazard exclusion. Because the pre-1966 policies combined the service and product aspects into one "Product-Completed Operations

---

[85] Friestad v. Travelers Indem. Co., 393 A.2d 1212 (Pa. Super. Ct. 1978), provides a well-reasoned decision that an installer of a product requires completed operation, but not products hazard, coverage.

Hazard,'' the majority of courts found that there was no coverage only if the work performed principally involved a product. As one court has pointed out, the effect of this majority view was that the completed operations hazard in pre-1966 policies was ''superfluous.''[86] Because the 1966 revisions clarified this ambiguity, courts in recent decisions have commented that earlier decisions involving pre-1966 language have ''little precedential value.''[87]

The post-1966 language separates the completed operations hazard from the products hazard and focuses on the principal issue surrounding the completed operations exclusion by defining when operations are deemed completed. The line between an uncompleted operation and a completed operation can be rather elusive. The Nebraska Supreme Court found that a welder's operations were completed and held the exclusion applicable, when a fire damaged equipment within 20 minutes after the welder left the scene.[88] In contrast, a North Carolina court affirmed a jury finding that an electrician had not completed his operations and that the exclusion was inapplicable when a fire was observed an hour and a half after the electrician's departure. The court disposed of the issue by commenting that the electrician's obligation ''was not 'completed' until the light fixtures were put in a *safe* working condition.''[89]

Operations are considered complete as soon as the particular portion of the work is put to its intended use by a third party, even though further service or maintenance work is required. Even this language, however, continues to lead to difficulties. The Colorado Supreme Court held that a contractor did not have coverage for injuries to a two-year-old who was badly burned by gasoline after entering a basement with a defective door latch. It reasoned that the door on which the contractor had not yet repaired the latch mechanism was complete, since its intended use was to provide access, rather than to act as a barrier.[90] In contrast, the South Carolina Supreme Court held that a stairway was not an excluded completed operation since it had not been put to its intended use, even though it had been used temporarily by the occupants of a new house. The court reasoned that the intended purpose was to afford passage to and from the attic ''indefinitely.''[91]

---

[86] *Id.* 1214.

[87] Martinez v. Hawkeye-Security Ins. Co., 576 P.2d 1017, 1018 n.1 (Colo. 1978); *see also* Tiano v. Aetna Cas. & Sur. Co., 102 Mich. App. 177, 301 N.W.2d 476 (1980) (''cases decided before 1966 are of scant use''); Abco Tank & Mfg. Co. v. Federal Ins. Co., 550 S.W.2d 193, 194 n.194 (Mo. 1977). An excellent discussion of the history of this exclusion is found in Henderson, *Insurance Protection for Products Liability and Completed Operations—What Every Lawyer Should Know,* 50 Neb. L. Rev. 415–446 (1971). Unfortunately, as the author notes in his conclusion, since no cases involving post-1966 language had been decided at the time the article was written, the principal value of the article is to provide historical perspective.

Pre-1966 interpretations have precedential value, of course, in continuing damage cases construing pre-1966 policies.

[88] Lavalleur v. State Auto. & Cas., 302 N.W.2d 362 (Nev. 1981).

[89] Woodard v. North Carolina Farm Bureau Mut. Ins. Co., 261 S.E.2d 43 (N.C. Ct. App. 1979).

[90] Martinez v. Hawkeye-Security Ins. Co., 576 P.2d 1017, 1020 (Colo. 1978).

[91] W.N. Leslie, Inc. v. Travelers Ins. Co., 215 S.E.2d 448, 451 (S.C. 1975).

Although historically the completed operations exclusions had a dramatic effect on the ''comprehensive'' coverage afforded to a contractor under a CGL policy, courts generally seemed willing to enforce the exclusion.[92] Pragmatically, courts were influenced by the fact that completed operations coverage was available to those contractors who agreed to pay the increased premium.

Even though the current CGL policy provides completed operations coverage, it is important that the contractor understand that the current CGL policy is written on an occurrence basis. There are a number of cases in which insureds had coverage for the period in which they were performing their operations, but failed to renew their coverage after operations were completed.[93] As discussed in **§ 9.5,** the trigger of coverage for policies written on an occurrence basis is the resulting damage or injuries, rather than the earlier negligent acts.

The proposed Insurance Services Office 1986 CGL revision simplifies the definition of *products-completed operations,* but makes no substantive changes in it. Exclusion l would exclude: '' 'Property damage' to 'your work' arising out of it or any part of it and included in the 'products-completed operations hazard.' This exclusion does not apply if the damaged work or the work out of which the damage arises was performed on your behalf by a subcontractor.'' This exclusion, in substance, tracks the current broad form property damage endorsement.

THE BUILDER'S RISK POLICY

§ 9.16 Introduction

Since the comprehensive general liability policy does not provide coverage for damage to the contractor's own work, the contractor can obtain additional coverage in the form of a first party *builder's risk policy.*[94] A builder's risk policy covers loss of or damage to the work which the contractor has contracted to perform.

---

[92] *See, e.g.*, Casey v. Employers Nat'l Ins. Co., 539 S.W.2d 181 (Tex. Civ. App. 1976) (affected area completed even though building only 75% complete); Meier v. Hardware Mut. Cas. Co., 281 So. 2d 793 (La. Ct. App. 1973) (construction of gin pole completed before collapse); Security Ins. Co. of Hartford v. Kay Mill Supply, Inc., 211 N.W.2d 519 (Minn. 1973) (emergency use of unfinished silo considered completed since put to intended use); Prieto v. Continental Ins. Co., 358 So. 2d 851 (Fla. Dist. Ct. App. 1978) (building completed before collapse); Adkins v. La Terre Dev. Corp., 387 So. 2d 652 (La. Ct. App. 1980) (insured failed to fill hole 11 months before claimant's injury); Tiano v. Aetna Cas. & Sur. Co., 102 Mich. App. 177, 301 N.W.2d 476 (1980) (exploding heating system considered completed).

[93] Singsaas v. Diederich, 238 N.W.2d 878 (Minn. 1976); Williams v. Aetna Cas. & Sur. Co., 376 A.2d 562 (N.J. Super. Ct. App. Div. 1977); Storms v. United States Fidelity & Guar. Co., 388 A.2d 578 (N.H. 1978).

[94] Coverage under builder's risk policies also are discussed in the following articles: Braude & Patin, *''All Risk'' Insurance*, Construction Briefings 1-12 (Nov. 1981); Annot., *Builder's Coverage Under Builder's Risk Insurance Policy*, 97 A.L.R.3d 1270 (1980).

The name of the policy comes from the common law rule that the damage or destruction of work in progress is a risk borne by the builder.[95] The policy insures against this risk. For example, the policy may cover damage to a building under construction, materials, and tools and equipment on the project site.

Depending on the language of the policy, the amount the contractor may recover for loss or damage under the builder's risk policy is either the actual value of the property at the time of loss or damage or the cost of repair or reconstruction.[96]

§ 9.17 Period of Coverage

Coverage under a builder's risk policy may commence on a specified date, but in many cases coverage attaches upon the commencement of construction and extends throughout the period of construction. Under a policy providing coverage "during construction," there is no coverage for losses occurring prior to actual physical commencement of construction. In one leading case, the court held that there was no coverage for damage to an old building, the component parts of which were to be used in constructing a new building, when the damage to the old building occurred prior to commencement of construction of the new building.[97]

Coverage may terminate on a specified date or upon some other occurrence specified in the policy. Termination of coverage under many builder's risk policies is dependent on completion of construction. Although *completion* usually means whatever the parties intend the term to mean, pursuant to the provisions of the construction contract itself,[98] in cases where the policy has not set forth a definition of *completion of the building,* courts have construed that term in a variety of ways. Some courts have held that coverage terminates when a building is ready to be put to the *full* use for which it was intended.[99] Other courts have held that the duration of coverage must be determined with reference to the object to be accomplished by the completed structure.[1]

A building is likely to be found to have been completed, and coverage thus terminated, even though minor repair work remains to be performed. In *Scottish Union & National Insurance Co. v. Encampment Smelting Co.,*[2] a builder's risk policy provided coverage for the building and machinery of an oremill which was destroyed

---

[95] *See, e.g.,* Ingle v. Jones, 2 Wall. 1, 17 L. Ed. 2d 763 (1864); *see also* Annot., *Builder's Risk Insurance Policy,* 97 A.L.R.3d 1270, 1274 n.5 (1980).

[96] Teeples v. Tolson, 207 F. Supp. 212, 215–16 (D.C. Or. 1962); Essex House v. St. Paul Fire & Marine Ins. Co., 404 F. Supp. 970 (D.C. Ohio 1975).

[97] Newman v. National Fire Ins. Co., 152 Miss. 344, 188 So. 295, 297 (1928).

[98] Philadelphia Facilities Management Corp. v. St. Paul Fire & Marine Ins. Co., 379 F. Supp. 780, 784 (E.D. Pa. 1974).

[99] Cuthrell v. Milwaukee Mechanics Ins. Co., 234 N.C. 137, 66 S.E.2d 649, 651 (1951).

[1] Scottish Union & Nat'l Ins. Co. v. Encampment Smelting Co., 166 F.2d 231, 238 (8th Cir. 1908).

[2] *Id.* 238.

after the building had been completed. The court held that, although the loss occurred after the building had been completed, the insured was allowed a reasonable amount of time "to test the composite structure, adjust its parts to each other and ascertain whether it (would) perform its function," and therefore the insured was covered under the policy during this period.[3]

Some policies provide for termination of coverage on the date a building becomes "occupied" or when the building is occupied for a certain period of time. When termination of coverage is dependent on occupancy of the building, the building will not be deemed to be "occupied" until it is put to a "practical and substantial use for the purpose for which it was designed."[4] "Transient" or "trivial" use will not terminate coverage.[5]

§ 9.18 Limitations on Coverage–Fortuitous Losses

Coverage under a builder's risk policy is subject to some important limitations in the form of implied conditions and specified exclusions. One fundamental implied limitation on builder's risk coverage is that only *fortuitous* losses are covered.[6] A *fortuitous loss* generally is one that is dependent on chance, is beyond human control, and is not certain to occur.[7] Accordingly, loss or damage resulting from depreciation or ordinary wear and tear would not be covered, because such events are usually certain to occur.[8] Many courts have held that *negligence* is a fortuitous event and have allowed contractors to recover for damage and loss caused by their own negligence and negligence of their agents.[9]

§ 9.19 –External Causes

Some policies limit coverage to losses resulting from an *external* cause. Courts distinguish between external *cause* and external *damage*: an *external cause* is one which has an external origin; *external damage* refers to the condition of the exterior part of an object. Coverage would only attach to damage to a structure caused

---

[3] *See also* Hendrix v. New Amsterdam Cas. Co., 390 F.2d 299, 304 (10th Cir. 1968); Fireman's Fund Ins. Co. v. Millers' Mut. Ins. Ass'n, 451 F.2d 1140, 1141 (10th Cir. 1971).

[4] Reliance Ins. Co. v. Jones, 296 F.2d 71, 74 (10th Cir. 1961).

[5] Cuthrell v. Milwaukee Mechanics Ins. Co., 234 N.C. 137, 66 S.E.2d 649, 651 (1951).

[6] Texas E. Transmission Corp. v. Marine Office-Appelton & Cox Corp., 579 F.2d 561, 564 (10th Cir. 1978); Avis v. Hartford Fire Ins. Co., 283 N.C. 142, 195 S.E.2d 545, 548 (1973).

[7] Texas E. Transmission Corp. v. Marine Office-Appelton & Cox Corp., 579 F.2d 561, 564 (10th Cir. 1978).

[8] Avis v. Hartford Fire Ins. Co., 283 N.C. 142, 195 S.E.2d 545, 548 (1973).

[9] C.H. Leavell & Co. v. Fireman's Fund Ins. Co., 372 F.2d 784, 789 (9th Cir. 1967); General Am. Transp. Corp. v. Sun Ins. Office Ltd., 369 F.2d 906, 908 (6th Cir. 1966).

by a force of external origin, and not to losses caused by damage inherent in the exterior of the structure.[10] Under policies containing an external cause requirement, when the cause of a loss cannot for some reason be explained or traced (e.g., collapse of building), courts generally do not require the insured to prove that the loss resulted from an external cause, and are willing to find coverage.

§ 9.20 —Exclusions

Coverage also may be limited by specific exclusions. Unless otherwise specified, these exclusions apply to subcontractors not named as insureds.[11] In one noteworthy case where a subcontractor was named as an insured,[12] a loss occurred as a result of the subcontractor's faulty workmanship. Even though the general contractor was not at fault, the court applied a *workmanship and materials* exclusion which made no distinction as to the source of the defective workmanship, and denied coverage.

Some of the more typical exclusions preclude coverage for loss or damage caused by faulty workmanship and materials; design defects, errors, and omissions; latent defects; inherent vice; equipment breakdown; and earth movement and subsidence. Damage resulting from fire, acts of nature, and war may also be excluded.

Contractors have been quite successful in avoiding the workmanship and materials exclusion. Most courts have held that the exclusion is limited to errors in the structure itself and not to errors in the methods of construction. These courts reason that *workmanship* refers only to work which affects the quality of materials used in the construction project, and does not necessarily refer to the defective construction work itself.[13] Thus, a contractor may avoid the exclusion by showing that the damage or loss was caused by its own negligence.

Contractors have avoided the *defective design* exclusion by showing (1) that the design was used successfully in other projects; (2) that the structure was successfully rebuilt using the same design; and (3) that an omitted element alleged to have caused damage or loss was not one customarily required for the particular design.[14]

In construing *latent defect* exclusions, courts have developed a general rule that a defect is latent if (1) it cannot be discovered by any known and customary test,

---

[10] Dubuque Fire & Marine Ins. Co. v. Caylor, 249 F.2d 162, 164–65 (10th Cir. 1957); City of Barre v. New Hampshire Ins. Co., 136 Vt. 484, 396 A.2d 121, 122 (1978); Connie's Constr. Co. v. Continental Ins. Co., 227 N.W.2d 204 (Iowa 1975).

[11] Kraemer Bros. Inc. v. United States Fire Ins. Co., 89 Wis. 2d 555, 278 N.W.2d 857 (1979).

[12] St. Paul Fire & Marine Ins. Co. v. Murray Plumbing & Heating Corp., 65 Cal. App. 3d 66, 135 Cal. Rptr. 120 (1976).

[13] Equitable Fire & Marine Ins. Co. v. Allied Steel Constr. Co., 421 F.2d 512, 514 (10th Cir. 1965); City of Barre v. New Hampshire Ins. Co., 136 Vt. 484, 396 A.2d 121, 122 (1978).

[14] Dow Chem. Co. v. Royal Indem. Co., 635 F.2d 379 (5th Cir. 1981).

or (2) it could not have been discovered by a proper or reasonable inspection.[15] In many cases, contractors have avoided the latent defect exclusion by showing that a particular loss or damage was caused by the contractor's own negligence.

An *inherent vice* exclusion excludes coverage for damage or loss caused by phenomena which occur over time due to the nature of the property itself (e.g., the rotting of wood or the fading of paint). Thus, the exclusion precludes coverage for losses caused entirely from the internal decomposition of the property which brings about its own injury or destruction.[16] Once again, however, the exclusion can be avoided by showing that the damage or loss was caused by the contractor's own negligence.

The *equipment breakdown* exclusion excludes coverage for loss or damage caused by mechanical breakdown, which is defined as a failure in the working mechanism of the machinery, i.e., a functional defect in the moving parts of the equipment which causes the equipment to cease functioning or to function improperly.[17] This exclusion applies only where the loss results from a breakdown caused by the inherent malfunctioning of the equipment, including instances where the mechanical failure is caused by negligence in the assembly or repair of the equipment.[18]

An *earth movement* exclusion precludes coverage for loss or damage caused by such things as earthquakes, volcanic eruptions, and landslides. To come within the exclusion, the insurer must show that the earth movement involved forces operating within the earth itself. Unless otherwise specified in the policy, superficial effects of external forces, such as erosion by runoff rainwater, do not come within the exclusion.[19]

Closely related to the earth movement exclusion is the *subsidence* exclusion. This exclusion applies where damage or loss is caused by a gradual settling which is the primary cause of collapse or other damage.[20]

§ 9.21 Burden of Proof

The builder's risk policy has a significant advantage over other forms of insurance policies with respect to burden of proof requirements. In the majority of jurisdictions, under an all-risk builder's risk policy, the insurer must prove that the policy

[15] General Am. Transp. Corp. v. Sun Ins. Office Ltd., 369 F.2d 906, 908 (6th Cir. 1966); Essex House v. St. Paul Fire & Marine Ins. Co., 404 F. Supp. 970, 992 (D.C. Ohio 1975).

[16] Employers Cas. Co. v. Holm, 393 S.W.2d 363 (Tex. 1965); Equitable Fire & Marine Ins. Co. v. Allied Steel Constr. Co., 421 F.2d 512 (10th Cir. 1965).

[17] National Investors Fire & Cas. Co. v. Preddy, 248 Ark. 320, 451 S.W.2d 457 (1970).

[18] Little Judy Indus., Inc. v. Federal Ins. Co., 280 So. 2d 14 (Fla. 1973).

[19] Peach State Uniform Serv. Ins. v. American Ins. Co., 507 F.2d 996, 1000 (5th Cir. 1975).

[20] Dow Chem. Co. v. Royal Indem. Co., 635 F.2d 379, 388–89 (5th Cir. 1981).

does not cover a particular loss or damage.[21] Although the contractor usually must prove that a particular loss or damage was caused by a fortuitous event, once this is shown the insurer must prove that the loss is not covered.[22]

§ 9.22 Sue and Labor Clause

Many builder's risk policies also contain *sue and labor clauses* which impose upon the insured a duty to "sue and labor" to safeguard the insured's property from actual or imminent loss or to minimize the loss once it occurs by taking immediate protective measures. Under a sue and labor clause, the insurer reimburses the insured for the expenses incurred in suing and laboring with respect to *covered* damages.[23]

§ 9.23 Subrogation

Under builder's risk policies, an owner's insurer may not subrogate to the owner's rights and sue either the general contractor[24] or a subcontractor[25] for causing damage to insured property where they are co-insureds of the owner under the policy.[26] When a subcontractor is not named as a co-insured of the general contractor or the owner, courts have held that if the subcontractor is not specifically excluded from coverage, it still is a co-insured immune from subrogation claims by the insurer.[27] One line of authority has rejected this rule, holding that the immunity of a subcontractor not named as an insured extends only to subrogation claims by the insurer for damage to the subcontractor's property, and does not protect the subcontractor against liability to others caused by its own negligence.[28]

---

[21] Redna Marine Corp. v. Poland, 46 F.R.D. 81, 86 (D.C.N.Y. 1969).

[22] Strubble v. United Servs. Auto. Assocs., 35 Cal. App. 3d 498, 504, 110 Cal. Rptr. 828 (1973); Texas E. Transmission Corp. v. Marine Office-Appelton & Cox Corp., 579 F.2d 561, 564 (10th Cir. 1978).

[23] Southern Cal. Edison Co. v. Harbor Ins., 83 Cal. App. 3d 747, 757, 148 Cal. Rptr. 106 (1978).

[24] Glen Falls Ins. Co. v. Globe Indem. Co., 214 La. 467, 38 So. 2d 139 (1948).

[25] Transamerica Ins. Co. v. Gage Plumbing & Heating Co., 433 F.2d 1051, 1055 (10th Cir. 1970).

[26] Harvey's Wagon Wheel, Inc. v. MacSween, 96 Nev. 215, 606 P.2d 1095, 1097–98 (Nev. 1980); Frank Briscoe Co., Inc. v. Georgia Sprinkler Co., 713 F.2d 1500, 1504 (11th Cir. 1983).

[27] Louisiana Fire Ins. Co. v. Royal Indem. Co., 38 So. 2d 807 (La. 1949) (the so-called "Louisiana Rule").

[28] Turner Constr. v. John B. Kelly Co., 442 F. Supp. 551 (E.D. Pa. 1976).

§ 9.24 Conclusion

Although the world of insurance may appear at first glance to be mysterious, there is a logic to it. If the contractor, its insurance broker, and its lawyer take the time and effort to understand the rationale and buy the right insurance package, they can avoid most of the problems which have set insureds against insurers in recent years.

The history of the standard form comprehensive general liability (CGL) policy is like the history of Holland: where the insureds have breached the dike, the insurers have shored it up. For every court decision which damaged the insurance industry, there is now an exclusion in the policy.

This policy continues today, primarily in the sensitive areas of toxic torts and pollution liability spanning many years of alleged continuing damage. As the courts have stretched the definition of *occurrence* to fit the coverage demanded by insureds, the insurance industry has responded with a decision to offer primarily claims-made liability insurance in the future.

Despite the major changes in the proposed 1986 CGL form, it seems inevitable that the tension between insureds and their insurers will last as long as the creativity of plaintiffs' lawyers creates new areas of liability for contractors. As big exposures to liability grow, contractors will continue to test the elasticity of the language of their policies. As they succeed, the insurance industry will tighten its policies.

CHAPTER 10

CONSTRUCTION LITIGATION INVOLVING CONTRACTORS AND THEIR SURETIES

Harvey C. Koch and James J. Myers*

Harvey C. Koch is the senior partner of the New Orleans law firm of Harvey C. Koch and Associates. Mr. Koch is a graduate of Tulane University and its School of Law and is a member of the Louisiana Bar. Since entering private practice in 1965, Mr. Koch has devoted a major portion of his efforts to litigation arising from all aspects of the construction process, with representation ranging from that of sureties to contractors and subcontractors to architects and engineers. Mr. Koch has authored numerous articles, lectured extensively, and served as chairman and co-chairman of various national programs on construction law and, in particular, on the relationship between contractors and sureties. Mr. Koch is currently a vice chairman of the Fidelity and Surety Committee of the International Association of Insurance Counsel, a member of the ABA Fidelity and Surety Committee and the Committee on International Construction Contracts of the International Bar Association, and has served as chairman of both the Engineers Selection Board and the Architects Selection Board of the State of Louisiana.

James J. Myers heads Gadsby & Hannah's construction law section at the firm's offices in Boston, New York City, and Washington, D.C. Mr. Myers is a graduate of the University of Pennsylvania and Boston University School of Law where he later served as a Lecturer of Law. He was admitted to the Massachusetts Bar in 1961, later became a member of the District of Columbia Bar, and has actively practiced law, concentrating on the construction industry, since that time. Mr. Myers currently serves as the Chairman-Elect of the Public Contract Law Section of the American Arbitration Association and is Chairman of the International Division of the American Bar Association's Forum Committee on the Construction Industry. Mr. Myers serves as a member of the Board of Advisors to the Bureau of National Affairs, Federal Contract Reports, and is a Fellow of the American College of Trial Lawyers.

* Sincere thanks are extended to R.D. Kuehnle, Of Counsel to Mr. Koch's firm, for his expert assistance in the preparation of this chapter.

SURETY BONDS GENERALLY

§ 10.1 Introduction

The modern way of obtaining security for the performance of public construction contracts is by surety bonds issued by a qualified surety corporation. Surety bonds are a form of assurance to the owner that the contractor will perform its work according to the contract and will pay for the labor and materials it uses in performing the contract. To a certain maximum amount, the surety guarantees to pay the cost of completing the work if the contractor defaults in its performance, and guarantees to pay for labor and materials for which the contractor does not pay.

Bonds are customarily required in private construction contracts, and are uniformly required by statute in federal and most state and municipal projects. The rationale for requiring payment bonds on public projects is that as a matter of public policy, liens cannot be filed upon public property. This is so because public policy opposes the sale of public property to satisfy liens, especially when such property is not available for sale under an execution following a judgment against a state or municipality.[1] Consequently, surety bonds have evolved as a substitute in public projects for the lien protection afforded to laborers and materialmen on private projects.

§ 10.2 Parties to Surety Bonds

The *surety,* an insurance company, is the party which obligates itself directly to an obligee by agreeing to perform the undertakings of the principal under its contract with the obligee. The bond is conditioned upon performance by the principal. In the event of failure or default of the principal, the obligee has a direct right against the surety to remedy the failure or default by performing or by responding in money damages.

The *obligee* is the party which receives the benefit, protection, and security of the surety bond. In essence, it receives the guarantee, by contract, of the surety

[1] *See, e.g.,* Friedman v. Hampden County, 204 Mass. 494, 90 N.E. 851 (1910).

that the obligations of the principal will be performed in accordance with the terms and provisions of the agreement between the obligee and the principal.

The *principal* is the party who enters into an agreement with the obligee and who has been required by the obligee to furnish bonds under which the contractual undertakings of the principal are guaranteed or bonded by a surety.

The payment bond involves the principal (the contractor), the obligee (the owner), the surety, and the beneficiaries of the bond, the *claimants.* Claimants are that class of persons defined in the bond who are effectively third-party beneficiaries of the bond and are entitled to enforce its terms.

SURETY BONDS ON PUBLIC AND PRIVATE PROJECTS

§ 10.3 Private (Common Law) Surety Bonds

There is no statutory requirement that private contracts be bonded either for performance or for payment of labor and materials. Parties to a private contract are at liberty to limit or condition their liability as long as they do not violate statutes or public policy. Considerable variety exists in form, type, and extent of protection afforded by private bonds. These private bonds are often referred to as *common law bonds.*

A common law bond is a contract for suretyship. The bond usually provides for joint and several liability of the principal and surety to pay up to the penalty sum of the bond in the event the obligee is damaged because of a default. The penalty sum is a fixed maximum sum for the surety's liability specified in the bond. Payment on the bond is subject to the condition that if the principal performs the obligation or undertaking for which the bond is given, the bond or obligation will be void. If the principal fails to perform, the bond is enforceable.

§ 10.4 Statutory Payment Bonds

Every state, as well as the federal government, requires that construction contractors for certain government contracts furnish bonds for the protection of laborers and materialmen. In general, these statutes follow the same suretyship principles as common law bonds, but require that certain procedural conditions precedent be met before a cause of action can be valid under their coverage. Since many of these statutes were modeled to a greater or lesser extent after the federal Miller Act,[2] these state statutes have become known as "little Miller Acts."

---

[2] 40 U.S.C. §§ 270a-270d.

A public contractor's bond given pursuant to such a statute includes the provisions of the statute by operation of law, whether or not such provisions are recited in the bond. This has come to be known as the *Christian doctrine,* established by the Court of Claims decision in *G.L. Christian & Associates v. United States.*[3] By virtue of this doctrine, the statutory provisions are, by law, read into the bond. Failure to include specific reference to the statute has been held to have no effect on the statutory validity of the bond. In *Martin Fireproofing Corp. v. Aetna Insurance Co.,*[4] the payment bond given to satisfy state law requirements contained no requirement for notice (which *was* required by the statute).[5] The court held that there could be no recovery based solely on the provisions in the bond and that the notice requirement in the statute governed.[6]

TYPES OF SURETY BONDS

§ 10.5 Bid Bond

Public agencies usually require bidders for public construction contracts to furnish surety bonds as a requisite to any final award of the contract. A contractor for public building or public work construction is ordinarily required to furnish bid security by a *bid bond.* The successful bidder is also usually required to furnish a *performance bond* for the performance of the contract and a *payment bond* for indemnification of the obligee against liability for the claims of laborers, materialmen, and subcontractors in the prosecution of the work.

A bid bond is a surety contract between the principal (bidding contractor), the obligee public authority and the surety. The bond provides that if the contractor is awarded the contract, it will, according to the contract provisions (which may include a specified time limit), enter into a formal contract with the obligee owner to perform the contract and also provide all required bonds for the proper performance of the contract.

§ 10.6 —Voiding of Bid Bond

While the contractor's compliance with the provisions upon award of the contract will render the bid bond void, failure of the contractor to enter into the contract or to provide the required surety bonds will result in the principal and surety

[3] 312 F.2d 418 (Ct. Cl.), *cert. denied,* 375 U.S. 954 (1963).

[4] 346 Mass. 498, 194 N.E.2d 101 (1963).

[5] Mass. Gen. Laws Ann. ch. 149, § 29 (West 1985).

[6] *See also* Metropolitan Pipe & Supply Co. v. D'Amore Constr. Co., 309 Mass. 380, 35 N.E.2d 211 (1941).

becoming liable to the obligee. The obligee (owner) is therefore secured against the default of the bidder for the contract, although the surety will be liable to indemnify the owner only up to the penal sum of the bond.[7]

§ 10.7 —Excuses

Refusal of the bidder to enter into the contract may be based on one of two grounds: change in condition, or mistake in the bid. Where mistake results from an honest error, the contractor is usually excused from its bid. Where the contractor is so excused, the bid bond surety is also relieved. For example, in *M.J. McGough Co. v. Jane Lamb Memorial Hospital,*[8] a contractor and its surety were found entitled to have the bid rescinded (and the surety therefore released from liability on the bond) when a clerical error caused the bid to be low by 10 percent of the total bid, and notification of the error, which was not caused by gross negligence, was received before the bid was accepted.

§ 10.8 —Form of Bid or Proposal Bond

Standard forms for bid or proposal bonds, such as American Institute of Architects Document A310 (1970 Edition) are available from most insurance agencies. Such bonds are conditioned upon the principal, if it is the successful bidder, executing a contract within a certain number of days after notice of award. Should the principal, after award, fail to execute the contract or fail to furnish a performance bond as required, then the condition of the obligation has not been met and the surety will be liable for the difference between the amount of the principal's bid and the amount of the bid selected, but not to exceed the penal sum of the bond.

§ 10.9 Completion Bond

Completion bonds are not frequently used today. Completion bonds were once commonly required of contractors who were developing real estate. The bond was conditioned upon the completion of the project within a specified time, free from liens.

---

[7] United States *ex rel.* Empire Plastics Corp. v. Western Cas. & Sur. Co., 429 F.2d 905 (10th Cir. 1970).

[8] 302 F. Supp. 482 (S.D. Iowa 1969).

§ 10.10 Payment Bond

A *payment bond* is a contract between a surety and the owner for the benefit of third-party laborers and material suppliers. The payment bond insures that persons furnishing labor and materials will be paid in full. This insulates the owner and insures payment to subcontractors and suppliers, regardless of what may be due to the prime contractor under its contract with the owner. Also, the payment bond is independent of the performance bond, under which additional monies may be due to the owner. A payment bond may be a standard form, such as American Institute of Architects Document A311 (1970 Edition), or (for public contracts) may be as prescribed by state statutes and regulations.

The payment bond and the performance bond are usually two separate bonds. Note, however, that a combination of payment and performance bond form in one instrument may involve a single penal sum. That single penal sum constitutes the limit of the surety's liability on both payment and performance obligations. Combination payment and performance bond forms, as well as straight payment bond forms, are available from most insurance agencies.

§ 10.11 Lien Bond

The contractor gives a *lien bond* to guarantee that it or its surety will indemnify the owner of the realty against losses resulting from the filing of liens against the owner's property. Lien bonds exist in two types: (1) to discharge a mechanic's lien; and (2) to indemnify against the filing of a mechanic's lien.

Bonds to discharge mechanics' liens are characterized by the undertaking of the prime contractor, with a surety, to assure that the obligee will not be encumbered by any mechanic's lien caused by debts to the contractor and/or the contractor's defaults in payment to third parties. An indemnity bond is given by the surety to provide that no lien will be filed. The bond substitutes the security of the bond for lien rights on the title to real estate. Forms for such a lien bond are available from most insurance agencies.

§ 10.12 Subcontractor's Bond

The *subcontractor's bond* is given by the subcontractor as principal, and its surety, to the general contractor; the general contractor and its surety are thus the obligees of the subcontractor's bond. Subcontractor's surety bonds do not run to other subcontractors or suppliers of the principal. Materialmen of the subcontractor ordinarily cannot recover on a subcontractor's bond.[9]

[9] Treasure State Indus., Inc. v. Welsh, 173 Mont. 403, 567 P.2d 947 (1977); McGrath v. American Sur. Co. of N.Y., 307 N.Y. 552, 122 N.E.2d 906 (1954).

Subcontractor's bonds are similar to the bonds furnished by the prime contractor (principal) to the owner (obligee). Payment and performance bonds furnished by the subcontractor guarantee that the subcontractor's surety will pay to the general contractor and its surety up to the amount of the penal sum of the bond if the subcontractor defaults on its contract obligations. Forms of subcontractor payment and performance bonds are available from most insurance agencies.

ROLE OF THE SURETY

§ 10.13 Generally

Payment and performance bonds in construction serve the purpose of assuring owners that contracts will be properly completed in accordance with the plans and specifications and that monies owed for labor and materials used in satisfaction of the contract will be paid. Issuance of contract performance and payment bonds by the surety signifies to the owner that the surety has prequalified—in effect guaranteed—the contractor's ability to complete the contract successfully. In other words, the surety has impliedly informed the owner that it has thoroughly investigated the contractor's ability to complete the contract successfully, and has determined that the contractor is capable of successful contract completion.

The product of the surety is the surety bond. Sureties, by assuming certain risks in their bonds, play an important and major role in assuring performance by their principals and in paying for losses resulting from nonperformance. One must recognize that surety companies are businesses whose objective is to be profitable. Consequently, unless it is found by a tribunal to be legally responsible for a loss, or concludes that it has substantial risk of liability, the surety is rarely a willing contributor to construction loss situations.

The most important principle to remember is that a bond is a contract. Just as contract provisions can differ and establish different rights between the parties, so too can bond provisions be written differently. Such provisions are enforceable in accordance with their terms.

§ 10.14 The Performance Bond

There are generally three parties to a performance surety bond. They are properly identified as the surety, the obligee, and the principal. The form of the performance bond may be that of American Institute of Architects Document A311 (1970 Edition), or as prescribed by state statutes and regulations. Such forms are available from most insurance agencies.

§ 10.15 Who Is Covered?

Under the performance bond, the obligations run solely between the prime contractor (principal), the owner (obligee), and the surety to insure proper compliance with and completion of the contract. Subcontractors or suppliers have no protection thereunder.[10]

Note that the relief available to the obligee of a performance bond is limited to a much narrower class of persons than under the payment bond. If a court finds, however, as it did in *Amelco Window Corp. v. Federal Insurance Co.*,[11] that the parties intended the performance bond to operate for the benefit of third parties, a subcontractor can maintain an action on the bond for payments due. Attempts have been made, mostly unsuccessfully, by subcontractors and suppliers to recover from performance bond sureties as third-party beneficiaries.[12] To avoid any unintended interpretations, performance bonds commonly include the provison that "no right of action shall accrue under this bond to or for the use of any person other than the obligee named herein."

§ 10.16 When the Principal Defaults

When a construction loss occurs on a project in which the surety is notified that the contractor defaulted or that the contractor was terminated, the performance bond surety must first conduct an investigation of the facts as to whether or not its principal is responsible in whole or in part for the alleged default. The surety must ascertain whether its principal, the contractor, is in default in order to determine whether it, as surety, has any obligations.

In situations other than those where the contractor has failed to perform because it ran out of money, the surety's principal vigorously (and most often, genuinely) asserts that it is not in default of its contract, that the owner's notice of termination was wrongful, and that the failure was not caused by the contractor's work but rather was due to design errors, owner-caused problems, and/or subcontractor errors. In large loss situations resulting in termination of its principal, a prudent surety will conduct an independent investigation to try to untangle the conflicting allegations of the parties with respect to responsibility for the failures that ultimately result in contract termination. Where a surety's exposure is relatively

[10] *See* United States *ex rel.* Warren v. Kimrey, 489 F.2d 339 (8th Cir. 1974), in which a subcontractor sought to recover against the contractor and its surety on a performance bond. The court of appeals ruled that a performance bond, furnished by the contractor's surety to the government, could not be converted into a payment bond for the benefit of labor and materialmen.

[11] 127 N.J. Super. 342, 317 A.2d 398 (1974).

[12] Continental Cas. Co. v. Hartford Accident & Indem. Co., 243 Cal. App. 2d 565, 52 Cal. Rptr. 533 (1966).

small, in an effort to economize, the surety will often make only a token investigation and accept the assertions of its principal.

A major consideration for the surety is the financial strength of its principal. If the principal is a financially strong contractor, with ample assets to satisfy a judgment if the principal is found to be in default, the surety will often take the principal's position as to lack of responsibility and permit the principal to defend the matter with its own resources, while the surety only monitors developments.

However, when the surety's principal is financially weak or questionable relative to the size of the risk, thereby exposing the surety's assets in the event a judgment is rendered against its principal, then the surety will normally assume a more active role. In this situation the surety's position tends to be more objective, and may not necessarily coincide with the position taken by its principal.

Sureties, like people, have different philosophies as to how to manage their obligations. Some sureties take a more active, positive, and dynamic role along with (and sometimes instead of) their principal when that principal is alleged to be in default of its contract. Other sureties are more passive, do little, and merely stay abreast of events as they occur.

In considering the financial strength of its principal, the surety often has the comfort of relying upon an indemnity agreement which it has obtained as a condition to furnishing the surety bond, whereby the principal (contractor) and the individual owners of the construction firm agree in a separate instrument with the surety that they will indemnify and hold the surety harmless should the surety sustain any costs, losses, or expenses, including attorneys' fees, in connection with the bond. While a corporate principal's resources may be weak or questionable, often the resources of the individual indemnitors are substantial, and the surety must then move to secure the indemnitor's assets so that they will be available to pay for such loss should a judgment be rendered against the principal. A surety will often take such action even before it is called upon to utilize its assets under the usual exoneration provisions of the indemnity agreement.

In order to gain insight into the legal principles which guide the surety's conduct, a basic understanding of the rights and obligations of the parties to the performance bond is necessary. When a performance bond surety is faced with the agonizing decision as to what course to pursue after default of its principal, the surety must consider the number and amount of subcontractors and suppliers who have not been paid and who may have valid claims against the surety's payment bond. There is one entire risk: part in the performance bond and part in the payment bond. The surety must assess both.

§ 10.17 Surety's Rights and Obligations upon Default

No obligation arises on the part of the surety until occurrence of the event upon which liability is conditioned. This is commonly referred to as the *default* of the principal. Under a performance bond, a default occurs when the contractor-principal fails to perform its contract obligations to the obligee under the construction contract. Under a payment bond, a default occurs when the principal fails to pay its debts to its subcontractors and suppliers.

When a default or claimed default of the contractor-principal occurs, then and only then does any obligation of the performance bond surety arise. The surety must be satisfied that its principal has in fact defaulted before it will undertake any obligations. If the owner claims that the contractor-principal is in default, and the contract requires positive action to terminate, the owner must place the contractor in default in accordance with the procedures set forth in the construction contract. For reasons that will be discussed later in this chapter, it is only in highly unusual situations that a surety will proceed to take over a completed job unless it is entirely convinced that its principal is in default of the contract, because of the serious negative consequences to the surety should it make the wrong decision.[13]

§ 10.18 —Events of Default

The contract between the obligee (owner) and the principal (contractor) usually defines those events which will constitute a default. Common *events of default* in construction contracts include: (1) substantial violation of the contract; (2) failure to proceed with or abandonment of the work; (3) persistent refusal to supply workmen to the job; (4) permitting liens to be filed against the real estate on which the work is being performed or against the contract funds; (5) insolvency proceedings; and (6) failure to pay for labor or material obligations incurred in the performance of the contract.

Typical of the termination for default provisions in construction contracts is the provision in American Institute of Architects Document A201, which provides as follows:

> 14.2 TERMINATION BY THE OWNER
>
> 14.2.1 If the Contractor is adjudged a bankrupt, or if he makes a general assignment for the benefit of his creditors, or if a receiver is appointed on account of his insolvency, or if he persistently or repeatedly refuses or fails, except in cases for which extension of time is provided, to supply enough properly skilled workmen or proper materials, or if he fails to make prompt payment to Subcontractors or for materials or labor, or persistently disregards laws, ordinances, rules, regulations or orders of any public authority having jurisdiction, or otherwise is guilty of a substantial violation of a provision of the Contract Documents, then the Owner, upon certification by the Architect that sufficient cause exists to justify such action, may, without prejudice to any right or remedy and after giving the Contractor and

[13] See § **10.22**.

his surety, if any, seven days' written notice, terminate the employment of the Contractor and take possession of the site and of all materials, equipment, tools, construction equipment and machinery thereon owned by the Contractor and may finish the Work by whatever method he may deem expedient. In such case the Contractor shall not be entitled to receive any further payment until the Work is finished.

14.2.2 If the unpaid balance of the Contract Sum exceeds the costs of finishing the Work, including compensation for the Architect's additional services made necessary thereby, such excess shall be paid to the Contractor. If such costs exceed the unpaid balance, the Contractor shall pay the difference to the Owner. The amount to be paid to the Contractor or to the Owner, as the case may be, shall be certified by the Architect, upon application, in the manner provided in Paragraph 9.4, and this obligation for payment shall survive the termination of the Contract.[14]

§ 10.19 —Notice and Investigation

The surety will normally receive notice of the principal's default from the obligee as a result of the obligee having declared the principal in default. This is usually because the obligee seeks the surety's participation to complete the contract. Unless the bond expressly provides otherwise, however, notice to the surety is not a condition of the surety's liability. Often a financially troubled principal voluntarily informs the surety of its inability to continue performance or to pay bills.

Upon receipt of notice, the surety normally investigates the status of the contract work and the financial condition of the principal. Assuming that the surety ascertains that the principal is in default, it must then determine whether any defenses exist which would discharge its obligations under the bond. If no such defenses exist, the surety must select one of a number of options in order to meet its bond obligation. Should the surety make the wrong decision, however, and take over its principal's contract when the principal is not in default, the surety is considered a volunteer and cannot recover its losses or expenses.[15]

§ 10.20 —Available Defenses

The surety may assert any defense that the principal could have asserted under the construction contract, and also any defense the surety may have as a result of the bond terms and conditions or the conduct of the obligee. The most common defense of the principal asserted by a surety defending against claimed liability is that the principal did not default or breach its contract with the obligee. Other less commonly available defenses of the principal which the surety might assert are:

---

[14] American Institute of Architects Document A201, General Conditions (1976 edition).

[15] Seaboard Sur. Co. v. Dale Constr. Co., 230 F.2d 625 (1st Cir. 1956).

1. Illegality of the construction contract
2. Impossibility of performance of the construction contract
3. Fraud or duress on the part of the obligee in obtaining the consent of the principal to the construction contract
4. Prior breach by the obligee of its obligations under the construction contract
5. Failure of consideration to support the construction contract
6. Material change in the principal's contract obligation without notice to or approval by the surety.

Defenses which arise as a result of the bond terms and conditions generally relate to the failure of the obligee to perform some condition precedent. Examples include failure of the obligee to bring an action within the time set forth in the bond and failure of the obligee to take any action required of it in the construction contract, such as to make payments to the principal in the manner and at the times specified in the construction contract.

§ 10.21 —Obligee's Conduct

Surety bonds normally contain *conditions precedent* to the enforceability of the bond. These conditions precedent must be satisfied in order for the obligee to recover on the bond.[16] Defenses of the surety which arise from the conduct of the obligee include: (1) fraud or misrepresentation of the obligee in inducing the surety to execute the bond; and (2) material alteration in the terms of the construction contract with respect to the scope of the work or the time and method of payment (which has not been waived by an express provision in the bond itself).

A condition precedent to the surety's obligation on the bond is that the obligee-owner perform all its obligations to the principal-contractor under the construction contract. The owner's primary obligation is to pay the contractor for its work in accordance with the contract. The surety normally has a defense if the obligee-owner pays the contractor too little, thereby causing the default, or pays the contractor too much, thereby reducing the contract balance available to the surety to complete and thus prejudicing the surety's rights. In this latter instance, however, the surety's defense is limited to the extent that the surety is monetarily prejudiced.

§ 10.22 Surety's Options upon Default of Performance Bond Principal

Assuming that the principal is in default and that no substantial defenses exist, the surety must select one of a number of ways to meet its performance bond obligation.

[16] Kanters v. Kotick, 102 Wash. 523, 173 P.2d 329 (1918).

First, the surety on the common form of performance bond may rest on its contract of suretyship by responding in damages to the consequences of the contractor's default, within the penal sum (limit) on the bond.[17] The surety may thus require the obligee to determine how to complete the project. Upon completion of the project, the obligee will be entitled to submit its claims to the surety for costs incurred in excess of the contract amount. The surety will thereafter be required to pay those claims as damages for the default, within the penal sum of its bond.

Second, the surety may enter into a *takeover agreement* with the obligee, whereby the surety agrees to complete the principal's contract and the obligee agrees to pay all retainages and contract proceeds, without reduction or setoff, to the surety. The surety then engages another contractor (one which is acceptable to the obligee) to complete the project. Sometimes the surety contracts with its own principal to complete, but conceptually it then becomes the surety's contract. In circumstances wherein the surety takes over and completes the contract, the surety loses its position as surety and assumes all the principal's contract obligations, thereby waiving the penal sum of the bond.[18] That is, when the surety takes over and assumes the obligation to complete, the surety must complete at any cost. By assuming the obligation to complete, unless the bond form expressly provides otherwise, the surety's obligation is not limited by the penal sum of the bond.

Third, the surety may select another contractor and persuade the obligee to enter into a contract with that new contractor for completion of the project. The surety will then pay (and its loss is generally limited in such agreements to) the difference between the remaining original contract funds and the contract price at which the substitute contractor has consented to complete the work.

Fourth, the surety may elect to finance the principal for the purpose of completing bonded work. This choice requires close supervision of the principal by the surety and joint control of all funds used for completion of the contract. Here again, the penal sum of the bond does not limit the surety's liability, and the surety's payments for or on behalf of its principal are not credited toward the bond penal sum.[19]

On rare occasions the surety will be able, by negotiating payment of a stipulated sum, to buy back its bond. While infrequently done, this may occur when work has not yet commenced or where other unusual circumstances exist.

§ 10.23 Damages Recoverable from Performance Bond Surety

Unless the bond itself places restrictions on the surety's liability, the surety's liability is coextensive with that of the principal.[20] Normally, however, the surety's liability

---

[17] Caron v. Andrews, 133 Cal. App. 2d 402, 284 P.2d 544 (1955).

[18] McWaters & Bartlett v. United States *ex rel.* Wilson, 272 F.2d 291 (10th Cir. 1959).

[19] Caron v. Andrews, 133 Cal. App. 2d 402, 284 P.2d 544 (1955).

[20] American Sur. Co. v. Wheeling Structural Steel Co., 114 F.2d 237 (4th Cir. 1940).

is limited to the penalty or penal sum of the bond.[21] Note, however, that where the surety undertakes to complete the contract following a default, the surety is generally deemed to have waived the penal sum limitation on its liability.[22]

If, however, the surety breaches its contract with the obligee to perform its obligations under the bond itself, then the penal sum of the bond still may not limit the liability of the surety. In *Continental Realty Corp. v. Andrew J. Crevolin Co.*,[23] wherein the bond required upon default that the surety make payments to the obligee or complete the work, and the surety did neither, the court found the surety liable for sums substantially in excess of the penal sum of the bond. These damages were assessed after the cost of completion had been determined. The surety totally failed in its contract obligation to the obligee, and thereby became liable to the obligee for a breach of contract separate from the obligation in the bond to indemnify the owner against the principal's default (recovery for which would be limited by the penal sum).

When the obligee completes the project, its recovery includes the amount in excess of the contract price actually and reasonably expended in completion. Special damages may be recoverable if they are within the contemplation of the parties to the contract.[24] Interest, while recoverable within the penal sum of the bond, is permitted to be recovered in excess of the penal sum of the bond only when there is an independent default of the surety.[25] Also, the surety will be liable for the obligee's attorney's fees only where the contract or a statute requires such payment.[26] Loss of profits and other consequential damages to the obligee resulting from refusal of the principal to complete its contract are not recoverable from the surety.[27]

§ 10.24 Right of Surety to Contract Funds

Controversy frequently arises between the completing surety of a defaulted contractor and the contractor's creditors (usually banks or lending institutions who have taken the contractor's receivables as security for a loan) as to who is entitled to the earned contract balance on the project. Courts have uniformly held that upon default of the principal, the surety is entitled to such contract balances and

[21] Chicago Bonding & Sur. Co. v. Augusta Savannah Navigation Co., 250 F. 616 (7th Cir.), *cert. denied,* 247 U.S. 509 (1918).

[22] McWaters & Bartlett v. United States *ex rel.* Wilson, 272 F.2d 291 (10th Cir. 1959).

[23] 380 F. Supp. 246 (D.W. Va. 1974).

[24] Miracle Mile Shopping Center v. National Union Indem. Co., 299 F.2d 780 (7th Cir. 1962).

[25] Cunningham v. Cunningham, 157 F.2d 859 (D.C. Cir. 1946).

[26] Maryland Cas. Co. v. Cunningham, 234 Ala. 80, 173 So. 2d 506 (1937).

[27] Terry v. United States Fidelity & Guar. Co., 196 Wash. 206, 82 P.2d 532 (1938).

retainages, provided the surety undertakes performance of the construction contract. This is so even though the surety has not perfected any security interest in the principal's receivables, because the surety's rights are based upon the theory that an assignee takes no greater rights than its assignor.[28]

Thus, courts generally hold that a completing surety is entitled to the contract balance after a default, even though other creditors, such as banks, may have perfected a security interest under the Uniform Commercial Code (U.C.C.).[29] The legal relationship of a surety in this situation is governed not by the U.C.C, but by equitable principles which are complementary to and not superseded by the U.C.C.[30] The surety's right to the unpaid balance of the principal's contract with the owner relates back to the time of the suretyship agreement, and is superior to any rights of the contractor or its assignees.[31] The surety does not, however, have priority to obtain, over other creditors of the principal, receivables from any project on which the principal is not in default and the surety is not completing the contract.

Note that progress payments or retainages already paid to the contractor or to other creditors can almost never be recovered by the surety. However, in *American Fidelity Fire Insurance Co. v. Pavia-Byrne Engineering Corp.*,[32] the court allowed recovery by a surety against a third-party engineer whose negligent issuance of payment certificates resulted in the owner's improper, but good faith, disbursement of retainage to the contractor.

§ 10.25 Subrogation

The prevailing rule of equitable subrogation is that a surety who pays the debt of its principal is thereby entitled to all the rights of the person it paid, to enforce the surety's right to be reimbursed. Thus, in one case, the rights of a surety who had paid the laborers and materialmen of its bankrupt principal, and thus stood in the shoes of those laborers and materialmen, were held to be superior to the rights of general creditors, which rights the receiver in bankruptcy represented.[33]

---

[28] Deer Park Bank v. Aetna Ins. Co., 493 S.W.2d 305 (Tex 1973); Travelers Indem. Co. v. Western Ga. Nat'l Bank, 387 F. Supp. 1090 (N.D. Ga. 1974).

[29] U.C.C. Art. 9.

[30] National Shawmut Bank of Boston v. New Amsterdam Cas. Co., 411 F.2d 843 (1st Cir. 1969); Canter v. Schlager, 358 Mass. 789, 267 N.E.2d 492 (1971).

[31] American Cas. Co. v. Shattuck, 228 F. Supp. 834 (W.D. Okla. 1964); Uptagraff v. United States, 315 F.2d 200 (4th Cir.), *cert. denied,* 375 U.S. 818 (1963).

[32] 393 So. 2d 830 (La. Ct. App.), *cert. denied,* 397 So. 2d 1362 (La. 1981).

[33] Segovia Dev. Corp. v. Constructora Maza, Inc., 628 F.2d 724 (1st Cir. 1980).

PARTIES AND CAUSES OF ACTION

§ 10.26 Introduction

This subchapter gives special treatment to those matters most frequently litigated between contractors and their sureties and related parties, with one exception. Litigation initiated by subcontractors and materialmen to enforce claims of such unpaid or otherwise aggrieved subcontractors and materialmen against contractors and their sureties is the subject of **Chapter 11.**

§ 10.27 Contractor versus Surety

The causes of action most frequently brought by contractors against their sureties are breach of contract, interference with performance of the contract, and breach of good faith. A contractor's charge of breach of contract on the part of the surety usually involves failure or refusal of the surety to pay the claims of subcontractors and materialmen against the contractor and/or to discharge liens against the project. As previously indicated, litigation of this nature is discussed in **Chapter 11,** with regard to actions brought against contractors by sureties. The elements of such surety claims, however, are being investigated as possible weapons for contractors in their surety-targeted attacks.

The latter two causes of action also used to be raised mostly as defenses, but are now being seen as offensive weapons.[34] An action for contract interference in construction cases usually arises after the surety notifies the owner that the contractor is in default and that no further contract funds should be released. This normally occurs where the contractor has not formally admitted default. The owner then usually holds the funds, depriving the contractor of the cash flow attributable to that specific contract. Subsequently, contractor sues surety for tortious interference with the prime contract.

This approach applies only to alleged *intentional* contract interference and is to be distinguished from *negligent* interference with a contract. Recovery for the latter is uniformly denied.[35] However, recovery for intentional interference with contractual relations is allowed in every jurisdiction except Louisiana.[36] The five elements of the tort are:

---

[34] Lambert v. Maryland Cas. Co., 418 So. 2d 553 (La. 1982). For a detailed analysis of this case and its significance, *see* Koch, *The Surety's General Indemnity Agreement: Current or Out of Date?,* in Proc. A.B.A. Fidelity & Surety L. Committee (New York 1984).

[35] Restatement (2d) of the Law of Torts § 766c, comment a (1977).

[36] P.P.G. Indus., Inc. v. Bean Dredging Corp., 447 So. 2d 1058, 1059 (La. 1984); Robins Dry Dock & Repair Co. v. Flint, 275 U.S. 303 (1927).

1. An enforceable contract (in Wyoming it applies also to an unenforceable contract)[37]
2. Knowledge of the contract's existence
3. Intentional inducement of a breach
4. Absence of justification
5. Resultant damages.[38]

However, one who in good faith asserts a legally protected, presently existing interest which may be impaired by performance of another contract is privileged to prevent that performance, and ill will toward the plaintiff will not defeat the privilege.[39]

Tortious interference with business expectancy or relations is a different cause of action.[40] A contractor's expectancy of good faith and fair dealing from its surety arises from the implied covenant of good faith and fair dealing inherent in any contract.[41] The breach of that implied covenant involves more than bad judgment, negligence, or insufficient zeal on the surety's part; it carries an implication of a dishonest purpose, conscious wrong, or breach of duty through motive of self-interest or ill will.[42]

Although the covenant of good faith and fair dealing is couched in contractual terms, it is largely a creation of tort law. Punitive damages will lie only if the insurer breaches the implied covenant and, in addition, acts with oppression, fraud, or malice.[43]

One commentator suggests that, although sureties have not yet been successfully attacked on the breach of good faith theory, by 1990 "all forms of insurance

---

[37] Kvenild v. Taylor, 594 P.2d 972 (Wyo. 1979).

[38] Phil Crowley Steel Corp. v. Sharon Steel Corp., 536 F. Supp. 429, 432 (E.D. Mo. 1982); Hurst v. Town of Shelburn, 422 N.E.2d 322 (Ind. Ct. App. 1981); Heheman v. E.W. Scripps Co., 661 F.2d 1115 (6th Cir. 1981).

[39] Bailey v. Meister Brau, Inc., 535 F.2d 982 (7th Cir. 1976); Brown v. Safeway Stores, Inc., 617 P.2d 704 (Wash. 1980); Alyeska Pipeline Serv. v. Aurora Air Serv., Inc., 604 P.2d 1090 (Alaska 1979); Wyatt v. Ruck Constr. Co., Inc., 571 P.2d 683 (Ariz. Ct. App. 1977).

[40] Sandolph v. P&L Hauling Contractors, Inc., 430 So. 2d 102 (La. Ct. App. 1983); Rickel v. Schwinn Bicycle Co., 192 Cal. Rptr. 732 (Ct. App. 1983); Friedman v. Edward L. Bakewell, Inc., 654 S.W.2d 367 (Mo. Ct. App. 1983); Leigh Furniture & Carpet Co. v. Isom, 657 P.2d 293 (Utah 1982); Joba Constr. Co. v. Burns & Roe, Inc., 329 N.W.2d 760 (Mich. Ct. App. 1982).

[41] Reinharcz v. Chrysler Motors Corp., 514 F. Supp. 1141 (C.D. Cal. 1981); Filner v. Shapiro, 633 F.2d 139 (2d Cir. 1980); Manchester Bank v. Connecticut Bank & Trust Co., 497 F. Supp. 1304 (D.N.H. 1980); Carlo C. Gelardi Corp. v. Miller Brewing Co., 502 F. Supp. 637 (D.N.J. 1980); Corbin on Contracts § 541, at p. 541 (1 vol. ed.1952); 17A C.J.S. *Contracts* § 328, n.40.5, citing cases from Maryland, Massachusetts, Oklahoma, Rhode Island, South Carolina, Utah, Vermont, and Washington. *See also* Miller, *Overview of the Problem of Bad Faith and Punitive Damages,* 15 Forum 194 (Fall 1979); Reinert, *The Duty of The Performing Surety to Its Bond Principal and The Indemnitors: Good Faith,* 15 Forum 1005 (Summer 1980).

[42] Hartford Accident & Indem. Co. v. Millis Roofing & Sheet Metal, Inc., 418 N.E.2d 645 (Mass. Ct. App. 1981).

[43] San Jose Prod. Credit Ass'n v. Old Republic Life Ins. Co., 723 F.2d 700, 703 (9th Cir. 1984).

relationships . . . [including] insured versus surety . . . will be vulnerable to some form of extra contract claim.''[44] Since that comment was written a surety has, in fact, been successfully sued under the breach of good faith theory. An Illinois appellate court found that the performance bond surety's actions did not ''demonstrate the good faith expected of a surety,'' and affirmed a judgment for more than double the face amount of the bond.[45]

§ 10.28 Contractor versus Owner

The possible actions a contractor might have against an owner from the bid process to the termination or completion of a project are myriad.[46] Two of the more common actions, however, *quantum meruit* and *damages for delay,* are exemplary of actions taken by contractors against owners.

The expression *quantum meruit* means literally ''as much as he deserves.'' It describes the extent of liability on a contract implied by law. This liability is predicated on the reasonable value of services performed.[47] The elements of this action are: (1) valuable services were rendered or materials furnished; (2) such services or materials were accepted, used, and enjoyed by defendant; and (3) circumstances exist charging defendant with notice that plaintiff expected to be paid for those services or materials.[48]

In the construction contract scenario, an action for quantum meruit usually arises after the formal contract has been terminated prior to completion[49] or when the required performance falls outside the terms of the original agreement and no supplemental agreement is evidenced.[50] A prerequisite for its application is that plaintiff have no remedy at law.[51]

The other common action arises from the theory that owner-caused delays which result in increased costs to the contractor or which prevent the contractor from completing the job on time can be held compensable by the owner.[52] As a direct reaction to a chain of large judgments in favor of contractors against owners, it

---

[44] Kornblum, *Extra-contract Actions Against Insurers: What's Ahead in the Eighties?,* 19 Forum 58, 60–61 (Fall 1983).

[45] Fisher v. Fidelity & Deposit Co., 466 N.E.2d 332, 339 (Ill. App. Ct. 1984).

[46] *See* Carrigan, Claims Against The Owner (Public and Private) (American Bar Ass'n Nat'l Inst. on Constr. Contract Claims, New York 1977).

[47] Nardi & Co., Inc. v. Allabastro, 314 N.E.2d 367, 370 (Ill. App. Ct. 1974).

[48] Montes v. Naismith & Trevino Constr. Co., 459 S.W.2d 691, 694 (Tex. Civ. App. 1970).

[49] Kinchen v. Gilworth, 454 So. 2d 1130 (La. Ct. App. 1984).

[50] Brookhaven Landscape & Grading Co. v. J.F. Barton Contracting Co., 676 F.2d 516 (11th Cir. 1982).

[51] Harrington v. Oliver, 459 So. 2d 111, 114 (La. Ct. App. 1984); Creely v. Leisure Living, Inc., 437 So. 2d 816 (La. 1983).

[52] *See* Carrigan, Claims Against The Owner (Public and Private) (American Bar Ass'n Nat'l Inst. on Constr. Contract Claims, New York 1977).

is now routine for owners to insert a *no damage for delay* clause in the construction contract.

This clause waives recovery for any damages arising from delays caused by the owner or another contractor.[53] Although the clause is a harsh measure which must be strictly construed, it is enforceable when it (1) is free from ambiguity, (2) reflects the intent of the parties, and (3) is reasonably interpreted.[54]

Exceptions not affected by the clause are delays not contemplated at the time the agreement is executed, unreasonably long delays, and delays resulting from fraud, bad faith, or active interference by owner.[55] Among events not considered exceptions are violation of scheduling provisions; inadequate storage space; labor problems; congestion at the job site; delay in arrival of equipment; weather delays; danger from falling objects; or rivalry among subcontractors.[56]

With these actions (and many others) available as recourse, many contractors, having dealt with public owners, tend to forget the basic requirement in dealing with a private owner, i.e., making certain that the party appearing to be the owner is the true owner of the property. A private owner in satisfactory financial condition at the start of a project may be insolvent before it is finished. Thus, the contractor's only security is its ability to place liens on the property—and liens are granted only to secure the obligations of the property owner.

Where the contractor is to work for a lessee, it is essential that the lease be scrutinized closely. Generally, an unpaid contractor can place a valid lien against the lessor's property if the lease requires such work to be done, if the lessor requests the work to be done, or if the lessor ratifies the work by its actions.[57] However, in Louisiana, for example, when a lessee builds at its expense and for its own use, even with the owner's consent, only the structure (not the land) can be encumbered with a lien.[58]

Private construction contracts may contain litigation traps not found in public works contracts. For example, one contract provided that the contractor had a "relationship of trust and confidence" with the owner. The court held that such a relationship required the contractor to inform the owner of incipient cost overruns, and failure to do so was a breach of contract.[59]

In another case, the owner was damaged through no fault of the contractor. The basic contract provided that the contractor would indemnify owner for all damages regardless of negligence. The American Institute of Architects indemnity provision, incorporated by reference, required indemnity only if the contractor was

[53] *See* Gontar, *The Enforceability of "No Damage for Delay" Clauses in Construction Contracts,* 28 Loy. L. Rev. 129 (1982). See also §§ **3.7, 3.10**.

[54] Annot., 74 A.L.R.3d 187, 207-13 (1950).

[55] *Id.* n.21, at 214-15.

[56] Gontar, *The Enforceability of "No Damage for Delay" Clauses in Construction Contracts,* 28 Loy. L. Rev. 129, 132–33 (1982).

[57] Christensen v. Idaho Land Developers, Inc., 660 P.2d 70 (Idaho 1983).

[58] La. Civ. Code Ann. arts. 464, 493 (West); P.H.A.C. Servs. v. Seaways Int'l, Inc., 403 So. 2d 1199 (La. 1981).

[59] Jones v. J.H. Hiser Constr., Inc., 484 A.2d 302 (Md. Ct. Spec. App. 1984).

negligent, but also provided that in any conflict between the two, the basic contract would prevail. The contractor escaped that trap by convincing the court that the difference between the two was not a conflict but merely a modification.[60]

§ 10.29 Contractor versus Miscellaneous Parties

A contractor possesses a cause of action against a subcontractor and its surety for breach of contract in the event of a default. It also has actions against a subcontractor on the same grounds that the owner has against the contractor, because the flow-down clause included in most subcontracts creates the same legal relationship between the subcontractor and the contractor as that existing between the contractor and the owner.[61] Most subcontracts also contain an indemnity agreement holding the prime contractor harmless from damages.[62]

Contractors may also have causes of action against architects, engineers, and others involved in the construction process for breach of their professional duties.[63] In a rather unusual case, a contractor also prevailed against its liability insurer for some $148,000 in attorneys' fees incurred by itself and its performance bond surety in defending a third-party demand in a negligence action.[64]

§ 10.30 Surety versus Contractor

Actions by a construction bond surety against its contractor/principal may be conveniently divided by reference to whether the surety's liability has become established and whether the surety has performed in accordance with its bond obligations. If the contractor has not defaulted, but has given the surety reason to be concerned about such matters as, for example, progress payments being applied to other, nonbonded projects, contract funds being diverted to the contractor's personal use, amounts due subcontractors and materialmen being understated, etc., an action in quia timet is appropriate.[65]

A legal concept developed under Roman law, *quia timet* means "because he fears," and it may be effectively used to protect the surety against a variety of reasonably anticipated future injuries. The action may be brought before the surety's liability has matured,[66] but the surety may not actually receive payments from the contract funds. Such an action is appropriate, for example, for requiring specific

[60] Sears Roebuck & Co. v. Shamrock Constr. Co., 441 So. 2d 379 (La. Ct. App. 1983).

[61] See § **8.4**.

[62] See § **8.8**.

[63] See § **10.33**.

[64] Merrick Constr. Co. v. Hartford Fire Ins. Co., 449 So. 2d 85 (La. Ct. App. 1984), *writ denied,* 450 So. 2d 958 (La. 1984).

[65] *See* Mann & Jennings, *Quia Timet: A Remedy for the Fearful Surety,* in II Surety Law (Proc. A.B.A. Fidelity & Surety Committee, Chicago 1984).

[66] 2 Story's Equity Jurisprudence §§ 825, 826 (1972).

performance of the contractor's duty under an indemnity agreement to deposit funds equal to any loss reserve established by the surety.[67]

A secondary benefit from the quia timet suit is that all the surety's actions against the contractor are taken in accordance with court orders. Thus, later claims by the contractor for domination, interference with contract, or breach of the implied covenant of good faith are prevented.

If the surety's liability has become absolute, it has available the remedy of *exoneration.* That remedy allows the surety to require the contractor to discharge the primary obligation before the surety has made any payments.[68] Where the contractor is financially unable to discharge its oligations, this equitable right to exoneration may be used to have any contract funds in the hands of the owner applied to the payment of materialmen or subcontractors.[69]

Occasionally, a project will be completed and accepted, with claims of materialmen and subcontractors still outstanding, when a contractor refuses to execute the close-out papers necessary to release the contract balances. In such a case, a surety can, under its exoneration right, have a receiver appointed to execute the necessary documents, thus releasing the funds for application by the surety to outstanding claims.[70]

After a surety has performed its bond obligations, the most frequent action brought against the defaulted contractor is for reimbursement of the surety's losses incurred in completing the job and/or paying the claims of subcontractors and materialmen. Such an action is usually brought under the provisions of a general indemnity agreement[71] executed by the contractor and individual indemnitors as part of the consideration for surety's issuance of the bonds. However, a performing surety is also entitled to indemnity from its contractor/principal by operation of law—its right of *equitable subrogation.*[72] The surety may bring this suit against one or more of the indemnitors without releasing those not joined as defendants, since the indemnity agreement specifies that their obligation is joint and several.[73] Usually, all indemnitors are joined.

Most suits against bond indemnitors are good candidates for summary judgment. Basically, the surety has only to attach copies of the indemnity agreements to its

---

[67] United Bonding Ins. Co. v. Stein, 273 F. Supp. 929 (D.C. Pa. 1967).

[68] Admiral Oriental Lines, Inc. v. United States, 86 F.2d 201, 204 (2d Cir. 1936).

[69] Glades County v. Detroit Fidelity & Sur. Co., 57 F.2d 449, 451-52 (5th Cir. 1932).

[70] Western Cas. & Sur. Co. v. Biggs, 217 F.2d 163 (7th Cir. 1954). *See generally* Downs, *A Surety's Basic Rights and Remedies,* 14 Def. L.J. 139 (1966); Auerbach, *Quia Timet: Relief by Summary Judgment on Indemnity Agreement,* 39 Ins. Couns. J. 536 (Oct. 1972); Babcock, *An Update—Suits for Exoneration and Other Special Relief,* 16 Forum 344 (Fall 1981).

[71] *See generally* Koch, *The Surety's General Indemnity Agreement: Current or Out of Date?,* in Proc. A.B.A. Fidelity & Surety L. Committee (New York 1984); McNamara & Mungall, Construction Contract Bond Indemnity Agreement Provisions (Defense Research Institute Monograph No. 3, 1975).

[72] Merrick Constr. Co. v. Hartford Fire Ins. Co., 449 So. 2d 85 (La. Ct. App. 1984), *writ denied,* 450 So. 2d 958 (La. 1984); United States v. Bellard, 674 F.2d 330, 335 (5th Cir. 1982); Hollabaugh, *Surety's Right to Equitable Subrogation: Overture for an Arrow,* 51 Ins. Couns. L.J. 547 (Oct. 1984).

[73] T.P.M.P.T. Employees Credit Union v. Charpentier, 376 So. 2d 592 (La. Ct. App. 1979).

complaint and exhibit a personal-knowledge affidavit of loss with the motion for summary judgment. Prayers for declaratory judgment fixing the ripening of surety's right of subrogation as of the bond date, and declaring the indemnity agreement assignments to be in effect as of the indemnity agreement date, are useful protection should subsequent third-party actions develop.

The indemnitors' answers will frequently deny the allegations of the complaint. In such cases, use of the request for admissions[74] will quickly establish the necessary undisputed facts.

The surety, of course, owes an indemnitor the duty of good faith—a duty which cannot be waived. The surety must also notify an indemnitor of its intention before using any assets which are available to reduce the indemnitor's liability, and must communicate to indemnitors all settlement offers.[75] Regardless of legal requirements, it is good practice promptly to notify all indemnitors of anything that might conceivably affect them, and invite their written objections, if any. Such practice can avoid obscure legal pitfalls. For example, if the surety has issued a statutory bond and the statutes specify the jurisdiction and venue of suits, settlement of a suit brought elsewhere may preclude recovery from an indemnitor.[76]

§ 10.31 Surety versus Bank

Conflicts between a construction bond surety and the contractor's bank frequently arise after the contractor defaults. Both assert claims to the contract funds remaining in the owner's hands: the bank by reason of assignments of all funds due or to become due the contractor, executed as security for interim financing; the surety by its right of subrogation and also through the assignments in the indemnity agreement.

The bank nearly always has its security interest perfected under the Uniform Commercial Code (U.C.C.); the surety seldom has anything publicly recorded. However, courts have uniformly held that the surety's right of subrogation is superior to the rights of the contractor/principal, its assigns, or a trustee in bankruptcy,[77] and relates back to the date of issuance of the bond.[78] This superiority

---

[74] Fed. R. Civ. P. Rule 36, or comparable state rule.

[75] New Amsterdam Cas. Co. v. Lundquist, 198 N.W.2d 543 (Minn. 1972).

[76] Safeco Ins. Co. v. J.L. Henson, Inc., 601 S.W.2d 183 (Tex. Civ. App. 1980).

[77] Prairie State Bank v. United States, 164 U.S. 227 (1869); Henningsen v. United States Fidelity & Guar. Co., 208 U.S. 404 (1908); Pearlman v. Reliance Ins. Co., 371 U.S. 132 (1962).

[78] Third Nat'l Bank in Nashville v. Highlands Ins. Co., 603 S.W.2d 730 (Tenn. 1980); Division of Employment Sec. v. Trice Constr. Co., 555 S.W.2d 65 (Mo. Ct. App. 1977); Finance Co. of Am. v. United States Fidelity & Guar. Co., 353 A.2d 249 (Md. Ct. Spec. App. 1976); Home Indem. Co. v. United States, 313 F. Supp. 212 (W.D. Mo. 1970). *See generally* Hallenbeck, *Surety's Remedies—Enforcing Rights to Contract Funds,* in Proc. I.A.I.C. Fidelity & Surety Committee (New York 1984).

applies to a financing bank's perfected security interest,[79] and that is so even if the security interest is perfected before the bond is issued.[80]

In the 1960s, some courts apparently felt that Article 9 of the U.C.C. superseded the construction bond surety's right of subrogation.[81] However, that contest was settled in the surety's favor by *Shawmut Bank.*[82]

The subrogation rights of the surety as to a defaulted contractor's materials are not as clear, but it has been held that an owner's right to such material is superior to that of the financing bank.[83] Consequently, since a performing surety is subrogated to the owner's rights,[84] and the owner would normally prevail in this scenario, so should the surety. Persuasive arguments in proper cases can also be made, by way of a surety's subrogation to rights of the contractor, subcontractors, and materialmen, that it has first claim to such materials.[85]

A bank cannot set off a debt against a contractor's account when it has information indicating that the contractor is de facto in default, even though no formal default has been declared. The bank is charged with knowledge that construction bond sureties routinely take indemnity agreements from their contractors.[86]

§ 10.32 Surety versus Owner

Most suits by sureties against owners involve the right to contract funds after the contractor has defaulted. These funds, depending on whether the default arises from nonperformance or nonpayment by the contractor, may be earned but unpaid progress payments, unearned contract funds, or the percentage of retained funds in the owner's hands after the job is completed.

Other claimants to such funds may be a financing bank, a contractor's trustee in bankruptcy, and the government. The performing surety's superior right as against the bank or trustee is well established, whether a performance or payment bond is involved.[87] However, where the government is the owner, the situation is more

---

[79] First Fed. State Bank v. Town of Malvern, 270 N.W.2d 818 (Iowa 1978); Alaska State Bank v. General Ins. Co. of Am., 579 P.2d 1362 (Alaska 1978); Schneider Fuel & Supply Co. v. West Alice State Bank, 236 N.W.2d 266 (Wis. 1978).

[80] Fidelity & Cas. Co. of N.Y. v. Central Bank of Birmingham, 409 So. 2d 788 (Ala. 1982).

[81] Whitney, Rubin & Stabbe, *Twenty Years of the Uniform Commercial Code and Fidelity and Surety Bonding— Some Random Observations,* 17 Forum 670, 676–80 (Summer 1983).

[82] National Shawmut Bank of Boston v. New Amsterdam Cas. Co., 411 F.2d 843 (1st Cir. 1969). *See also* Hallenbeck, *Surety's Remedies— Enforcing Rights to Contract Funds,* in Proc. I.A.I.C. Fidelity & Surety Committee (New York 1984).

[83] Howard Dodge & Sons, Inc. v. Finn, 391 N.E.2d 638 (Ind. Ct. App. 1979); Graves Constr. Co., Inc. v. Rockingham Nat'l Bank, 28 U.C.C. Rep. Serv. (Callaghan) 588 (Va. 1980).

[84] First Fed. State Bank v. Town of Malvern, 270 N.W.2d 818 (Iowa 1978).

[85] *See* Sutton, *Contractor's Default: Can the Surety Take the Construction Materials?,* 4 Construction Law. 5 (1983).

[86] INA v. Northhampton Nat'l Bank, 708 F.2d 13 (1st Cir. 1983).

[87] *See* Prairie State Bank v. United States, 164 U.S. 227 (1869); Henningsen v. United States Fidelity & Guar. Co., 208 U.S. 404 (1908); Pearlman v. Reliance Ins. Co., 371 U.S. 132 (1962).

complex. The government cannot set off unpaid taxes against retained funds when the surety is acting under a performance bond,[88] but it can do so when the surety is acting under a payment bond.[89] Liquidated damages can be set off against retained funds regardless of the nature of the bond issued.[90] A performing surety is subrogated to the government's right of setoff, even against unbonded jobs.[91]

Where, in a suit for possession of the retainage, the court indicates it will grant surety's motion for summary judgment, but the owner's counterclaim remains to be tried, there is a useful mechanism for putting the retained funds to work: ask the court to certify a summary judgment under Federal Rules of Civil Procedure 54(b) but to stay it under Rule 62(h) until the counterclaim is decided. Meanwhile, the owner deposits the funds with the court, where they are invested in short-term, high-yield government obligations. The result is a better return than legal prejudgment interest would have provided.[92]

Another fairly common action against the owner is an action to recover contract funds paid to the contractor after default. This situation arises when the contractor is completing or has completed the job satisfactorily but subcontractors and materialmen remain unpaid. Some owners do not realize that failure to pay for labor and materials is as much a default as abandoning the job.[93] Consequently, even though the surety has formally notified the owner of the default and requested that no further payments be made, the owner may still release the funds to the contractor.

Such a payment constitutes an unlawful conversion of funds to which the performing surety is entitled.[94] The surety's rights in this situation apply to the United States as well as to state or private owners, and to progress payments as well as retainage.[95] However, a government contracting officer is allowed considerable discretion with respect to progress payments, and it is essential for the surety to be able to establish when notice of default was actually received.[96] Like the suit against indemnitors, this action is usually a good candidate for summary judgment.

The Bankruptcy Reform Act of 1978 (hereafter Bankruptcy Code)[97] substantially affected surety rights and must be considered in any litigation involving

---

[88] Trinity Universal Ins. Co. v. United States, 382 F.2d 317 (5th Cir. 1967), *cert. denied,* 390 U.S. 906 (1967).

[89] Royal Indem. Co. v. United States, 371 F.2d 462 (Ct. Cl. 1967), *cert. denied,* 389 U.S. 833 (1967).

[90] Great Am. Ins. Co. v. United States, 492 F.2d 821 (Ct. Cl. 1974). *See generally* the Supplementary Analysis No. 1 in American Fidelity Fire Ins. Co. v. Construcciones Werl, 407 F. Supp. 164, 191 (V.I. 1975).

[91] District of Columbia v. Aetna Ins. Co., 462 A.2d 428, 432 (D.C. 1983).

[92] Hartford Accident & Indem. Co. v. Boise Cascade Corp., 489 F. Supp. 855 (N.D. Ill. 1980).

[93] Martin v. National Sur. Co., 300 U.S. 588 (1937); United States Fidelity & Guar. Co. v. Triborough Bridge Auth., 74 N.E.2d 226, 228 (N.Y. 1974).

[94] Claiborne Parish School Bd. v. Fidelity & Deposit Co. of Md., 40 F.2d 577, 578–79 (5th Cir. 1930).

[95] National Sur. Corp. v. United States, 319 F. Supp. 45, 48–49 (N.D. Ala. 1970).

[96] *See* United States Fidelity & Guar. Co. v. Orlando Utils. Comm'n, 564 F. Supp. 962 (M.D. Fla. 1983); § **10.35**.

[97] 11 U.S.C. § 101 *et seq.*

sureties.[98] A prefiling analysis of the Bankruptcy Code's possible effects may be important even though the contemplated suit's defendant cannot take bankruptcy.

For example, assume suit is to be filed against a state to recover retained funds released to a defaulted contractor after notice and request to stop payment, and another suit is to be filed against that contractor and its individual indemnitors for reimbursement of the surety's losses in paying subcontractors and materialmen. The contractor is in precarious financial condition and might well file for bankruptcy as soon as the indemnity suit is served. State law provides that all parties to a contract are indispensable parties in a suit arising from that contract.

In such a situation, the strategy must be to bring suit against the state first. If the reverse order is used and the contractor files for bankruptcy, the surety may be left without a forum in which to recover the released retainage. This is so because the Bankruptcy Code's automatic stay would prevent the surety from joining the contractor as a defendant in the state suit, failing which the state court would dismiss. The surety cannot alternatively pursue its claim for the retainage in the bankruptcy court because the Constitution's sovereignty clause does not allow a state to be sued in a federal court.[99]

§ 10.33 Surety versus Miscellaneous Parties

Professionals whose services are involved in a construction project owe a duty to others involved in the same project without regard to privity of contract.[1] This duty has been expressed as that ''which rests upon any one to another where such person pretends to possess some skill and ability in some special employment, and offers his fitness to act in the line of business for which he may be employed.''[2] It is also expressed as the duty ''of exercising that degree of professional care and skill customarily employed by others of the same profession in the same general area.''[3]

---

[98] *See* Haug & Haug, *Bankruptcy 1984 v. The Surety's Right to Contract Proceeds,* in Proc. A.B.A. Fidelity & Surety L. Committee (New York 1985); Griffin, *General Contractor in Proceedings under the Bankruptcy Code: A Surety's Perspective,* in American Bar Association Joint Program on Bankruptcy, Crisis in the Construction Industry (New York 1983); Cushman & Miller, *Effect of the New Bankruptcy Law on Traditional Surety Rights,* 16 Forum 1011 (Summer 1981); Duree, *The Effect on the Surety of Bankruptcy Reorganizations of Bond Principals, Obligees and Claimants under the Bankruptcy Reform Act of 1978,* in Proc. A.B.A. Fidelity & Surety L. Committee (New Orleans 1981).

[99] U.S. Const. amend. XI; Hans v. Louisiana, 134 U.S. 1 (1890); Governor v. Madrazo, 26 U.S. (1 Pet.) 110, 122-23 (1828).

[1] *See generally* Chapman, *The Liability of Design Professionals to the Surety,* in Proc. A.B.A. Fidelity & Surety L. Committee (New York 1985); Hirsch & Karpowitz, Liability of Architects and Engineers (Defense Research Inst. No. 1, 1982).

[2] Coombs v. Beede, 36 A. 104, 104–05 (Me. 1896).

[3] Calandro Dev., Inc. v. R.M. Butler Contractors, Inc., 249 So. 2d 254, 265 (La. 1971).

This duty is owed by architects,[4] engineers,[5] appraisers/inspectors,[6] and surveyors.[7] The duty is owed not only to owners but also to contractors,[8] subcontractors,[9] sureties,[10] users of the project,[11] and the guarantor of the bank loan financing the construction.[12] Thus, a performing surety may have a right of action against those parties either through its right of subrogation[13] or by a direct action for negligent breach of that duty to perform professionally.[14] Among the breaches for which damages have been held recoverable by the surety are negligent release of retainage;[15] negligent certification of progress estimates;[16] failure to detect improper construction techniques;[17] failure to detect misuse of contract funds;[18] false certification of work performed;[19] negligent interpretation of plans and specifications;[20] and negligent inspection of the work.[21]

Through its right of subrogation, the prime contractor's surety can also reach a subcontractor's surety, despite a clause in the subcontractor's bond declaring that it "is one of indemnity only and does not inure to the benefit of or confer any right of action upon any person other than (the general contractor)."[22] The crucial distinction leading to that result is that the prime's surety is not "another person" but is the successor to the general contractor.[23] Under proper factual conditions, a surety can also recover from its principal's liability insurer either by

---

[4] General Trading Corp. v. Burnup & Sims, Inc., 523 F.2d 98, 101 (3d Cir. 1975).

[5] American Fidelity Fire Ins. Co. v. Pavia-Byrne Eng'g Corp., 393 So. 2d 830 (La. Ct. App.), *cert. denied,* 397 So. 2d 1362 (La. 1981).

[6] Alley v. Courtney, 448 So. 2d 858 (La. Ct. App. 1984).

[7] Jenkins v. Krebs & Sons, Inc., 322 So. 2d 426 (La. Ct. App. 1975).

[8] United States v. Rogers & Rogers, 161 F. Supp. 132 (S.D. Cal. 1958).

[9] E.H. Leavell & Co. v. Glantz Contracting Corp. of La., 322 F. Supp. 779 (E.D. La. 1971); Gurtler, Hebert & Co., Inc. v. Weyland Mach. Shop, Inc., 405 So. 2d 660 (La. Ct. App. 1981).

[10] American Fidelity Fire Ins. Co. v. Pavia-Byrne Eng'g Corp., 393 So. 2d 830 (La. Ct. App.), *cert. denied,* 397 So. 2d 1362 (La. 1981).

[11] Emond v. Tyler Bldg. & Constr. Co., Inc., 438 So. 2d 681, 685 (La. Ct. App. 1983).

[12] Alley v. Courtney, 448 So. 2d 858 (La. Ct. App. 1984).

[13] Peerless Ins. Co. v. Cerny & Assocs., Inc., 199 F. Supp. 951, 954 (D. Minn. 1961).

[14] State *ex rel.* National Sur. Corp. v. Malvaney, 72 So. 2d 424 (Miss. 1954); Unity Tel. Co. v. Design Serv. Co., Inc., 201 A.2d 177 (Me. 1964).

[15] State *ex rel.* National Sur. Corp. v. Malvaney, 72 So. 2d 424 (Miss. 1954).

[16] Peerless Ins. Co. v. Cerny & Assocs., Inc., 199 F. Supp. 951 (D. Minn. 1961).

[17] Aetna Ins. Co. v. Hellmuth, Obata & Kassabaum, Inc., 392 F.2d 472 (8th Cir. 1968).

[18] *Id.*

[19] Westerhold v. Carroll, 419 S.W.2d 73 (Mo. 1967).

[20] Givoh Assocs. v. American Druggist Ins. Co., 562 F. Supp. 1346 (E.D.N.Y. 1983).

[21] Calandro Dev., Inc. v. R.M. Butler Contractors, Inc., 249 So. 2d 254 (La. 1971).

[22] Financial Indem. Co. v. Steele & Sons, Inc., 403 So. 2d 600 (Fla. Dist. Ct. App. 1981).

[23] Argonaut Ins. Co. v. Commercial Standard Ins. Co., 380 So. 2d 1066 (Fla. Dist. Ct. App. 1980), *review denied,* 389 So. 2d 1108 (Fla. 1980).

subrogation, directly as the real party in interest, or as a third-party beneficiary of the liability policy.[24]

§ 10.34 Owner versus Contractor and/or Surety

Just as the other participants in the construction project have direct rights of action against the owner for any breach of a contractual or other duty owed them, so the owner has similar rights. Upon default of the contractor, the owner can proceed against both the contractor and the surety guaranteeing the performance of the work and payment of those who furnish labor or material to the project. For breaches of contract short of default, such as defective workmanship,[25] the owner has a similar right.

In theory, the owner is also entitled to delay damages from a contractor. In practice, however, the owner, being in the stronger bargaining position, is able to insist on a *liquidated damages clause* being included in the contract. This clause usually provides for a fixed sum per day to be paid the owner by the contractor for each day the project remains unfinished beyond the time period or date specified in the contract.

Such a clause is valid and enforceable if the stipulated amount represents a reasonable estimate, at the time the contract was executed, of actual damages.[26] On the other hand, if the amount is excessive, considering the terms, nature, and purpose of the contract, it will be treated as a penalty, regardless of how it is named.[27] Whether the clause is a legitimate attempt to assess liquidated damages or the imposition of a penalty is a question of law for the court.[28] A liquidated damages clause will not be enforced when: (1) it is held to impose a penalty;[29] (2) no actual damages have been sustained;[30] (3) the delay was caused in part by the owner;[31] or (4) the owner's acts or silence constitute an election not to invoke the clause.[32]

---

[24] *See* Crowder, *Surety's Remedies—Rights Other than Those to Contract Funds,* in Proc. I.A.I.C. Fidelity & Surety Committee (NewYork 1984). For a nonconstruction case, *see* Western World Ins. Co. v. Travelers Indem. Co., 358 So. 2d 602 (Fla. Dist. Ct. App. 1978).

[25] Pickeys v. Clovis, 431 So. 2d 880 (La. Ct. App. 1983).

[26] National Coop. Refinery Ass'n v. Northern Ordnance, Inc., 258 F.2d 803 (10th Cir. 1956).

[27] Pembroke v. Caudill, 37 So. 2d 538 (Fla. 1948).

[28] Yarborough Realty Co. v. Barar, 67 So. 2d 853 (Ala. 1953).

[29] National Coop. Refinery Ass'n v. Northern Ordnance, Inc., 258 F.2d 803 (10th Cir. 1956).

[30] Northwest Fixture Co. v. Kilbourne & Clark Co., 128 F. 256, 261 (9th Cir. 1904).

[31] United States v. United Eng'g & Contracting Co., 234 U.S. 236 (1913).

[32] 5 S. Williston, A Treatise on the Law of Contracts § 688 (3d ed. 1961); Autrey v. Williams & Dunlap, 343 F.2d 730 (5th Cir. 1965).

An owner may also include in the prime contract an *indemnification agreement* requiring the contractor to hold the owner harmless from all injuries and damages, however caused. Such a clause has been held valid, even to indemnify the indemnitee against its own negligence.[33] However, in some jurisdictions, a contract to indemnify another against his or her own negligence is against public policy.[34]

In seeking indemnification from the contractor's surety, the owner must be particularly cautious. A performance bond cannot be substituted for liability insurance.[35] Thus, where an owner incurred attorney's fees in defending a subcontractor's suit, and subsequently sued the prime's performance bond surety to recover those fees, its failure to allege and prove that the prime contractor was the owner's indemnitor barred recovery.[36]

A troublesome source of litigation between these parties is when the owner discovers defects in the project after it has been certified complete, accepted, and paid for. The contractor and surety then refuse either to remedy the defects or to pay damages. Their reasons might be that an acceptance and occupancy of the project waives any claims for defective work and/or that applicable guaranty periods or statutes of limitation periods have expired. The owner then files suit.

The key determination in such litigation is whether the defects were *patent* or *latent*, that is, whether they were discoverable upon a reasonable inspection (patent defects)[37] or whether they were not so discoverable (latent defects).[38] Acceptance, payment, and occupancy waive any action for patent defects,[39] but the owner still has recourse against the contractor and surety for latent defects.[40] Since in the latter situation, applicable statutes of limitation do not begin to run until the latent defect is discovered, contractors and sureties have been subjected to what amounts to open-ended liability.[41]

---

[33] Employer Ins. of Wausau v. Cajun Contractors & Eng'rs, Inc., 459 So. 2d 610 (La. Ct. App. 1984); Frank Corrente v. Conforti & Eisele Co., Inc., 468 A.2d 920 (R.I. 1984); Farmington Plumbing & Heating Co. v. Fischer Sand & Aggregate, Inc., 281 N.W.2d 838 (Minn. 1980); Dorothy L. Draper v. Airco, Inc., 580 F.2d 92 (3d Cir. 1979).

[34] Cal. Civ. Code § 1782 (West 1972); Ga. Code § 13-8-2 (Harrison 1982); Ill. Rev. Stat. ch. 29, § 61 (West 1985); Mich. Comp. Laws § 691.991 (West 1968); N.Y. Gen. Oblig. Law § 5-324 (West 1978).

[35] Long v. City of Midway, 311 S.E.2d 508 (Ga. Ct. App. 1983).

[36] State Sur. Co. v. Lamb Constr. Co., 625 P.2d 184 (Wyo. 1981).

[37] Renown, Inc. v. Hensel Phelps Constr. Co., 201 Cal. Rptr. 242 (Ct. App. 1984); Mattingly v. Anthony Indus., Inc., 167 Cal. Rptr. 292 (Ct. App. 1980).

[38] City of Osceola v. Gjellefald Constr. Co., 279 N.W. 590 (Iowa 1938); Town of Tonawanda v. Stapell, Mumm & Beals Corp., 270 N.Y.S. 377 (N.Y. 1934). *See* Warin, *The Surety's Liability for Latent Defects,* in Proc. A.B.A. Fidelity & Surety L. Committee (New York 1985).

[39] Stearns Constr. Corp. v. Carolina Corp., 217 N.W.2d 291 (Wis. 1974); Salem Towne Apartments, Inc. v. McDonald & Sons Roofing Co., 330 F. Supp. 906 (E.D.N.C. 1970); Fitzgerald v. LaPorte, 40 S.W. 61 (Ark. 1897).

[40] Aubrey v. Helton, 276 Ala. 134, 159 So. 2d 837 (1964).

[41] School Bd. of Pinellas County v. St. Paul Fire & Marine Ins. Co., 449 So. 2d 872 (Fla. Dist. Ct. App. 1984), citing authorities from Iowa, New Jersey, New York, North Carolina, Oregon, and Texas; Milana, *The Performance Bond and the Underlying Contract,* 12 Forum 187 (Fall 1976).

To counteract this situation, some 40 states have passed special statutes of limitations to cover participants in the construction industry.[42] Some of these statutes provide for minimum and maximum periods within which actions for latent defects may be brought; others set only the outer limits. In either case, they provide some relief to contractors and sureties, and indirect incentives to those professionals employed to supervise and inspect construction projects for giving increased attention to the building process.[43]

NOTICE AND LIMITATION OF ACTIONS

§ 10.35 Notice Periods

The notice and limitation periods which are most often raised in construction litigation are those controlling claims and suits by subcontractors and materialmen against the contractor and/or its surety. These matters, as previously noted, are covered in **Chapter 11.**

There are, however, less well-known matters of notice and limitations which can seriously affect the contractor, owner, and surety when litigation arises. For example, the language of a performance bond creates an implied duty in the bond obligee/owner to notify the surety of the contractor's delays. Failure to do so releases the surety from liability for resulting damages to the extent that the surety could have remedied those delays.[44] Nevertheless, if an owner's late notice is not shown to prejudice the surety, no release is effected.[45] Although the foregoing is the generally accepted rule, one recent case holds that actual prejudice is not required for an insurer to disclaim liability when notice was not timely given.[46]

A detailed discussion of the Miller Act[47] and its jurisprudence is beyond the scope of this chapter. However, it should be noted that the only safe way to serve notices such as letters of default, direction, and stop-payment on the government is by hand delivery to and personal receipt by the contracting officer of copies of the documents. Otherwise, that officer may later testify that he or she did not actually see those notices. Such testimony, coupled with the broad discretion granted a contracting officer in making progress payments, can result in a substantial portion of contract funds, which should have been available to the surety for payment

---

[42] *See* Witherwax, *Special Statutes of Limitations Against the Contractor—A Defense to the Surety,* 16 Forum 1057 (1980).

[43] *See* Hinchey & Lord, *Liability of Surety, Contractor and Subcontractor after Final Completion and Acceptance* (A.B.A. Joint Program on Trouble Areas in Construction, New York 1985).

[44] Blackhawk Heating & Plumbing Co. v. Seaboard Sur. Co., 534 F. Supp. 309 (N.D. Ill. 1982).

[45] United States Fidelity & Guar. Co. v. Orlando Utils. Comm'n, 564 F. Supp. 962 (M.D. Fla. 1983).

[46] Kosanovich v. Meade, 463 N.E.2d 257 (Ind. 1984).

[47] 40 U.S.C. §§ 270a-270d.

of subcontractors and materialmen, disappearing into the defaulted contractor's pocket.[48]

§ 10.36 Limitation Periods

In a suit by the owner against the contractor and surety, different limitation periods may apply. Such was the holding in a suit for defective work brought 11 years after final certification and delivery of the job. Because the bond was under seal while the contract was not, a 12-year limitation applied to the surety, whereas a three-year period applied to the contractor. The court held that although a valid cause of action existed against the surety, the surety was entitled to indemnification and exoneration from the contractor since the contractor's defense of the statute of limitations was a personal defense not available to the surety.[49]

Carelessness in filling in blank spaces in contracts can be costly, particularly when that space sets the period for limitation of actions. Such a failure may create open-ended liability for the surety.[50] Sureties issuing construction bonds to subcontractors on government projects should also be aware of the fact that their exposure to suit may be for a much longer period than they expect—longer than the periods provided for by the Miller Act,[51] longer than those provided for by states' public and private works acts, and longer than those specified in the bonds.

This possibility of longer exposure arises, as to the Miller Act, because that act does not require a subcontractor's bond and, therefore, such a bond does not automatically incorporate the notice and limitation periods of the act.[52] Similarly, since a government project does not belong to a state or any agency or subdivision thereof, a state's public works law limitation periods would not apply. A state's private works statutes would also not apply, because they are predicated on the assumption that the bonded work is lienable, and government property is not lienable. Finally, in states which do not allow general statutory limitation periods to be shortened by contract, but only by specific statutory limitation periods, a shorter period provided for in the bond would have no legal effect. Thus, what is left is the limitation period for actions on a contract, which is usually longer than that for claims arising out of construction projects. In Louisiana, for example, the period is 10 years.[53]

The automatic stay provisions of the Bankruptcy Code[54] do not toll the bond limitation period for an action by a Chapter 11 debtor/contractor against a

[48] United States Fidelity & Guar. Co. v. United States, 676 F.2d 622 (Ct. Cl. 1982).

[49] President, Georgetown College v. Madden, 660 F.2d 91 (4th Cir. 1981).

[50] Peters Township School Auth. v. United States Fidelity & Guar. Co., 467 A.2d 904 (Pa. 1983).

[51] 40 U.S.C. §§ 270a-270d.

[52] United States *ex rel.* Wheeling-Pittsburgh Steel Corp. v. Algernon Blair, Inc., 329 F. Supp. 1360 (D.C.S.C. 1971).

[53] La. Civ. Code Ann. art. 3499 (West 1985).

[54] 11 U.S.C. § 362, Rules 11-44(a) and 12-33(a).

subcontractor and its surety. The reason is that those stay provisions apply only to actions brought *against* the debtor, not *by* the debtor.[55]

Is all lost for the surety when it files suit against the owner beyond the two-year limitation period provided for in the bond? Not necessarily. Why not argue that the suit was brought, not under the bond, but under the indemnity agreement's assignment from the contractor, to whose rights the surety is subrogated? At least one court has agreed.[56]

CONSTRUCTION OF DOCUMENTS AND PAROL EVIDENCE

§ 10.37 Construction of Contracts

Generally, public works statutes are construed liberally to effect the intent of the legislative body to protect those whose labor and materials go into public projects.[57] Contracts, if they contain ambiguities, are construed against the drafter. Compensated surety bonds are construed by most courts as requiring the use of both rules.[58] Obviously, such construction rules strongly affect litigation involving contractors and sureties, usually adversely to the surety. Only occasionally will a court apply the rules evenhandedly and, for example, hold that if the surety bond follows statutory language and contains an ambiguity it must be strictly construed against the state.[59]

Further complicating surety bond construction is the fact that at least one court has specifically declared such a bond to be an insurance contract,[60] and most others treat it as such. Consequently, the doctrines of reasonable expectations[61] and unconscionability could be applied to such bonds.[62]

The Arizona Supreme Court may have provided some light at the end of this tunnel. The court noted that applying the *four corners* rule of contract interpretation to standardized contracts has such mischievous consequences that courts are

---

[55] 1616 Reminc Ltd. Partnership v. Atchison & Keller Co., 14 Bankr. L. Rep. (CCH) ¶ 484 (Bankr. 1981).

[56] Balboa Ins. Co. v. W.C.B. Assocs., Inc., 390 So. 2d 172 (Fla. Dist. Ct. App. 1980).

[57] Clifford F. MacEvoy Co. v. United States *ex rel.* Calvin Tomkins Co., 332 U.S. 102 (1944).

[58] *See* Frederick v. Electro-Coal Transfer Corp., 548 F. Supp. 83 (E.D. La. 1982), and cases cited therein.

[59] Florida *ex rel.* Consolidated Pipe & Supply Co. v. Houdaille Indus., 372 So. 2d 1177 (Fla. Dist. Ct. App. 1979).

[60] Austin v. Parker, 672 F.2d 508, 519 (5th Cir. 1982).

[61] Keeton, Basic Text on Insurance Law 360 (1971); Carpenter v. Suffolk Franklin Sav. Bank, 346 N.E.2d 892 (Mass. 1976).

[62] *See* Mittelman, *Reasonable Expectations and Bad Faith in the Settlement of a Health Insurance Claim,* 19 Forum 340 (Winter 1984); Koch, *The Surety's General Indemnity Agreement: Current or Out of Date?,* in Proc. A.B.A. Fidelity & Surety L. Committee (New York 1984).

required to invent ambiguities and then construe those ambiguities against the drafter to reach a just result. Declaring that "the time has come to remove insurance law from the land of make-believe," the court proceeded to revise the entire body of Arizona insurance law.[63]

The rules which the *Darner* case propound are fair and recognize the realities of modern business practices. They may well become the cutting edge of a new trend in the construction of standardized contracts, which of course include surety bonds and indemnity agreements.

§ 10.38 Parol Evidence

Oral change orders are a source of frequent dispute among owners, contractors, and subcontractors, and some of these disputes result in litigation.

When a contractor follows an oral change order, it is not liable for resulting damage.[64] The rationale in most jurisdictions is that a written contract may be waived or modified by a subsequent parol agreement.[65] This is so even though the contract contains a capitalized provision stating that: "No extra work shall be paid for unless authorized in writing from owner's office. Verbal authorization shall not be recognized."[66] Louisiana law allows parol evidence of a subsequent agreement to modify or revoke a written agreement unless that written agreement is one required by law to be in writing.[67]

The problem, of course, is proving the oral agreement. Immediately sending a confirmation letter containing one's understanding of the agreement to the other party is a recommended procedure.

§ 10.39 Conclusion

At the heart of construction litigation involving contractors and sureties is detail. Probably to no other field of law is the "horseshoe nail" progression more applicable.[68] In the cases cited throughout this chapter, the losing parties either did something prohibited or failed to do something required by the statutes, contract

---

[63] Darner Motor Sales, Inc. v. Universal Underwriters Ins. Co., 682 P.2d 388 (Ariz. 1984).

[64] Mannella v. Pittsburgh, 6 A.2d 70 (Pa. 1939).

[65] 30 Am. Jur. 2d *Evidence* § 1063 (1967).

[66] W.E. Garrison Grading Co. v. Piracci Constr. Co., Inc., 221 S.E.2d 512 (N.C. Ct. App. 1975), *cert. denied,* (Feb. 3, 1976); Son-Shine Grading, Inc. v. ADC Constr. Co., 315 S.E.2d 346 (N.C. Ct. App. 1984). *See also* Ekco Enters., Inc. v. Remi Fortin Constr., Inc., 382 A.2d 368 (N.H. 1978); D.K. Meyer Corp. v. Bevco, Inc., 292 N.W.2d 773 (Neb. 1980).

[67] Torrey v. Simon-Torrey, Inc., 307 So. 2d 569 (La. 1974).

[68] "For want of a nail, the shoe was lost; For want of a shoe, the horse was lost; For want of a horse, the soldier was lost; For want of a soldier, the battle was lost; For want of a battle, the war was lost."

law, or tort law of the jurisdiction. Reading between the lines in some of those cases, it seems probable that the losing party did perform properly but simply could not prove the necessary facts. As aptly put by Judge Murnaghan, "there are very few cases indeed in which ultimately the facts do not control the outcome. Arguments impeccable in their abstract reliance upon broadly phrased principles of law tend to evaporate when the equities point to a different result. Purely linguistic considerations should not be permitted to outweigh substance."[69]

The filing of any of these two-party suits could, under modern pleading practice, rather quickly expand through counterclaims, crossclaims, third-party demands, interpleaders, answers, and consolidations to involve most, if not all, of the parties and actions listed.[70] How then can a contractor, surety, owner, architect, engineer, subcontractor, or materialman minimize the litigation risks of participating in a construction project? Only by being aware of, and incorporating into its operating procedures, the current law of the situs applicable to the contract documents, the bonds, and the acts of the parties.

This means documenting contemporaneously (not trying to remember and reconstruct at a later date) all facts that would be relevant to the causes of action which might be brought by or against the participant. While such documentation may be inconvenient at times, the rewards are substantial—a greatly increased probability of (1) avoiding disputes with other participants, (2) reaching a favorable settlement if a dispute does arise, and (3) winning when litigation becomes unavoidable. The good news is then, that, given the fact that many construction projects present complex legal questions before, during, and after completion of the project, one who meticulously follows the guidelines inherent in the jurisprudence discussed in this chapter can and should surely win.

---

[69] Campbell v. Virginia Metal, 708 F.2d 930, 930 (4th Cir. 1983).

[70] *See,* for example, Barr/Nelson, Inc. v. Tonto's, Inc., 336 N.W.2d 46 (Minn. 1983), in which contractor sued the owner and architect for breach of contract, interference with business relationships, and malpractice. The owner and the architect counterclaimed and cross-claimed. The owner third-partied surety for breach of contract, breach of good-faith covenant, fraud, and punitive damages. The surety counterclaimed and cross-claimed for subrogation and indemnity.

CHAPTER 11

PROSECUTING AND DEFENDING PAYMENT BOND CLAIMS

Edward Graham Gallagher

Edward Graham Gallagher is a graduate of Williams College and George Washington University Law School and practices law in Washington, D.C. with the firm of Thompson, Larson, McGrail, O'Donnell & Harding. Mr. Gallagher is active in the Fidelity and Surety Committee of the American Bar Association and the International Association of Insurance Counsel, and has been a speaker at numerous programs on surety law. His articles on surety topics have been published by the Tort and Insurance Practice Section of the American Bar Association.

§ 11.1 Introduction

Suretyship is a contractual relationship in which one party agrees in writing to answer for the debt or default of another.[1] Every surety contract involves three parties: a principal, a surety, and an obligee. The obligee is the party for whose benefit and protection the surety contract is written. The principal is the party primarily liable to the obligee, and the surety is secondarily liable to the obligee if the principal fails to pay the debt or perform the obligation which underlies the surety contract.

In a modern labor and material payment bond, the principal is the contractor or subcontractor who provided the bond, the surety is usually an insurance company, and the named obligee is the owner, lender, or prime contractor who required the bond. Although not named as obligees, persons who furnish labor or material to the principal for use in the performance of the contract work are explicitly or implicitly additional obligees entitled to sue on the bond.[2]

§ 11.2 Purpose of Payment Bond

On a private (i.e., nongovernment) construction project, persons who furnish labor or material for use in the performance of the work may file mechanic's liens against the property. In order to protect the property from such liens, an owner can require the contractor to provide a bond conditioned on payment of persons who have furnished such labor or material.[3] If the contractor fails to make payment, the surety on the bond will have to do so, and the owner's property will be protected.

Although government property is usually exempt from mechanic's liens,[4] the government has a moral obligation to see that those whose work has gone into

---

[1] Stearns, Law of Suretyship § 1.1, at 1 (5th ed.); 74 Am. Jur. 2d *Suretyship* § 1, p. 2.

[2] *See,* for example, 40 U.S.C. § 270b(a); American Institute of Architects Document A311, Performance Bond and Labor and Material Payment Bond (Feb. 1970 ed.).

[3] There are also "no lien" bonds which require only that the surety protect the owner's property from mechanics' liens.

[4] F.D. Rich Co. v. United States *ex rel.* Industrial Lumber Co., 417 U.S. 116 (1974); Stauffer Constr. Co., Inc. v. Tate Eng'g, Inc., 407 A.2d 1191 (Md. Ct. Spec. App. 1979).

the construction of a public project are paid. In order to meet this moral obligation,[5] and to make it easier for public contractors to obtain material on credit, Congress and the state legislatures enacted statutes requiring public works contractors to post payment bonds.[6]

§ 11.3 Typical Payment Bond Provisions

The basic condition of a payment bond is that the principal promptly make payment for labor and material furnished for use on the project. Each bond form, however, has slightly different phrasing and may impose conditions or restrictions on the surety's obligation. Anyone concerned with prosecuting or defending a bond claim should first and foremost *read the bond.*

The United States government requires use of a standard form payment bond[7] on all federal projects. The most common form on private jobs has been American Institute of Architects (A.I.A.) Document A311,[8] although a new A.I.A. form, Document A312, has recently been issued.[9]

Some state agencies draft their own forms, which are usually patterned after the federal one, but most will accept any reasonable payment bond. Most surety companies have printed bond forms which their agents will use if the obligee does not require a specific form. Many of these company forms are merely the A.I.A. bond reprinted on the company's letterhead, but others vary from company to company.

The coverage, notice, and suit limitations commonly included in payment bonds are discussed later in this chapter. It is important to remember that a surety is not an insurer of its principal[10] and is liable only if a claimant has strictly complied with the conditions of the bond, unless, of course, a restriction in the bond is itself contrary to law.

---

[5] United States *ex rel.* Mississippi Road Supply Co. v. H.R. Morgan, Inc., 542 F.2d 262 (5th Cir. 1976), *cert. denied,* 434 U.S. 828 (1977); R.C. Stanhope, Inc. v. Roanoke Constr. Co., 539 F.2d 992 (4th Cir. 1976).

[6] *See,* for example, 40 U.S.C. § 270a; Md. Ann. Code art. 21, § 3-501 (1957); Va. Code § 11-58.

[7] General Services Administration Standard Form 25-A (June 1964 ed.).

[8] American Institute of Architects Document A311, Performance Bond and Labor and Material Payment Bond (Feb. 1970 ed.). This form, and all A.I.A. documents referred to in this chapter, are available from the American Institute of Architects, 1735 N.Y. Avenue, N.W., Washington, D.C. 20006.

[9] American Institute of Architects Document A312, Performance Bond and Payment Bond (Dec. 1984 ed.).

[10] Pearlman v. Reliance Ins. Co., 371 U.S. 132 (1962); United States *ex rel.* Getz Bros. & Co. v. Markowitz Bros. (Delaware), Inc., 383 F.2d 595 (9th Cir. 1967); State Highway Admin. v. Transamerica Ins. Co., 367 A.2d 509 (Md. 1976).

§ 11.4 Statutory Requirements

At common law there was no requirement of a payment bond. When Congress and various state legislatures enacted laws compelling government contractors to provide such bonds, they also specified certain terms and conditions which the bond would have to incorporate. These terms and conditions are read into the bond by operation of law, whether or not they are actually stated in the bond.[11]

For example, the federal *Miller Act*[12] provides that a claimant must file suit within one year of the date he or she last furnished labor or material for use in the performance of the contract. Although the bond itself may contain no such provision, the one-year limitation of the statute is read into the bond, and a claimant who lets the one-year period run without filing suit will lose his or her right to collect on the payment bond.

In order to construe a bond given by a public works contractor, therefore, it is essential to read the statute requiring the posting of the bond as well as the bond itself. The text of the Miller Act is reproduced as **Appendix 11-1**. Almost all state bond statutes are patterned after that Act, and, in the absence of controlling state decisions, the state courts will look to Miller Act precedents for guidance.[13]

The results, however, are by no means uniform, and state bond statutes and the decisions interpreting them vary from state to state. One should never assume that the rights of a claimant on a state bond are identical to those of a claimant under the Miller Act. There is simply no substitute for reading the bond itself and the statute, if any, which required that it be furnished.

§ 11.5 Resolution of Conflicts between Bond and Statute

In an ideal world there would be no conflict between a public works contractor's bond and the statute requiring that it be posted. The contractor and its surety's agent would be sure that the bond form they used conformed to the statutory requirements; if they failed to do so, the government's contracting officer would reject the bond and require that a proper one be furnished. In the real world, however, mistakes are made and the resolution of a conflict between the bond and statute occasionally determines the validity of a claim.

If the bond purports to provide less than the minimum coverage required by the statute, the statute will be read into the bond on the theory that the surety entered into its contract with knowledge of the law. The surety will be held to have provided

---

[11] Mayor of Baltimore v. Fidelity & Deposit Co. of Md., 386 A.2d 749 (Md. 1978).

[12] 40 U.S.C. § 270a *et seq.*

[13] Solite Masonry Units Corp. v. Piland Constr. Co., 232 S.E.2d 759 (Va. 1977); Montgomery County Bd. of Educ. v. Glassman Constr. Co., 225 A.2d 448 (Md. 1967).

the minimum coverage specified by law.[14] On the other hand, most courts hold that the surety is free to contract to provide more than the required coverage, and if a claim is valid under the bond the claimant will usually prevail even though the minimum requirements of the statute would not have covered the claim.[15]

Many bond forms, including the American Institute of Architects bonds, provide that if the suit limitation period provided in the bond is prohibited by law, then the limitation shall be deemed amended to the minimum period permitted by the statute. At least one court interpreted its state payment bond statute to "prohibit" a limitations period different from that specified in the statute, and so construed the bond to provide only the statutory period for suit even though the period stated in the bond was longer. The court recognized that the surety could contract to provide greater coverage than required by the statute, but held that it had not done so because of the savings clause in the bond.[16]

§ 11.6 Persons Who May Make a Bond Claim

On any construction project, a great many persons play some part, directly or indirectly, in the performance of the work. Employees of the contractor perform some of the work using material purchased from distributors or manufacturers. Other parts of the work are subcontracted to firms which in turn purchase materials, hire laborers, and, possibly, re-subcontract parts of their portion of the work. Tools and equipment to be used on the project are purchased or rented. Consumables such as electric power, fuel, telephone service, and portable sanitation facilities are necessary even though they are not incorporated into the building, and even intangibles such as insurance and working capital are vital to the successful completion of the work.

Potentially, an almost unlimited number of persons could claim to have contributed something to the performance of the contract. That does not mean, however, that each of them can make a claim under the contractor's payment bond. The surety is not an insurer of the principal's obligations.

Employees of the contractor who actually worked on the project are certainly within the coverage of the bond, but it is almost never necessary for them to resort to the surety for payment. If a contractor does not pay his employees one week, they are unlikely to continue to work. A significant arrearage of back wages almost never accrues.

A building cannot be constructed with labor alone, however, and the contractor must also acquire materials. These are normally supplied on credit, and a person

---

[14] Mayor of Baltimore v. Fidelity & Deposit Co. of Md., 386 A.2d 749 (Md. 1978).

[15] Reliance Ins. Co. v. Trane Co., 194 S.E.2d 817 (Va. 1971); American Cas. Co. of Reading v. Irvin, 426 F.2d 647 (5th Cir. 1970).

[16] Joseph F. Hughes & Co. v. George H. Robinson Corp., 175 S.E.2d 413 (Va. 1970).

who merely sells material is called a materialman.[17] A materialman who sells directly to the bond principal is covered by the bond,[18] but a materialman who sells to another materialman, such as a wholesaler or distributor, is not covered under the Miller Act bond or under most other bonds.[19]

A materialman who sells to a sub-subcontractor may be covered if it gives written notice as required by the terms of the bond. Most bonds require notice to the bond principal,[20] but some provide for notice to any two of the principal, owner, or surety.[21] A materialman who sells to a sub-contractor or more remote subcontractor is usually not covered.[22]

A *subcontractor* is generally defined as one who takes over performance of a specific part of the work from the prime contractor.[23] A subcontractor must be distinguished from a *materialman,* who supplies material but does not have responsibility for any specific part of the work, and from a *sub-subcontractor* (or *second-tier subcontractor*), who performs a specific part of the work but does not have a direct contractual relationship with the prime contractor. Confusion can result if the definition of *subcontractor* in a case addressing the distinction between a subcontractor and a materialman is applied in a case involving a sub-subcontractor.[24]

A subcontractor (as defined above) is covered under the prime contractor's payment bond. A sub-subcontractor can be within the coverage of the prime contractor's

---

[17] As opposed to a *subcontractor* who takes over a specific portion of the contract work.

[18] Such a materialman, of course, has a direct contractual relationship with the bond principal. *See* 40 U.S.C. § 270b; American Institute of Architects Document A311 (Feb. 1970 ed.), and Document A312 (Dec. 1984 ed.).

[19] Clifford F. MacEvoy Co. v. United States *ex rel.* Calvin Tomkins Co., 322 U.S. 102 (1944); United States *ex rel.* Hasco Elec. Corp. v. Reliance Ins. Co., 390 F. Supp. 158 (E.D.N.Y. 1975); United States *ex rel.* Bryant v. Lembke Constr. Co., 370 F.2d 293 (10th Cir. 1966); United States *ex rel.* Wellman Eng'g Co. v. MSI Corp., 350 F.2d 285 (2d Cir. 1965).

[20] Under the Miller Act (40 U.S.C. § 270a *et seq.*), American Institute of Architects Document A312 (Dec. 1984 ed.)., and most state statutes. *See* J.W. Cooper Constr. Co. v. Public Hous. Admin. *ex rel.* Rio Grande Steel Prods. Co., 390 F.2d 175 (10th Cir. 1968); Travelers Indem. Co. v. United States *ex rel.* Western Steel Co., 362 F.2d 896 (9th Cir. 1966); F.D. Rich Co. v. United States *ex rel.* Industrial Lumber Co., 417 U.S. 116 (1974).

[21] Under American Institute of Architects Document A311, ¶ 3(a): "No suit or action shall be commenced hereunder by any claimant: a) Unless claimant, other than one having a direct contract with the Principal, shall have given written notice to any two of the following: the Principal, the Owner, or the Surety . . . ."

[22] J.W. Bateson Co. v. United States *ex rel.* Board of Trustees, 434 U.S. 586 (1978), but *compare* Md. Ann. Code art. 21, § 3-501 (1957), which requires such coverage.

[23] Clifford F. MacEvoy Co. v. United States *ex rel.* Calvin Tomkins Co., 322 U.S. 102 (1944); F.D. Rich Co. v. United States *ex rel.* Industrial Lumber Co., 417 U.S. 116 (1974); J.W. Bateson Co. v. United States *ex rel.* Board of Trustees, 434 U.S. 586 (1978).

[24] *See,* for example, the decision of the United States Court of Appeals for the District of Columbia Circuit in United States *ex. rel.* Board of Trustees v. J.W. Bateson Co., Inc., 551 F.2d 1284 (D.C. Cir. 1977), which was reversed by the Supreme Court in J.W. Bateson Co. v. United States *ex rel.* Board of Trustees, 434 U.S. 586 (1978).

bond if it gives the required notice.[25] A third-tier or lower subcontractor is not within the coverage of a Miller Act bond[26] or A.I.A. bonds[27] but may be covered by some state statutory bonds.[28]

A supplier to a materialman of any tier, including a materialman who has a direct contractual relationship with the prime contractor, is not normally covered by the prime contractor's bond.[29]

§ 11.7 Types of Claims Covered by Bond

A payment bond is conditioned on payment for labor or material furnished for use in the prosecution of the contract work.[30] Labor performed on the job site in connection with the contract is covered and usually causes no difficulty. Similarly, there is usually no contest over coverage of material which is actually incorporated into the finished project.

A more difficult question is presented by a claim for material ordered for use on the bonded job but diverted by the contractor to some other use. The claimant can still recover if it can show that it had a reasonable, good faith belief that the material was intended to be used on the bonded job.[31] The material does not have to have been delivered to the job site.[32]

Some tangible items which are consumed during the prosecution of the contract work are not actually incorporated into the job. At one time there was considerable litigation over whether such consumables were covered by the bond, but now, either by statute, explicit bond language,[33] or court decision, they are generally

---

[25] *See* Clifford F. MacEvoy Co. v. United States *ex rel.* Calvin Tomkins Co., 322 U.S. 102 (1944); F.D. Rich Co. v. United States *ex rel.* Industrial Lumber Co., 417 U.S. 116 (1974); J.W. Bateson Co. v. United States *ex rel.* Board of Trustees, 434 U.S. 586 (1978); United States *ex rel.* Benkart Co. v. John A. Johnson & Sons, Inc., 236 F.2d 864 (3d Cir. 1956).

[26] United States *ex rel.* Powers Regulator Co. v. Hartford Accident & Indem. Co., 376 F.2d 811 (1st Cir. 1967).

[27] *See* American Institute of Architects Document A311, ¶ 1 (Feb. 1970 ed.), and Document A312, ¶ 15.1 (Dec. 1984 ed.).

[28] *See, e.g.,* Md. Ann. Code art. 21, § 3-501 (1957).

[29] *See* Clifford F. MacEvoy Co. v. United States *ex rel.* Calvin Tomkins Co., 322 U.S. 102 (1944); United States *ex rel.* Hasco Elec. Corp. v. Reliance Ins. Co., 390 F. Supp. 158 (E.D.N.Y. 1975); United States *ex rel.* Bryant v. Lembke Constr. Co., 370 F.2d 293 (10th Cir. 1966); United States *ex rel.* Wellman Eng'g Co. v. MSI Corp., 350 F.2d 285 (2d Cir. 1965).

[30] 40 U.S.C. § 270b; American Institute of Architects Documents A311 (Feb. 1970 ed.) and A312 (Dec. 1984 ed.).

[31] Glassell-Taylor Co. v. Magnolia Petroleum Co., 153 F.2d 527 (5th Cir. 1946); United States *ex rel.* Westinghouse Elec. Supply Co. v. Endebrock-White Co., 275 F.2d 57 (4th Cir. 1960).

[32] United States *ex rel.* Carlson v. Continental Cas. Co., 414 F.2d 431 (5th Cir. 1969).

[33] *See, e.g.,* American Institute of Architects Documents A311 (Feb. 1970 ed.) and A312 (Dec. 1984 ed.).

covered. Thus, claimants who supplied fuel,[34] groceries (if they are not otherwise generally available in the area of the job site),[35] utilities,[36] transportation,[37] and tires[38] have all been held to have valid payment bond claims on the theory that the items supplied were used or consumed during the performance of the work.

If a contractor *purchases* equipment and happens to use it on a bonded job, the seller of the equipment does not thereby acquire a claim against the surety for any unpaid part of the purchase price.[39] The equipment can be used on many jobs, and the surety should not be held liable for its cost. If the contractor *rents* equipment for use on a particular job, however, its payment bond surety will be liable for any unpaid obligations under the rental contract. This liability includes the cost of maintaining or repairing the equipment, if the rental contract puts these obligations on the contractor, as well as the unpaid rent.[40]

§ 11.8 —Taxes, Contributions, and Insurance

Contractors who encounter financial difficulty often fail to pay withheld payroll taxes to the Internal Revenue Service (I.R.S.). The employees still get credit on their tax returns for the withheld monies, and the I.R.S. has to try to collect from the delinquent employer. The I.R.S. often looked to the payment bond and claimed that withholding taxes for employees on the bonded job were monies due for labor furnished in the prosecution of the work. The sureties argued that the employees had been paid in full and that the I.R.S. was simply a general creditor of the contractor and had not supplied labor or material for use on the bonded job.

On federal jobs, this controversy was decided by a 1966 amendment to the Miller Act making the performance bond liable for withheld payroll taxes if the government gives the surety written notice of its claim within certain time limits. On non-Miller Act projects, unless an applicable statute or the bond explicitly provides coverage for withheld taxes, the surety should not be liable for such tax claims.[41]

---

[34] Glassell-Taylor Co. v. Magnolia Petroleum Co., 153 F.2d 527 (5th Cir. 1946).

[35] Brogan v. National Sur. Co., 246 U.S. 257 (1918); Equitable Cas. & Sur. Co. v. Helena Wholesale Grocery Co., 60 F.2d 380 (8th Cir. 1932), but *compare* Delaware Dredging Co. v. Tucker Stevedoring Co., 25 F.2d 44 (3d Cir. 1928).

[36] American Institute of Architects Documents A311 (Feb. 1970 ed.), A312 (Dec. 1984 ed.); Va. Code § 11-58.

[37] Standard Accident Ins. Co. v. United States *ex rel.* Powell, 302 U.S. 442 (1938).

[38] United States *ex rel.* J.P. Byrne & Co. v. Fire Ass'n of Philadelphia, 260 F.2d 541 (2d Cir. 1958).

[39] United States *ex rel.* Eddie's Sales & Leasing, Inc. v. Federal Ins. Co., 634 F.2d 1050 (10th Cir. 1980); Ibex Indus., Inc. v. Coastline Waterproofing, 563 F. Supp. 1142 (D.D.C. 1983).

[40] Illinois Sur. Co. v. John Davis Co., 244 U.S. 276 (1917); United States *ex rel.* Mississippi Road Supply Co. v. H.R. Morgan, Inc., 542 F.2d 262 (5th Cir. 1976), *cert. denied,* 434 U.S. 828 (1977).

[41] United States v. Commonwealth of Pa., Dep't. of Highways, 349 F. Supp. 1370 (E.D. Pa. 1972); United States v. Maryland Cas. Co., 323 F.2d 473 (5th Cir. 1963); United States v. Crosland Constr. Co., 217 F.2d 275 (4th Cir. 1954).

If the contractor fails to pay wage-based contributions to a union trust fund, the surety will be liable for that portion of the delinquent payment attributable to the bonded job, and, in a suit by the trustees of the fund, the surety will also be liable for auditing fees, interest, and attorneys' fees to the extent they are provided for in the union contract signed by the principal. The courts have accepted the unions' argument that such benefit payments are part of the amount due for work performed on the job by the contractor's employees.[42]

Insurance premiums are not labor or material and thus are not within the coverage of the payment bond even though insurance is necessary to the performance of the contract. This is true even for workmen's compensation insurance, for which the premium is based on wages paid by the contractor.[43]

§ 11.9 —Delay Damages and Lost Profits

Payment bonds typically give a right of action for sums "justly due." Claimants argue that they are justly due an amount which will make them whole for the contractor's default, including delay damages, lost profits, interest, and attorneys' fees. They want to be in the same financial position as if the contractor had never defaulted. The surety, on the other hand, argues that it is liable only for the actual cost of labor or material furnished for use on the bonded job, and that such items are neither labor nor material.

There is a split of authority as to whether a surety is liable for delay damages. Some courts reason that excess costs incurred by a subcontractor as a result of a compensable delay are part of the amount justly due for the labor or material supplied. These courts accept the argument that such items as escalation or disruption costs and extended overhead are part of the cost of the labor or material furnished.[44]

Other courts, however, reject such arguments and hold that a claimant's recovery is limited to the price stated in the original contract with the surety's principal. These courts reason that the parties themselves set the price of the labor and material and that delay damages are, as the name implies, due to the subcontractor to compensate for delay rather than for supplying labor or material. Thus, there are cases holding that the surety is not liable for delay damages.[45]

---

[42] United States *ex rel.* Sherman v. Carter, 353 U.S. 210 (1957); J.W. Bateson Co. v. United States *ex. rel.* Board of Trustees, 434 U.S. 586 (1978).

[43] United States *ex rel.* West v. Peter Kiewit & Sons' Co., 235 F. Supp. 500 (D. Alaska 1964); United States *ex rel.* Bordallo Consol., Inc. v. Markowitz Bros. (Delaware), Inc., 249 F. Supp. 610 (D. Guam 1966).

[44] United States *ex rel.* Mariana v. Piracci Constr. Co., Inc., 405 F. Supp. 904 (D.D.C. 1975); United States *ex rel.* Heller Elec. Co. v. William F. Klingensmith, Inc., 670 F.2d 1227 (D.C. Cir. 1982); General Fed. Constr., Inc. v. D.R. Thomas, Inc., 451 A.2d 1250 (Md. Ct. Spec. App. 1982).

[45] McDaniel v. Ashton-Mardian Co., 357 F.2d 511 (9th Cir. 1966); United States *ex rel.* Pittsburgh-Des Moines Steel Co. v. MacDonald Constr. Co., 281 F. Supp. 1010 (E.D. Mo. 1968); United

The trend in the law, however, seems to be toward finding liability for delay damages. The payment bond statutes are universally admitted to be remedial legislation designed to protect those whose work goes into government construction projects, and they are liberally construed to effect this legislative intent. Delay damages are an attempt to quantify the increased costs incurred by a contractor or subcontractor as a consequence of a wrongful act of another party. If the bond claimant's contract entitles it to a price increase because of a compensable delay, then the surety should logically be liable for that increase as an amount justly due in return for the labor or material the claimant has furnished.[46]

At one time sureties argued that they should not be liable for a claimant's profit, since profit is neither labor nor material, but now everyone recognizes that a reasonable profit is part of the amount justly due for the labor or material furnished to the principal (but not, of course, lost profit not earned because of the principal's breach of contract).[47] Similarly, it would seem that in the future increased costs due to compensable delays will be recognized as a proper part of the surety's liability.

§ 11.10 —Attorneys' Fees and Interest

Claimants routinely demand interest and attorneys' fees, but seldom receive them. This poor record of success is partly because claimants usually make the wrong arguments to bolster their position. If a claimant's only argument in support of an award of interest or attorneys' fees is that they are necessary to make it whole for the principal's default, then it is unlikely to prevail, since the surety did not insure it against damage from its mistake in contracting with an insolvent principal. The surety is liable for the amount justly due for labor or material furnished for use on the bonded job, and interest or attorneys' fees are not labor or material. In order to have a chance to recover attorneys' fees, the claimant must point to a statute, a contract provision, or an action by the surety itself which would justify their award. The mere fact that the principal did not pay the claimant is not sufficient.

The *American Rule* is that a prevailing party is not entitled to attorneys' fees as an item of damages. The Supreme Court explicitly applied this general rule to Miller Act bonds,[48] and the logic of the Court's holding would be equally relevant

---

States *ex rel.* E&R Constr. Co., Inc. v. Guy H. James Constr. Co., 390 F. Supp. 1193 (M.D. Tenn. 1972), *aff'd,* 489 F.2d 756 (6th Cir. 1974).

[46] United States *ex rel.* Mariana v. Piracci Constr. Co., Inc., 405 F. Supp. 904 (D.D.C. 1975); United States *ex rel.* Heller Elec. Co. v. William F. Klingensmith, Inc., 670 F.2d 1227 (D.C. Cir. 1982); General Fed. Constr., Inc. v. D.R. Thomas, Inc., 451 A.2d 1250 (Md. Ct. Spec. App. 1982).

[47] Arthur N. Olive Co. v. United States *ex rel.* Marino, 297 F.2d 70 (1st Cir. 1961).

[48] F.D. Rich Co. v. United States *ex rel.* Industrial Lumber Co., 417 U.S. 116 (1974).

to other bonds if there is no controlling state statute or decision. There are four recognized exceptions to the American Rule. One allows attorneys' fees to a prevailing plaintiff who has acted in the public interest. This so-called *private attorney general exception* would not seem to be relevant to a payment bond claim.

A second exception allows attorneys' fees if a statute authorizes them to be awarded. The Freedom of Information Act, for example, provides that a plaintiff who substantially prevails shall be awarded a reasonable attorney's fee. The Miller Act and most state bond statutes are silent on the question of attorneys' fees. Some state statutes, however, do provide for fees, and a prevailing claimant on a bond under such a statute can recover. A claimant on a Miller Act bond, however, cannot recover attorneys' fees even if the state within which the project is located has such a provision in its statute.[49]

A third exception is if the surety's conduct in the handling of the claim violates an applicable statute which authorizes attorneys' fees or if the court finds the surety's conduct to have been so outrageous that attorneys' fees should be awarded as, in effect, punitive damages.[50]

A claimant's best chance of recovering attorneys' fees is under the fourth exception: fees authorized by contract. If the bond itself (either explicitly or by incorporating the contract by reference)[51] or the claimant's contract with the surety's principal[52] provides that the prevailing party will be entitled to attorneys' fees, then the claimant can argue that the attorneys' fees are part of the bargained-for price of the labor and material furnished and are justly due under the bond.

Interest is awarded to compensate one party for another party's delay in payment. If the claimant's contract with the bond principal calls for interest, it can argue that such interest is part of the amount justly due for labor or material.[53] If the contract does not provide for interest, the claimant can still obtain it if he or she would be entitled to recover it under state law. Many states, by statute or court decision, provide for interest if payment of a liquidated amount is delayed.[54]

---

[49] *Id.*

[50] United States *ex rel.* General Elec. Supply Co. v. Minority Elec. Co., 537 F. Supp. 1018 (S.D. Ga. 1982).

[51] Dale Benz, Inc. Contractors v. American Cas. Co., 303 F.2d 80 (9th Cir. 1962), *reh'g denied,* 305 F.2d 641 (9th Cir. 1962).

[52] Travelers Indem. Co. v. United States *ex rel.* Western Steel Co., 362 F.2d 896 (9th Cir. 1966); United States *ex rel.* Sherman v. Carter, 353 U.S. 210 (1957); United States *ex rel.* Carter Equip. Co. v. H.R. Morgan, Inc., 544 F.2d 1271 (5th Cir. 1977), *reh'g granted,* 554 F.2d 164 (5th Cir. 1977).

[53] D&L Constr. Co. v. Triangle Elec. Supply Co., 332 F.2d 1009 (8th Cir. 1964); United States *ex rel.* Billows Elec. Supply Co., Inc. v. E.J.T. Constr. Co., Inc., 517 F. Supp. 1178 (E.D. Pa. 1981), *aff'd,* 688 F.2d 827 (3d Cir.), *cert. denied,* 103 S. Ct. 126 (1982).

[54] Arnold v. United States *ex rel.* Bowman Mech. Contractors, 470 F.2d 243 (10th Cir. 1972); United States *ex rel.* A.V. DeBlasio Constr., Inc. v. Mountain States Constr. Co., 588 F.2d 259 (9th Cir. 1978); United States *ex rel.* Astro Cleaning & Packaging Corp. v. Jamison Co., 425 F.2d 1281 (6th Cir. 1970).

If interest is proper, it will run from the date the payment should have been made.[55] Virtually all bonds provide that the surety is liable only if the principal has failed to make payment within 90 days of the date upon which the claimant last furnished labor or material, and interest should not commence to run against the surety until that 90-day period expires. In addition, the surety cannot be expected to pay until a demand is made against it.[56] Some courts, however, have awarded prejudgment interest against the surety from the date its principal should have made payment.[57]

§ 11.11 Notice Requirements

Virtually all bonds require a claimant who does not have a direct contractual relationship with the bond principal to give written notice of its claim. Since a supplier to a materialman is not covered by most bonds,[58] the practical effect of the notice requirement is that a sub-subcontractor or a supplier to a subcontractor must give written notice of its claim or lose its rights against the prime contractor's bond.

The purpose of this notice requirement is to enable the prime contractor to protect itself by withholding payments otherwise due the subcontractor until the subcontractor pays its debt.[59] It must be remembered that a prime contractor and its surety can be liable on the payment bond to such a supplier or sub-subcontractor even though the prime contractor has paid the subcontractor in full for the work or material for which the claim is being made. The notice requirement is designed to mitigate the unfairness of imposing such double liability on the prime contractor while still giving the supplier or sub-subcontractor an opportunity to be paid under the bond.

There is no reason to require that notice be given by a materialman or subcontractor who is in a direct contractual relationship with the bond principal, since the principal already knows that such a materialman or subcontractor is, or claims to be, unpaid. Such a notice requirement would serve no purpose and could be a trap for the unwary or unsophisticated.

---

[55] United States *ex rel.* Schaefer Elec., Inc. v. O. Frank Heinz Constr. Co., 300 F. Supp. 396 (S.D. Ill. 1969).

[56] American Auto Ins. Co. v. United States *ex rel.* Luce, 269 F.2d 406 (1st Cir. 1959); United States *ex rel.* Caldwell Foundry & Mach. Co. v. Texas Constr. Co., 237 F.2d 705 (5th Cir. 1955).

[57] *See, e.g.,* Baker & Ford Co. v. United States *ex rel.* Urban Plumbing & Heating Co., 363 F.2d 605 (9th Cir. 1966); J.F. White Eng'g Corp. v. United States *ex rel.* Pittsburgh Plate Glass Co., 311 F.2d 410 (10th Cir. 1962); United States *ex rel.* Astro Cleaning & Packaging Corp. v. Jamison Co., 425 F.2d 1281 (6th Cir. 1970).

[58] See **§ 11.6.**

[59] United States *ex rel.* Honeywell v. A&L Mechanical Contractors, Inc., 677 F.2d 383 (4th Cir. 1982); United States *ex rel.* Kinlau Sheet Metal Works, Inc. v. Great Am. Ins. Co., 537 F.2d 222 (5th Cir. 1976).

Under the Miller Act, A.I.A. Document A312,[60] and most state bond statutes, notice must be given to the prime contractor. Under A.I.A. Document A311, however, notice must be given to only two out of three among the owner, the contractor, and the surety.[61] If notice is given to the owner or surety, they will presumably inform the contractor, but since the purpose of requiring notice is to alert the contractor to a default by its subcontractor, it does not seem logical to rely on such indirect notification.

The Miller Act and both A.I.A. bonds require that the notice be in writing and that it ''state with substantial accuracy the amount claimed and the name of the party to whom'' the labor or material was furnished. At least part of the notice must be in writing, but it can be supplemented by information communicated orally. Thus, once some written communication is established, the court will consider other contacts to decide whether all the elements of notice have been furnished.[62]

The courts also give a liberal interpretation to a requirement that notice be sent by registered or certified mail. The general rule is that such a requirement is merely to provide proof that the notice was received; if receipt is admitted or otherwise proven, notice will not fail because it was not sent in the prescribed manner.[63]

In judging the required contents of the notice, however, the courts are only slightly less liberal. The contractor should be informed of the approximate amount owed[64] and the subcontractor for whom the work was performed.[65] Most importantly, the notice must tell the contractor that the unpaid supplier or sub-subcontractor is looking to the contractor for payment.[66] The sufficiency of any given notice will be judged in light of the purpose of the notice requirement, but the prime contractor must be advised that the claimant is looking to it for payment so that it can withhold the amount claimed from any payments otherwise due the defaulting subcontractor.

§ 11.12 —Time Period for Giving Notice

The courts strictly enforce the requirement that notice must be given within 90 days of the last date upon which the claimant furnished labor or material for which

[60] American Institute of Architects Document A312 (Dec. 1984 ed.).

[61] American Institute of Architects Document A311, ¶ 3(a) (Feb. 1970 ed.).

[62] United States *ex rel.* Kelly-Mohrhusen Co. v. Merle A. Patnode Co., 457 F.2d 116 (7th Cir. 1972).

[63] United States *ex rel.* Hillsdale Rock Co., Inc. v. Cortelyou & Cole, Inc., 581 F.2d 239 (9th Cir. 1978); Fleisher Eng'g & Constr. Co. v. United States *ex rel.* Hallenbeck, 311 U.S. 15 (1940).

[64] United States *ex rel.* Pool Constr. Co. v. Smith Rd. Constr. Co., 227 F. Supp. 315 (D. Okla. 1964).

[65] Apache Powder Co. v. Ashton Co., Inc., 264 F.2d 417 (9th Cir. 1959).

[66] United States *ex rel.* Charles R. Joyce & Son, Inc. v. F.A. Baehner, Inc., 326 F.2d 556 (2d Cir. 1964); United States *ex rel.* Jinks Lumber Co. v. Federal Ins. Co., 452 F.2d 485 (5th Cir. 1971); Bowden v. United States *ex rel.* Malloy, 239 F.2d 572 (9th Cir. 1956), *cert. denied,* 353 U.S. 957 (1957); United States *ex rel.* Bailey v. Freethy, 469 F.2d 1348 (9th Cir. 1972).

claim is made. The determination of that date is complex, and the criteria to be considered are discussed at some length in connection with the commencement date of the suit limitations period in **§ 11.13**.

One problem unique to the 90-day notice requirement, however, is this: what happens if 90 days elapse following delivery of material, with no notice given, and then additional material is delivered, followed by timely notice? The claimant will argue that the timely notice resurrects the barred claim, while the prime contractor and surety will argue that they can be liable only for the additional material.

The Miller Act provides that the notice must be given within 90 days of the last delivery for which claim is made.[67] Thus, if the additional material is not included in the claim (if it was delivered C.O.D. or delivered directly to and paid for by the prime contractor, for example), then it will not resurrect the barred claim.[68] There is also authority holding that notice given within 90 days of delivery of material under a separate purchase order or contract on the same project will not revive a barred claim.[69]

On the other hand, if the claimant has a single contract with a subcontractor to supply material on the bonded job, a 90-day gap in deliveries will not prevent the date of the later delivery from being the last date upon which material for which claim is being made was supplied, and the additional delivery triggers a new 90-day period during which notice can be given for the entire claim.[70]

If a claimant finds that its right of action on the payment bond is or may be barred by the 90-day notice requirement, it can do one of two things. It can try to convince the court that it was a subcontractor or materialman with a direct contractual relationship with the bond principal (and thus not required to give the notice), or it can argue that the principal waived the notice requirement or is estopped to assert it.

One way of establishing a direct contractual relationship with the prime contractor is to show that the purported subcontractor with whom the claimant dealt was really an alter ego of the prime. The courts will not permit a contractor to reduce the protection afforded a subcontractor under the Miller Act by creating a "straw" or dummy subcontractor and then arguing that persons who have dealt with this subcontractor must give the 90-day notice.[71] That is not to say that a prime contractor cannot subcontract work to a related but separate entity. The courts

---

[67] 40 U.S.C. § 270b(a).

[68] United States *ex rel.* Harris Paint Co. v. Seaboard Sur. Co., 437 F.2d 37 (5th Cir. 1971); United States *ex rel.* Olmsted Elec., Inc. v. Neosho Constr. Co., Inc., 599 F.2d 930 (10th Cir. 1979).

[69] United States *ex rel.* J.A. Edwards & Co. v. Peter Reiss Constr. Co., 273 F.2d 880 (2d Cir. 1959), *cert. denied,* 362 U.S. 951 (1960); United States *ex rel.* I. Burack, Inc. v. Sovereign Constr. Co., 338 F. Supp. 657 (S.D.N.Y. 1972). But *compare* Noland Co. v. Allied Contractors, Inc., 273 F.2d 917 (4th Cir. 1959).

[70] United States *ex rel.* J.A. Edwards & Co. v. Peter Reiss Constr. Co., 273 F.2d 880 (2d Cir. 1959), *cert. denied,* 362 U.S. 951 (1960).

[71] National Sur. Corp. v. United States *ex rel.* Way Panama, S.A., 378 F.2d 294 (5th Cir.), *cert. denied,* 389 U.S. 1004 (1967).

will consider factors such as whether the prime and purported subcontractor have separate facilities, equipment, and employees, and whether they work on separate jobs; in the final analysis, the claimant will have to convince the court that in reality, if not on paper, the prime and purported subcontractor are one and the same.

If a subcontractor encounters financial difficulties, the prime contractor will often take steps to insure a smooth flow of material to the job. If the prime merely issues joint checks payable to the subcontractor and the supplier, it should not be found to have entered into a direct contractual relationship with the supplier.[72] On the other hand, if the prime contractor makes a separate agreement with the supplier, such as by guaranteeing payment or agreeing to pay the supplier directly or by taking over the subcontract work itself and requesting that the supplier or sub-subcontractor continue to work, then the court will find a direct contractual relationship and no 90-day notice will be required.[73]

If a subcontractor is in financial difficulty, the prime contractor should be very careful in any dealings with the suppliers or sub-subcontractors. There is no reason to risk incurring liability for a past-due debt barred by the supplier's failure to give the 90-day notice if an item is readily available from another source or can just as well be purchased by a replacement subcontractor.

The prime contractor should also be very careful not to waive the 90-day notice requirement or estop itself from asserting it. Anything the prime does which would *reasonably* lead an unpaid supplier or sub-subcontractor to believe it will be paid without having to give the 90-day notice invites a court to find that the notice has been waived or that the prime contractor is estopped from asserting the defense of failure to give the notice.[74] Vague assurances that the matter is being taken care of or promises of future payments are especially dangerous. Prime contractors can avoid handing claimants such waiver or estoppel arguments if they make no promises about past-due accounts and explicitly reserve all rights under the payment bond and applicable statutes. Contact with potential claimants should be in writing or, at least, followed promptly by a letter confirming the reservation of rights.

§ 11.13 Suit Limitations

The most common, serious, and easily avoided mistake committed by payment bond claimants is the failure to file suit against the surety within the time limits

[72] United States *ex rel.* State Elec. Supply Co. v. Hesselden Constr. Co., 404 F.2d 774 (10th Cir. 1968); United States *ex rel.* Fordham v. P.W. Parker, Inc., 504 F. Supp. 1066 (D. Md. 1980).

[73] United States *ex rel.* Billows Elec. Supply Co., Inc. v. E.J.T. Constr. Co., Inc., 517 F. Supp. 1178 (E.D. Pa. 1981), *aff'd,* 688 F.2d 827 (3d Cir.), *cert. denied,* 103 S. Ct. 126 (1982); United States *ex rel.* Greenwald-Supon, Inc. v. Gramercy Contractors, Inc., 433 F. Supp. 156 (S.D.N.Y. 1977); American Cas. Co. of Reading v. Southern Materials Co., 261 F.2d 197 (4th Cir. 1958).

[74] United States *ex rel.* Franklin Paint Co. v. Kagan, 129 F. Supp. 331 (D. Mass. 1955).

set by the bond or statute. Virtually all bonds require that suit be filed within one year, but the event with which that one-year period commences varies from bond to bond.

The Miller Act and most state payment bond statutes require that suit be filed within one year of the last date upon which the claimant furnished labor or materials for use in the prosecution of the contract work.[75] A.I.A. Document A311 has the one-year period run from the date upon which the principal ceased work on the bonded project, unless such a limitation is prohibited by law, in which event the limitation is the minimum permitted by law.[76] Under A.I.A. Document A312, the one-year period commences on the earlier of the dates on which the claimant gave the contractor written notice of the claim or the last date upon which anyone furnished labor or material under the construction contract.[77] Other bonds start the one-year period from final completion or final acceptance of the contract work, or from the making of final payment to the contractor.

Since the various payment bond statutes, including the Miller Act, create a remedy unknown at common law, the courts hold that suit within the one-year period is a *condition precedent* to an action on the bond: it is a limitation on the right itself, not merely on the remedy. That is, the one-year limitation is not merely a statute of limitations which will be waived if not properly pled, but a jurisdictional prerequisite which can be raised at any time, even on appeal.[78] In most cases, this distinction is one of semantics, with no practical consequences. In Miller Act cases, however, the surety can argue that a United States District Court has no subject matter jurisdiction because of the claimant's failure to file within the one-year period.

---

[75] 40 U.S.C. § 270b(b).

[76] American Institute of Architects Document A311, ¶ 3(b) (Feb. 1970 ed.):

> ¶ 3. No suit or action shall be commenced hereunder by any claimant: . . .
> b) After the expiration of one (1) year following the date on which Principal ceased Work on said Contract, it being understood, however, that if any limitation embodied in this bond is prohibited by any law controlling the construction hereof such limitation shall be deemed to be amended so as to be equal to the minimum period of limitation permitted by such law.

[77] American Institute of Architects Document A312, ¶ 11 (Dec. 1984 ed.):

> **11** No suit or action shall be commenced by a Claimant under this Bond . . . after the expiration of one year from the date (1) on which the Claimant gave the notice required . . . or (2) on which the last labor or service was performed by anyone or the last materials or equipment were furnished by anyone under the Construction Contract, whichever of (1) or (2) first occurs. If the provisions of this Paragraph are void or prohibited by law, the minimum period of limitation available to sureties as a defense in the jurisdiction of the suit shall be applicable.

[78] United States *ex rel.* Celanese Coatings Co. v. Gullard, 504 F.2d 466 (9th Cir. 1974); United States *ex rel.* Soda v. Montgomery, 253 F.2d 509 (3d Cir. 1958); United States *ex rel.* Statham Instruments, Inc. v. Western Cas. & Sur. Co., 359 F.2d 521 (6th Cir. 1966); American Masons' Supply Co. v. F.W. Brown Co., 394 A.2d 378 (Conn. 1978).

§ 11.14 —Commencement of the Limitation Period

Since construction projects often take a long time to complete even if the contractor is not in financial difficulty, a limitation period which commences upon completion of the project gives such an extended period within which to file suit that few claimants find themselves barred. Those that do typically argue that the contract is not complete (or the principal has not ceased work, or the project is not accepted) until final payment is made or even until all warranties have expired.

Many claimants, however, run afoul of the one-year limitation if it commences on the last date they furnished labor or material. Although a year would seem to be plenty of time in which to sue, many claimants are unaware of their bond rights and delay seeking legal advice. For whatever reason, a great deal of litigation has been devoted to determining the last date upon which a claimant furnished labor or material.

There is no minimum amount of work which must be done to qualify as furnishing labor or material, but the work must have been required under the prime contract.[79] A visit to the job by the claimant's employees for inspection purposes or to prepare a final estimate of labor and material supplied or to correct work previously performed but found to be defective will not count as furnishing labor or material.[80] Work performed or material delivered as a sham to gain coverage under the bond will not be considered.[81]

If the claimant is a materialman, delivery slips will almost always establish the date upon which material was furnished. If the claimant is a subcontractor, certified payrolls, daily reports, inspectors' reports, and invoices are useful in determining the nature of the work which was performed. A surety will often find that the subcontractor's own invoices show its work to be 100 percent complete long before the date it now claims to have furnished labor or material. If the claimant rented equipment, the last day upon which the equipment was available for use on the job (rather than the last day it was actually used) will start the one year running, but the claimant cannot extend the period by leaving unused equipment on the job site after a request to remove it.[82]

---

[79] United States *ex rel.* Joseph I. Richardson, Inc. v. E.J.T. Constr. Co., Inc., 453 F. Supp. 435 (D. Del. 1978).

[80] United States *ex rel.* Billows Elec. Supply Co., Inc. v. E.J.T. Constr. Co., 517 F. Supp. 1178 (E.D. Pa. 1981), *aff'd,* 688 F.2d 827 (3d Cir.), *cert. denied,* 103 S. Ct. 126 (1982); United States *ex rel.* H.T. Sweeney & Son, Inc. v. E.J.T. Constr. Co., Inc., 415 F. Supp. 1328 (D. Del. 1976); General Ins. Co. of Am. v. United States *ex rel.* Audley Moore & Son, 406 F.2d 442 (5th Cir.), *on reh'g,* 409 F.2d 1326 (5th Cir.), *cert. denied,* 396 U.S. 902 (1969); United States *ex rel.* Greenwald-Supon, Inc. v. Gramercy Contractors, Inc., 433 F. Supp. 156 (S.D.N.Y. 1977); United States *ex rel.* Altman v. Young Lumber Co., 376 F. Supp. 1290 (D.S.C. 1974); United States *ex rel.* T Square Equip. Corp. v. Gregor J. Schaefer Sons, Inc., 272 F. Supp. 962 (E.D.N.Y. 1967).

[81] United States *ex rel.* First Nat'l Bank of Jackson v. United States Fidelity & Guar. Co., 240 F. Supp. 316 (N.D. Okla. 1965).

[82] Mike Bradford & Co. v. F.A. Chastain Constr., Inc. 387 F.2d 942 (5th Cir. 1968).

The most common dispute is over labor or material supplied after substantial completion of the claimant's work. The surety will argue that any additional items are maintenance or guaranty work, while the claimant will say it is original contract work which was not done previously.

The crucial question is whether the item was never supplied, or was furnished but proved defective. If it was omitted entirely, then it is original work when it is furnished for the first time.[83] If a defective item was furnished, and is then corrected or replaced, the corrective work is not considered in computing the one-year period.

Once the commencement date of the one-year period is determined, the one year is computed according to Rule 6(a) of the Federal Rules of Civil Procedure. The day upon which the last work was performed is not included, but the final day of the one-year period is included. Thus, if January 11, 1985, is the last day upon which labor or material was furnished, suit must be filed on or before January 12, 1986.[84]

For the first 90 days of the year following the date upon which the claimant last furnished labor or material, the surety cannot be sued. The surety's liability commences if the claimant has not been paid within this 90-day period, and, theoretically, premature suits should be dismissed. If such dismissal takes place after the year has expired, the claimant loses its bond rights.

Under modern pleading, however, the complaint can be amended, or a supplemental complaint filed, to allege that the 90-day period has expired, and the amended complaint should relate back to the date of the original filing. No modern court is going to let a surety escape liability because of a premature suit or pleading mistake, if the claimant acts promptly to correct its error.[85] Although a more difficult case is presented if a claimant files a timely suit but names the wrong surety, it can probably save itself after the one-year period has expired by amending the complaint to name the proper defendant and arguing that the amended complaint relates back to the date of filing of the original complaint, if the requirements of Rule 15(c) of the Federal Rules of Civil Procedure are met.[86]

---

[83] United States *ex rel.* Lank Woodwork Co., Inc. v. CSH Contractors, Inc., 452 F. Supp. 922 (D.D.C. 1978).

[84] United States *ex rel.* Pre-Fab Erectors, Inc. v. A.B.C. Roofing & Siding, Inc., 193 F. Supp. 465 (S.D. Cal. 1961).

[85] United States *ex rel.* Capitol Elec. Supply Co., Inc. v. C.J. Elec. Contractors, Inc., 535 F.2d 1326 (1st Cir. 1976); Security Ins. Co. v. United States *ex rel.* Haydis, 338 F.2d 444 (9th Cir. 1964); United States *ex rel.* Pittsburgh-Des Moines Steel Co. v. Dauphin Steel & Eng'g Co., 53 F.R.D 382 (S.D.N.Y. 1971).

[86] United States *ex rel.* Statham Instruments, Inc. v. Western Cas. & Sur. Co., 359 F.2d 521 (6th Cir. 1966) (dissenting opinion); Travelers Indem. Co. v. United States *ex rel.* Construction Specialities Co., 382 F.2d 103 (10th Cir. 1967); 3 Moore's Federal Practice ¶ 15.15 [4.-2].

§ 11.15 —Estoppel

Whatever the outcome of the unusual cases in which a timely suit is filed against the wrong defendant or a premature suit is filed, the one-year limitation is a very powerful defense which time and time again allows sureties to defeat otherwise valid claims. If the one-year period has expired, a claimant loses its rights against the payment bond unless it can convince the court that the surety should be estopped from asserting the limitations defense.

Although rare, such an estoppel has been found where the surety misled the claimant into reasonably believing that its claim would be paid without the necessity of filing suit.[87] If the surety represents to a claimant that it will be paid, but asks for more time in which to investigate, it is inviting an estoppel argument. For this reason anyone defending against a claim should be sure to make no promises and to reserve all rights under the bond and any applicable statute. Such a reservation should always be made or confirmed in writing, and many sureties make it a standard practice to write each claimant suggesting that the claimant take whatever steps it deems necessary to protect its position.

§ 11.16 Jurisdiction and Venue Requirements

The Miller Act provides that suit must be brought "in the United States District Court for any district in which the contract was to be performed and executed and not elsewhere . . . ."[88] Almost all other bonds require that suit be filed in the state court for the political subdivision in which the project is located or in the federal court for the district in which the project is located. The Miller Act gives exclusive jurisdiction to the federal courts without regard to the amount in controversy or the citizenship of the parties.[89] The parties cannot confer jurisdiction on a state court by contract or subsequent agreement, and a suit filed in a state court is ineffective even to toll the running of the one-year limitation period.[90]

The federal court in which the suit is filed, however, is only a venue requirement, and can be waived by the parties.[91] If suit is erroneously filed in a federal district other than one where the contract is to be performed, the defendants may

---

[87] United States *ex rel.* Nelson v. Reliance Ins. Co., 436 F.2d 1366 (10th Cir. 1971); United States *ex rel.* E.E. Black, Ltd. v. Price-McNemar Constr. Co., 320 F.2d 663 (9th Cir. 1963); United States *ex rel.* Humble Oil & Ref. Co. v. Fidelity & Cas. Co., 402 F.2d 893 (4th Cir. 1968).

[88] 40 U.S.C. § 270b(b).

[89] *Id.*

[90] United States *ex rel.* Harvey Gulf Int'l Marine, Inc. v. Maryland Cas. Co., 573 F.2d 245 (5th Cir. 1978).

[91] United States *ex rel.* Caswell Equip. Co. v. Fidelity & Deposit Co., 494 F. Supp. 354 (D. Minn. 1980); United States *ex rel.* Capolino Sons, Inc. v. Electronic & Missile Facilities, Inc., 364

waive the improper venue by failing to make a timely objection; if objection is made, the suit may be transferred to the proper district.[92] Some courts have even held that the bond principal may contract to have venue in another district and that the surety is bound by the principal's contract.[93]

If a Miller Act bond is given for a contract to be performed outside the United States, a claimant will not be deprived of a remedy for lack of proper venue. Even over the objection of the defendants, the suit will be heard.[94]

§ 11.17 Possible Remedies if Payment Bond Claim Is Unavailable

Even though a statute or contract may have required a payment bond, it is possible for a person who has supplied labor or material to an insolvent contractor not to have bond protection. Claims against the bond can exceed the bond penalty, or a claimant can lose its bond remedy by failing to file a timely suit against the surety. Although rare, there are even cases in which the required bond was never furnished.

A claimant who has the dual misfortune of selling to an insolvent contractor and finding that its payment bond remedy is unavailable very likely will not be paid. In some circumstances, however, such a claimant can still recover all or part of what it is owed.

The first possible source of payment is any contract balance still in the hands of the owner. Once the contract is complete, the owner is a stakeholder of any remaining contract funds, and persons who have supplied labor or material for use in the performance of the contract have an equitable claim to be paid out of

---

F.2d 705 (2d Cir.), *cert. denied,* 385 U.S. 924 (1966); United States *ex rel.* Angell Bros., Inc. v. Cave Constr., Inc., 250 F. Supp. 873 (D. Mont. 1966); but *compare* United States *ex rel.* Coffey v. William R. Austin Constr. Co., 436 F. Supp. 626 (W.D. Okla. 1977).

[92] 28 U.S.C. § 1406(a). *See* Griners' & Shaw, Inc. v. Federal Ins. Co., 234 F. Supp. 753 (D.S.C. 1964); United States *ex rel.* Caswell Equip. Co. v. Fidelity & Deposit Co., 494 F. Supp. 354 (D. Minn. 1980); United States *ex rel.* Essex Mach. Works, Inc. v. Rondout Marine, Inc., 312 F. Supp. 846 (S.D.N.Y. 1970); United States *ex rel.* Harvey Gulf Int'l Marine, Inc. v. Maryland Cas. Co., 573 F.2d 245 (5th Cir. 1978).

[93] United States *ex rel.* Fireman's Fund Ins. Co. v. Frank Briscoe Co., 462 F. Supp. 114 (E.D. La. 1978); United States *ex rel.* Caswell Equip. Co. v. Fidelity & Deposit Co., 494 F. Supp. 354 (D. Minn. 1980); but *compare* United States *ex rel.* Vermont Marble Co. v. Roscoe-Ajax Constr. Co., 246 F. Supp. 439 (N.D. Cal. 1965).

[94] United States *ex rel.* Bryant Elec. Co. v. Aetna Cas. & Sur. Co., 297 F.2d 665 (2d Cir. 1962); United States *ex rel.* Bailey-Lewis-Williams of Fla., Inc. v. Peter Kiewit Sons Co., 195 F. Supp. 752 (D.D.C. 1961), *aff'd,* 299 F.2d 930 (D.C. Cir. 1962).

the contract balance.[95] If the owner is not protected by sovereign immunity, the unpaid subcontractor or supplier can sue to recover the amount due.[96] If the owner is the United States, sovereign immunity will bar a direct suit,[97] but the unpaid claimant can prevent anyone else from obtaining the funds. The court may authorize the United States to pay the claimant, but cannot compel the government to do so.[98]

A second possible source of recovery is any other bond which the contractor may have furnished. That is, if a payment bond was not written but a bid bond or performance bond was, the unpaid subcontractor or supplier can argue that it was a third-party beneficiary of the bid or performance bond.[99]

§ 11.18 Practical Aspects of a Payment Bond Claim

If a contractor is solvent and appears to be generally meeting its obligations, its surety is very unlikely to become involved in a dispute between the contractor and its subcontractors or suppliers. The surety is a financial guarantor, not a judge or referee, and if the principal denies owing a claim the surety will normally support it. A claimant will do better to put its efforts into reaching an agreement with the contractor than into trying to use the surety to pressure the contractor into paying a disputed claim.

Once it becomes apparent that the contractor is in financial difficulty, however, the unpaid subcontractor or materialman should immediately present a claim to the surety. The first step is to get a copy of the bond from the owner[1] and be sure of the identity of the surety. The bond and any applicable statute should be carefully studied for any notice or limitation requirements, and a detailed, supported claim sent to the surety.

---

[95] American Sur. Co. of N.Y. v. Westinghouse Elec. Mfg. Co., 296 U.S. 133 (1935); Pearlman v. Reliance Ins. Co., 371 U.S. 132 (1962); United States Fidelity & Guar. Co. v. United States, 475 F.2d 1377 (Ct. Cl. 1973).

[96] Kennedy Elec. Co., Inc. v. United States Postal Serv., 508 F.2d 954 (10th Cir. 1974).

[97] United Elec. Corp. v. United States, 647 F.2d 1082 (Ct. Cl. 1981).

[98] United States Fidelity & Guar. Co. v. United States, 475 F.2d 1377 (Ct. Cl. 1973); North Denver Bank v. United States, 432 F.2d 466 (Ct. Cl. 1970).

[99] United States *ex rel.* Empire Plastics Corp. v. Western Cas. & Sur. Co., 429 F.2d 905 (10th Cir. 1970); United States *ex rel.* Victory Elec. Corp. v. Maryland Cas. Co., 213 F. Supp. 800 (E.D.N.Y.), *further proceedings*, 215 F. Supp. 700 (E.D.N.Y. 1963); 17 Am. Jur. 2d *Contractors' Bonds* § 19; *cf.* Johns-Manville Sales Corp. v. Reliance Ins. Co., 410 F.2d 277 (9th Cir. 1969).

[1] The Miller Act (40 U.S.C. § 270c) and most state bond statutes require the government to furnish a certified copy of the payment bond to any potential claimant.

Most sureties will want to see delivery slips, invoices, copies of any purchase orders or subcontracts, and a summary of any running account showing purchases for and payments made on account of their bonded job. If the claimant has a written admission by the bond principal of the amount due, a copy should be sent to the surety along with an unequivocal demand for immediate payment.

Sureties are generally large, bureaucratic organizations, and it is easy for letters to be filed away and forgotten. If a satisfactory response is not received promptly, the surety itself (not its local agent) should be telephoned. A claimant should, if necessary, call the bond claim department in the surety's home office and ask about the status of its claim. One should ask for the claim number and the name and mailing address of the person handling the file, and address future correspondence to that person. If the surety has not set up a claim file and assigned a claim number, it should be a warning that they are not planning to take any action in the near future.

A surety will certainly need time to obtain information from its principal and consider its position, but if a commitment to pay an undisputed claim has not been made within a reasonable time, the claimant should write a letter threatening to file suit for the undisputed amount, plus interest and attorneys' fees, if a commitment to pay the claim is not made by a date certain in the near future. If the surety does not respond satisfactorily to the threatening letter, file suit. Some suits may be filed on claims which the surety would have paid anyway, but it is better to file suit than to join the legion of claimants who delayed so long that their claim was barred.

The surety is entitled to information and supporting documents, and they should be supplied promptly, but once it has the facts the surety should be able to make a decision. In surety claims, as in much else in life, the squeaky wheel gets the grease.

§ 11.19 Conclusion

Almost every court considering a payment bond claim notes that the Miller Act, or applicable state statute, is highly remedial legislation designed to protect those whose work goes into public works projects, and that the bond should receive a liberal interpretation in order to carry out its purpose. The courts usually add, often in the same sentence, that a liberal interpretation does not justify ignoring plain words of limitation in order to impose wholesale liability on sureties.

The 90-day notice requirement and one-year suit limitation will be enforced by the courts, and a contractor should be aware of the provisions of the bond and statute before it starts work on the project. There is no reason why a person who has supplied labor or material to a contractor or subcontractor on a public project should go unpaid, but many do because they fail to take the steps necessary to preserve their bond rights.

APPENDIX 11-1

THE MILLER ACT, 40 U.S.C. §§ 270a-270c

§ 270a.

(a) Before any contract, exceeding $2,000 in amount, for the construction, alteration, or repair of any public building or public work of the United States is awarded to any person, such person shall furnish to the United States the following bonds, which shall become binding upon the award of the contract to such person, who is hereinafter designated as "contractor":

(1) A performance bond with a surety or sureties satisfactory to the officer awarding such contract, and in such amount as he shall deem adequate, for the protection of the United States.

(2) A payment bond with a surety or sureties satisfactory to such officer for the protection of all persons supplying labor and material in the prosecution of the work provided for in said contract for the use of each such person. Whenever the total amount payable by the terms of the contract shall be not more than $1,000,000 the said payment bond shall be in a sum of one-half the total amount payable by the terms of the contract. Whenever the total amount payable by the terms of the contract shall be more than $1,000,000 and not more then $5,000,000, the said payment bond shall be in a sum of 40 per centum of the total amount payable by the terms of the contract. Whenever the total amount payable by the terms of the contract shall be more than $5,000,000 the said payment bond shall be in the sum of $2,500,000.

(b) The contracting officer in respect of any contract is authorized to waive the requirement of a performance bond and payment bond for so much of the work under such contract as is to be performed in a foreign country if he finds that it is impracticable for the contractor to furnish such bonds.

(c) Nothing in this section shall be construed to limit the authority of any contracting officer to require a performance bond or other security in addition to those, or in cases other than the cases specified in subsection (a) of this section.

(d) Every performance bond required under this section shall specifically provide coverage for taxes imposed by the United States which are collected, deducted, or withheld from wages paid by the contractor in carrying out the contract with respect to which such bond is furnished. However, the United States shall give the surety or sureties on such bond written notice, with respect to any such unpaid taxes attributable to any period, within ninety days after the date when such contractor files a return for such period, except that no such notice shall be given more than one hundred and eighty days from the date when a return for the period was required to be filed under the Internal Revenue Code of 1954. No suit on such bond for such taxes shall be commenced by the United States unless notice is given as provided in the preceding sentence, and no such suit shall be commenced after the expiration of one year after the day on which such notice is given.

§ 270b.

(a) Every person who has furnished labor or material in the prosecution of the work provided for in such contract, in respect of which a payment bond is furnished under section 270a of this title and who has not been paid in full therefor before the expiration of a period of ninety days after the day on which the last of the labor was done or performed by him or material was furnished or supplied by him for which such claim is made, shall have the right to sue on such payment bond for the amount, or the balance thereof, unpaid at the time of institution of such suit and to prosecute said action to final execution and judgment for the sum or sums justly due him: *Provided, however,* That any person having direct contractual relationship with a subcontractor but no contractual relationship express or implied with the contractor furnishing said payment bond shall have a right of action upon the said payment bond upon giving written notice to said contractor within ninety days from the date on which such person did or performed the last of the labor or furnished or supplied the last of the material for which such claim is made, stating with substantial accuracy the amount claimed and the name of the party to whom the material was furnished or supplied or for whom the labor was done or performed. Such notice shall be served by mailing the same by registered mail, postage prepaid, in an envelope addressed to the contractor at any place he maintains an office or conducts his business, or his residence, or in any manner in which the United States marshal of the district in which the public improvement is situated is authorized by law to serve summons.

(b) Every suit instituted under this section shall be brought in the name of the United States for the use of the person suing, in the United States District Court for any district in which the contract was to be performed and executed and not elsewhere, irrespective of the amount in controversy in such suit, but no such suit shall be commenced after the expiration of one year after the day on which the last of the labor was performed or material was supplied by him. The United States shall not be liable for the payment of any costs or expenses of any such suit.

§ 270c.

The Comptroller General is authorized and directed to furnish, to any person making application therefor who submits an affidavit that he has supplied labor or materials for such work and payment therefor has not been made or that he is being sued on any such bond, a certified copy of such bond and the contract for which it was given, which copy shall be prima facie evidence of the contents, execution, and delivery of the original. Applicants shall pay for such certified copies such fees as the Comptroller General fixes to cover the cost of preparation thereof.

CHAPTER 12

EFFECTIVE PREPARATION FOR A CONSTRUCTION ARBITRATION

Robert S. Peckar

Robert S. Peckar is a partner in the law firm of Peckar & Abramson, with offices in New Jersey, New York, and Florida, specializing in representation of members of the construction industry. Mr. Peckar, who is general counsel to the Building Contractors Association of New Jersey, has authored numerous articles on construction problems and has lectured to organizations representing contractors, design professionals, sureties, and other industry groups. He represents major contractors throughout the United States. He is a graduate of Rutgers College (AB) and Columbia Law School (JD). He is former chairman of the Public Contract Law Committee of the New Jersey State Bar Association and is presently Chairman of Region III of the Public Contract Law Section of the American Bar Association.

§ 12.1 Introduction

Thorough preparation is essential in all manner of dispute resolutions. It is not necessary for this chapter to advocate the necessity for preparation generally; rather, this chapter provides a practical guide for the unique preparation advisable for a construction arbitration.

For the purposes of providing a meaningful context for this chapter, it is assumed that an arbitration has already been commenced in accordance with the filing of a demand for arbitration under the auspices of the American Arbitration Association, pursuant to the American Arbitration Association's Construction Industry Arbitration Rules,[1] or pursuant to a submission by the parties to some other arbitration procedure. It is also assumed that the disputes involve substantially complex issues and/or substantial sums.

[1] As amended and in effect February 1, 1984. All references to the A.A.A. Rules or the Construction Industry Arbitration Rules are to these rules.

DIFFERENCES BETWEEN PREPARATION FOR LITIGATION AND PREPARATION FOR ARBITRATION

§ 12.2 Overview

Those who have made the tactical error of coming before a panel of arbitrators inadequately prepared are usually the victims of common misperceptions and erroneous generalizations concerning arbitration. The root of the problem is attitudes such as:

> "Arbitration is informal."
>
> "Arbitration is where I sit down and tell my story to my fellow industry members and they will recognize the validity of my case. That will be that."
>
> "It does not matter what you do or say, the arbitrators are going to split your demand in half anyhow."
>
> "The law has nothing to do with arbitration."
>
> "Why pay my lawyer to prepare for the case or my testimony? In arbitration I can say what I want without any judge to stop me."

Some of these misperceptions have even been afforded respectability by inclusion in published works. For example, in one textbook for construction management students, the author states:

> The advantages of arbitration are the following:
>
> 2. Legal costs are reduced, both because the sessions are shorter and because the parties concerned can state their own case without the difficulties in admission of evidence they would encounter in Court. The lawyer, if used, spends less time in preparation of the case. In a technical [Court] case, an attorney must spend a great deal of time preparing the matter so it is apparent to a lay jury; this preparation is unnecessary in arbitration.[2]

As the above are in fact misconceptions, it is worthwhile briefly to review what the arbitration process is and why that process requires a very high level of preparation.

§ 12.3 Informality

Arbitration is not a formal court proceeding; however, most arbitrations are far from informal. The claimant presents its case first and the respondent usually has

---

[2] R. King, The Construction Manager (Prentice-Hall, Inc. 1974).

the opportunity to cross-examine the claimant's witnesses before having the opportunity and burden to present its own case.[3] Testimony is usually offered in the identical question-and-answer format as in a court trial. As there is often at least one attorney sitting as an arbitrator, a reasonable amount of objections by counsel to the evidence offered by his or her adversary will be tolerated by the panel of arbitrators. These objections will be heard, if not to apply court rule restrictions, then for the instructive value to the arbitrators of the explanations and arguments accompanying these objections. Adding to this formalized informality is the practice of using court stenographers to aid the arbitrators and parties in overcoming the time lapse between hearings.[4] With transcribed hearings, the level of formality is of course heightened.

The inapplicability of the rules of evidence which apply in court proceedings is a major factor in the need for extraordinary preparation. The Construction Industry Arbitration Rules of the American Arbitration Association provide that: "The arbitrator shall be the judge of the relevance and the materiality of the evidence offered, and conformity to legal rules of evidence shall not be necessary."[5]

While adherence to the rules of evidence in court imposes a somewhat cumbersome and time-consuming ritual, adherence to these rules also affords predictable protection against certain types of evidence which have proved to be unreliable. For example, a court would not permit testimony by a subcontractor's carpenter foreman that he knew that the general contractor used inadequate concrete by his statement that "one of the guys on the job mentioned a problem with concrete to a buddy of mine and he then told me." This statement would be inadmissible upon the grounds that such a statement constitutes hearsay. In an arbitration, the statement would likely be heard by the arbitrators. Therefore, the well-prepared attorney and his or her client must be able to handle the unpredictable. To do so requires imaginative and thorough preparation.

§ 12.4 Role of Law

It is true that the body of court cases and the legal principles and doctrines which have developed from them are technically not binding upon arbitrators. However, that fact does not serve as a limitation upon the right or prerogative of a party in an arbitration to bring supportive court cases to the attention of the arbitrators as guidance in resolving the dispute. Since an attorney's skill is evidenced by applying the facts of a case to existing law and urging a consistent result, it should come as no surprise that judicial decisions are often cited by lawyers to arbitrators, and legal briefs are both offered to and requested by arbitrators.

There is a danger in this practice, though. Judges are trained (and usually experienced enough) to be able to discern inappropriately cited cases, to distinguish

---

[3] A.A.A. Rule 29, "Order of Proceedings."

[4] See § **12.10**.

[5] A.A.A. Rule 31.

between cited cases and the case before them, and to go to their libraries to verify and further research the cases cited to them. The value of contractor and architect/engineer arbitrators is their technical knowledge and familiarity with the construction industry. These industry arbitrators are not trained to evaluate legal arguments formulated upon judicial decisions. Thus, the shifting of issues in an arbitration from facts to cited legal cases should be considered dangerous.

Purposefully ignored legal decisions and statutory law in the presentation of a case to arbitrators, or prevailing upon the arbitrators that they should ignore legal precedent, is not without danger either. While most courts refuse to invade the realm of arbitrator discretion even upon a determination that the arbitrators ignored or misapplied a legal principle, there have been cases where courts vacated arbitration awards upon such grounds.[6] The role of the law, therefore, is not at all outside the parameters of many arbitrations.

§ 12.5 Role of Lawyers

In virtually every significant arbitration, the parties are represented by lawyers—and wisely so. Lawyers are trained to present cases, and in the substantial arbitration the services of the competent attorney are invaluable. However, the training and competence of that lawyer must express itself in keeping with the tone and setting of a construction industry arbitration. Arbitrators readily perceive and react to the differences between attorneys who act in a manner which appears consistent with the arbitrators' desire to receive evidence in an orderly, expeditious, competent, and fair way and those who impose courtroom behavior and tactics upon the intimacy of the arbitration table.

§ 12.6 Prehearing Discovery

One of the major features of litigation is the ability to learn about the other party's case through the use of depositions of witnesses, the propounding of interrogatories in writing, the inspection and copying of documents, and other pretrial discovery procedures. The primary purposes of discovery in civil litigation are to avoid surprise during trial, to avoid the enormous consumption of court time while parties go on "fishing expeditions" during trial, to permit the parties to learn relevant information about the dispute that they otherwise would be unable to learn without access to their adversary's files and witnesses and, most importantly, to cause the parties to be so aware of each other's strengths and weaknesses as to generate

[6] Under the Uniform Arbitration Act and most arbitration statutes, errors of "law" by arbitrators are not grounds to vacate or modify an arbitration award. Yet, in some reported and unreported court decisions, courts have in fact vacated awards on the grounds that the arbitrators misapplied the law. *See, e.g.*, Lewis v. Erie Ins. Exch., 421 A.2d 1214 (Pa. Super. Ct. 1980).

the likelihood of a settlement. In recent years, there has been substantial discussion concerning abuse of the discovery process, as the cost, both in time and dollars, may exceed a reasonable relationship with the magnitude of the dispute and the sums involved in the litigation. One of the features of arbitration which may be advantageous is the absence of automatic prehearing discovery.[7] There is no provision in the A.A.A.'s Construction Industry Arbitration Rules which specifically permits prehearing discovery. However, the statutes of some jurisdictions permit discovery in aid of an arbitration upon the direction of the arbitrators or upon application to a court.[8]

In most instances the first hearing of arbitration will proceed within six months after the demand for arbitration has been filed; thus, it is more likely than not that the parties will not have the opportunity for discovery of any kind unless they mutually agree to conduct expeditious discovery. The American Arbitration Association now appears to be advocating the utilization of discovery in the complex commercial case. In an article by Allen Poppleton, which the American Arbitration Association reprinted for guidance of arbitrators in expediting large and complex commercial cases,[9] a new role for prearbitration discovery is advocated, even though the Construction Industry Arbitration Rules do not provide for the discovery which is promoted. That article states:

> **Exchange of Information**
>
> The Arbitrator should advise that the parties will be required to cooperate with each other in permitting, conducting and completing the exchange of information concerning each other's documents and witnesses. It should be made clear that any questions as to whether something is or is not within the scope of a request to exchange should be resolved in favor of production rather than concealment.
>
> If, despite this mandate to cooperate, the parties cannot resolve the difference of opinion, the problem should be referred to the arbitrator for an immediate ruling on the issue . . . .[10]

In another section of the Poppleton article, reference is made to "[t]estimony from a witness whose deposition has been taken. . . ."[11] It appears that the arbitration

---

[7] McDonald, *Major Construction Claims/Arbitration or Litigation,* Construction Specifier 62 (Oct. 1984).

[8] The Uniform Arbitration Act of 1955, which has been adopted by a majority of states, specifically permits the issuance by the arbitrators of subpoenas to compel the attendance of witnesses at hearings or depositions, as well as the production of documents and other evidence. However, the availability and issuance of subpoenas are not the equivalent of broad pretrial discovery available in the court action setting.

In New York, for example, the civil practice rules specifically provide that a party may seek a court order to provide a party with discovery in aid of an arbitration. N.Y. Civ. Prac. Rules § 3102 (McKinney). *See also* Willenhen, *Discovery in Aid of Arbitration,* 6 Litigation 16 (1980).

[9] Poppleton, *The Arbitrator's Role in Expediting the Large and Complex Commercial Case,* 36 Arb. J. 6 (Dec. 1981).

[10] *Id.* 7.

[11] *Id.* 8.

of large and complex commercial cases will undergo substantial changes which are likely to include a prehearing discovery phase.

There is another issue related to prehearing discovery which presents a compelling basis for permitting discovery. Under the Construction Industry Arbitration Rules, the arbitration may be initiated upon the filing of a demand for arbitration, which is required to contain specified information as follows: "[T]he notice shall contain a statement setting forth the nature of the dispute, the amount involved and the remedy sought . . . ."[12] In practice, some claimants and respondents, in filing of their claims in arbitration, offer no more insight into the nature of the disputes to be arbitrated and the amount in controversy than the cryptic statements set forth in the following example:

Nature of Dispute:

Breach of Construction Contract

The Amount Involved:

The amount in controversy has not yet been fully determined, but is greater than the amount of $10,000.

The Remedy Sought:

An award against the respondent for claimant's damages plus interest, costs of suit, and reasonable attorneys' fees.

As preposterous as it might sound, some parties actually find themselves attending the first hearing of arbitration knowing nothing more about the other party's case than the above-stated "one liners." Such a situation is intolerable. A method for resolving this problem is discussed in **§§ 12.11-12.24.**

§ 12.7 Arbitration Awards

Despite the gross and inaccurate generalization that arbitrators "split the baby," it is more common for arbitrators to render a carefully considered award following study of the evidence and lengthy deliberation among the arbitrators. Major victories and major losses upset many parties to arbitration, as some claimants harbor false expectations of receiving at least half of whatever sum they demanded and some respondents foolishly expect to avoid having to pay at least half of a valid debt. Some arbitration panels have even awarded a claimant more than the sum demanded when convinced of the respondent's bad faith. Hence, a party entering an arbitration should expect a favorable award if its position is valid and effectively presented to the arbitrators.

---

[12] A.A.A. Rule 7, "Initiation under an Arbitration Provision in a Contract."

§ 12.8 Arbitrators and Their Conduct

Perhaps the greatest distinction between litigation and arbitration is the identity of the triers of fact. In litigation, the trier of fact is either a judge or a jury. Aside from the specialized courts and administrative tribunals that hear construction disputes, very few judges have any background at all in construction or in the law relating to the resolution of construction disputes. Fewer judges have any background in the technical elements of engineering and construction which are inherent aspects of most construction disputes. Certainly, a jury selected from the community is not likely to have an understanding of engineering or construction. Therefore, an alluring feature of arbitration from the perspective of a contractor is the ability to present his or her case to a group of individuals who are knowledgeable about construction and construction disputes.

Another distinction between arbitration and litigation is the ability of the parties to a dispute to select the individuals who will be the triers of fact. While it is true that the jury selection process affords some degree of control to the attorney, for the most part that process merely affords the opportunity to eliminate obviously biased or unqualified jurors.

Parties rarely are able to have any say in the court's decision as to which judge will try their case. In arbitration, the parties may select the individuals either by stipulation or in accordance with the applicable rules governing the arbitration. In A.A.A. arbitration, arbitrators are proposed by the A.A.A. with biographical data provided and the parties are afforded the opportunity to select from that list of proposed arbitrators.

The selection of arbitrators is of critical importance to the resolution of a dispute and should be afforded extremely close attention by the attorney. Aside from the obvious desire to eliminate arbitrators who would have a conflict or some personal interest in the outcome of the arbitration, the attorney must carefully evaluate the proposed arbitrators from the standpoint of their experience in construction and their experience as an arbitrator, from the standpoint of the discipline in which they have experience or training, and otherwise to assure that an arbitrator who is likely to understand and appreciate the party's position will be included on the panel.

A typical three-member panel includes an attorney, a contractor, and an architect or engineer.[13] There is no rule or protocol which determines which member of the panel will serve as its chairman, yet the attorney member almost always assumes the chairman's seat. Some such chairmen are extremely deferential to their industry member co-arbitrators, while others preempt the meaningful participation

[13] The A.A.A. Rules refer to a singular "arbitrator" (*see* A.A.A. Rule 13, "Appointment from Panel"). However, the Rules also provide that the A.A.A. may, in its discretion, direct that a "greater number of arbitrators be appointed," if the arbitration agreement does not so provide (*see* A.A.A. Rule 17, "Number of Arbitrators"). Accordingly, a *request* for the appointment of a three-member panel must be made in the absence of a contractual requirement. The A.A.A. usually appoints a three-member panel in complicated cases and in cases involving a significant sum which exceeds a threshold set by the A.A.A.

of the nonlawyer arbitrators. In those few instances when no attorney sits on a panel, or where a nonattorney serves as chairman, hearings range from well-conducted business sessions to formless meetings in which abuse of arbitration's weaknesses is rampant.[14]

Arbitrators often interrupt the parties and their attorneys with questions and think aloud about the issues before them. Arbitrators usually accept virtually every document placed before them as evidence, even if the document is obviously self-serving. However, they do so with the condition that in "admitting the document as evidence" (note the formality of this procedure!), they will afford it "whatever weight it deserves." Whatever that statement means, the arbitrators are going to read the document.

Unlike judges who have the benefit of pleadings, motions, and other proceedings before trial, all of which educate the court as to the nature of the case, arbitrators arrive at the first hearing with nothing more than the demand and the answering statement. Those documents are usually so cryptic as to be virtually useless as an informational tool.[15] Therefore, arbitrators respond very favorably to a well-prepared and well-presented introduction at early sessions and a continuation of that level throughout the hearings. The American Arbitration Association promotes early presentation of an overview of the case to the arbitrators at procedural conferences held in advance of hearings.[16]

§ 12.9 Finality of the Decision

It is a prudent assumption that the arbitration award rendered by the arbitrators will be final and binding. The grounds upon which most state and federal courts will vacate an arbitration award are restricted to circumstances which hardly ever occur.[17]

---

[14] McDonald, *Major Construction Claims/Arbitration or Litigation,* Construction Specifier 63 (Oct. 1984).

[15] See **§ 12.6**.

[16] Poppleton, *The Arbitrator's Role in Expediting the Large and Complex Commercial Case,* 36 Arb. J. 6, 7 (Dec. 1981):

> Immediately upon being appointed in a large and complex case the arbitrator should assume control of the case by calling for a procedural conference in advance of the evidentiary hearing. At that conference counsel for each party should be required to give a brief outline of the case and the issues involved, the information that must be exchanged before the evidentiary hearing, the anticipated length of the evidentiary hearing, and any particular scheduling problems.
>
> Having thus gained a preliminary overview of the case, the arbitrator should thereupon (a), advise the parties that he or she intends to conduct the case in an expeditious manner and (b), then proceed to establish the ground rules to be followed before, during and after the evidentiary hearing.

[17] The grounds for vacating an arbitration award are extremely limited, to such circumstances as:

(1) Egregious and prejudicial conduct by the arbitrators exceeding the bounds of their authority
(2) An undisclosed conflict of interest of an arbitrator and a party

§ 12.10 Time and Cost

Disputes reach hearings before arbitrators far sooner than comparable cases reach the courtroom, as long as both parties seek an expedited result. Unfortunately, one party can easily and successfully delay an arbitration for quite some time by using dilatory tactics. Once the arbitration hearings commence, they usually occur with lengthy intervals between hearings over a prolonged period. It is not at all uncommon for 20 arbitration hearings to span several years. This intermittent procedure results from the understandable difficulty in coordinating the business and vacation commitments of three arbitrators, the parties, the attorneys, and the witnesses—from as few as seven people to as many as a dozen or more—just to select one day for a hearing. An obvious adjunct of this situation is loss of memory, loss of familiarity with detail and evidence, and the need to backtrack at each hearing to overcome the impact of these interruptions upon the arbitrators and the parties.

A METHOD OF PREPARATION FOR CONSTRUCTION ARBITRATION

§ 12.11 Introduction

The foregoing is a description of the arbitration process as it really is. A description of the particular preparation that should be accomplished by the serious party and its counsel in view of that description follows. Every attorney and each client has his or her own style and method. The method set forth below is the author's and it has been used with significant success.

§ 12.12 Knowledge of the Adversary's Claim

As stated earlier, a party may commence an arbitration or file a counterclaim in arbitration without affording the other party very much information about the claims to be presented. Indeed, the example given in **§ 12.6** was drawn from significant cases where one of the parties attempted to avoid allowing the other party even to understand the degree of significance of the claim asserted in the arbitration. There are several ways to compel the other party to disclose what its claims are, and the exercise of one or more of these methods is essential to the proper preparation of the case. Without information as to the magnitude of the claims being asserted against a party, and without a clear understanding of the matters in dispute, it is

---

(3) Actual fraud by the arbitrator

(4) The arbitrator's failure to follow the procedures under the applicable arbitration statute.

impossible to prepare properly for the arbitration. In fact, without this information, a party is deprived of a fair understanding of the seriousness of the impending arbitration.

§ 12.13 —Application to the American Arbitration Association for Relief

In the event that the opposing party files a cryptic nondisclosing demand or counterclaim which fails to comply with the obvious intent of A.A.A. Rule 7 (that the nature and amount of the dispute and the remedy sought be reasonably described), a letter should be directed to the American Arbitration Association's regional director insisting that the rule has not been complied with and contending that, absent such compliance, the arbitration may not be administered by the American Arbitration Association. The A.A.A. is usually not inclined to *direct* a party to do anything; it is the essential nature of the A.A.A. to serve merely as a valuable administrator in the arbitration process and not to substitute itself for the judgment of the arbitrators.

However, there are some arguments which nonetheless motivate the American Arbitration Association to take action. One of those arguments is based upon the fact that A.A.A. Rule 7 also requires that the appropriate filing fee (as provided in § 48 of the A.A.A. Rules) be submitted to the American Arbitration Association together with the demand or answering statement. Those fees are set forth in a schedule contained within the A.A.A. Rules. The A.A.A.'s fees are computed upon a percentage of the amount of claim or counterclaim. Although the Rules do not clearly enable the A.A.A. to avoid the administration of a case due to the reluctance of a party to disclose the amount of its claim and pay the required fee over and above the $200 initiation fee, in practice the AAA can be prodded to exert substantial pressure upon a party to express the dollar amount of the claim and to file an appropriate filing fee.

In the event that the foregoing method is unsuccessful, a party can request a prehearing conference with the regional director pursuant to A.A.A. Rule 10. At that conference, the regional director will attempt to mediate between the parties and suggest that the amount in dispute and the nature of the dispute be more adequately disclosed. A.A.A. Rule 10 specifically contemplates that "an exchange of information" will be accomplished as a result of the conference. If the opposing party is purposely avoiding the disclosure of its claim, the prehearing conference will be of no particular value, because the A.A.A.'s case administrator has no authority to impose any sanctions or to involve himself or herself in discussions with the arbitrators about its perceptions of the behavior of one party or the other.

The ultimate step available within the context of the A.A.A. is a conference with the arbitrators, at which time application can be made for an order compelling the other party to disclose its claims in a sufficient manner. Rule 53 of the

Construction Industry Arbitration Rules provides that the ''Arbitrator shall interpret and apply these Rules insofar as they relate to the arbitrator's powers and duties.'' Rule 31 requires that the parties produce ''such additional evidence as the arbitrator may deem necessary to an understanding and determination of the controversy.'' It is clear that the arbitrators may compel, by the issuance of an order, the furnishing of information which will enable the other party to understand the nature and scope of the claims asserted against it.

It is very likely that the arbitrators would compel such a result and that any attempts by the other party to avoid cooperating will be unsuccessful and severely prejudice the arbitrators' view of that party. However, it should be noted that the first time the Construction Industry Arbitration Rules provide a circumstance which could result in the other party being *compelled* to come forward with a reasonable statement of the value and nature of its claims is after the arbitrators have been appointed, in a prehearing conference with the arbitrators. Many months will likely have passed between the initial filing of the claim and the time the arbitrators require the filing of this additional information. Furthermore, once the arbitrators meet as a panel of arbitrators, they tend to be anxious to expedite the commencement of evidentiary hearings. Therefore, waiting for a prehearing conference with the arbitrators to compel the other party to provide such basic information runs the risk of severely limiting the time within which a party will be able to prepare its own case in response to that information. Accordingly, access to the courts may be of value.

§ 12.14 —Application to Court

It should be possible in virtually every jurisdiction to make an application to court to compel the other party to disclose the nature of its claims. Many judges will appropriately remind the complainant that the courts look favorably upon arbitration, in large part because it removes such specialized commercial disputes from the dockets of the courts. Those judges, therefore, are not delighted at the prospect of being involved in the procedural aspects of a contractually stipulated arbitration. However, when basic due process considerations, arising from a lack of awareness of the claims that a party is defending against, are at issue, courts are more inclined to become involved and to compel a reasonable adherence to the arbitration rules (just as the court could compel the performance of any other contractual obligation).

The benefit of commencing a court proceeding in such an instance is severalfold. First, access to the court may be accomplished on an expedited basis, so as to obtain an earlier knowledge of the other party's claims. Second, a party who intended to abuse the arbitration process is less likely to continue its attempts to do so once it has had to appear before a court which can hold that party in contempt for failure to comply with a court order. Third, the party complaining to the court will leave with a judicial determination in hand that says that the other party was not "playing fair" from the first moment. That judicial determination may be placed before the arbitrators at an appropriate time if the other party continues to play games once the arbitration commences.

In determining to apply to a court, the attorney should evaluate the local arbitration statutes and compare them with the Federal Arbitration Act,[18] as there may be an available election to apply either to the United States District Court or to the local courts, provided that jurisdiction is otherwise established. The attorney should also evaluate any time limitations applicable to such relief under prevailing statutes.[19]

§ 12.15 Preliminary Matters

Who are the arbitrators? What are their fields of expertise? Have they served on arbitration panels before? These and other such questions should be answered *before* the first day of arbitration. Having that information aids later decisions regarding the manner in which the case will be presented. This information, together with the attorney's demonstrated familiarity with each arbitrator's name and background, permits and facilitates that attorney's development of a level of effective communication with the arbitrators.

§ 12.16 Maintaining the Proper Focus

The attorney and his or her client both come before the arbitrators with the purpose of convincing the arbitrators of the appropriateness of that client's position. The other party arrives with the same goal. The arbitrators enter upon the arbitration process with the goal of reaching a fair, correct, and just result. The focus of the attorney and his or her client should be to convince the arbitrators of the validity of that client's story in a way that demonstrates that they share the arbitrators' goal and that they are doing everything they can to present fair and truthful evidence of what really happened.

Remembering that the arbitrators probably know absolutely nothing about the project, the dispute, or the parties, the first order of business is the education of

---

[18] United States Arbitration Act, 9 U.S.C. § 1 *et seq.*

[19] In addition to a concern about timing, the attorney should be careful to avoid filing a lawsuit upon the claims pending before the arbitrators in order to obtain the benefit of discovery, as such action may constitute a waiver of the right to arbitrate in some jurisdictions. *See* Burton-Dixie Corp. v. Timothy McCarthy Constr. Co., Inc., 436 F.2d 405 (5th Cir. 1971) (holding that waiver occurred); E.C. Ernst, Inc. v. City of Tallahassee, 527 F. Supp. 1141 (D. Fla. 1981) (holding that no waiver occurred where the adverse party did not change its position by reason of the "waiving" conduct); Hudik-Ross, Inc. v. 1530 Palisades Corp., 131 N.J. Super. 159 (1974) (holding that right to arbitrate was not waived where party in litigation preserved arbitration as a pleaded defense).

the arbitrators about those critical facts. Arbitrators delight in a party's obvious desire for the arbitrators to understand the case. In the process, a party offering that insight to the arbitrators will find that its credibility is immediately enhanced. Such a presentation is best accomplished by a combination of: (1) the written submission of a factual summary of the history of the dispute which is accurate and essentially objective; (2) the submission of a copy for each arbitrator of all critical contract documents and other undisputed important documents; and (3) an oral presentation by the attorney of the important elements of the dispute, using graphic aids. Aside from the good public relations value of this presentation, the arbitrators will benefit from an early understanding of the scope of the disputed issues and will therefore be better able to appreciate an effectively presented case.

Regardless of the particular format of this presentation, at least eight items should be addressed; these elements are separately discussed in §§ **12.17-12.24**.

§ 12.17 —Description of the Project

The presentation of the site plan and elevations is usually sufficient to describe the project physically. Too much detail at this early stage may be counterproductive. However, an overview of the structure's components and the identity of the contractor or other party responsible for each major portion of the work should be described. Brief reference to areas of the project which are the subject of or related to the issues in dispute should be identified, and their relevance explained without zealous advocacy.

§ 12.18 —Description of the Project Time Frame

A graphic representation of the time periods applicable to the project should be presented to each arbitrator. This chart will serve as a fast reference tool for the arbitrators during the balance of the arbitration. The chart should include, at a minimum:

1. The bid date
2. The contract date
3. The date work commenced
4. The scheduled project completion date
5. The actual substantial completion date
6. The actual project completion date.

If the parties hold different positions regarding important dates, such as substantial completion, it may be advisable to indicate the dates contended by each party on the chart.

The benefit to the party of a demonstration of its objectivity in this early presentation is that party's enhanced credibility from the very start of the arbitration. A detriment may occur, however, if the adverse party takes issue with its adversary's apparent attempt to commit it to a position with which it disagrees, thereby affording the adversary a platform to criticize its opponent's integrity, honesty, and fairness. In other words, in the absence of a formal, written, and demonstrable position by the adverse party, that party should be left to announce its own positions.

§ 12.19 —Contract Price/Payment Analysis

If the dispute involves money, it is extremely helpful for the arbitrators to know from the first day of arbitration how the money damages sought, or sought to be voided, relate to the contract, change orders, backcharges, and payments. In preparing such a submission, each party should present this analysis from its own perspective. That is, the claimant seeking a money recovery should indicate those contract price adjustments which are bilateral, and then simply list the other adjustments it seeks, without seeking to indicate its adversary's position. This presentation should be approached as if the arbitrators might simply convert the statement of account into the arbitrators' award. Therefore, a schedule similar to the example below should be presented to the arbitrators:

(a)	Original contract price		$10,000,000
(b)	Change orders:		
	C.O.1 (5/19/82) Add		1,000,000
	C.O.2 (7/30/82) Deduct		(500,000)
	(list all change orders signed and recognized by both parties)		
(c)	Accepted backcharges		
	8/1/82 Electric bill Deduct		(10,000)
	10/1/82 Additional cleanup Deduct		(50,000)
(d)	Entitlement to additional compensation		
	(i) 8/10/82 claim for additional pumping	Add	100,000
	(ii) 9/30/82 claim for additional cost of two block wall system at east face of garage	Add	50,000
	(list all claims which are intended to be proved)		
(e)	Claimed adjusted contract price		10,590,000
(f)	Payments to date		(9,000,000)
(g)	Amount sought in this arbitration		1,590,000
(h)	Interest from 8/10/82 to date of award		
(i)	Costs of arbitration		
(j)	Reasonable attorneys' fees		

The above chart format should, of course, be modified as appropriate to the particular case. For the party defending a claim or seeking an affirmative award that the amount it owes to the other is less than claimed, item (d) should be called "Entitlement to credits," or some similar terminology. Under that category, matters such as backcharges, credits, claims, and other such price-reducing matters should be set forth. Additionally, if the retainage should be withheld, an item indicating withheld retainage should be inserted after item (f) in this chart.

Many parties make the mistake of placing only those items they perceive as contested claims before the arbitrators, omitting items such as the contract balance which they perceive as not contested. An omission on that basis is dangerous and inappropriate. The contract balance and other uncontested matters should always be included in the claims before the arbitrators. The adversary's acknowledgement of such matters can always be graciously accepted.

It is customary for most parties in arbitration to include a demand for reasonable attorneys' fees. In item (j) in the example, reasonable attorneys' fees are included in the presentation of damages and claims. However, there is a risk in seeking and receiving an award for reasonable attorneys' fees. In most jurisdictions, attorneys' fees are not awarded as a matter of course to the prevailing party. Therefore, an award of attorneys' fees by arbitrators will not be recognized as consistent with the law unless there is a contractual provision in the underlying contract between the parties to the arbitration which permits such an award.

In a case where such a contract clause exists, the arbitrators are merely fulfilling their responsibilities to carry out the terms of the contract. However, in those instances where there is no contract provision permitting the award of attorneys' fees, but where the arbitrators are sufficiently moved by the egregious conduct of one party so as to award attorneys' fees anyway, that award may be an unfortunate victory. The inclusion of attorney's fees in an award without a contract clause permitting it may jeopardize the validity of the entire award, and may subject the prevailing party to a court challenge on the basis that the arbitrator's award, on its face, demonstrates that the arbitrators exceeded their authority. Accordingly, lacking contractual entitlement, the attorney should consider whether it is worth the risk to include a request for reasonable attorneys' fees.

§ 12.20 —Scope of Work

Virtually every construction arbitration involves issues concerning the scope of work which the contractor was required to perform. Before informing the arbitrators of the particular items of work in dispute,[20] it is wise to inform the arbitrators of the general scope of work of the contract in issue. A simple and fast way to accomplish this is to present the arbitrators with copies of a highlighted table of contents from the technical specifications in conjunction with an oral description

---

[20] See § 12.22.

of the items of the work which are part of the subject contract. In multiple prime contractor situations, this clarification, presented early in the arbitration, may avoid confusion in later hearings. At least one complete set of the specifications should be presented to the arbitrators as well. Every effort should be made to present a *clean* set of specifications which has been conformed to addenda, bulletins, and other amendments.

If at all possible, a clean set of the plans should also be presented to the arbitrators, together with an oral presentation as to which drawings (by category or list) are particularly relevant to the arbitration. However, with diplomacy and tact, the attorney should urge the arbitrators to be patient and avoid getting into the details and contentions of the case while viewing the drawings during this preliminary phase. Nonetheless, once the drawings are on the table, it is common for the arbitrators to leaf through them and ask questions. It is important that the attorney and his or her client be fully prepared to answer those questions. This situation is precisely what was meant by the earlier statement that construction arbitration requires extraordinary preparation. The dividends from being able to demonstrate that level of preparation from the outset are self-evident, and the arbitrators will communicate their approval and appreciation accordingly.

§ 12.21 —Schedule

If delay is an element of the dispute, the initial presentation should include an as-planned/as-built comparative chart. This chart may be in bar graph form or in network form. The extent of detail, the breakdown of trades and areas of work, and other such considerations should be determined *after* the case has been thoroughly prepared, so that the chart highlights the areas of strength in the case and provides a reference tool for the arbitrators that is consistent with the testimony and exhibits offered in the arbitration.

It is important to emphasize the fact that the presentation of a well-thought-out comparative schedule in the preliminary phase also benefits the attorney in his or her closing arguments after he or she has produced testimony and documentary evidence which is consistent with the graphic presentation offered at the commencement of the hearings. However, this chart is not intended to be the evidentiary submission of the actual project critical path method (CPM), bar chart, or other such scheduling chart.

§ 12.22 —Overview of the Claims

The presentation of an overview of the claims is akin to the opening argument before a court. Many experienced trial attorneys would counsel their colleagues to offer as brief and nonspecific an opening as possible, so as not to commit to specific assertions of facts to be proved which may haunt that attorney at the end of the trial. There is merit to that position. However, arbitration is different from court litigation in important respects, and this author believes that those differences bespeak the advisability of a full explanation of the claims to the arbitrators on the first day of hearings. The key reasons for such a presentation, by way of summary, are:

1. The likely substantial time gaps between hearings
2. The arbitrator's lack of any knowledge of the case before the first hearing
3. The substantial possibility that the arbitrators may insist upon such a presentation if not voluntarily offered. No advocate worth his salt wants to be placed in the position of saying "I am not prepared to do that today," or just shooting from the hip to avoid that embarrassing admission.

The presentation should take advantage of and build upon the submissions discussed in the previous sections. However, there is a particular format which industry arbitrators will relate to and understand well. Indeed, the advocate, by using this presentation format, will come as close as he or she may to the ultimate dream of presenting the case without the risk of testimonial bloopers by his or her client.

Taking each major element of the claim in whatever order is appropriate to the case plan, the attorney should establish the following:

1. What does the contract require? (Cite the sections of the specifications and/or the drawings, details, shop drawings, etc.)
2. What work was actually performed or had to be performed?
3. How does the actual condition of the work differ from the contractual scope?
4. Why did the work have to be performed as it was?
5. Who is responsible for the need to perform the work differently from that set forth in the original contract scope?
6. That the work was performed (or that the performance of it is necessary)
7. The cost of the disputed work
8. The impact of the disputed work on time, sequences, productivity, or other features of the project.

§ 12.23 —Prehearing Written Submission

In some instances it may be both possible and advisable to present the arbitrators with a prehearing written submission which essentially provides the same information as the presentation outlined in § **12.22**. An advantage of a prehearing written submission is having the arbitrators understand the case before the first meeting. If the other party does not provide a similar submission, the arbitrators may even arrive with a favorable bias. However, the attempt to furnish the arbitrators with a prehearing submission is not always successful, and in the process the other party may become motivated to prepare such a presentation which it otherwise would not have thought to do.

In the setting of the American Arbitration Association, it is A.A.A. practice to require that the other party be furnished a copy of the proposed submission and that the other party consent to the submission before the A.A.A. will send it on to the arbitrators. If the other party refuses, the Association asks the arbitrators if they wish to receive the submission before the hearings in view of the objection by the other party. Even if the arbitrators decide to receive the submission, although they will often decline to do so under such circumstances, the edge (which was the goal of the submission) may be lost, and the other party receives a full education in the process. Therefore, planning for a presentation at the first arbitration session is most advisable and preferable.

§ 12.24 —Summary

A review of the recommended preceding suggestions should convince the attorney and client of the need for comprehensive preparation *before the first meeting* with the arbitrators. That effort may not be easy or inexpensive; however, its successful performance will translate into the most advantageous setting for a successful arbitration result.

§ 12.25 Preparing for Hearings

Having planned the education of the arbitrators, the prosecution of the claims must be properly planned before preparation for testimonial and documentary evidence may proceed. The backdrop of this planning must include an awareness of the fact that industry arbitrators, not a judge, will decide the dispute. The other differences between arbitration and litigation highlighted in this chapter must also be considered.

§ 12.26 —Who Will Tell the Story?

Arbitrators prefer to hear a story from someone who was involved in the circumstances in dispute. They also enjoy a well-presented summary presentation by someone (attorney, client, consultant, or expert) who places the actual evidence in a perspective which is meaningful to experienced industry arbitrators. The attorney's determination of which witnesses will present the factual and summary evidence is critical to the arbitration strategy.

When an attorney is planning the witnesses in contemplation of a jury trial, he or she might well fear that a poorly spoken field superintendent will make an unfavorable impression before the jury, will be easily abused by a skilled adversary, and will do great harm to the case in the process. If arbitration has one particularly redeeming feature, it is the understanding of construction and construction projects that industry arbitrators bring to the hearing. That understanding and appreciation for people involved in construction provides a setting in which that very same field superintendent may be rather comfortable and enjoy a heightened level of credibility, notwithstanding his or her lack of finesse, as long as he or she is perceived as honest and sincere.

The arbitrators will often assist such a witness while he or she testifies to assure that the questions asked are generally understandable and sensible in the context of the dispute. If there is an obvious attempt to confuse the witness with questions that the arbitrators readily recognize as being unfair, tricky, or ridiculous, solely for the purpose of confusing this field worker, the arbitrator will often caution the attorney to limit cross-examination to relevant questions having to do with the issues in the particular dispute. Consequently, there should be much less concern about utilizing a field worker before a panel of industry arbitrators. Indeed, it is often best to utilize such individuals as, given this industry setting, the field worker can present the most comprehensive and credible explanation of what actually occurred.

In contrast, many attorneys prefer to place the college-educated officer or supervisory person on the stand before a jury, to project a very credible, refined, and businesslike demeanor to the jury. Experienced arbitrators are capable of making insightful distinctions between what that individual knows from his or her personal knowledge and what he or she has learned from reading the documented records of the individuals who were actually involved in the construction of the project.

Many arbitrators will confirm that some of the most effective testimony in arbitration is that given by the field superintendent, the staff engineer, the owner's clerk of the works, or other such individual whose involvement in the disputed project is at a level which permits that witness to speak definitively about matters from his or her experience, rather than relying upon second- and third-hand references to letters and other documents. Therefore, the attorney should carefully consider the appropriate scope of the fact witness's testimony with the arbitration setting in mind.

The desire of arbitrators to get to the heart of a dispute is in part a foreseeable feature of their not being afraid to delve into very technical aspects of a construction dispute. This feature is a direct result of the arbitrators' construction industry background. Few judges would willingly embark upon a journey into highly technical matters, such as the inherent stresses in certain types of concrete structures, in order to evaluate the credibility of the witnesses. Arbitrators, on the other hand, who are engineers or architects by profession or who are contractors with substantial experience will not only desire to get into such technical matters but will insist that those matters be evaluated. In that respect, a mere recitation of the self-serving positions taken in letters, telegrams, and memoranda to file will usually not suffice in an arbitration.

§ 12.27 —Use of Experts

It is important to realize that the role of an expert in an arbitration may be very different from that role in a court case. A clear distinction between arbitration and litigation is the likelihood that a claims consultant will be permitted to present testimony in an arbitration which is tantamount to the advocate's dream of a professional witness, even if that spokesman is doing little more than presenting the writings of other people in an organized and convincing fashion. In planning the preparation and presentation of a case in arbitration, the attorney should consider the different perspective of arbitrators and judges very carefully before making the assumption that the testimony or report of a claims consultant or other expert testifying before the arbitrators will be given much weight. There is great danger in that assumption. However, claims consultants and experts may be able to offer substantial value in the preparation and presentation of an arbitration, even if their testimony is not advisable.

The role of an expert in a court trial is supposed to be that of an advisor to the court. The expert is permitted to testify because the subject matter is beyond the technical understanding of the judge and the jury. As a result, very few construction cases do not involve expert testimony. Considering that some arbitrators consider themselves to be experts and in fact testify as experts in judicial proceedings, the selection of an expert becomes a critical matter, as does the intended use of his or her testimony.

There is little question that in highly technical matters the testimony of an expert can be very helpful. If that expert is counselled to testify in a manner which demonstrates to the arbitrators that he or she is objective and fair, the testimony may have a far greater effect before industry arbitrators than might otherwise be achieved in court. The industry arbitrators will be able to recognize and independently conclude that the expert's testimony is correct, based upon their technical understanding of the subject matter. Accordingly, in cases where expertise is important, the use of a qualified expert is recommended.

However, it is critical for the attorney to establish the factual predicates for the expert's testimony by using witnesses who were involved in the construction of the project. Although a claims consultant's testimony may be valuable, some attorneys make the mistake of believing that a claims consultant's testimony or attractively bound report will serve as a substitute for the factual testimony of individuals involved in the project. To be certain, many claims consultants employ individuals who are experienced witnesses and who afford the attorney and the client a sense of security that this very well-spoken person will present their claims.

However, when the attorney uses the claims consultant as the client's spokesman, he or she runs a substantial risk of convincing the arbitrators that his or her client is trying to cleverly package a factually unsupportable claim. The greater value of claims consultants is their ability to assemble the documentation of a substantial dispute in a workable order and to call upon the expertise and experience of the members of their firm to analyze all those records in a way which identifies for the attorney the critical elements of the dispute for which testimony is needed.

Is the intended witness truly an expert, or is he or she just a consultant? If the proposed expert witness is going to attempt to tell the arbitrators what occurred in a dispute by reciting what the parties said to one another in letters, by making characterizations of and judgments about the other party's behavior based on reading those letters, and by then graphically demonstrating how he or she has concluded from the written record that his or her client should be the prevailing party, then that expert is likely to cause a very unfavorable impression before the arbitrators. If, on the other hand, the expert has been brought in to render technical valuations on technical issues which do require expertise, and if he or she is qualified and thoroughly prepared, the arbitrators may actually hold a discussion with that expert and genuinely form opinions relying upon his or her expertise.

When an expert is put on the stand in an arbitration merely to characterize a job's history with an advocate's twist, the arbitrators may react very unfavorably. However, the same expert or consultant can render valuable assistance in the preparation of the proper fact witnesses. They may also build upon the testimonial foundation laid by the fact witnesses, in which case the expert's testimony will then be valuable in the arbitration.

§ 12.28 —Preparing Exhibits

Once a determination has been made as to which documents will be used as exhibits, a sufficient number of copies should be prepared so that each arbitrator and each adverse party will receive a copy. Exhibits should be marked by the offering party with exhibit numbers. Having a sufficient quantity of pre-marked exhibits will expedite proceedings enormously. Since many arbitration hearings commence at 10:00 a.m. and terminate at 4:00 p.m., with an hour or so for lunch, it is extremely important that the attorney make the most of the available hearing time. Any effort to streamline the procedure results in fewer interruptions and enables the attorney to present far more testimony and evidence during a day.

With respect to exhibits, there are some practical considerations which should be taken into account by the attorney. More often than not arbitration hearings are held in hotel rooms, conference rooms, or other locations to which the arbitrators must travel. Therefore, the arbitrators must take with them whatever exhibits are offered in the arbitration. Arbitrators do not appreciate being asked to carry around voluminous documents and heavy packages of exhibits. Therefore, in planning the presentation of evidence, it is often wise to offer pages from specification books and individual drawings, or copies of details on drawings as single sheets, while marking and having custody of the full volumes of specifications and drawings available at each hearing for the arbitrator's reference.

§ 12.29 —Use of Affidavits

Under the Construction Industry Arbitration Rules, a party may offer an affidavit in lieu of oral testimony before the arbitrators.[21] Such an offer is certain to result in the strenuous objection of the adversary, as it is impossible to cross-examine an affidavit. If a party intends to offer an affidavit, it is important to be prepared to explain the unique circumstances which justify the use of that affidavit, and to offer the adverse party some opportunity for examination or rebuttal testimony which will demonstrate to the arbitrators that the proposed use of an affidavit was not intended to obtain an unfair advantage. Before offering evidence by affidavit, the attorney should weigh the value of such an offer against the likely resistance by the adverse party and perhaps the arbitrators. It is not at all uncommon for arguments and discussion about the appropriateness of accepting the affidavit to consume far more time than the presentation of testimony by the proposed affiant in the arbitration hearing. Additionally, the adverse party will almost always urge the arbitrators to believe that the motive behind use of an affidavit was to avoid the obvious embarrassment that the offering party would suffer if the witness appeared and was cross-examined. There are few instances in the complicated arbitration when testimony by affidavit accomplishes anything meaningful.

§ 12.30 —Creative Evidence

The ''informality'' of the arbitration process offers the imaginative attorney any number of opportunities to effectively communicate his or her client's case to the arbitrators. The Construction Industry Arbitration Rules specifically allow the arbitrators to visit the construction site with the parties.[22] In an appropriate case, a party may gain a substantial advantage by having the arbitrators visit the site and inspect the work which is in dispute. The timing of that request, the timing of the visit, the manner in which that visit should be handled, and the careful preparation for that visit to avoid unpleasant surprises are matters which must be addressed fully by the attorney.

Photographs have long been a mainstay of construction dispute proceedings, in accordance with the adage that one picture is worth a thousand words. In this era, the ability to communicate a situation photographically has been greatly enhanced by the advent of the video cassette recording. A two- or three-hour video

---

[21] A.A.A. Rule 32, ''Evidence by Affidavit and Filing of Documents.''

[22] A.A.A. Rule 33, ''Inspection or Investigation.''

cassette presentation could very well communicate circumstances which might take many days of testimony. Another benefit of the video cassette presentation is the ability to have the witness narrate as the video is made. By so doing, and by having that witness available for cross-examination at a hearing, it is possible for the witness to effectively communicate images and facts which may be extremely difficult to convey simply through testimony.

The two preceding suggestions are only examples of the type of procedures that a creative attorney may propose to the arbitrators as a method for expediting the presentation of his or her case. Perhaps the greatest attribute of arbitration from the attorney's perspective is the ability to use the most effective way of communicating the client's case, even if that presentation involves methods or procedures not customarily employed in litigation before the courts.

§ 12.31 Summary

To many attorneys, the arbitration process is something of a paradox. On the one hand, it offers unique opportunities to present a case in the most effective manner, even if that manner would not be acceptable in litigation. On the other hand, arbitration presents uncertainties and risks which many attorneys are not accustomed to. The method presented in this chapter provides an overview to the preparation of the complicated construction arbitration, so that the attorney may obtain the benefits of the process without having to fear those uncertainties and risks.

CHAPTER 13

PREPARING THE CASE FOR LITIGATION

Laurence M. Scoville, Jr., and Philip R. Placier

Laurence M. Scoville, Jr. is the head of the Litigation Department of Clark, Klein & Beaumont, a general practice law firm in Detroit, Michigan. He is also a member of the firm's Executive Committee. He graduated from Dartmouth College in 1958 and the University of Michigan Law School in 1961. He has primary responsibility for a diversified litigation practice with particular emphasis on complex matters involving engineering and construction.

Philip R. Placier is Chairman of the Management Committee of Thelen, Marrin, Johnson & Bridges, which has offices in San Francisco, Oakland, San Jose, Los Angeles, Newport Beach, and Houston. A 1958 graduate of the University of Michigan Law School, he has had over 20 years of litigation experience in handling trials and arbitrations on construction industry disputes. In recent years, he has handled a number of complex cases involving nuclear power plants and other major construction projects.

§ 13.1 Introduction

This chapter presents a number of guidelines for pretrial preparation of a contractor's claim in a typical construction litigation setting involving multiple claims, and usually multiple parties as well. Effective pretrial preparation is important because contractor claims are almost always factually complex and based on subjective or judgmental assumptions. A successful claim or defense requires adequate investigation, thorough research, and presentation in a cogent, organized manner.

INITIAL INVESTIGATION AND PLANNING

§ 13.2 Start the Investigation Early

Whether the dispute casts the contractor defending against owner or subcontractor claims, or prosecuting its own claims, the initial approach is substantially the same. At the earliest opportunity, begin to familiarize yourself with the dispute and start your investigation to identify the legal and factual issues most relevant to the contractor, depending upon its role in the project and the nature of the dispute.

The first step is to ascertain what the crux of the dispute is, where the major problem areas are, and what damages and/or exposures are involved. Try to assume as little as possible. If at all practical, arrange to talk to the people directly involved to get their first-hand perspective of the dispute.

Equally important is the need to begin marshalling the people and resources which will be involved in resolution of the dispute. It is not too early to begin developing basic litigation strategy and the decisionmaking process best suited for litigating this particular dispute. Establish the lines of communication for distribution of pertinent information and periodic litigation status reports, and the procedures for identifying and making important policy decisions regarding development, prosecution, and perhaps settlement of the action.

Establish an initial plan; this should be regularly reviewed and updated as the case is prepared for trial. The legal and factual complexities involved in today's construction litigation require planning to avoid duplication and wasted effort. A preliminary identification and outline of the major questions or issues involved in the dispute will determine how to organize the tremendous volume of incoming information, and how to extract that portion which will eventually be needed for discovery or at trial.

Although the initial investigation will typically be confined to in-house interviews, the overall inquiry into the factual circumstances underlying the dispute should cover three sources of available information: in-house investigation or interviews; outside investigation involving third parties or persons not directly involved in the litigation; and discovery proceedings in the litigation itself. Do not rely on any one of the three to the exclusion of any other. Each should be used to complement the overall investigation, because each has its own advantages and disadvantages.

Two questions must be answered promptly. Who is involved in the dispute? What work is involved? Because of the interlocking legal relationships among the various parties involved in a construction project, it is important to determine promptly the party or parties contributing to the cause or circumstances giving rise to the dispute. It is also important to get an early sense of the alignment and degree of responsibility of each of the parties already involved in the case, and those who are likely to become involved in the case, so that decisions can be made regarding whether and which other parties need to be brought into the dispute.

If the contractor had only a peripheral role in the project, some prompt initial discovery should be considered to learn more about the roles and responsibilities of the various parties involved. A brief set of interrogatories and a simple request for documents may be all that is either required or desirable.

There is little to be gained and much to be lost by waiting to start your investigation until you are responding to an opposing party's discovery. Memories fade, recollections become selective, and persons with pertinent information become unavailable. Moreover, it is human nature for people to lose interest as they become concerned with today's problems rather than yesterday's.

§ 13.3 —Contractor's Representatives

The initial investigation stage is also the time to select responsible management representatives of the contractor to assume an active and continuing role in the management of the litigation. A team approach which combines the respective resources of legal counsel and knowledgeable contractor personnel is usually the most effective and mutually satisfactory approach for both the contractor and counsel.

The team approach has several distinct advantages which more than compensate for the modest extra demands it may place on the contractor. It promotes the best possible utilitization of the resources of both legal counsel and the contractor, thereby maximizing the likelihood of success in the litigation. Counsel should not overlook the potential wealth of information and experience relevant to the dispute which the contractor possesses. By the same token, the contractor must not expect counsel alone to simply take care of the matter and then expect a cost-effective defense. Regular meetings should be scheduled between the contractor's management and counsel to monitor the progress of the case.

The contractor's active involvement with the litigation team and the decision-making process in the litigation also insures that major decisions are consistent with the contractor's actual experience and in line with the company's general policies. As a less direct benefit, the contractor's responsible management will necessarily become educated regarding the advantages and disadvantages of various practices and procedures and can begin shaping future policies to avoid repeating prior mistakes.

§ 13.4 Develop a Theory of the Case and a Litigation Plan

A convincing theory of the case and a well-thought-out litigation plan are necessary to litigate today's complex construction dispute effectively with a satisfactory net result to the contractor. A theory of the case maximizes both the likelihood and extent of recovery, and a litigation plan helps minimize the costs of litigation in general and discovery proceedings in particular.

Keep in mind that litigation is ultimately a question of persuasion. A theory of the case should therefore strive to pull together all the necessary elements of the claim in a manner which is most persuasive to the trier of fact, whether it is an arbitrator, a contracting officer, a judge, or a jury. Regardless of the forum, the trier of fact will be called upon to decide in the context of an adversarial proceeding which side is right. It will reach its decision in the relatively abbreviated time period of a trial, on the basis of only a fraction of the evidence available to the parties.

It is unlikely that the trier of fact will have any intuitive sense as to the merits of the dispute. Its knowledge of the facts surrounding the claim is confined to those presented by the parties to the dispute. Hence, the task of education is crucial. The trier of fact must be able to understand the claim to be persuaded of its merits.

In sifting through the facts and circumstances of the dispute to develop the theory best suited for the particular case, it is helpful to keep certain general guidelines in mind. To be persuasive, the evidence to be presented must be both understandable and believable. Therefore, whatever the theory, it must have a solid basis in facts which can be put in evidence, and conversely, should not be flatly contradicted by other strong evidence. Any basic recurring themes in the litigation should be emphasized whenever possible.

Next, the theory should appeal to both common sense and reason. The trier of fact will have neither training, experience, nor familiarity with the general background of the dispute, and will instead have to rely upon common sense in reaching a decision. Once the trier of fact has been shown what the important facts are and how to tie those unfamiliar facts and circumstances together in a logical way, the task of persuasion becomes easier.

Lastly, the theory of the case should seek to utilize the most impressive evidence available which is favorable, and at the same time seek to explain away as much of the unfavorable evidence as possible. There will often be key facts ranging from a damaging admission by the adversary to the uncontroverted testimony of a neutral witness. Integrate these key facts into theory to maximize their impact on the trier of fact.

The contractor's theory of the case will therefore determine what type of facts and what evidence are needed to prove its claims, and hence should dictate the foundation of the litigation plan based on those facts. As information is developed, the theory of the case must remain flexible to account for both the positive and negative points which are developed.

§ 13.5 —Litigation Plan and Trial Book

As a general rule, at least a preliminary litigation plan should be prepared as soon as counsel has a general understanding of the dispute. The timetable will vary with each dispute and the particular needs of the contractor. The general rule is the earlier, the better.

The preparation of the litigation plan itself provides several benefits. It has the immediate effect of forcing counsel and the litigation team to take the time to organize their thoughts and gives the contractor's management a clear picture of what to expect ahead. A litigation plan should typically include provisions for all major aspects of the litigation, including:

1. Investigation and research
2. Pleadings
3. Motions
4. Experts
5. Discovery (includes document requests, interrogatories, depositions, and other items)
6. Settlement
7. Pretrial
8. Trial.

An incidental benefit of developing and implementing a litigation plan is that it can promote settlement. Management should get a realistic assessment of its claim or exposure as soon as practical and can make efforts to settle the valid claims early.

One effective tool in organizing and implementing your litigation plan is to start a trial book early. The *trial book* is usually a looseleaf binder, subdivided into various categories. The categories should include:

1. Things to do
2. Issues
   —Factual
   —Legal
3. Pleadings
4. Discovery
5. Witnesses
6. Exhibits
7. Trial preparation.

The trial book should include a sufficient number of further categories or subcategories to include all the important issues involved in the case. Note in writing everything you have done, and insert your notes into your trial book as you go. If you do so, there will be no question later as to what you have previously done, and more importantly, you will be able to organize your ideas in a convenient way to manage the incoming information.

COLLECTING ALL AVAILABLE INFORMATION

§ 13.6 General

One of the first tasks of a litigation plan should be to plot out the initial investigation to effectively match previously identified general areas of inquiry with the sources of information available at the time. Relative priorities should be established in order to obtain the most important material promptly.

Certain types of files and other sources of information are almost always needed and should therefore be obtained promptly. The contract documents, the plans and specifications concerning all disputed items of work, the change orders file if extra work is in issue, the correspondence file, and other similar records are the types of documents generally needed to understand the circumstances behind the dispute.

Because the documents generally take time to assemble, you should promptly begin collecting all the various records, files, documents, and other needed materials. Contract documents and those files and documents relating to the essence of the claims should be critically reviewed as soon as possible. Similarly, key witnesses should be identified and initial contacts made at the earliest opportunity; any experts needed should also be contacted early.

§ 13.7 Documents

Project records, correspondence, and other contemporaneous documents often play a pivotal role in construction litigation, with both sides attempting to reconstruct their respective versions of the dispute using the existing documentation. A prompt initial compilation and review of the contractor's own documents provide a head start toward understanding the crux of the dispute. Despite their importance, however, both initial review and subsequent document production during discovery are all too often done routinely and hence not in a very efficient or cost-effective manner. It is therefore desirable to develop an efficient initial document review system which will not require duplication of effort during subsequent discovery proceedings.

To a large extent, the complexity of the case and the number of documents involved will determine both the level of effort and type of approach best suited to the particular case. A small mechanic's lien case will of course be treated much differently from a multiparty multimillion-dollar engineering/construction dispute. However, despite the range of differences, several general rules apply.

First, take the time to inventory what documents are available, where they are physically located, and who maintains them or is knowledgeable about their contents. Attempt to determine approximately how many documents are involved.

Once a rough inventory is made, an initial inspection to identify the most important documents can begin.

It is generally useful to prepare a comprehensive index of the files or other documents which are or may become relevant to the dispute. If the indexing is done by or under the direction of counsel, the index will probably be protected from subsequent disclosure by the attorney work-product privilege.

The time and effort saved through using the index during subsequent review or production will more than pay for its initial cost. There are also other indirect advantages to using a file index system approach. It permits a more orderly initial review and affords a way to keep track of the files which have been reviewed. Additionally, if a large number of documents is involved, an index system facilitates a staggered or phased production of documents to opposing counsel which is less burdensome to the contractor. For example, if part of the documents requested include current or working files, you can arrange to have them produced on non-business hours or on scheduled days, to minimize the inconvenience to the contractor's staff. Similarly, if bulky files are located in other places or offices, as is frequently the case with large contractors who have home, regional, and project offices as well as a field office, separate arrangements can be made. As long as the index is updated and file integrity is maintained, subsequent production, inspection, and copying may be done according to a prearranged schedule with minimum supervision by counsel and a minimum of disruption to the contractor.

§ 13.8 —Selection, Form, and Numbering

Once a rough inventory of the documents has been made, the next step is to develop a procedure for extracting the most important documents. One of the primary functions of the initial inspection is to determine which documents need to be copied, and in what form or format copies should be made. The economy of scale factors previously mentioned also apply to this second stage of initial document review. It is probably more efficient to copy the entire file of a small case than to sift through the file to decide what needs to be copied.

The big case presents a decidedly different situation. Unless there is a compelling reason to do so, do not routinely decide to "copy all the documents." Copy only those documents reasonably needed, consistent with the time frame in which they will be needed. If an index exists, additional copies of files or documents can be obtained merely by designating the appropriate file title from the index.

If large numbers of copies are to be made, consider using microfilm. Hard copies of all the important documents will be needed, but not of the vast majority of the documents involved in the dispute. Some principal advantages of microfilm include ease of storage of the microfilm, reasonable cost per copy, a ready-made sequential numbering system of documents by reel and frame and an index of documents by document number, portability of microfilming equipment for on-site document copying, and subsequent ease of access to specific documents for the reviewer.

Disadvantages of microfilm include the need for (and hence cost of) the microfilm reader, office space needed for the reader, decreased efficiency in reviewing documents, and generally less desirable quality hard copies. In many locales, independent litigation support service companies, including large copying services, can be used for this work.

If hard copies of all or substantially all the documents are to be made for a working file, consider having an extra set made at the same time for subsequent production during discovery. It is usually considerably easier to have two (or more) sets of copies made at the same time than it is to return to recopy all the documents later. Maintaining a master set from which all subsequent copies are made will cut down on unnecessary recopying, which usually results in deterioration of copy quality.

Sequential numbering of a master set of the pertinent documents, either continuously from any starting point or staggered by groups (e.g., all document numbers preceded with a P are produced by plaintiff or all documents in the P10000 series are job files) may prove to be useful later on. Additionally, the numbering provides a reliable system for keeping track of what has been produced and by whom.

The actual mechanics of affixing the document number to each of the documents presents an important logistical consideration. Defacing original documents in current working files by stamping numbers on them is generally not greeted with enthusiasm by contractors, although it does help to ensure file integrity. On the other hand, because it is desirable to retain a numbered set of the documents to be produced to opposing parties, copying the documents and then numbering a set of the copies introduces extra trouble and requires extra copying.

§ 13.9 —Review

Once the initial inventory and screening is accomplished, the review must commence. While often the sheer volume and technical nature of the documents require that an initial screening be done by others, any substantive review should be done by someone with an understanding of the issues, both legal and factual, involved in the litigation.

If separate and unrelated claims are involved, physical segregation of the documents by issue can be useful. As long as there is not a large number of overlapping issues which require excessive cross-indexing or an excessive number of copies of the same document, this may be the simplest way. Placing some or all the documents in chronological order is another useful system.

If the litigation is exceptionally complex, consider the feasibility of utilizing a computerized document retrieval system. This approach is not appropriate for every case, primarily because of its sharply higher cost, but it can be invaluable in the right circumstances. Large numbers of documents can be cross-referenced by topic, author, witness, or any other issue and accessed at once. The system should be designed with a certain amount of flexibility so that developments in the nature of the case and the evidence can be assimilated. An outside litigation support service consultant can often provide the appropriate software and systems.

§ 13.10 Witnesses

It is important in construction litigation to identify promptly the witnesses with knowledge of the dispute, and to ascertain their future availability. Hourly personnel, including foremen, often change jobs or leave the area when the project is completed, making it more difficult to locate them later and making subsequent interviews inconvenient and expensive. Likewise, salaried personnel may be transferred to other projects in other areas, again making it inconvenient and more costly to interview them.

To the extent reasonably practical, keeping in mind that opposing parties may not be contacted by counsel after the litigation has commenced, attempt to interview all the witnesses who have knowledge of the dispute, whether friendly, neutral, or hostile. Start by identifying the contractor's personnel who know the most about the dispute. There will usually be a small number or group of individuals who know the most about the dispute, either because of their position in project management, their tenure on the job, or a combination of both. It is advisable to contact these individuals first, and to make arrangements for continued access to them.

Because many construction disputes are complex, the desire for promptness in interviewing witnesses must be balanced with the need to conduct sufficient preparatory work to be able to identify the relevant issues and orient and refresh the witness's recollection, particularly when dealing with nonparty witnesses. At times, general observations and impressions are all that may be desired, while at other times a detailed reconstruction of a disputed sequence of events is critical. The particular need will determine both the timing and the level of preparatory work necessary.

Generally speaking, witness interviews are best conducted one-on-one in a comfortable setting. Let the witness do the talking and tell the story in his or her own words. Take notes as necessary so you can come back to areas which you want to cover in greater detail or want to ask further questions about. Write up your interview notes and your general impressions of the witness as soon as possible, and insert them in the trial book under the witness's name.

Consider whether it is advisable to obtain written or recorded witness statements from contractor employees or from neutral witnesses. If witness statements are desired, they should probably be obtained by counsel to preserve the attorney work-product privilege.

§ 13.11 Experts

As the construction industry becomes increasingly specialized, so too do the nature and complexity of the technical issues underlying disputes within the industry. It is becoming increasingly common for experts to become involved early to assist in investigating, quantifying, and assembling the claim or the lawsuit, as well as in their more traditional roles of expert consultant and trial witness.

If the dispute involves technical issues or work of a specialized nature, it will often be desirable (if not necessary) to use the services of one or more consulting experts to assist in preparation of the case. Consultants can be used to educate counsel and other members of the litigation team, as well as to help identify the critical issues in the case. Depending upon the expertise and the level of sophistication of the contractor, the technical nature and complexity of the problems, and the size, posture, and importance of the case, counsel may choose to rely on knowledgeable individuals already employed by the contractor or may instead decide to retain independent experts. Cost is obviously a significant factor to be considered.

Independent experts should be retained early. If the area of technical specialty needed is unique, locate and retain an expert as soon as possible. The best experts may also be sought by the opposition, particularly if there are multiple parties involved in the litigation. Retaining the best experts early, regardless of what their ultimate opinion is or whether they eventually testify or not, helps educate the entire litigation team to the strengths and weaknesses of the case and may also forestall highly damaging expert testimony against the contractor.

In order to preserve the attorney work-product privilege, counsel, not the contractor, should retain the expert. It is also a good practice to insist that the expert sign a secrecy agreement preventing disclosure or subsequent use of the information provided or developed in connection with the case.

If it is anticipated that the expert will be used as a witness at trial, and hence deposed by an opposing party, several basic rules should be kept in mind when communicating with the expert. First, everything the expert receives or relies upon may have to be revealed to the opposition. Therefore, to the extent practical, the basic background information about the dispute given to the expert should be provided by way of pleadings or other documents already available to the opposition in the litigation.

Second, once it has been decided to designate the expert to testify at trial, communicate with him or her primarily by telephone or in meetings and not in writing, as correspondence may be discoverable. Resist the temptation to ask for written reports. Reports in draft form and initial conclusions are not necessarily protected from disclosure and often provide a fertile source of impeachment material.

Lastly, make sure that the expert is timely provided with all the underlying information, both favorable and adverse, necessary to get a complete picture and to reach an informed opinion. If, on the basis of his or her initial review of material provided, the expert feels there is or may be additional information relevant to the dispute or to his or her opinion, take steps to provide it in a timely manner.

Your expert will likely be asked during deposition, and perhaps at trial, to identify all the information both reviewed by him or her and relied upon as the basis of his or her opinion. It reflects adversely on the expert's opinion if important material has not been reviewed and evaluated or if other information was requested or needed and not furnished. Furthermore, although the expert's opinion need not be based on admissible evidence, as long as it is based on the type of material reasonably relied upon by other experts in the field in forming their opinions, the underlying material should be credible or it will undermine the validity of the expert's opinion.

DISCOVERY

§ 13.12 Developing the Discovery Plan

Proper planning is the key to effective discovery. Once a litigation plan has been developed, documents and witnesses located, and expert technical assistance secured, the next step is to formulate a discovery plan. This plan will be your road map through the discovery process.

The single most important rule for this initial and every other phase of discovery is to think through the problem thoroughly before taking any action. Review your theory of the case and litigation plan to determine what types of evidence will be needed, then decide how to go about obtaining that evidence. The initial investigation may already have produced much of the necessary proof, in which case discovery requires only filling in the gaps. Comprehensive discovery in such a case may only serve to tip your hand to opposing counsel. Conversely, it may be necessary in some cases to use discovery to elicit nearly all the relevant facts. In any event, the time spent thinking about how discovery should be utilized before actually beginning the process is invariably time well spent.

A thorough review of the applicable court rules should be undertaken at the outset. This is especially important where the forum is an unfamiliar one, of course, but knowledge of the current rules is essential in any context. Do not assume that you have an adequate command of the rules simply because you have conducted discovery before. The federal *Manual for Complex Litigation*[1] can also be a valuable source of information on marshalling and conducting discovery in a complicated case.

Be sure you have made adequate provision for the processing, storage, and retrieval of information and documents to be produced during discovery. The indexing system established during the initial investigation phase should be expandable to include items disclosed by discovery. A system which permits cross-indexing between different types of information (e.g., between deposition summaries and documents produced) is highly desirable. As noted, in a complex case you may

[1] Manual for Complex Litigation (Fed. Judicial Center 2d ed. 1982).

wish to use a computer-based system of information storage and retrieval. Keep in mind, however, that such a system will be far more costly than simple manual storage.

If an independent expert has been retained for consulting services, or if the contractor can provide expert personnel from its own staff, use the expert in developing the discovery plan. He or she can be helpful in analyzing documents and in preparing witnesses and counsel for depositions. The expert can also help formulate discovery requests to insure that they focus on the right type of evidence, the right witnesses, and the right technical and engineering standards.

Explore the possibilities of cooperation with opposing counsel. Discovery will proceed more smoothly and effectively if agreement can be reached as to discovery schedules or even limitations on discovery requests. If trade secrets or confidential or privileged documents are likely to be involved, it may be possible to stipulate to a reciprocal form of protective order without the necessity of intervention by the court. In fact, a more favorable protective order can usually be obtained through negotiation with the opposition than can be elicited from a court.

All these considerations apply with equal force to arrangements with counsel for third parties or co-parties, including third parties not directly involved in the case. Your client may have claims, defenses, and strategies in common with co-parties (although the opposite may be equally true). Do not let the opposition use the technique of "divide and conquer." In the appropriate case, pool discovery plans and schedules with co-parties to take maximum advantage of common strengths.

Finally, remember that discovery is the most expensive part of litigation, and that the contractor will not intuitively understand the need for all the discovery you deem desirable for adequate preparation of the case. Use the team approach to keep the contractor involved in, and in ultimate control of, the services it is paying for. At the same time, develop a sense of perspective about the importance of discovery in the particular case relative to its cost. The contractor engages in cost-benefit analysis every day in doing its job, and you should do the same in doing yours.

§ 13.13 Written Discovery Requests

The three most widely used types of written discovery requests are interrogatories, requests for production of documents, and requests for admission. A construction case of any complexity is likely to require the use of at least one of these devices, and perhaps all of them.

Written discovery devices are not only useful, they are indispensable in most cases. They are extremely flexible, and with careful drafting can perform a wide variety of useful functions. They can bring to light both the factual and legal contentions of the party to whom they are addressed. They can be broadly drafted, to elicit the entire factual background of a case, or directed narrowly at particular facts or issues, and can serve the ends of both investigation and advocacy.

Nevertheless, written discovery requests have a number of disadvantages. The most important of these is that preparation of the response to the request will undoubtedly be supervised by the attorney representing the responding party. Responses are thus likely to be carefully crafted, and there is no opportunity to cut through an incomplete or evasive answer with an immediate follow-up question. Also, written discovery procedures are slow. The responding party may be permitted as many as 30 days to answer, and extensions are routinely sought. If answers are not timely filed, a motion to compel must be brought, which entails further delays. Furthermore, written requests by definition are limited to a question-and-answer format, and as such are not suited to elicitation of narrative factual descriptions. Many of these disadvantages can be avoided or minimized, however, by thoughtful and careful preparation of discovery requests.

§ 13.14 —Drafting Discovery Requests

In the area of drafting, the guiding principle is again to *think before you write.* Consider precisely what you hope to elicit with your request. Boiler-plate interrogatories, document requests, and requests for admissions developed in other cases, or taken from the "form file," are no substitute for discovery requests tailored to the specific facts of the case. Form requests almost always produce too much of what is not needed and too little of what is needed. Of course, a form can be useful to outline and establish the basic structure of the discovery device, the logical order of questions or requests, and the like. Nevertheless, the actual interrogatories, document requests, or requests for admission should be as narrowly and thoughtfully directed to the particulars of the case at hand as possible.

A well-drafted discovery request should start with a preface which puts the party or person to whom it is directed on notice as to what is being requested and what the penalties are for noncompliance. The preface should also make it clear that information sought includes matters within the knowledge of not only the person addressed, but also its attorneys,[2] agents, employees, officers, and representatives. The rules concerning supplementation of discovery requests[3] should also be alluded to.

Definitions and instructions are a second important feature of any written discovery device. Using a defined term to encompass a long list of details allows concise and efficient drafting of the actual discovery requests, and facilitates ex-

---

[2] A discovery request which includes information in the hands of the responding party's attorneys is, of course, likely to be objected to on the grounds of privilege or work product. Do not automatically acquiesce to such an objection. Demand identification of any information withheld on the ground of privilege, and a detailed statement of the basis for the claim. The courts have interpreted privilege and work-product claims in a surprisingly narrow fashion in many cases. Not as much information can be protected on these grounds as many attorneys would like to think.

[3] Fed. R. Civ. P. 26(e).

pansion of the scope of particular questions. Common examples are defining *document* to include all imaginable media through which information can be stored or transmitted, or *person* to include all conceivable legal entities. In interrogatories, the term *identify* is often defined to require a listing of specified information concerning the person or thing identified. Terms such as *the project* or *the accident* can be used to refer to specific aspects of the case. Comprehensive definitions encompassing critical issues, such as what documents are included in the *contract,* help avoid evasive objections and allow for considerable economy of language in drafting broad-reaching questions.

The form of the questions or requests themselves depends upon the particular type of discovery device being used, but some basic rules apply to all. For example, it usually makes sense for the sequence of questions to track those pleadings that set out the answering party's claims and defenses (e.g., the complaint, answer, or affirmative defenses). Also, the litigation plan should be consulted to make sure that discovery requests cover all areas that should be explored at that particular stage of litigation.

Lastly, principles of writing in plain English should not be overlooked. If the opponent can rightfully claim that a discovery request is ambiguous or otherwise unanswerable because of bad grammar, poor sentence structure, or typographical errors, it is unlikely that he or she will be required to respond to it. Make every effort to draft with both clarity and precision. Do not try to get too much out of any one question or request. Break the subject matter down into its elements and cover each element separately. The information elicited through this approach will be more useful, and the responding party will find such requests more difficult to evade.

Always remember that discovery is a two-edged sword. A common tactic is for opposing counsel to change the names on a discovery request from his or her client's to that of the party propounding it, and send it right back. If a document you have drafted is capable of this treatment, be sure you are in a position to respond to what is in essence your own discovery request.

§ 13.15 —Interrogatories

Interrogatories will undoubtedly be used in all but the simplest cases. They are relatively easy to prepare, and when well prepared may be very difficult to answer in a way which avoids disclosing the information requested. Furthermore, under the Federal Rules of Civil Procedure and in the practice of most states, interrogatory answers must be signed under oath by the party or its representative. Thus, while prior inconsistencies in deposition testimony may sometimes be explained away as a result of faulty recollection or a failure to comprehend the question, a party's written interrogatory answers (which presumably it had the opportunity to investigate and contemplate) are much more likely to be viewed by the trier of fact as firmly binding.

Under the Federal Rules of Civil Procedure, and in most states, interrogatories may inquire into any discoverable matter. They are, however, most useful to elicit general and background information and to narrow issues. In a reasonably complex case, interrogatories should be divided into successive sets, each of which performs a distinct function. The first set should identify the entities, persons, and documents involved in the case, as well as their relationship in a general sense. The second set, perhaps in conjunction with witness depositions, should establish the basic facts which will form the foundation of the case to be presented to the trier of fact. Additional sets, possibly in conjunction with requests for admission, should be used to fill in any holes in the evidence prior to trial. Keep in mind, however, that state or local court rules may limit the number of interrogatories (including subquestions) which may be propounded.

Interrogatories used to identify witnesses should inquire specifically about the identity of expert witnesses and the subject matter of expert testimony, and should request a summary of any expert testimony to be offered and the grounds therefor. Federal R. Civ. P. 26(b)(4)(A) and some state rules expressly allow such discovery. Some states, however, have separate and distinct procedures for mutual disclosure and discovery of expert witnesses and their opinions.

Interrogatories should ask for the location and the identity of the custodian of the opponent's documents, and for identification of all documents the opponent intends to use to support every element of its claims or defenses. The manner in which the opponent keeps its records, the types of documents that are available, and the types of information they contain can also be learned through interrogatories. This information will make specific requests for document production much easier to draft.

Specific inquiry should always be made into the nature and amount of each element of any claimed damages. Establishing the parameters of the contractor's exposure in this manner is vital to an intelligent evaluation of the advisability of settlement as opposed to litigation.

Finally, interrogatories can also be used to explore a party's legal theories. Federal R. Civ. P. Rule 33(b) permits interrogatories to seek "an opinion or contention that relates to fact or the application of law to fact. . . ." Thus, to the extent possible within the written-question format, the opponent may be compelled to disclose its legal, as well as factual, theories. This can in turn lay the groundwork for a motion for summary judgment or other pretrial disposition.

§ 13.16 —Requests for Production of Documents

Construction projects are often voluminously, if not meticulously, documented. This often makes a complete document review a practical impossibility in a contract case. Nevertheless, documents will nearly always be central to the issues involved in the litigation. Top priority must be given to identifying the documentary evidence the other side will rely upon, obtaining copies of those documents, and learning how to interpret them. Identification, inspection, and copying of documents in the possession of parties are accomplished under Fed. R. Civ. P. 34 and parallel state rules through a *request for production of documents*. Documents of nonparties must be obtained through voluntary cooperation or a subpoena duces tecum. Some state rules provide that document production can be obtained only by court order after motion and hearing.

Document production efforts are most frequently derailed when, in response to an overly broad document request, the other side simply offers to open all of its files for inspection. The party requesting production is then put to the task of sorting through and attempting to understand immense amounts of material, much of which may be irrelevant to the case.

It is possible to avoid, or to at least minimize, this inundation. First of all, as with all other discovery requests, materials seeking document production should be precisely drafted. Do not mechanically request ''all documents'' relating to the dispute. To the extent possible, only specified *types* or categories of documents should be requested. Your consulting expert or the client's employees will likely be able to assist in identifying documents which should be sought.

It is possible that the opposing party has a file or document indexing system already in use. If the system was devised and put into use by the party, as opposed to counsel, this index should be discoverable. If so, it will make identification of relevant documents much simpler. It may also be advisable at an early point to schedule the deposition of the custodian of the opponent's documents to ascertain whether all available sources have been checked and prior document requests fully complied with.

It is also important to insist that all documents withheld under a claim of privilege or on some other basis be identified, with a full explanation of the basis of the claim. As noted in **§ 13.14**, courts generally do not interpret the doctrine of privilege as broadly as most lawyers do, and a motion to compel may be indicated. Again, however, discovery is a two-edged sword, and you must be prepared to disclose and defend your own claims of privilege on the same basis.

The types of documents which should be requested will vary with the issues of each case. Nevertheless, in most construction litigation at least some of the following types of documents will be sought:

1. Contract documents, including documents on which the opposing party intends to rely, all copies which bear notes or interpretive markings, and any drafts, working papers, and notes or memos used in contract negotiation
2. Drawings and specifications
3. Change orders
4. Construction logs and foremen's diaries
5. Quality control/quality assurance reports
6. Reports to owners; progress and payment schedules

7. Reports to regulatory agencies and authorities
8. Accident or incident reports
9. Insurance policies and bonds
10. All correspondence, including interoffice memoranda.

These categories, and any other categories relevant to the particular case, should be reflected in the document indexing system previously devised during the investigatory phase. If a mechanized or computerized system is in use, be sure it can be expanded or adapted to accommodate new categories disclosed during discovery.

§ 13.17 —Requests for Admission

Requests for admission under Fed. R. Civ. P. 36 are a powerful discovery tool and an effective strategic device as well. They cast on the opponent the burden of taking a specific position on identified issues which may well go to the very heart of the case. Admissions elicited early in the case may lead to a motion for full or partial summary judgment, or to settlement, allowing the client to attain its objectives short of trial. Nevertheless, attorneys by and large do not appear to make sufficient use of requests for admission in construction litigation.

The degree to which admissions obtained from an opposing party will be useful depends directly on the clarity and specificity of the requests themselves. The objective is to have the opponent admit exactly what is needed to prove or facilitate proof of the client's claim or defense, neither more nor less. Each individual request should deal with a single fact, and should be phrased so as to leave the other side with no practical alternative other than a yes or no answer or its equivalent. Vague terminology or incomplete descriptions that give the other side the opportunity to be nonresponsive should be avoided. Denial of requested admissions should always be followed up with interrogatories directed to the basis of the denial.

The subject matter of requests for admission will vary with the facts of the particular case. Nevertheless, there are recurring features of construction lawsuits which lend themselves particularly well to this discovery device:

Authentication of or foundation for documents. Because construction cases are often lengthy and complex, requests for admissions are frequently used to establish the authenticity of or evidentiary foundation for documents. This avoids tedious trial testimony on authenticity, and leaves counsel flexibility in witness scheduling. Generally, the opponent is simply asked to admit the authenticity of specified documents, copies of which are attached to the request.[4] Other bases for the evidentiary admissibility of documents, such as elements of the business record exception to the hearsay rule, can also be established through this type of request for admission.

---

[4] Such copies *must* be attached under Fed. R. Civ. P. 36.

Basic facts. The factual building blocks of the case, such as dates, times, addresses, dollar amounts, the accuracy of numerical computations, and other basic matters are also well suited for inclusion in requests for admission. Witnesses often display uncertainty on such subjects, particularly numbers, at trial. Obtaining admissions concerning hard-to-remember but basically uncontested details facilitates a smooth presentation of evidence.

Application of law to fact. Requests for admission concerning opinions or the application of law to fact are permitted under the Federal Rules, even if they involve questions of ''ultimate'' law or fact. This can be an extremely effective aspect of this discovery device in the proper case.

§ 13.18 Obtaining Responses to Discovery Requests

Service of a discovery request unfortunately does not automatically ensure that the information sought will be forthcoming. The party served with the request may try to avoid answering through delay, or by objecting to a particular request, or by intentionally or negligently failing to provide all the information sought. Effective use of discovery requires conscientious follow-up and a swift reaction to any attempt to avoid or delay responding.

Almost inevitably, opposing counsel will request an extension of time to respond. Whether to grant such a request depends on a variety of factors well known to most practitioners, including the degree to which professional courtesies exist or are honored in the locality. Any extension should, however, be memorialized in a written stipulation and order specifying the date on which discovery is to be forthcoming. This precludes any misunderstanding and facilitates a motion to compel if the extended deadline is not met.

Meritless objections to discovery requests should be addressed through a motion to compel. Some states require, in fact, that the party *objecting* to the discovery arrange to have the objections brought before the court. Most lawyers have found, however, that judges by and large dislike discovery motions, and prefer that the parties work out their differences about discovery without the court's intervention. One manifestation of this reluctance to become involved in discovery disputes is that motions to compel are frequently the subject of local court rules and thus subject to special requirements and procedures. Rejection of a motion to compel on purely procedural grounds simply compounds the delay inherent in the discovery process.

All possible efforts should be made to obtain compliance informally before bringing a motion to compel, even if such efforts are not strictly required by court rule. If a motion does in fact become necessary, the moving party will wish to be able to demonstrate that the other party was given a fair chance to respond.

A corollary of this proposition is that motions to compel can be used to educate the court, both about the issues and about the parties. Choose your opportunity with care. *Do not* run to the court for relief from trifling omissions or on matters which can be obtained through other means. *Do,* however, ask the court to deal firmly with parties or counsel who are acting in palpable bad faith.

The range of valid grounds for objecting to discovery is actually fairly limited. One such objection is relevance. Many courts construe relevance broadly for discovery purposes, however, and will allow discovery in all but the most egregious circumstances. Burdensomeness is a frequent objection to discovery in construction litigation, and is best avoided by careful preparation of discovery requests in the first instance. The requesting party should be able to demonstrate that, even though the discovery request is extensive, everything sought is important to the parties' *and the court's* understanding of the case. Confidentiality is also a frequent objection to discovery in construction cases. Offering to submit to a protective order providing for limited disclosure of information, in camera inspection limited to the court and attorneys for the parties, or other precautions is an effective response.

§ 13.19 Responding to Discovery Requests

Discovery, of course, works both ways in construction litigation as elsewhere. Many of the same general guidelines applicable to the preparation of discovery requests should also be followed in the preparation of responses. Interrogatory answers, answers to requests for admission, and the production of certain documents may be construed as admissions of the party at trial. Accordingly, careful analysis of the request and a thorough evaluation and understanding of the legal and tactical situation are essential. For example, if a party is asked to produce all documents upon which it intends to rely to prove a particular point, it may be precluded at trial from using other documents not brought forward in response to the requests. Similarly, counsel in receipt of discovery requests should keep in mind the duties of supplementation imposed by court rules such as Fed. R. Civ. P. 26(e).

Responding to interrogatories and requests for admission is much like preparing a witness for cross-examination. It is important to answer only what the question asks and not to build the other side's case. Objections to discovery requests should be carefully considered, keeping in mind the judicial attitude toward discovery disputes previously discussed in **§ 13.18**. The Federal Rules of Civil Procedure permit interposition of an objection, in lieu of an answer to an interrogatory, so long as the grounds for the objection are clearly stated. In some cases, however, it may be preferable to state both the objection and the answer, thereby preserving both the objection and the court's good graces.

Promptness is also important; the indulgence of counsel should not be relied upon to excuse untimely responses. If an extension of time to respond is requested and granted, it should be confirmed in writing.

§ 13.20 Depositions

The oral deposition is one of the most flexible (and certainly one of the most expensive) forms of discovery available. Because of this flexibility, depositions are useful in following up on interrogatory answers and document productions and in framing requests for admission.

Persons having direct knowledge of the operative facts of the case should be identified as soon as possible. Witnesses who will testify for the opposing party should be deposed early in the discovery process, both to tie them to a particular version of the facts and to preserve their testimony for potential impeachment use at trial. Each witness should be asked in the course of the deposition to identify other persons who may have knowledge of the pertinent facts.

The depositions of experts and employees of the owner or subcontractors who do not have direct knowledge of the operative facts but who will have knowledge of the circumstances, procedures, methods of operation, and other relevant information, may be deferred until the later phases of the discovery process. This allows time for adequate preparation and facilitates the authentication of documents needed for trial.

It is not always necessary to know in detail the management hierarchy of an organization from which discovery will be sought in order to begin taking depositions. Federal R. Civ. P. 30(b)(6) permits the organization itself to be named as the deponent in a notice of taking deposition or subpoena which describes "with reasonable particularity" the matters to be inquired into. The organization must then itself designate one or more persons to testify in response to the notice or subpoena (a subpoena must notify a nonparty organization of this duty). This presents a useful way to get a foot in the door of an unfamiliar organization for discovery purposes.

Recent technological advances can make depositions more useful. If numerous depositions are contemplated (as is usually the case in large construction cases), and if evidence is being compiled and stored in a computerized data base, look for a court reporting service whose recording equipment can interface directly with the computer. If key deponents will be unavailable for trial, videotape depositions are the best substitute for live testimony, especially if the demeanor of the witness is important. If video depositions are contemplated, local court rules relative to their taking and admissibility should be carefully analyzed and followed. Preparation for videotape depositions requires attention to a number of factors not pertinent to the normal transcribed deposition.[5]

---

[5] Especially careful preparation is required for videotape depositions, since shuffling papers and long pauses can prove distracting, and will detract from the effectiveness of the presentation. In many cases, hiring a reputable firm to produce the deposition is the best means of assuring a quality presentation. Different camera angles should be used to eliminate the "talking head" effect of one stationary camera focused on the witness.

§ 13.21 —Conduct of Deposition

No two depositions are quite alike, and no two lawyers will conduct a deposition in quite the same way. Nevertheless, some guiding principles have universal applicability:

> Prepare for the deposition as you would for the testimony of that witness at trial, even if you are not conducting the examination. After all, the witness may become unavailable, in which case the deposition testimony may actually become the witness's testimony. Moreover, if important questions go unasked or improper material is not objected to, the damage may be difficult (if not impossible) to repair later.
>
> If you are taking the deposition, know what you want to accomplish. Is this merely an exploratory, fact-finding inquiry, or do you plan to elicit particular facts or admissions for use at trial? Are there documents that should be authenticated by this witness, and if so, are your copies adequate and legible? Is there an applicable adverse witness statute that will let you proceed by leading questions?
>
> If your witness is being deposed, make sure he or she is thoroughly prepared. The best preparation is for you to take the role of opposing counsel, conducting the deposition as if the witness were adverse. You will have done your job if opposing counsel asks no questions that you did not. Be every bit as hard on your witness as you expect opposing counsel to be. The witness will be better prepared, and you will have an opportunity to assess the strengths and weaknesses of your own case.
>
> Have a plan, but be flexible. You should almost always have a written list of subjects you want to cover with the witness, but be prepared to abandon the list temporarily to follow unexpected developments in the testimony. Conversely, do not rely on a written list of *questions*, except for particular critical points, as this practice often stifles spontaneity and creativity.

Additional considerations apply to depositions of expert witnesses. Your consulting expert or the contractor's expert personnel should be involved in preparation for the deposition and in formulating hypothetical questions. Find out as much as you can about the deponent prior to the deposition. Review his or her curriculum vitae in advance, and read any of his or her published works that may touch on the subject matter of the testimony—the expert may have taken a position in an article or publication which is at variance with the anticipated testimony. Answers to interrogatories relating to the expert's proposed testimony will also be helpful in planning the scope and direction of your examination. Do not let the witness get away with conversing in technical jargon; make him or her explain answers that would not otherwise be clear to the trier of fact.

In a complex case, it is likely that numerous depositions will be taken, and that each deposition will be quite lengthy. This makes it essential that a system of transcript summarization be instituted. Actual preparation of summaries can often be delegated to a paralegal, but the lawyer who attended the deposition should review the summary as soon as possible after it is prepared to be sure that essential points are not missed. If the attorney who took the deposition will not be conducting the actual trial, an early conference between the deposing attorney and the trial attorney is essential so that the witness's strengths, weaknesses, and overall demeanor can be discussed and analyzed before trial.

§ 13.22 Other Discovery Devices

There are two other related discovery devices which are useful more often in construction cases than in cases of other types: the request to inspect, test, or sample physical things in the possession of another party, and the request to go upon the land of another party to measure, survey, photograph, test, and the like. Both are governed by Fed. R. Civ. P. 34.

As a rule, if these devices are likely to be useful at all, they should be used early in the case. The physical condition of real and personal property relevant to the case may change dramatically over the course of discovery, which often takes several years. The admissibility of evidence relating to physical condition may turn directly on the amount of time that has elapsed between the events giving rise to the litigation and the inspection or testing of the item.

There are a number of ways in which these devices can be useful. In a case involving failure of a building component, physical testing may reveal whether the failure resulted from faulty design, faulty materials, or faulty workmanship. Soil tests may reveal whether building subsidence resulted from subsoil conditions or from oversight by the architect.

Where construction on the project continues during litigation, the ability to photograph the site in the condition existing at the time of the dispute may be invaluable. Likewise, measurements and surveys may be relevant to disputes involving claims of architectural malpractice.

TRIAL PREPARATION

§ 13.23 General Preparation

As the trial date approaches, but before discovery cutoff, the litigation plan must be reevaluated to make sure it satisfactorily accounts for new or additional facts and contentions disclosed through discovery. Experts should have concluded their investigations and reached their opinions, and will probably have been deposed. The basic framework of the dispute has been set.

If permissible under the local rules of the forum, it is often useful to propound a last set of clean-up interrogatories to tie up any loose ends which may remain. These should include an interrogatory asking if any previous interrogatory answers need to be updated or changed to reflect subsequently discovered information or documents.

At the beginning of this period, attention should be devoted to the precise proof needed to present the case, and how the proper foundation for its admissibility will be established. Agreeing to reasonable stipulations as to the foundation for or admissibility of routine documents will save time at trial.

Determine what demonstrative evidence is necessary or desirable to the presentation of the case. Photographs, models, or other tangible exhibits are usually necessary to tie together the testimony of various witnesses and aid in focusing the attention of the trier of fact. Graphs, charts, or summaries may help simplify, clarify, or explain otherwise unmanageable or overly complicated supporting documentation, such as cost accounting records.

To be admissible, demonstrative evidence must be a fair and accurate representation of what it purports to represent, it must be relevant, and it must help the trier of fact understand the evidence. Standards for admissibility vary from state to state, court to court, and even judge to judge. It is therefore prudent to inquire as to the local rules and practice before taking any substantial steps towards developing any particular form of demonstrative evidence.

The earlier efforts to organize the documents will pay handsome dividends during trial preparation. Determine which documents are needed and in which order. Unless the foundation has already been agreed to with the opposition, be sure all the foundational requirements are met for each exhibit, given the sequence of presentation of witnesses.

§ 13.24 Conclusion

Most lawyers would agree that cases are more often won or lost in the preparation stages than at trial. This is particularly true of construction litigation. These cases are for the most part not decided on the basis of intangibles like credibility of witnesses or the demeanor of the parties; they turn, rather, on complex factual determinations and inferences. For this reason, it is almost invariably the best-organized lawyer, the one with the most complete mastery of the facts, who will prevail. In this arena, effective advocates are made, not born, and they are made on the basis of effective preparation.

CHAPTER 14

PROVING LIABILITY TO AND DAMAGES OF THE CONTRACTOR

Peter Goetz*

Peter Goetz is a founder and senior member of the New York City law firm of Goetz and Fitzpatrick, P.C., specializing in construction and engineering matters throughout the United States and abroad. Mr. Goetz is a graduate civil engineer from Rensselaer Polytechnic Institute and practiced for a number of years as an engineer in the field of construction supervision, including major highway, bridge, and airport projects. Since admission to the New York Bar, he has specialized in the practice of construction law, including arbitration and litigation of construction claims throughout the United States in both the private and public sectors of the construction industry. In 1967, Mr. Goetz established his present firm and since its formation has represented all facets of the construction industry, including construction managers, contractors, subcontractors, bonding companies, owners, architects, engineers, and manufacturers of building systems and materials. Mr. Goetz writes a monthly article on construction law in the Associated Builders & Contractors' national magazine *Builder & Contractor.*

Thanks and credit are due Donald J. Carbone, Esq., a senior associate of Goetz & Fitzpatrick, P.C., for his invaluable assistance in the preparation of this chapter.

* This chapter is dedicated to the memory of Gerard E. Fitzpatrick, my partner, friend, and outstanding construction attorney.

§ 14.1 Introduction

On a factual basis, there are as many variations in the substantive nature of claims made by contractors in the courts as there are differences in the construction techniques used in erecting or constructing buildings, airports, bridges, and highways. Since there is so much diversity in the subject matter of this chapter, the format chosen was that which covers the claim areas having the biggest economic impact on the contractor's costs. In terms of financial considerations, delay-related claims are the big money items in the field of litigated construction matters. Acceleration, suspension, and interference actions likewise are generally claims with significant financial repercussions.

As is shown in this chapter, a delay claim may involve a myriad of separate subsidiary claims which in and of themselves fall under separate and distinct categories. In **§§ 14.3-14.12**, examples of these claims are given in terms of such subsidiary claim items as defective plans of specifications, failure to obtain site access, unforeseeable subsurface conditions, etc. Each of these items gives the contractor the right to recover its extra costs, yet each may also give rise to a more significant claim, i.e., a delay claim.

This chapter focuses on *time of performance* claims and shows how these time-related claims are caused by (or at least frequently influenced by) various satellite claims which orbit the principal claim. Put another way, the sum proves to be greater than total of the individual numbers. Delay, acceleration, and other time-related claims are accentuated because the success or failure of a project may hinge on the amount recovered by the contractor for these claims.

DELAY CLAIMS

§ 14.2 General

By far the most commonly and hotly litigated claims of contractors involve delays caused by the owner or by persons for whom the owner is responsible. As a general rule, if a contractor agrees to do certain work within a specified time, but is prevented from performing the contract by the act or default of another party, or by the acts of persons for whose conduct that other party is responsible, the contractor is entitled to the economic loss sustained as a result of the delay.[1] Of course, this is somewhat of a simplification, because the contract language in the owner-contractor documents controls the liabilities of these parties.

§ 14.3 Owner's Failure to Perform an Affirmative Act

In situations where the owner is under a contractual obligation to perform an affirmative act, its failure to perform will give rise to a claim for damages caused by such failure. For example, in *Glassman Construction Co., Inc. v. Maryland City Plaza, Inc.*,[2] the owner's failure to timely obtain building permits and locate sewer lines (along with a continuous flow of revisions from the owner) prevented the contractor from performing its work. The contractor was entitled to recover its losses due to delay.

Likewise, in *George A. Fuller Co. v. United States,*[3] the government was held liable for damages sustained by the contractor which resulted from government delays in furnishing models as required by contract.

§ 14.4 Defective Plans and Specifications

If the furnishing of defective plans and specifications is the cause of delay, the contractor, in addition to recovering its extra costs, also has a viable cause of action against the owner for delay damages. An example where extra work costs were allowed in a delay situation is *Jasper Construction v. Foothill Junior College District,*[4] which involved a claim by the contractor that inadequate and defective plans and specifications caused it to suffer delays and added expenses. The court reiterated the general rule that "a contractor of public work who, acting reasonably, is misled by incorrect plans and specifications issued by public authorities as the basis for bids and who, as a result, submits a bid which is lower than he would have otherwise made may recover in a contract action for extra work or expenses necessitated by the conditions being other than represented."[5] However, in qualifying this general rule, the court stated that the contractor must demonstrate an affirmative misrepresentation or concealment of material fact in the plans and specifications.

In *Laburnum Construction Corp. v. United States,*[6] the delay was caused by deficient specifications being furnished to the contractor by the government. The

---

[1] *See* 13 Am. Jur. 2d *Building and Construction Contracts* § 48 (1964).

[2] 371 F. Supp. 1154 (D. Md. 1974), *aff'd,* 530 F.2d 968 (4th Cir. 1975).

[3] 69 F. Supp. 409 (Ct. Cl. 1947).

[4] 91 Cal. App. 3d 1, 153 Cal. Rptr. 767 (1979).

[5] 91 Cal. App. 3d at 8, 153 Cal. Rptr. at 770.

[6] 325 F.2d 451 (Ct. Cl. 1963).

court directed that the contractor could recover its increased costs from the owner. The owner alone caused the delay, and it was therefore a compensable delay.

§ 14.5 Owner's Failure to Provide Access to the Site

Unfortunately, it is common for a contractor to be denied timely access to the project site. The courts have historically entertained contractors' claims for delay based on such access denial, because the law assumes that timely access was anticipated by the parties. For example, in *E.C. Ernst, Inc. v. Koppers Co.*,[7] the follow-on contractor, Ernst, which could only perform its work after other contractors had prepared the work area, was awarded damages as a result of being delayed from commencing its work.

However, in *Burgess Construction Co. v. M. Marrin & Son Co., Inc.*,[8] the court found that a general contractor's failure to provide access to the site on the dates stated in the contract was not a breach of its implied promise not to hinder or delay its subcontractor's performance. After examining the contract language and surrounding circumstances, the court determined that the general contractor did not have an unconditional obligation to complete work necessary for subcontractor's site access on the specified date. The court found dispositive the fact that the contract provided for an extension of the subcontractor's time in the event of a delay by the general contractor.[9]

§ 14.6 Unforeseeable Subsurface Conditions

As a general rule, the contractor can recover additional costs when it encounters unforeseeable subsurface conditions. In *L.G. DeFelice & Sons, Inc. v. State*,[10] the court stated that, ordinarily, delays in performance of contracts are compensated only by an extension of time for completion; such extensions are routinely granted when site availability is delayed or there is interference between the contractor and other contractors or utility owners.[11] The court further held that for delays in contracts to be actionable, they must be more than ordinary delays, and in fact should be ''extraordinary'' and not reasonably contemplated by the parties. The *DeFelice* court found that an increase of almost 400 percent in the amount

---

[7] 626 F.2d 324 (3d Cir. 1980).

[8] 526 F.2d 108 (10th Cir. 1976).

[9] See §§ **14.13** and **14.22-14.25**, for treatment of contract clauses which exculpate one party from its otherwise ordinary obligations.

[10] 63 Misc. 2d 357, 313 N.Y.S.2d 21 (Ct. Cl. 1962).

[11] See § **14.7**.

of topsoil required constituted an extraordinary delay and was not reasonably contemplated by the parties at the time of entering into the contract. Thus, the contractor was entitled to recover additional sums expended.

§ 14.7 Substantial Changes in Scope of Contract

The courts have held that a contract may validly provide that a contractor shall be entitled to no relief for delay except an extension of time of performance, even though such delay does in fact increase the costs.[12] In most jurisdictions, though, where the increased costs result from a substantial change in the scope of the contract, such a disclaimer is ineffective.[13]

In *Peter Kiewit Sons' Co.*,[14] the prime contractor, instead of permitting the subcontractor to backfill unimpeded as agreed to by contract, required the backfilling to be done in three phases after pipe had been laid. This increased the difficulty of the backfill operation and resulted in a 300 percent increase in backfill costs. The court determined that the prime contractor's conduct constituted a substantial breach since it substantially changed the scope of the work. The court expressly found that a disclaimer providing for an extension of time while prohibiting additional costs was not applicable in a situation where the increase in the subcontractor's costs was caused primarily by a change in the scope of the contract.[15]

§ 14.8 Extension of Time Clauses

Customarily, a time extension provision neither bars nor allows per se the right to recover compensation for delays. In the strictest sense, a time extension clause simply provides that under general or specified circumstances, whichever the case, if a delay is beyond the control of the contractor, the owner is obligated to extend the time of completion by the amount of the excusable delay period.

Time extension clauses are not to be confused with other contract provisions which relate to compensation for extra work or forfeiture of monies (liquidated damages) caused by delays in performance of the work. Its only function is to provide extended completion time in the event the contractor is delayed for causes either within or beyond the owner's control.

---

[12] Lichter v. Mellon-Stuart Co., 305 F.2d 216 (3d Cir. 1962).

[13] Peter Kiewit Sons' Co. v. Summit Constr. Co., 422 F.2d 242 (8th Cir. 1969).

[14] *Id.*

[15] *Id.* 265.

§ 14.9 Changes

Virtually all modern contracts contain a changes clause. In fact, without such a clause, the owner would be unable to alter or add work without being liable for breach of contract. A typical changes clause is found in Article 12 of A.I.A. Document A201.[16]

The changes clause generally delineates, in substantial detail, the specific manner in which changes are to be effectuated. These clauses usually require that notice of changes be made in writing, and specify the means for adjusting the contract price.

The courts have allowed contractors in some cases to recover costs of delays under the changes clause. The disputes that generally arise from these clauses, however, are whether the changes are within or beyond the contract scope of work.[17]

§ 14.10 —Directed Changes

If the owner directs the contractor to make changes, and the subject changes are covered by the changes clause of the contract, the contractor may also be entitled to damages for delay. In *Brooks Towers Corp. v. Hunkin-Conkey Construction Co.*,[18] changes in the structural specifications are covered under the changes clause caused changes in the sequence of work and resulted in delays. The court allowed recovery of delay damages under the changes clause.

§ 14.11 —Changes Clause Strictly Construed

Courts will generally strictly construe the notice provisions of the changes clause, on the rationale that the owner should be made aware that the modification constitutes a change for which the contractor seeks extra compensation. In *State v. Omega Painting, Inc.*,[19] a contractor's claim for delays under a changes clause was denied where it failed to strictly adhere to notice requirements found in that clause.[20]

Most changes clauses require the change to be issued to the contractor in writing, in the form of an extra work or change order. The contractor should insist on strict adherence by the owner to such a provision. Typically, the changes clause

---

[16] American Institute of Architects Document A201, General Conditions, Art. 12, ''Changes in the Work'' (1976 ed.).

[17] See **§ 14.7**.

[18] 454 F.2d 1203 (10th Cir. 1973).

[19] 463 N.E.2d 287 (Ind. Ct. App. 1984).

[20] *See also* J.A. Ross & Co. v. United States, 115 F. Supp. 187 (Ct. Cl. 1953).

also will provide for alternate methods of payment (i.e., agreed or contract unit price, lump sum, or, in the absence of an agreement, a time and material basis).

Always check the language of the changes clause to see the extent to which the owner can authorize changes. In public works, by contract or by legislative enactment, the owner's representative may not have authority to issue change orders beyond a certain magnitude or make changes that are beyond the scope of the contract.

Similarly, in public contracts, the contractor should be extremely careful to determine what person or persons have authority to issue changes or change orders. Without contract authority, regardless of good-faith reliance by the contractor who acted on the directed changes, it is unlikely that recovery can be had for unauthorized changes. In one case,[21] the contractor was denied payment for extra work, despite the fact that the owner's representative authorized the extra work. (The contractor also failed to strictly adhere to the written notice requirements.) The court reasoned that since only the entire school board could authorize changes, the representative's authorization was not controlling. Obviously, if delays are part of the change's extra cost, then that claim will likewise be denied if the extra work claim fails to qualify.

§ 14.12 —Constructive Changes

Another source of change found under law is the *constructive change.* These are changes which are not acknowledged by the owner or its representative at the time they are issued, but are treated by the courts as changes to the same extent as if the owner had issued a formal order.

Various types of claims are allowed under the changes clause on the legal theory of constructive changes. In instances where defective or incomplete plans and specifications required extra work to accomplish the intention of the contract, recovery by the contractor has been allowed on the theory that the owner warrants the sufficiency of the plans and specifications.[22] Also allowed are claims for extra work caused by: (1) performance in a different manner due to impossibility of performance as specified or designed; (2) enforcement of extraordinarily rigid performance requirements; (3) superior knowledge by the owner of work requirements requiring performance in a more costly manner; and (4) ambiguous plans and specifications under which the contractor contemplated a less costly means of performance.[23]

---

[21] Nether Providence Township School Auth. v. Thomas M. Durkin & Sons, Inc., 476 A.2d 904 (Pa. 1984).

[22] United States v. Spearin, 248 U.S. 132 (1918). See also **§ 14.4**.

[23] United States v. Spearin, 248 U.S. 132 (1918).

§ 14.13 Restrictions on Contractor's Recovery

A common means by which the owner attempts to restrict a contractor's ability to recover for delay damages is to insert a *no damage for delay clause* into the contract. Essentially, the no damage clause is an exculpatory provision which provides that regardless of what one party does to delay the work of the other party, even if caused through neglect, fault, or omission, it will not be liable for delay suffered by the injured party. A typical no damage clause provides: "No payment or compensation of any kind shall be made to the contractor for damages because of the hindrance or delay from any cause in the progress of the work, whether such hindrance or delays be avoidable or unavoidable."[24]

As a general rule, such clauses are enforceable when they are clear and unambiguous and reflect the intent of the parties.[25] Due to the harsh results fostered by such clauses, they are strictly construed by the courts.[26] However, the mere fact that the result is a harsh one will not prevent the application of a no damage clause. They are effective to prevent liability for unreasonable delays as well. It is not the harshness of the results which is the touchstone of the effectiveness of such clause, but rather the scope of the delay contemplated by the parties at the time the contract was made.

§ 14.14 —Delays Contemplated by the Parties

If the delays encountered are found to be within the contemplation of the parties at the time they entered into the contract, the no damage clause will be enforced.

In *McDaniel v. Ashtown-Mardian Co.*,[27] a subcontractor brought an action against the general contractor and its bonding company under the Miller Act.[28] The subcontract was expressly made subject to the prime contract and the subcontract recognized within its own provisions that changes and additions might delay the work. The court held that the subcontractor could not recover damages from the general contractor for delays occasioned by proper change orders of the government where the general contractor and subcontractor both knew when they entered into the subcontract that it might very well not be completed within the time specified in the general contract. The court also held that changes might be made by the government which would necessarily extend the performance period, and that if any such

[24] Psaty & Fuhrman, Inc. v. Housing Auth., 76 R.I. 87, 68 A.2d 32 (1949) (cited in 10 A.L.R.2d 789, 795 (1950)). *See* 74 A.L.R.3d 187, *Delay in building or construction contract* (1976).

[25] Note, *The Enforceability of "No Damage for Delay Clauses" in Construction Contracts,* 28 Loy. L. Rev. 129 (1982); Cunningham Bros., Inc. v. City of Waterloo, 117 N.W.2d 46 (Iowa 1962).

[26] Cunningham Bros., Inc. v. City of Waterloo, 117 N.W.2d 46 (Iowa 1962).

[27] 357 F.2d 511 (9th Cir. 1966).

[28] 40 U.S.C. § 270a *et seq.*

changes authorized by the government should result in delay damages to the general contractor, no compensation for such damages could be recovered by the general contractor from the government.

In *Wes-Julian Construction Corp. v. Commonwealth,*[29] a highway contractor petitioned to recover for extra work and increased costs and expenses from the State of Massachusetts. The court held that the contract provision giving the state power to delay commencement of the work "if it shall deem it best for its interests to do so" exculpated the state from liability for delay even if its actions were negligent, unreasonable, due to indecision, or arbitrary, willful, and capricious. In this case, the contractor had no claim for damages, but would have been entitled to an extension of time.

Western Engineers Inc. v. State Road Commission[30] involved a situation in which plaintiffs were to complete their performance in nine months, but did not complete until approximately three and one-half years had elapsed. The court held that the no damage clause precluded recovery for damages. The contract clause provided that "no changes or claims for damages will be made by them for any delay or hindrances, of any cause whatsoever, during the progress of any portion of services specified in the agreement; such delays or hindrances, if any, shall be compensated for by an extension of time for such reasonable periods as the road commission may decide."[31]

§ 14.15 Exceptions to No Damages for Delay Clauses

There are various exceptions which have been held to preclude the application of the no damage for delay provisions found in construction contracts. They can be broadly classified as follows:

1. A delay that is the result of wrongful conduct, particularly fraud, bad faith, or active interference, on the part of one seeking the benefit of the provision
2. A delay which has extended such an unreasonable length of time that the party delayed would have been justified in abandoning the contract
3. A delay not within the specifically enumerated delays to which the "no damage" clause is to apply
4. A delay which, in light of the relationship of the parties and objective and attendant circumstances, was not intended or contemplated by the parties to be within the purview of the no damage provision.[32]

---

[29] 351 Mass. 588, 223 N.E.2d 72 (1967).

[30] 20 Utah 2d 294, 437 P.2d 216 (1968).

[31] 20 Utah 2d at 296, 437 P.2d at 217.

[32] 10 A.L.R.2d 801, *No Damage Clause—Delay* (1950); 74 A.L.R.3d 187, *Construction Delay—No Damage Clause,* at 214-15 (1950); E.C. Ernst, Inc. v. Manhattan Constr. Co., 551 F.2d 1026 (5th Cir.), *cert. denied,* 434 U.S. 1067 (1977).

§ 14.16 —Active Interference

In *Peter Kiewit Sons' Co. v. Iowa Southern Utilities Co.*,[33] the court, in addressing the concept of active interference, stated:

> It, therefore, would appear that to be guilty of "active interference" in the instant action, the defendants herein would have to have committed some affirmative, willful act, in bad faith to unreasonably interfere with plaintiffs' compliance with the terms of the construction contract. This court concludes that the use of the term "active" to modify "interference" by the Court in *Cunningham* was intentional, is significant, and clearly implies that more than a simple mistake, error in judgment, lack of total effort or lack of complete diligence is needed for plaintiff to prove the "active interference" necessary to render unenforceable an otherwise clear and unambiguous "no damage" clause [citations omitted].[34]

The *Peter Kiewit* court declined to find active interference when a general contractor engaged in an electric power plant construction brought suit against the owner and engineering contractor for expenses incurred because of interruptions in the construction. The court noted that in order for a plaintiff to recover due to construction delays of both a tortious and contractual nature, it must "prove that the delays charged in the complaint were the product of fraud, evil intent, improper motive or willfulness amounting to bad faith."[35]

Johnson v. State[36] provides an example where active interference was found. Over the protests of the contractor, the state opened the roadway on which the contractor was working to traffic for the use of other contractors. It also actively interfered with the contractor's seeding and planting, which caused it to miss the planting season. The court found that the extra work and costs to which the contractor was subjected by reason of the state's action went far "beyond the cooperation required under the terms of the contract" and allowed recovery for damages by reason of the state's active and indirect interference with the performance of the work.

In *Grant Construction Co. v. Burns*,[37] where a contractor's work depended upon the removal of utility poles by third parties, the failure of the state Highway Department to schedule removal of the poles was held not to have been within the contemplation of the parties, and thus it constituted active interference. It is interesting to note that the court found the delay in the removal of the poles, once scheduled, *was* contemplated by the parties.

---

[33] 355 F. Supp. 376 (S.D. Iowa 1973).

[34] *Id.* 399.

[35] *Id.* 398.

[36] 24 N.Y.2d 958, 302 N.Y.S.2d 489 (1969).

[37] 92 Idaho 408, 443 P.2d 1005 (1968).

Bad faith was demonstrated in *United States Steel Corp. v. Missouri Pacific Railroad Co.,*[38] where the court found active interference and held the railroad liable for damages suffered by a bridge builder while it waited for 120 days past its original startup date to gain access to the job site. The court found the railroad committed an affirmative willful act in bad faith, which unreasonably interfered with the contractor's compliance with the terms of the contract, when it issued a notice to proceed to the contractor knowing that necessary substructure work would not be completed on time and knowing of the large numbers of commitments which contractor would make on the basis of the start up order.

§ 14.17 –Active Interference Not Sufficient

New York's highest court recently decided that active interference by the owner, standing alone, no longer warrants an exception to a no damage for delay clause. In *Kalisch-Jarcho, Inc. v. City of New York,*[39] the heating, ventilating, and air conditioning (HVAC) contractor on a new New York City police headquarters building sought damages as a result of the job's completion being extended for an additional 28 months. The active interference alleged by the contractor included "endless" revisions of plans and drawings, the city's failure to coordinate the activities of the prime contractors, and other acts of omission or commission interfering with the sequence and timing of the work.

The court focused on the contract language, the intent of the parties, and the purported public policy surrounding no damage clauses. It found that the unambiguous contract language provided that the contractor was to make no claims for delay damages caused by *any* act or omission by the city. It determined that the clause's "unmistakable intent" was that the contractor was to absorb damages occasioned by contractee-caused delay, whether it be a reasonable or unreasonable delay.[40] Finally, the court acknowledged that such no damage clauses are commonly found in construction contracts and are enforced as a matter of public policy except in exceptional circumstances, such as fraud, malice, bad faith, or willful or gross negligence.

Finally, the court held that active interference alone was insufficient to show the gross negligence necessary to invalidate the no damage clause. The court found that, to meet that standard, the contractor would have to prove "the City acted in bad faith and with deliberate intent delayed the [contractor] in performance of his obligation."[41]

Kalisch-Jarcho is important because it indicates that the courts, at least in New York, will not make an exception to the no damage clause even when the fact-

---

[38] 668 F.2d 435 (8th Cir.), *cert. denied,* 459 U.S. 836 (1982).

[39] 58 N.Y.2d 377, 448 N.E.2d 413, 461 N.Y.S.2d 746 (1983).

[40] 58 N.Y.2d at 384, 448 N.E.2d at 416, 461 N.Y.S.2d at 749.

[41] 58 N.Y.2d at 386, 448 N.E.2d at 418, 461 N.Y.S.2d at 751.

finder determines that active interference caused the delays and caused the contractor unexpected additional costs. In the writer's opinion, the *Kalisch-Jarcho* decision is hard law and should be restricted to the facts of that case only.

Three interesting cases decided after *Kalisch-Jarcho* are *Slattery Associates v. City of New York,*[42] *Bradley Environmental Contractors v. Village of Sylvan Beach,*[43] and *Corinno Civetta Construction Corp. v. City of New York.*[44] In *Slattery,* the court failed to find an exception to a no damage clause where the only contention that the city was guilty of intentional misconduct was its alleged failure to provide adequate police protection for the job site.

In *Bradley,* the contractor sued the village for breach of contract in connection with a water pollution abatement project and water pollution control plant. The contractor sought damages resulting from the village's delay in, among other things, approving shop drawings, materials, and vendors. The appellate division reversed the lower court's dismissal of these causes of action, which was based on its finding that the actions were barred by a clause in the contract precluding recovery by the contractor for damages occasioned "by any delay on the part of the owner in doing the work or furnishing the material to be done or furnished by it." In reversing, the appeals court stated that "a no damage for delay clause must be strictly construed against the party seeking exemption from liability because of its own fault."[45]

In *Corrino,* a sewer contractor sued the city for, among other things, delays resulting from the city's alleged interference with the contractor's work, failure to perform its own work in a punctual manner, and failure to coordinate properly the project, such that the contractor was forced to work outside of the normal sequence of construction. The appellate court dismissed these causes of action because it concluded that the contractor failed to allege conduct on the city's part amounting to bad faith or gross negligence.

It will take some time for the full impact of *Kalisch-Jarcho* to be realized. The *Kalisch* court appeared to sidestep the issue of what would happen if the delay was beyond the anticipation of the parties, although a footnote in the decision[46]

---

[42] 98 A.D.2d 686, 469 N.Y.S.2d 758 (1st Dep't 1983).

[43] 98 A.D.2d 973, 470 N.Y.S.2d 214 (4th Dep't 1983).

[44] N.Y.L.J., Jan. 22, 1985, at 14, col. 1.

[45] Bradley Environmental Contractors v. Village of Sylvan Beach, 98 A.D.2d 973, 470 N.Y.S.2d 214, 215 (4th Dep't 1983).

[46] Kalisch-Jarcho, Inc. v. City of N.Y., 58 N.Y.2d 377, 448 N.E.2d 413, 461 N.Y.S.2d 746, 750 n.7 (1983):

> In its brief on this appeal, Kalisch in part now argues that "the City's failure to disclose [that the contract drawings were defective] coupled with its positive misrepresentation of the fact at the pre-bid meeting, amounted to actual or constructive fraud." Intent is, of course, an essential element of fraud. But, even if one were to assume that such an inference can be drawn from the record, there was no request that such a theory, which would have been within the realm of willful misconduct, be charged to the jury. Concomitantly, there is nothing in the interrogatories to suggest that such a claim was to be considered.

seems to indicate that active interference could under certain circumstances give rise to a finding that the no damage clause would be unenforceable. In the past, the New York courts rendered numerous decisions holding that clauses which are totally one-sided, and which may be considered unconscionable where a party would benefit from its own wrongdoing, are not sanctioned by the law. Likewise, it can be projected that an owner who causes a contractor serious delays with little or no concern for the contractor's rights may not be able to hide behind the shield of a no damage clause.

§ 14.18 —Delays Not Contemplated by the Parties

In most states, if the courts determine that the delay was caused by an occurrence which was not within the contemplation of the parties, the no damages for delay clause will be held inapplicable.

U.S. Industries Inc. v. Blake Construction Co.[47] is a case in which the court, noting that exculpatory clauses generally are not favored, strictly construed a no damage clause to determine that the delay was not intended or contemplated by the parties to be within the purview of the no damage provision. The court held that the provision in a mechanical subcontract for construction of a hospital, which exculpated the contractor from liability for delay on the part of other contractors or subcontractors involved in the work or furnishing of materials, did not immunize the prime contractor from damages for delay in erecting the steel structure of the building and its outer shell, which was allegedly caused by failure of the prime contractor's supplier to timely deliver the steel to the job site. The no damage clause in pertinent part stated that "contractor shall not be liable for any damages that may occur from delays or other causes on the part of other contractors or subcontractors involved in work, or furnishing of materials pertaining to the project."[48] The court found that if the parties had intended to immunize the prime contractor from liability for delay caused by its suppliers and vendors, they easily could, and presumably would, have so provided.

In *Johnson v. City of New York,*[49] the court determined that under a contract for sewer work (providing that the borough president might suspend work, without compensation to contractor, if he should deem it in the city's interest), the borough president was not authorized to suspend work because the city had made no appropriation to pay for the engineering and inspection it was bound to furnish. The court found dispositive the fact that the city suspended the work because it failed in one of its obligations under the contract.

---

[47] 671 F.2d 539 (D.C. Cir. 1982).

[48] *Id.* 543.

[49] 191 A.D. 205, 181 N.Y.S. 137 (2d Dep't 1920).

§ 14.19 —Unreasonable or Excessive Period of Delay

If the period of delay is found to be unreasonable or excessive, courts do not allow no damages for delay clause to bar recovery, especially if the delay would have enabled the contractor to justifiably abandon the contract. In *John T. Brady & Co. v. Board of Education,*[50] a landmark case in New York prior to *Kalisch-Jarcho,* a contractor engaged to construct a public school in the City of New York within 350 days sued because of the city's unreasonable neglect in removing certain structures from the site of the proposed school. The court, despite the existence of a no damage clause, found the city's delay of eight months "sufficiently unreasonable to strike at the heart of the contract and justify the contractor in abandoning it."[51]

In *People ex rel. Wells & Newton Co. v. Craig,*[52] a heating and ventilating contractor was delayed in excess of two years on a contract intended to be completed within 120 days. The court, notwithstanding a no damage claim, granted damages. The court ruled that such a clause "must be construed as inapplicable to delays so great or so unreasonable that they may fairly be deemed equivalent to an abandonment of the contract . . . ."[53]

It should be noted that unreasonable delays were addressed by the court in *Kalisch-Jarcho,*[54] but that court held that unreasonable delays were covered by the no damage clause. Therefore, the applicability of no damage clauses to unreasonable delays is in a state of flux, at least in New York.

§ 14.20 —Bad Faith or Fraud

If the delay is caused by the owner's misrepresentation, bad faith, or fraud, the resulting delay claims are not immunized by a no damages for delay clause.[55] Courts nullify the effects of a disclaimer where they find active misrepresentation on behalf of the party seeking the benefits of such clauses.[56] In *Warner Construction Corp. v. City of Los Angeles,*[57] the court, in sustaining a claim for breach of warranty and fraudulent concealment, stated that a cause of action for nondisclosure of material facts arises when: (1) the defendant makes representations but does not disclose facts which materially qualify the facts disclosed, or which render the disclosure likely to mislead; (2) the facts are known or accessible only to defen-

[50] 222 A.D. 504, 226 N.Y.S. 707 (1st Dep't 1928).

[51] 226 N.Y.S. at 709.

[52] 232 N.Y. 125, 133 N.E. 419 (1921).

[53] 232 N.Y. at 143.

[54] Kalisch-Jarcho, Inc. v. City of N.Y., 58 N.Y.2d 377, 448 N.E.2d 413, 461 N.Y.S.2d 746, 749 (1983).

[55] *See* 74 A.L.R.3d 187, *Construction Delay—No Damage Clause* (1976).

[56] County Asphalt, Inc. v. State, 40 A.D.2d 26, 337 N.Y.S.2d 415 (3d Dep't 1972).

[57] 2 Cal. 3d 285, 466 P.2d 996, 85 Cal. Rptr. 444 (1970).

dant, and defendant knows they are not known to or reasonably discoverable by plaintiff; or (3) the defendant actively conceals facts from discovery by the plaintiff. The court found the city liable in all three instances.

An owner's extreme negligence, which the court found almost became willful negligence, was held actionable by a contractor. In *Ozark Dam Constructors v. United States,*[58] a joint venture contracted to build a dam for which the government was to supply the cement. The government, faced with a strike by the railroad which transported the cement, made absolutely no provisions for alternate transportation until three days after the strike had begun. The court found determinative that "the possible consequences were so serious, and the action necessary was so slight, that the neglect was almost wilful. It showed a complete lack of consideration for the interest of the [contractor]."[59]

§ 14.21 —Concurrent Delays Attributable to Both Parties

Frequently, the responsibility for the delay in a construction project cannot be attributed solely to either one of the parties, but rests in some measure with both. In some cases, where both the owner and the contractor proximately contribute to the same delay, neither party is allowed to recover. For example, in the case of *J.A. Jones Construction Co. v. Greenbriar Shopping Center,*[60] the contractor contributed to the delay since its subcontractor failed to perform on time. The owner contributed to the delay with repetitive design changes and late delivery of the architect's drawings. The court refused to allow the owner to collect liquidated damages and denied the contractor the delay costs.[61] This is the so-called *rule against apportionment,* which states that under a liquidated damage provision against delay, where the owner has contributed to delays on the project, it may not apportion the fault, and forfeits all right to recovery under the provision.

This rule has been adopted by some jurisdictions but rejected by others.[62] In *E.C. Ernst, Inc. v. Manhattan Construction Co. of Texas,*[63] the Fifth Circuit, applying Alabama law, held that an owner was entitled to liquidated damages for delay in completion of a construction contract despite the fact that it had contributed to such delays itself.

In a well-reasoned opinion presenting the modern trend, the court in *United States v. William F. Klingensmith, Inc.*[64] apportioned damages between a subcontractor

---

[58] 127 F. Supp. 187 (Ct. Cl. 1955).

[59] *Id.* 189.

[60] 332 F. Supp. 1336 (N.D. Ga. 1971), *aff'd,* 461 F.2d 1269 (5th Cir. 1972).

[61] *See* 17A C.J.S. *Contracts* § 502 (1963).

[62] 152 A.L.R. 1349, *Building Contract—Liquidated Damages* (1944).

[63] 551 F.2d 1026 (5th Cir.), *cert. denied,* 434 U.S. 1067 (1977).

[64] 670 F.2d 1227 (D.C. Cir. 1982).

and contractor when both were found to have contributed to the cause of the delay. The court determined that the general contractor had violated its contractual obligations to its subcontractor on a parking garage by its failure to adequately coordinate the work of a subcontractor and failing to provide sufficient supervisory personnel on the job, notwithstanding the fact that the subcontract contained no completion date. The court also determined that the subcontractor breached the subcontract by contributing to the delays.

The court held that "where each party delays the other, it follows that each should be able to recover to the extent of the injury caused by the other's delay."[65] The court felt that such a rule protects each party from losses due to the delay of the other throughout the period of performance. It also induces each party to avoid imposing such losses on the other at any time during the period of delay.

Addressing itself to the rule against apportionment, the court stated that:

> a rule precluding a party from recovering damages for delay, once the party itself delays, would leave the parties to a contract unnecessarily vulnerable to delay by the other. We see no wisdom in, nor authority for such a rule of preclusion. Therefore, when both parties to a contract breach their contractual obligations by delaying performance, a court must assess the losses attributable to each party's delay and apportion damages accordingly.[66]

Where there are concurrent delays, i.e., by both owner and contractor, the latter is still entitled to a time extension.

§ 14.22 Excusable Delays of the Contractor

Excusable delays are those delays created by events beyond the control of either the contractor or the owner for which neither can be held responsible. Where a delay is excusable, neither party can recover damages, and the contractor may not be held in default. Examples of excusable delays would be those caused by unusually severe weather and acts of God such as tornadoes, floods, or earthquakes.

§ 14.23 —Unforeseeability

An essential element of all excusable delays is that they are caused by an unforeseen occurrence. In *United States v. Brooks-Calloway Co.*,[67] the contractor sought to

---

[65] The court noted that neither party claimed that its own delay was caused by the other party's delays. The court indicated that if such a claim was asserted, it would relate both to the claimant's responsibility for its own delay and the other party's responsibility for the delay. *Id.* 1230. *See* 6A Corbin, Contracts § 1264 (2d ed. 1964).

[66] 670 F.2d at 1231.

[67] 318 U.S. 120 (1943).

be relieved of liquidated damages based upon his defense that his work on a levee had been delayed by floods. The lower court found the contractor not at fault. However, the Supreme Court reversed and sent the case back to the lower court for further consideration, holding that there was no finding that the floods were unforeseeable (in fact, high water was found to have been predicted), which was necessary if the contractor was to be free of liability.

§ 14.24 —Unusually Severe Weather

In a government Board of Contract Appeals case, *Herman H. Neumann,*[68] the board considered the question of a delay caused by weather, and granted a time extension when it found that the delay was caused by unforeseeable and unusually severe weather.

The key to determining whether severe weather will give rise to a time extension is the unusual nature of the weather for the particular geographical area involved. *Acts of God* include unusually severe weather. By definition they are unforeseeable and beyond the control of the contractor. Therefore, the contractor cannot be held responsible for its failure to perform caused by such disturbances. Natural disasters such as earthquakes, unforeseeable floods, and tornadoes would all be classified as *acts of God.*

§ 14.25 —Labor Disputes

Labor disputes or strikes which are unforeseeable may excuse the delay, but usually only if the contract contains a provision protecting against a delay caused by a strike. If there is a *no strike* reference in the time extension clause, the contractor will be liable for delay damages to the owner.

In *Fritz-Rumer Cooke Co. v. United States,*[69] the court considered a ''construction type'' contract between the contractor and the United States, covering the removal and loading of railroad track from an area near a gaseous diffusion plant. The contract contained no provision protecting against delay caused by strike. The contractor was required to complete its undertaking without right of recovery against the United States for any damages that might be sustained by reason of a strike at the plant; during the course of that strike, a picket line was set up, which the contractors' employees observed. The government's position was that it provided the contractor with access to the job site, which remained available during the picketing. The court held: ''It has long been a well-settled rule of law that if a

[68] 70-1 B.C.A. (CCH) ¶ 8216 (1970).

[69] 279 F.2d 200 (6th Cir. 1960).

party by his contract charges himself with an obligation possible to be performed, unforeseen difficulties, however great, will not excuse him, unless the performance is rendered impossible by Act of God, the law or the other party."[70]

ACCELERATION CLAIMS

§ 14.26 Acceleration Defined

Many contractors' claims are based upon the theory of *acceleration.* Although relatively old, perhaps one of the clearest descriptions of these costs is found in a Controller General's report to the Congress:[71]

> Acceleration costs are described as those additional costs incurred by a contractor to overcome excusable delays, including those caused by the addition of extensive modifications work, within the time limits originally established in the contract to meet the relatively inflexible completion dates set by the [owner]. Acceleration costs are paid for multiple shift operations, and other increased operating expenses related to the performance or the work on an accelerated time schedule.

Another good definition is provided by the court in *Contracting & Material Co. v. City of Chicago*,[72] which stated:

> Acceleration is the process by which the material or ordinary progress of events is quickened. In the case of a contract, acceleration occurs when the contractor speeds up his work so that he is performing the job at a faster rate than prescribed in the original contract. Commonly acceleration is achieved by working overtime or working double shifts. Acceleration costs, as claimed by plaintiff, are costs incurred by virtue of acceleration.

§ 14.27 Actual and Constructive Acceleration

There are basically two types of acceleration, *actual* and *constructive.* The former is easy to identify: it occurs when the contractor is directed to accelerate its work by adding manpower, working weekends, evenings, etc.

---

[70] *Id.* 201.

[71] Controller General's Report to the Congress, *Review of the Administration of Construction of Certain Launch Facilities for the Atlas and Titan Intercontinental Ballistic Missile at Selected Air Force Bases* 31-32 (Jan. 1963).

[72] 20 Ill. App. 3d 684, 692, 314 N.E.2d 598, 604 (1974), *rev'd on other grounds,* 64 Ill. 2d 21, 349 N.E.2d 389 (1976).

Constructive acceleration does not result from a direct order from the owner, but results from the owner's refusal to allow a rightfully entitled time extension. With constructive acceleration, no order or directive is given, but the contractor is required to accelerate its work in order to keep pace with the job or to complete its work by the contract completion date. A typical constructive acceleration claim situation arises when the contracting officer or architect refuses to act on the contractor's time extension request, requiring the contractor to accelerate, when in fact the contractor is entitled to the extension.

Another common scenario is when the owner's agent tells the contractor that it will consider the time extension request at the end of the job. This benevolent gesture places the contractor in the predicament of not knowing what to do. If it waits until job completion for the owner's agent's decision, then it runs the risk of exposing itself to liquidated damages or other damage claims by the owner if the time extension request is rejected. If it chooses to accelerate, it will normally be entitled to recover its costs on a constructive acceleration claim, provided it notifies the owner it is accelerating the work.

In both these situations, however, the problem for the contractor is the same. When a time extension request is rejected, the contractor has only two choices: it can finish its work late or it can accelerate to meet the schedule imposed upon it.

In *Wallace Process Piping Co. v. Martin-Marietta Corp.*,[73] the court recognized acceleration costs while determining equitable adjustments in government contracts. The court was careful to state that acceleration costs are allowable only for efforts to overcome excusable delays. However, in this case the court did not grant acceleration costs for Wallace's claim for its excess labor costs, since Wallace did not establish by a preponderance of the evidence that its directions to accelerate were the sole cause, or even the primary cause, for costs in excess of those for which it had already received compensation.

The *Wallace* court then held that an order to complete additional work, without an extension of time, which made necessary a work week in excess of the 50 hours upon which the subcontract price was based, was a change within the meaning of the subcontract. Therefore, the subcontractor was entitled to be compensated for overtime or premium pay resulting from orders to accelerate the work.

Acceleration costs have been recognized by the courts in equitable adjustments of private contracts as well. In *Natkin & Co. v. George A. Fuller Co.*,[74] the court held that the insistence by the owner and the prime contractor (implemented by threats and withholding of the subcontractor's progress payment) that subcontractor add manpower or go on overtime to complete the work, including change order work at a time when certain delays were known to have occurred, constituted a change under the contract. Thus, the subcontractor was entitled to recover additional costs incurred for and as a result of such acceleration.

Acceleration claims were also allowed in *Nat Harrison Associates, Inc. v. Gulf States Utilities Co.*,[75] where the defendant indicated to plaintiff-contractor that

---

[73] 251 F. Supp. 411 (E.D. Va. 1965).

[74] 347 F. Supp. 17 (W.D. Mo. 1972).

[75] 491 F.2d 578 (5th Cir. 1974).

no extensions of time for delay would be permitted and directed it to provide additional manpower and equipment to maintain the original work schedule (Harrison had agreed to construct approximately 158 miles of transmission line in Louisiana for Gulf States Utilities). Although the contract allowed for time extensions for delays attributable to ''strikes and other unavoidable casualties,'' and Harrison was delayed due to inclement weather, strikes, and other unavoidable causes, Harrison's requests for extensions were either ignored or refused by Gulf States. Thus, the court allowed Harrison's acceleration damages claim.

In *Contracting & Material Co. v. City of Chicago*,[76] the contractor claimed it was forced into an abbreviated performance schedule resulting in overtime expenses when the owner wrongfully and inequitably refused to extend the contract performance date to reflect delays caused by a 46-day suspension order and a 51-day strike. The court held that the owner's refusal to recognize that an excusable delay had occurred, its refusal to grant a legitimate time extension, and its insistence that the contractor was obligated to complete the work on the original completion date constituted a constructive acceleration.

§ 14.28 Disallowance of Acceleration Damage Claims

Acceleration claims are disallowed where contract performance is behind schedule due to the contractor's own action or delay. In *Tri-Cor, Inc. v. United States*,[77] although the contracting officer notified the contractor that contract performance was behind schedule, and directed the contractor to take whatever steps were necessary to meet the completion dates, the contractor was not entitled to a price adjustment based on additional acceleration costs, because it was not entitled to a time extension. Likewise, in *Siefford v. Housing Authority of City of Humboldt*,[78] acceleration was denied where contractor was substantially and continually behind schedule under the contract, and the delay was not caused by breach of contract on the part of the owner or willful hindrance of the contractor's work by the owner.

In *Johnson Controls, Inc. v. National Valve & Manufacturing Co.*,[79] a case decided under Oklahoma law, a subcontractor was denied claims for additional expense it incurred for acceleration, because it failed to make written application for an extension of time and for compensation due to acceleration, although such was required by the contract.

In this case, the subcontractor contracted with the prime contractor to provide certain instrumentation work on a power plant project located near Hugo, Oklahoma. The terms of the prime contract between owner and contractor were incorporated into the subcontract. The prime contract contained a rather common clause which allowed the owner to make changes to the work and required the contractor to

---

[76] 64 Ill. App. 2d 21, 349 N.E.2d 389 (1976).

[77] 458 F.2d 112 (Ct. Cl. 1972).

[78] 192 Neb. 643, 223 N.W.2d 816 (1974).

[79] 569 F. Supp. 758 (E.D. Okla. 1983).

give written notice to the owner of time extensions resulting from the changes and estimates of any additional costs resulting therefrom. This clause also provided for a waiver of the contractor's right to additional compensation or a time extension if the contractor failed to give written notice in advance of performance of the changed work.

The prime contractor's actual performance fell substantially behind schedule, thus delaying the subcontractor's work. In order to meet the mandatory deadline, the subcontractor accelerated its performance by tripling its work force, working overtime and weekends, and adding a night shift. It then sought additional compensation for the acceleration and extra work it had performed. The contract provided and the parties had agreed that Oklahoma law would govern. The subcontractor argued, however, that since no existing Oklahoma cases dealt with acceleration of work claims, federal government contract law should be applied. Specifically, the subcontractor urged the court to follow *Norair Engineering Corp. v. United States,*[80] wherein a prima facie case for increased costs of acceleration under a changes clause was established when the contractor demonstrated: (1) that any delays giving rise to the order of acceleration were excusable; (2) that the contractor was ordered to accelerate; and (3) that the contractor in fact accelerated performance and incurred extra costs. A specific advance request for an extension of time or for additional compensation was not required, however, in *Norair.*

The court rejected the use of federal law in the matter, holding that Oklahoma indicated clearly its desire for strict compliance with the unambiguous terms of a contract, that the changes provision was not ambiguous, and that the parties had agreed that Oklahoma law was to govern this contract. The court cited Oklahoma case law and concluded that: "If a contract unambiguously requires written application for additional compensation, then that requirement is a condition precedent to recovering additional compensation . . . When a party makes a contract and reduces it to writing, he must abide by its plainly stated terms."[81]

The *Johnson Controls* case points out the necessity for a contractor to scrutinize its contract and follow the notice requirements contained in these provisions.

OTHER DAMAGE CLAIMS

§ 14.29 Changed Conditions

Another common ground for contractors' claims for damages is *changed conditions,* a problem which usually also involves delays. Changed condition clauses are common to many public and private contracts, and are intended to provide the contractor with a means of protecting itself against unexpected subsurface condi-

---

[80] 666 F.2d 546 (Ct. Cl. 1981).

[81] Johnson Controls, Inc. v. National Valve & Mfg. Co., 569 F. Supp. 758, 761 (E.D. Okla. 1983).

tions while at the same time protecting the owner from inflated prices needed to cover the unexpected conditions. Changed condition clauses provide for a price adjustment and time extension if the contractor encounters materially different subsurface or latent physical conditions than anticipated. In the absence of a changed condition clause or a quality variation clause,[82] most jurisdictions will deny the contractor an adjustment in price for substantial quantity changes unless there has been fraud or misrepresentation by the owner.

In *Depot Construction Corp. v. State,*[83] an excavation contractor had agreed to a unit price for the removal of what was expected to be 1,000 cubic yards of rock. When the job was finished, 3,000 cubic yards had been excavated, the work had stretched into winter, and additional as well as different equipment had been required. The court held that the contractor was limited to the contract unit price even in the face of a lower court's findings that the test borings were clearly inadequate. The court reasoned that the existence of the unit price contract, with clear language stating there were no representations as to the amount of work to be performed, conclusively demonstrated that the parties, each of whom had the same information at their disposal, considered all the contingencies.

In *P.J. Maffei Building Wrecking Corp. v. United States,*[84] a demolition contractor prepared its bid on the demolition and removal of the United States Pavilion from the 1964 World's Fair site by relying on drawings referred to in the invitation to bid as "for information only," which were specifically not made part of the contract. The invitation to bid specifically stated: "the quantity, quality, completeness, accuracy and availability of this drawing are not guaranteed." The court denied the contractor's claim under the changes provision of the contract, noting that the contractor's reliance on the documents was not reasonable where the invitation to bid was replete with caveats to the effect that documents in question were not in the possession of the federal government and would not be part of the contract documents.

In *Fattore Co. v. Metropolitan Sewerage Commission Co.,*[85] the court awarded substantial damages of $1.5 million to a sewer contractor under a changed condition and equitable adjustment provision of the contract. The court reasoned that when a government entity seeks to reduce its contract costs by making borings and providing the information to bidders, it is deemed to warrant the adequacy of its plans and specifications to the extent that compliance therewith will result in satisfactory performance.

In *Borell v. Municipality of Metropolitan Seattle,*[86] a contractor brought suit against a municipality for extra compensation for laying pipe at locations which had unexpected soil conditions. The contract did not contain a changed conditions

---

[82] A quality variation clause is generally provided in unit price contracts which acts to set percentage limits for both increases and decreases, beyond which percentage adjustments are not permitted.

[83] 278 N.Y.S.2d 363 (1st Dep't 1969).

[84] 732 F.2d 913 (Fed. Cir. 1984).

[85] 505 F.2d 1 (7th Cir. 1974).

[86] 698 F.2d 1229 (9th Cir. 1983).

clause. In fact, it contained express *disclaimers* with respect to any representations concerning subsurface conditions. Moreover, it expressly provided that no payment would be made over contract price on account of differences between contract soil information and conditions disclosed at the site. The court denied recovery.

§ 14.30 Low Bidder Claims

In government contract cases, the unmistakable trend appears to be that a low bidder who was wrongfully denied the contract can recover its bid preparation costs. In *Scanwell Laboratories, Inc. v. Thomas,*[87] the court reiterated that suits by disappointed bidders for bid preparation costs "are founded in an implied contract between the government and its bidders under which the government agrees to consider each bid fairly and honestly in return for the bidder's submission of an eligible bid."[88] While it noted that the Court of Claims consistently denies recovery of lost profits,[89] the *Scanwell* court, without accepting or rejecting the contractor's tort claim for lost profits, held that the claim fell within one of the exceptions to government liability of the Tort Claims Act.

In *Keco Industries, Inc. v. United States,*[90] the court held that the ultimate standard in deciding whether an unsuccessful low bidder is entitled to be compensated for the expense of undertaking the bidding process is whether the government conduct toward the bidder-claimant was arbitrary and capricious. In determining whether the government's behavior was arbitrary and capricious, the court addressed four general criteria:

> One is that subjective bad faith on the part of procuring officials, depriving a bidder of the fair and honest consideration of his proposal normally warrants recovery of bid preparation costs [citation omitted]. A second is that proof that there was "no reasonable basis" for the administrative decision will also suffice at least in many situations [citation omitted]. The third is that the degree of proof of error necessary for recovery is ordinarily related to the amount of discretion entrusted to the procurement officials by applicable statutes and regulations [citation omitted]. The fourth is that proven violation of pertinent statutes or regulations can but need not necessarily be a ground for recovery [citation omitted]. The application of these four general principles may well depend on (1) the type of error or dereliction committed by the government and (2) whether the error or dereliction occurred with respect to the claimant's own bid or that of its competition.[91]

---

[87] 521 F.2d 941 (D.C. Cir. 1975), *cert. denied,* 425 U.S. 910 (1976).

[88] 521 F.2d at 946.

[89] *See, e.g.,* McCarty Corp. v. United States, 499 F.2d 633 (Ct. Cl. 1974).

[90] 492 F.2d 1200 (Ct. Cl. 1974).

[91] *Id.* 1204.

ASSERTING A DAMAGE CLAIM

§ 14.31 Strict Adherence to Contract Provisions

A contractor or subcontractor should be extremely mindful of any and all contract provisions which restrict its rights to pursue claims for damages. Such clauses as discussed in §§ **14.32–14.34**, must be strictly adhered to in order for a claim to be considered.

§ 14.32 Delay and Time Extension Clauses

All modern contracts spell out the time of commencement, contract calendar days allotted for performance of the work, and the dates for substantial and final completion. More sophisticated contracts will designate the method of scheduling to be used (e.g., CPM). Generally, these contracts also provide for specific periodic updating and revisions to progress schedules. Failure to comply with these updating provisions can negate or void an otherwise viable delay claim.

Time extension clauses generally set out the causes for which the owner agrees to allow an extension of contract time. These causes are for acts and circumstances beyond the control of the contractor, i.e., strikes, acts of God, unusually severe weather, etc.[92] When presenting a time extension request or delay claim, the contractor must be sure that it is cast in terms that track the contractual agreements.

§ 14.33 Notice and Deadline Clauses

Customarily, the changes, changed conditions, disputes, and time extension clauses all contain notice requirements. Notice requirements are usually mandatory, and failure to comply may result in a waiver of the claim. Frequently, in public contracts, there are also *follow-up notice clauses* which require even more; for instance, a detailed statement of all claims, to be submitted with the final payment requisitions, is often mandated.

A contractor should consider deadline clauses to be the same as contract limitations. As an example, in certain federal contracts containing a 30-day period to appeal from a contracting officer's final decision, the courts have held that failure to comply with this deadline provision was fatal to the contractor's claim, regardless of the merit of the appeal.

---

[92] See §§ **14.22–14.25**.

§ 14.34 When Time Is of the Essence

As a general rule, time is not of the essence in a construction contract in the absence of a provision in the contract making it such. When time is of the essence, the party not performing on time is liable for damages resulting from the delay. However, an express statement, using those exact words, is not really necessary to indicate that time is of the essence in a contract. As expressed by Professor Corbin:

> Again, the agreement may be in such terms that the first party promises performance by a stated time and the second party expressly makes its return promise conditional on the exact performance; in such a case, failure by the first to perform on time gives to the second a right to damages and also the legal privilege of not performing his own promise. And still further, although the second party's return promise may not be in terms conditional, it may be made so by construction of law because the first party's failure to perform on time will deprive the second of the substantial benefit for which he bargained.[93]

There is little doubt that time was of the essence in *General Insurance Co. of America v. Hercules Construction Co.*[94] Here the general contractor contracted with the owner to construct a parking structure in St. Louis, Missouri. That general contractor signed a subcontract whereby the subcontractor agreed to supply it with custom precast concrete components, which were to comprise a substantial portion of the parking structure. The subcontract specifically directed attention to the fact that the target date for completion of the entire project was August 1, 1962, and that completion of the project was contingent upon adherence to the progress schedule for the phase of the work covered by the subcontractor. Delivery of the precast components was to commence approximately January 6, 1962, and be completed approximately April 15 of that year. The subcontractor was made aware, by express provisions in the subcontract and by subsequent dealings with the general contractor, that timely delivery in proper sequence of the precast components was an overriding feature of the agreement.

The subcontractor breached the subcontract. Delivery was one month late to start and almost five months late on final delivery. Furthermore, almost 25 percent of the precast units delivered were defective. The general contractor thereby incurred extra costs in the area of labor, erection equipment, costs, premium time, and additional construction costs. The court held that ultimate acceptance of the parking structure by the owner without penalty to the general contractor for a two-month delay in completion was irrelevant to whether the general contractor was entitled to recover on the subcontractor's bond for extra costs it incurred because of the breach, since time was of the essence.

---

[93] 3A Corbin, Contracts § 713 (1951) (discussion of "When Is Time of the Essence").

[94] 385 F.2d 13 (8th Cir. 1967).

This case is noteworthy because it refers to a surety's liability for delay damages and also because it clearly illustrates that a subcontractor can be liable for delay damages even though the general contractor is exonerated for delay damages by the owner.

PROOF OF DAMAGES

§ 14.35 Actual Damages

Once the contractor or subcontractor has established its claim, the next step is the proof of damages to sustain the claim. Various legal theories of damages are regularly sustained by the courts.

A contractor may recover damages for breach of contract only for such losses as were reasonably foreseeable at the time of execution of the contract.[95] In other words, the contractor must demonstrate either that such losses are those which usually flow from the type of contract which was breached or that the breaching party had reason to know, through some form of communication or otherwise, that such special losses would occur.[96]

Generally, damages need not be proved with mathematical certainty but rather by "whatever degree of certainty the case permits."[97] Proof of damages may consist of reasonable possibilities and inferences[98] but cannot be based on "mere speculation, guess or conjecture."[99] There must be a logical nexus between the amount of damages demanded and the proof presented.[1] Basically, a contractor must prove its damages with a reasonable degree of certainty.

A concise and clear test on the law of proof of damages as it relates to the aspect of certainty is set forth in *M&R Contractors & Builders v. Michael.*[2] There the court said:

> Courts have modified the "certainty" rule into a more flexible one of "reasonable certainty." In such instances, recovery may often be based on opinion evidence,

---

[95] Hadley v. Baxendale, 156 Eng. Rep. 145 (1854).

[96] *See also* Restatement of the Law of Contracts § 330 (1932); Restatement (2d) of the Law of Contracts § 351 (1981).

[97] John E. Green Plumbing & Heating Co., Inc. v. Turner Constr., 742 F.2d 965 (6th Cir. 1984), quoting Shakin v. Buskirk, 454 Mich. 490, 93 N.W.2d 293 (1958).

[98] E.C. Ernst, Inc. v. Koppers Co., 626 F.2d 324 (3d Cir. 1980).

[99] Zirin Laboratories Int'l v. Mead-Johnson & Co., 208 F. Supp. 633 (E.D. Mich. 1962).

[1] Manshul Constr. Co. v. Dormitory Auth. of N.Y., 79 A.D.2d 383, 436 N.Y.S.2d 724 (1st Dep't 1981).

[2] 215 Md. 340, 138 A.2d 350 (1958).

in the legal sense of that term, from which liberal inferences may be drawn. Generally, proof of actual or even estimated costs is all that is required with certainty.

Some of the modifications which have been aimed at avoiding the harsh requirements of the ''certainty'' rule include: (a) if the fact of damage is proven with certainty, the extent or the amount thereof may be left to reasonable inference; (b) where a defendant's wrong has caused the difficulty of proving damage, he cannot complain of the resulting uncertainty; (c) mere difficulty in ascertaining the amount of damage is not fatal; (d) mathematical precision in fixing the exact amount of damage is not required; (e) it is sufficient if the best evidence of the damage which is available is produced; and (f) the plaintiff is entitled to recover the value of his contract as measured by the value of his profits.[3]

§ 14.36 —Total Cost Method

One method which has been used to calculate damages is the *total cost method,* whereby the contractor takes actual job costs less anticipated bid costs or contract price to arrive at actual damages. Courts have sanctioned the use of this method when they feel no other method is available;[4] however, they acknowledge that it is not the preferred method of calculating damages.[5]

The majority view rejects this method since, to accept it, one must assume that claimants' original bid estimate was reasonable, that claimants' costs were reasonable, and that claimants were not responsible for any of the increased costs.[6] The courts realize that such is not always the case, and there are too many unknowns which creep into this theory of damages.

Sections 14.37–14.42 cover various measures of damages recoverable when a contractor or subcontractor demonstrates that each is applicable to its particular theory of damages.

§ 14.37 —Labor Costs

The contractor may experience increased labor costs, both because of additional man-hours expended on the job and as a result of escalated wages needed for premium pay for overtime and/or increased union wages. Additionally, if out-of-sequence work is performed, and added manpower or overtime is used to accelerate the work, the contractor will most likely experience diminished productivity from its labor force.

---

[3] 138 A.2d at 355.

[4] *See* Joseph Pickard's Sons Co. v. United States, 532 F.2d 739 (Ct. Cl. 1946).

[5] *See* Moorehead Constr. Co., Inc. v. City of Grand Fortis, 508 F.2d 1008 (8th Cir. 1973).

[6] F.H. McGraw & Co. v. United States, 130 F. Supp. 394 (Ct. Cl. 1955); Oliver-Finnie Co. v. United States, 279 F.2d 498 (Ct. Cl. 1960); Fattore Co. v. Metropolitan Sewerage Comm'n, 505 F.2d 1 (7th Cir. 1974).

§ 14.38 —Material Costs

When a contractor is delayed, more often than not its material costs increase, sometimes sharply, especially in inflationary periods. Three methods used to demonstrate additional material costs are:

1. Comparison of the material costs on the days it should have been purchased with actual costs
2. The use of indices
3. The cost of replacing vendors who have defaulted or refused to honor quoted prices.

If the contractor had materials on hand or specially fabricated for the project, it is entitled to demurrage and storage costs.

§ 14.39 —Job Site Indirect Costs

Job site indirect costs are the costs of maintaining the job site, and include supervisory and personnel costs, field engineer costs, field office costs, costs of rental equipment, watchmen services, etc. These costs usually can be allocated to a specific project without difficulty.

§ 14.40 —Home Office Overhead

Costs for extended home office overhead have been the subject of much litigation. Since it was first introduced in 1960, the *Eichleay*[7] formula and its variations have become very popular in construction contract litigation, but in the past few years the formula has drawn increasing criticism regarding its true usefulness as an exclusive and singular means for computing delay damages. Even though *Eichleay* per se may be fast on its way out, no discussion of delay damages is complete without some explanation of the formula which, up to recent time, was used in almost all delay claims. Moreover, the appropriate use of an *Eichleay*-type formula coupled with supplemental evidence of a relationship between home office expenses and the delay should still prove effective in proving delay damages. The basic *Eichleay*-type approach can be reduced to three simple steps in equation form:

---

[7] Eichleay Corp., 60-2 B.C.A. (CCH) ¶ 2688 (A.S.B.C.A. 1960).

Step 1:

Contract billing / Total company billing for original contract period	×	Total home office overhead for contract period	=	Home office overhead allocable to contract

Step 2:

Allocable overhead / Days of contract performance	×	Daily home office overhead allocable to project

Step 3:

Daily overhead rate	×	Number of days of compensable project delay	=	Total compensable extended home office overhead

Over the years, the courts and the boards have criticized contractors' sole and exclusive reliance on the *Eichleay* formula on the grounds that it does not provide a logical nexus between what is proved and the damages the factfinder is requested to determine.

In *Berley Industries, Inc. v. City of New York,*[8] the highest court in New York State held that the *Eichleay* formula was insufficient insofar as it was offered as a substitute for direct evidence, to establish that the plaintiff sustained delay damages for increased home office overhead expenses during the period it was unable to start on the project after receiving the notice to proceed. There was no showing that delay caused an increase in home office overhead activity or expense of any kind and no attempt was made to show that the formula was logically calculated to produce a fair estimate of actual damages.

In *Savoy Construction Co., Inc. v. United States,*[9] and *Capital Electric Co.,*[10] the courts, while applying *Eichleay,* noted the formula's shortcomings, i.e., its lack of mathematical precision and the failure to relate compensable delay to alleged increased home office overhead. In *Capital* the court in a terse statement said "[Eichleay] . . . neither associates cause with effect nor allocates costs that cannot be so associated to a specific accounting period."[11]

Therefore, the contractor should use an approach which clearly demonstrates a direct cause/effect correlation between delay and home office overhead. In New York, the courts in cases subsequent to *Berley* have approved of the use of various formulas provided the plaintiff first demonstrates that: (1) the defendant was responsible for the delay; (2) such delays caused delays in completion of the contract;

[8] 45 N.Y.2d 683, 385 N.E.2d 281, 412 N.Y.S.2d 589 (1978).

[9] 80-1 B.C.A. (CCH) ¶ 392 (A.S.B.C.A. 1980), *motion for reconsideration denied,* 80-1 B.C.A. (CCH) ¶ 14,724 (A.S.B.C.A. 1980).

[10] 65 B.C.A. (CCH) Nos. 5313, 5316 (1965).

[11] *Id.*

and (3) the plaintiff suffered damages as a result of such delays resulting in increased home office overhead. In these cases, the courts approved the use of the formulas because they were critical in furnishing some rational basis for the court to estimate such damages, since precise measure was not possible.[12]

§ 14.41 —Finance Costs

Finance costs are those costs the contractor encounters as a result of not receiving its money in a timely manner. If the contractor is forced to borrow money to finance construction, the interest costs of borrowing represent a finance cost. If the contractor is substantial and can prove it would have invested the monies withheld, the lost investment revenues or lost opportunity costs constitute another financing cost. Note, though, that the federal government does not recognize interest as a cost of performing extra or delayed work.[13]

§ 14.42 Disruption Damages

Disruption damages are those suffered by the contractor as a result of the owner's breach of implied obligation not to hinder the contractor in its performance. The courts, depending on the cause, often distinguish between delay and disruption activities.

In *Gardner Displays Co. v. United States,*[14] a manufacturer of rubber terrain maps sued for the increased cost of latex during a period of government-caused delay in inspection and approval of the preproduction model where, despite the delay, the contractor still managed to perform within the contract time limits. The court found the government unreasonable in its dilatory and inconclusive inspection and approval procedure which hindered the contractor's performance and awarded money damages for the disruption.

In *Metropolitan Paving Co. v. United States,*[15] the court stated that if the contractor suffered damages as a result of "deliberate harrassment and dilatory tactics" employed by the government's employees, it could recover.

In *D'Angelo v. State,*[16] the contractor sued the state on the ground that it failed to provide the contractor with designs for culverts shown on the plans until it was

---

[12] *See* Fehlhaber Corp. v. State, 69 A.D.2d 362, 419 N.Y.S.2d 773 (3d Dep't 1979); Manshul Constr. v. Dormitory Auth. of N.Y., 79 A.D.2d 383, 436 N.Y.S.2d 724 (1st Dep't 1981).

[13] Dravo Corp. v. United States, 594 F.2d 842 (Ct. Cl. 1979); Framlau Corp. v. United States, 568 F.2d 687 (Ct. Cl. 1977); Bell v. United States, 404 F.2d 975 (Ct. Cl. 1968).

[14] 346 F.2d 585 (Ct. Cl. 1965).

[15] 325 F.2d 241 (Ct. Cl. 1963).

[16] 41 A.D.2d 77, 341 N.Y.S.2d 84 (3d Dep't 1973), *modified,* 46 A.D.2d 983, 355 N.Y.S.2d 377 (1974).

too late to construct them during the winter season as planned. Despite the fact that contractor finished the job within 19 days of the scheduled completion date, the court said that damages could be awarded, since it noted that the contract may have been completed within such a short time only as a result of costly overtime expense.

A contractor may also recover delay damages where it can prove that it planned on performing the work in a shorter time than the specified completion time but, due to interference by the owner, it completed its work only by the contract completion date.

§ 14.43 Lost Profits

When a contractor or a subcontractor is prevented from performing because of a breach by the owner or a person for whose action the owner is responsible, it is entitled to lost profits.[17] As the *Restatement* provides, the builder is to be put "in as good a position as he would have been had the contract been fully performed."[18] In order to do this, the contractor should be entitled to the full contract price less any amount it saves as a result of the owner's breach.[19]

The courts generally refuse to award damages under the theory of lost profits for a contractor's alleged unearned profits on other jobs on which it was unable to bid due to the owner's breach. The courts apparently feel that it is far too speculative to grant recovery for such claims when it is not clear that the contractor would have bid on other jobs, would have been awarded the contracts, and, most importantly, would have earned any profits on the prospective jobs.[20]

---

[17] Restatement of the Law of Contracts § 346, comment G (1932).

[18] *Id.*

[19] *See also* Sampley Enters., Inc. v. Laurilla, 404 So. 2d 841 (Fla. Dist. Ct. App. 1981); Fostic v. United Home Improvement Co., 428 N.E.2d 1351 (Ind. Ct. App. 1981); Ferrell Constr. Co. v. Russell Creek Coal Co., 645 P.2d 1005 (Okla. 1982).

[20] Manshul Constr. Corp. v. Dormitory Auth. of N.Y., 111 Misc. 2d 209, 444 N.Y.S.2d 792 (1981).

CHAPTER 15

REPRESENTING THE CONTRACTOR IN DEFENSE OF OWNER CLAIMS

Mark L. Fleder and Peter J. Smith

Mark L. Fleder, a graduate of Princeton University and the Rutgers University School of Law, is a partner in the Newark, New Jersey law firm of Connell, Foley & Geiser, Esqs. Mr. Fleder has limited his practice during the past 15 years to construction matters, principally in the negotiation, litigation, and arbitration of complex technical disputes. He is past chairman of the New Jersey State Bar Association's Public Contract Law Section and a member of the American Bar Association's Forum Committee on the Construction Industry as well as the Public Contract Law Section. Mr. Fleder also lectures widely to attorney and nonattorney audiences on various subjects in construction law and has authored recent publications on selected topics.

Peter J. Smith is a senior associate at the Newark law firm of Connell, Foley & Geiser, Esqs. He graduated from the University of Notre Dame and, in 1981, from the Rutgers University School of Law. He specializes in the litigation of construction and related commercial matters and is a member of the Public Contract Law Sections of the New Jersey Bar Association and the American Bar Association.

§ 15.1 Available Defenses

Regardless of the size of the project and dollar amount of a contractor's agreement with an owner, a contractor is always vulnerable to claims of an owner for any of the multitude of problems which may afflict the project. Most of the rights and duties with which the law is concerned and about which the contractor should be concerned may be classified generally as arising either out of contract or tort. Contract rights and duties are voluntarily assumed by agreement of the parties and are specially tailored for the particular owner, contractor and project. Assuming that the terms of the contract are not violative of public policy, each party is entitled to performance by the other, and if there is breach by one of the parties which causes damage to the other, the law provides a remedy.

A *tort* is a civil wrong committed by one to the damage of another, arising not from any contract between them but rather by reason of their relationship in society. Thus, for example, the driver of an automobile has a duty to exercise reasonable care in the operation of the automobile so as not to interfere unreasonably with the right of another to operate a car or to walk across the street. The important consideration is always the relationship between the parties, because rights and duties spring only from that relationship.

This chapter gives an overview of the defenses available to a contractor when an owner presents a claim against it. Legal principles set forth herein are for general reference and should be considered in light of any applicable local, state, or federal statute or regulation, as well as the terms of any contract between the contractor and the owner.

This general review of the defenses available to a contractor against the claims of an owner deals with the primary (and most common) owners' claims: claims of defective construction, claims of delay in completion of the project, and claims in connection with the termination of a contractor prior to completion of the project.

§ 15.2 Compliance with Plans and Specifications

The rule is well settled in nearly every jurisdiction that where a contractor is hired to build according to plans or specifications furnished by the owner, and those plans are followed, the contractor will not be responsible for damages which result solely from defects in the plans or specifications themselves, absent any negligence on the contractor's part or any express warranty by the contract that the plans and specifications are free from defects.[1] This is true whether the plans were actually furnished by the owner itself or by an architect employed by the owner.[2] The rule has been noted by courts as particularly applicable where a contractor becomes aware of defects in plans and warns the owner of the defects, but the owner refuses to make any changes.[3]

Although it is probably not a reliable precedent today,[4] one case went so fai as to hold that even where the contractor drew up the plans, which were subsequently accepted and adopted by the owner after review, the contractor was not liable for damages arising solely from defects in the plans.[5] There is, however, some authority for the proposition that the contractor is liable for damages resulting from defective plans furnished by the owner if the damage occurs prior to completion of the contractor's performance.[6] These cases were decided on the theory that where one contracts to build a building according to plans which are open to inspection, one cannot be relieved of one's contractual duty in the absence of fraud or duress. The cases that stand for this proposition, however, are not recent, and their rationale was particularly appropriate to a time when construction design was simpler. Even in jurisdictions which adhered to this theory, contractors have been held not liable where the owner impliedly or expressly warranted the plans as free from defect.[7]

---

[1] 13 Am. Jur. 2d *Building and Construction Contracts* § 28 (1964); Annot., 6 A.L.R.3d 1394, 1397 (1966); United States v. Spearin, 248 U.S. 132 (1918).

[2] 13 Am. Jur. 2d *Building and Construction Contracts* § 28 (1964); Staley v. New, 56 N.M. 745, 250 P.2d 893 (1952) (plans furnished by owner's architect); United States v. Spearin, 248 U.S. 132 (1918) (plans furnished by United States government/owner).

[3] Ridley Inv. Co. v. Croll, 192 A.2d 925 (Del. 1963) (cited in 6 A.L.R.3d 1389 (1966)); Passaic Valley Sewerage Comm'rs v. Tierney, 1 F.2d 304 (3d Cir. 1924).

[4] Barraque v. Neff, 202 La. 360, 11 So. 2d 697 (1942); Baerveldt & Honig Constr. Co. v. Szombathy, 365 Mo. 845, 289 S.W.2d 116 (1956).

[5] Roberts v. Sinnot, 55 Mont. 369, 177 P. 252 (1918).

[6] Lonergan v. San Antonio Loan & Trust Co., 101 Tex. 63, 104 S.W. 1061 (1907), *motion for reh'g overruled,* 101 Tex. 81, 106 S.W. 876 (1908); Brasher v. Alexandria, 215 La. 887, 41 So. 2d 819 (1949); N.J. Magnam Co. v. Fuller, 222 Mass. 530, 111 N.E. 399 (1916).

[7] Pittman Constr. Co. v. Housing Auth. of New Orleans, 169 So. 2d 122 (La. Ct. App. 1964), *writ refused,* 237 La. 343, 170 So. 2d 865, *writ refused,* 247 La. 346, 170 So. 2d 866 (La. Ct. App. 1965); Penn Bridge Co. v. New Orleans, 222 F. 737 (5th Cir. 1915).

Generally, there are two situations is which a contractor may be held liable for damages caused by defects in the plans and specifications furnished by an owner. One is where the contractor makes significant departures from the plans or specifications.[8] Notwithstanding this, where the contractor deviates from the plans with the owner's knowledge and the owner does not object, the contractor will not be held liable unless the consequences of such deviation are so well known that the contractor should have been aware of the danger.[9] In a case where the contractor originally deviated from the plans, but made corrections pursuant to directions from the owner's architect, the contractor was also found to be not liable.[10] In another case, where the contractor deviated from plans and failed to correct deviations despite the owner's protest, the contractor was found liable.[11]

The second situation in which a contractor may be held liable for damages caused by defects in plans furnished by the owner is where the contractor expressly or impliedly warrants the sufficiency of the owner's plans or clearly contracts to erect a building free of defect.[12] The existence of this warranty should be determined from a close examination of both the contract and the relationship of the contractor and the owner on the project. Notwithstanding boiler-plate warnings and disclaimers in bidding documents with regard to plans and specifications, the very limited time available to bidders to estimate cost does not permit any substantial investment in making thorough independent studies of site conditions. The contractor should thus be careful not to sign a contract expressing any warranty of site conditions without the opportunity to thoroughly inspect those conditions.

Based on the conventional wisdom that the best defense is a good offense, the contractor should consider the potential liability of the design professional for design shortcomings. The Restatement (Second) of Torts provides that:

> (1) One who, in the course of his business, profession or employment, or in any other transaction in which he has a pecuniary interest, supplies false information for the guidance of others in their business transactions, is subject to liability for pecuniary loss caused to them by their justifiable reliance upon the information, if he fails to exercise reasonable care or competence in obtaining or communicating the information.
>
> (2) Except as stated in Subsection (3), the liability stated in Subsection (1) is limited to loss suffered
>
> (a) by the person or one of a limited group of persons for whose benefit and guidance he intends to supply the information or knows that the recipient intends to supply it; and

---

[8] Drummond v. Hughes, 91 N.J.L. 563, 104 A. 137 (1918); Filbert v. Philadelphia, 181 Pa. 530, 37 A. 545 (1897); 13 Am. Jur. 2d *Building and Construction Contracts* § 28 (1964).

[9] Mann v. Clowser, 190 Va. 887, 59 S.E.2d 78 (1950).

[10] Fuchs v. Parsons Constr. Co., 172 Neb. 719, 111 N.W.2d 727 (1961).

[11] Otto Misch Co. v. E.E. Davis Co., 241 Mich. 285, 217 N.W. 38 (1928).

[12] 13 Am. Jur. 2d *Building and Construction Contracts* § 28 (1964); Cameron-Hawn Realty Co. v. Albany, 207 N.Y. 377, 101 N.E. 162 (1913); Miller v. Broken Arrow, 660 F.2d 450 (10th Cir.), *cert. denied,* 455 U.S. 1020 (1981); Southern New England Contracting Co. v. State, 165 Conn. 644, 345 A.2d 550 (1974); Mayville-Portland School Dist. v. C.I. Linfoot Co., 261 N.W.2d 907 (N.D. 1978).

(b) through reliance upon it in a transaction that he intends the information to influence or knows that the recipient so intends or in a substantially similar transaction.
(3) The liability of one who is under a public duty to give the information extends to loss suffered by any of the class of persons for whose benefit the duty is created, in any of the transactions in which it is intended to protect them.[13]

While the above statement is not necessarily representative of the law in any given state, it appears to be the general direction in which the law is moving in this field. It has been held that a design professional is answerable in tort to a contractor who claims damage as the result of reliance upon plans prepared by that professional, even in the absence of privity of contract.[14] In a Texas case, however, the court of appeals for that jurisdiction reached a different conclusion, holding that a contractor could not maintain a suit against an architect/engineer charging negligence in failing to properly supervise and administer the project.[15] There was a vigorous dissent by the chief justice, who announced that he would have followed the view discussed above and allowed the suit.

While the Texas decision is probably inconsistent with the law in most other jurisdictions, it is possible to reconcile the case. The Texas case alleged negligence only in the performance of duty during the construction phase, as opposed to negligent design. Construction supervision duties spring entirely from the contract between the architect/engineer and the owner, and thus it can be argued that the contractor, who is a stranger to that agreement, should not be permitted to complain over the fact that the architect/engineer may have breached its contract with the owner.

§ 15.3 Waiver

Generally, when a contract is breached in a material respect, the injured party is accorded a choice of remedies for damage incurred. However, when the injured party treats the contract as still in effect after knowledge of the breach by the other party, its conduct is incompatible with an intention to consider the contract ended. Therefore, the otherwise injured party is deemed to have elected not to assert the breach as a cause for refusing to continue the contract.[16] This is the legal doctrine of *waiver*.

In the context of a construction project, there is no difference in principle from the situation where there is a claimed defect under a contract for the sale of goods. Generally, if a defect in the work is or ought to be known, its acceptance imposes a duty to pay for it, and if no protest or complaint about the quality of work is

---

[13] Restatement (2d) of the Law of Torts § 552 (1977).

[14] Conforti & Eisele, Inc. v. John C. Morris Assocs., 175 N.J. Super. 341, 418 A.2d 1290 (1981).

[15] Bernard Johnson, Inc. v. Continental Constructors, Inc., 630 S.W.2d 365 (Tex. Civ. App. 1982).

[16] 5 S. Williston, A Treatise on the Law of Contracts § 700 (3d ed. 1961).

promptly made, any right to claim damages for defects in performance is also discharged.[17] For example, it has been held that the supervision of an entire sidewalk construction project and the acceptance of the sidewalk upon completion of the project amounted to waiver.[18] Even though knowledge of the defects was not expressly found, such knowledge was imputed. Likewise, it has been held that daily inspections of a road under construction, and acceptance of the road upon completion, amounted to waiver of a claim against the contractor.[19]

The waiver rule is often invoked where the defective performance is a failure to complete construction in accord with contract requirements, followed by acceptance of the late performance by the owner. In such cases, the defect is obvious and the only question is whether acceptance was intended by the owner. A number of cases have held that waiver as to the time of performance was an effective defense available to the contractor.[20] There are, however, two major exceptions in the construction contract context in which there can be no waiver by an owner of the right to remedy for defective performance. The first such situation is where the defects are latent and are thus unknown and unknowable by the owner at the time of acceptance. Acceptance by the owner in such situations has been held not to amount to waiver.[21]

The other significant exception to the waiver doctrine in the construction contract context is where the contractee/employer is the owner of the land upon which the building or construction work is to be done. In this case, the owner may be obliged to accept the work in order to take or retain posession of his land. Accordingly, there "is no legal presumption in this type of contract that acceptance will discharge any right of damages for defects in performance, unless a length of time unreasonable under the circumstances lapses without complaint."[22] The question is one of fact: in some cases, acceptance and occupancy without objection, under circumstances which make it clear that the owner accepts the building as is, will amount to waiver.

---

[17] *Id.* § 724; 13 Am. Jur. 2d *Building and Construction Contracts* § 55 (1964). *See also* Johnson v. Fenestra, Inc., 305 F.2d 179 (3d Cir. 1962).

[18] Katz v. Bedford, 77 Cal. 319, 19 P. 523 (1888).

[19] City of St. Claire Shores v. L.&L. Constr. Co., 363 Mich. 518, 109 N.W.2d 802 (1961). *See also* Kandalis v. Paul Pet Constr. Co., 210 Md. 319, 123 A.2d 345 (1956); K&G Constr. Co. v. Harris, 223 Md. 305, 164 A.2d 451 (1960).

[20] *See, e.g.,* Phillips & Colby Constr. Co. v. Seymour, 91 U.S. 646 (1875); Smither & Co. v. Calvin Humphrey Corp., 232 F. Supp. 204 (D.D.C. 1964); Commercial Contractor, Inc. v. United States Fidelity & Guar. Co., 524 F.2d 944 (5th Cir. 1975).

[21] 13 Am. Jur. 2d *Building and Construction Contracts* § 55 (1964). *See also* Clear v. Patterson, 80 N.M. 654, 459 P.2d 358 (1969); Salem Realty Co. v. Batson, 256 N.C. 298, 123 S.E.2d 744 (1962); Cantrell v. Woodhill Enters., Inc., 273 N.C. 490, 160 S.E.2d 476 (1968).

[22] 5 S. Williston, A Treatise on the Law of Contracts § 724 (3d ed. 1961). *See also* Restatement (2d) of the Law of Contracts § 246(d), illustrations 6 & 7 (1981); Aubrey v. Helton, 276 Ala. 134, 159 So. 2d 837 (1964); Weinberg v. Wilensky, 26 N.J. Super. 301, 97 A.2d 707 (App. Div. 1953); 13 Am. Jur. 2d *Building and Construction Contracts* § 56 (1964).

In a case where the owner had use of a building long enough to discover defects, the waiver doctrine was held to apply.[23] Payment, whether partial or in full, is merely one fact to be considered in determining whether an owner has accepted the building as is and waived exception. Payment is not in itself conclusive on the issue of waiver.[24]

§ 15.4 Statutes of Limitation

Statutes of limitation govern the time period within which a contractor may be held accountable for its activity in relation to a project. An action against a contractor may be based on any number of theories, all of which may have distinct statutes of limitation.

Generally, statutes of limitation for actions based in contract are longer than those for actions based in tort. Typically, an owner's claims against a contractor include counts based on tort theory and separate counts based on allegations of breach of contract. Apart from the distinction between contract and tort actions, some jurisdictions have separate statutes of limitation covering actions for injuries to real property[25] and actions on unwritten contracts.[26] Some jurisdictions even have a statute of limitation which applies specifically to actions brought against contractors, engineers, architects, builders, developers, etc., for inadequate performance under a construction contract; some such statutes apply to both contract and tort actions.[27]

In situations where a reviewing court is faced with a choice between a statute of limitation for a contract action and a statute of limitation for a tort action, the cases generally hold that the longer contract statute applies.[28] Notwithstanding this general rule, if the action does not sound in both contract and tort, but only tort, a court will of course limit the action to the shorter tort limitation period.

For example, in *Lexington Insurance Co. v. Abarca Warehouses Corp.*,[29] the court held that an owner of property which had suffered serious damage in a fire was bound by a one-year tort limitation statute rather than a fifteen-year contract limitation statute. The court found the action to be based upon the negligence of an employee of the contractor and not upon the failure to install a cooling tower which apparently could have prevented the fire.

---

[23] Steltz v. Armory Co., 15 Idaho 551, 99 P. 98 (1908). *See* 13 Am. Jur. 2d *Building and Construction Contracts* § 56 (1964); Restatement (2d) of the Law of Contracts § 246(d), illustration 7 (1981).

[24] 13 Am. Jur. 2d *Building and Construction Contracts* §§ 58–59 (1964).

[25] *See* 34 Am. Jur. *Limitation of Actions* § 73 (1941).

[26] *See id.* §§ 92-94.

[27] *See, e.g.*, N.J. Stat. Ann. § 2A:14-1.1 (West 1952).

[28] *See* Annot., 1 A.L.R.3d 916 (1965).

[29] 337 F. Supp. 902 (D.C.P.R. 1971), *aff'd*, 476 F.2d 44 (1st Cir. 1973).

Accordingly, counsel defending a contractor from the claim of an owner should carefully examine the allegations of the complaint and, if appropriate, raise the defense that the action is barred by the applicable statute of limitation.

§ 15.5 Supervening Impracticability

The doctrine of supervening impracticability is set forth in the *Restatement of Contracts* as follows:

> Where, after a contract is made, a party's performance is made impracticable without his fault by the occurrence of an event the non-occurrence of which was a basic assumption on which the contract was made, his duty to render that performance is discharged, unless the language or the circumstances indicate the contrary.[30]

It has been held that where a contractor contracts to repair an owner's building, and before the contractor has finished repair to the building, it is destroyed by causes outside the responsibility of the contractor, the contractor's performance is discharged.[31] If the contractor had begun repairs at the time the building was destroyed, the contractor may recover the value of the services already rendered, again assuming that the building was destroyed through no fault of the contractor.[32] Should the parties contractually allocate the risk to the contractor, however, the allocation of that responsibility will be given effect;[33] that is, a contractor may not be able to recover in quantum meruit for services and materials rendered where it retains control over the materials used in the construction.[34]

In a contract for construction of a building, in general, impossibility of performance due to an unanticipated supervening event is the only excuse upon which a contractor may rely to excuse nonperformance of its contractual obligations. Mere impracticability in terms of hardship to the contractor in performing its obligations due to adverse weather[35] or higher than anticipated construction costs[36] will not excuse performance. It has been held that the risk of abnormal weather is

---

[30] Restatement (2d) of the Law of Contracts § 261 (1981).

[31] *Id.* §§ 263; 261 comment D, illustration 6; 13 Am. Jur. 2d *Building and Construction Contracts* § 68 (1964). *See also* Annot., 28 A.L.R.3d 788, 792 (1969).

[32] 28 A.L.R.3d 788, 795 (1969); 13 Am. Jur. 2d *Building and Construction Contracts* § 68 (1964). *See also* Bell v. Carver, 245 Ark. 30, 431 S.W.2d 452 (1968) (cited in 28 A.L.R.3d 781 (1969)); Trip v. Henderson, 158 Fla. 442, 28 So. 2d 857 (1947); Matthews Constr. Co. v. Brady, 104 N.J.L. 438, 140 A. 433 (1928).

[33] 28 A.L.R.3d 788, 809 (1969).

[34] Schaeffer Piano Mfg. Co. v. National Fire Extinguisher Co., 148 F. 159 (7th Cir. 1907).

[35] 18 S. Williston, A Treatise on the Law of Contracts § 1964 (3d ed. 1978); White v. Mitchell, 123 Wash. 630, 213 P. 10 (1923).

[36] Restatement (2d) of the Law of Contracts § 261, comment B (1981). *See also* Chouteau v. United States, 95 U.S. 61 (1877); Frank T. Hickey, Inc. v. Los Angeles Jewish Community Council, 128 Cal. App. 2d 676, 276 P.2d 52 (Dist. Ct. App. 1954); Barnard-Curtis Co. v. United States, 301 F.2d 909 (Ct. Cl. 1962).

assumed by the construction contractor except where a provision in the contract states otherwise.[37] Likewise, a mere change in the degree of difficulty or expense due to such causes as increased wages, prices of raw materials, or costs of construction, unless well beyond normal range, do not amount to impracticability, since this is the sort of risk that a fixed-price contract is intended to cover.[38]

The impracticability which excuses nonperformance must be due to an act of God or other unforeseeable, drastic change in a material condition upon which the contract is made. It is not enough that the change only makes it impracticable for the particular contractor to perform: performance itself must be impossible.[39] Notwithstanding this, a drastic change in economic conditions may be an excuse for nonperformance when caused by the occurrence of an event the nonoccurrence of which was a fundamental assumption of the contract.[40] Likewise, where an unanticipated supervening event makes construction difficult—as, for example, where zoning laws change after a contract to build is entered into—the contractor is obliged to ''use reasonable efforts to surmount obstacles to performance and a performance is impracticable only if it is so in spite of such efforts.''[41] It has been held that the rule excusing a contractor from performance of a contract for impracticability when performance will violate zoning regulations is not applicable where the zoning laws permit a variance and the contractor has not made a bona fide attempt to obtain a variance.[42]

As can be seen, the doctrine of supervening impracticability, in effect, affords a contractor a very narrow defense in only the most unusual of circumstances. It can be anticipated that these circumstances will be viewed strictly by any reviewing body.

§ 15.6 Pre-Existing Impracticability

If, subsequent to entering into a contract, the contractor and the owner become aware of facts which render performance impracticable (or which substantially frustrate the principle purpose of the contract) and of which neither party knew or had reason to know at the time of making of the contract, then discovery of such facts is treated as if it were a supervening event which rendered the contract impractical to perform.[43] For example, in *Boston Plate & Window Glass Co. v. John Bowen Co.*,[44] a contractor who won a public bid entered into a subcontract

---

[37] Associated Eng'rs & Contractors, Inc. v. State, 58 Hawaii 322, 568 P.2d 512 (1977).

[38] Restatement (2d) of the Law of Contracts § 261, comment D (1981).

[39] *Id.* § 261, comment E, illustration 14.

[40] *Id.* § 261, comment D.

[41] *Id.* § 261.

[42] Pennsylvania State Shopping Plazas, Inc. v. Olive, 202 Va. 862, 120 S.E.2d 372 (1961).

[43] Restatement (2d) of the Law of Contracts § 266(1), (2) (1981); 18 S. Williston, A Treatise on the Law of Contracts § 1933 (3d ed. 1978). *See also* Annot., 84 A.L.R.2d 12, 32 (1962).

[44] 335 Mass. 697, 141 N.E.2d 715 (1957).

with the plaintiff. Before the subcontractor began work, a court declared the general contractor's contracts with the state void for reasons of which the contractor had no knowledge. The court ruled that the contractor was not liable to its subcontractor for breach of contract because its performance was impossible due to pre-existing impossibility. Likewise, in *Partridge v. Presley*,[45] a contract for the sale of real estate included a provision that the seller would deliver a permit allowing a modification of the structure to become a two-family house. Because the zoning laws prohibited such a modification, the court declared the contract void due to pre-existing impossibility.

Some authorites state that the defense of pre-existing impossibility/impracticability, if proven, renders a contract void ab initio, rather than discharging the duty of the contractor to perform.[46]

As with supervening impracticability, the pre-existing fact which renders performance impracticable must be a fact the *nonexistence* of which was a basic assumption of the contract. Whether something was a basic assumption of the contract is a question of fact and depends upon all circumstances surrounding the contract.[47] Occasionally, for instance, a court will find that soil conditions suitable to the intended purpose of the contract are a basic assumption of the contract. This assumption is usually made, however, in mining cases. For example, in *Mineral Park Land Co. v. Howard*,[48] the court held that where soil conditions under a contract to remove dirt made the contract impracticable of performance (performance would cost 10 times as much as anticipated), the contract was held to be void.[49]

In construction contracts, it is generally held that the contractor agrees unconditionally to perform in building a house, building, road, laying pipe, etc., and unanticipated conditions of the soil or other factors rendering performance impracticable is no excuse. It has been held, for instance, that where soil conditions made building a house on a proposed site impracticable, the contractor who failed to build the house was liable for breach of contract.[50] Also, in a case where a contractor could not lay water mains under a river according to the terms of the contract, it was held liable for breach of contract when another contractor subsequently did the job.[51] Subjective impossibility is no excuse.

However, if the owner prepares the plans, chooses the site, or does other acts which amount to a warranty that the plans for the building on that site are sound,

---

[45] 189 F.2d 645 (D.C. Cir.), *cert. denied,* 342 U.S. 850 (1951).

[46] Restatement (2d) of the Law of Contracts § 266(1), (2) (1981). *See also* Annot., 84 A.L.R.2d 12, 32 (1962); Varia v. Southwick, 81 Idaho 68, 337 P.2d 374 (1959).

[47] *See* Restatement (2d) of the Law of Contracts §§ 261, comment B; 266, comment A (1981).

[48] 172 Cal. 289, 156 P. 458 (1916).

[49] *See also* Virginia Iron, Coal & Coke v. Graham, 124 Va. 692, 98 S.E. 659 (1919).

[50] Stees v. Leonard, 20 Minn. 494 (1874).

[51] B's Co., Inc. v. B.T. Barber & Assocs., Inc., 391 F.2d 130 (4th Cir. 1968). *See also* 13 Am. Jur. 2d *Building and Construction Contracts* § 64 (1964); 18 S. Williston, A Treatise on the Law of Contracts § 1964 (3d ed. 1978).

the builder may be excused from performance.[52] Of course, the contract itself may either explicitly or by inference allocate the risk of impracticability, in which case the terms of the contract control.[53]

If only part of the contractor's performance is impracticable, its duty to render the remaining part is unaffected if it can render substantial performance of the contract, or if the owner agrees to accept the partial performance in full satisfaction of the contract.[54] If the contractor renders partial substantial performance according to the terms of the contract, it must render substitute performance for the part which is impracticable to the extent that such substitute performance is a reasonable alternative.[55] Where the contractor has rendered in good faith a performance which does not strictly comply with the contract requirements, but which amounts to substantial performance, the contractor is entitled to recovery on a quantum meruit basis.[56] In determining what amounts to *substantial performance*, it is helpful to consider the following definition:

> In the common case where performances are to be exchanged under an exchange of promises, performance would be substantial if the failure of performance would not be material . . . . [B]oth parties then remain bound to complete the exchange, subject to discharge of the duty to perform the impracticable part and a compensatory claim for restitution.[57]

The same rule applies where, instead of a part of the contractor's performance being impracticable, a condition of the contract does not occur. That is, "impracticability excuses the non-occurrence of a condition if the occurrence of the condition is not a material part of the agreed exchange and forfeiture would otherwise result."[58] For example, in *Greiner v. Compratt Construction Co.*, the contractor completed road construction but the city engineer refused to exercise his judgment as to the adequacy of the work. Because the contractor could not obtain a letter of certification from the city engineer, as required by the contract, the court held for the contractor, noting that the failure to obtain the letter of certification did not materially affect the bargain since the contractor had fulfilled the purpose of the condition by another device.[59]

---

[52] Annot., 6 A.L.R.3d 1384, 1409-12 (1966). *See also* Ridley Inv. Co. v. Croll, 192 A.2d 925 (Del. 1963); Miller v. Broken Arrow, 660 F.2d 450 (10th Cir.), *cert. denied,* 455 U.S. 1020 (1981).

[53] Restatement (2d) of the Law of Contracts § 265, comment C (1981). *See also* Philadelphia Hous. Auth. v. Turner Constr. Co., 343 Pa. 512, 23 A.2d 426 (1942) (contractor warranted sufficiency of plans supplied by the owner, and thereby assumed the risk of impracticability).

[54] Restatement (2d) of the Law of Contracts § 270, comment B (1981). *See also* Grenier v. Compratt Constr. Co., 189 Conn. 144, 454 A.2d 1289 (1983).

[55] Restatement (2d) of the Law of Contracts § 270(a) (1981).

[56] 13 Am. Jur. 2d *Building and Construction Contracts* §§ 41–42 (1964).

[57] Restatement (2d) of the Law of Contracts § 270, comment B (1981).

[58] *Id.* § 271.

[59] 189 Conn. 144, 454 A.2d 1289, 1292 (1983).

§ 15.7 Warranties and the Limitations Thereof

While the Uniform Commercial Code (U.C.C.) provides an effective body of law for construing contracts for the sale of goods, in general, the provisions of the U.C.C. covering warranties and limitations of warranties do not apply, except by analogy, to construction contracts, insofar as the latter deal exclusively with supplying labor, design, or other services on the construction project. It is not surprising, therefore, that courts have held that a contract for the waterproofing of a basement, for example, is not a sale of goods;[60] neither is a contract to build a house[61] or construct an in-ground swimming pool.[62]

However, the provisions of the U.C.C. relating to expressed warranties,[63] implied warranties,[64] and the limitations thereof,[65] will apply where construction, design, or installations services provided with materials are merely incidental. Thus, a subcontract for the supply and installation of an air-conditioning unit has been held to be a *sale of goods*,[66] supporting the application of the U.C.C.; a contract to supply and pour concrete has also been held to be a sale of goods.[67]

Even though buildings and the like are generally not considered goods, if the finished product itself is found to be a good, in many instances courts will find construction contracts to be, in reality, contracts for the sale of goods. For instance, a grain silo[68] and a diving board installed with a built-in-pool[69] were both held to be "goods"; a sale of prefabricated modular buildings,[70] the sale and installation of an electrical floor,[71] the sale and construction of a one-million-gallon water tank,[72] the design, construction, and delivery of a sewage processing plant,[73] and the installation of a pulp mill boiler and equipment[74] were all held to be "sales

---

[60] Peltz Constr. Co. v. Dunham, 436 N.E.2d 892 (Ind. Ct. App. 1982).

[61] G-W-L, Inc. v. Robichaux, 643 S.W.2d 392 (Tex. 1982).

[62] Chlan v. K.D.I. Sylvan Pools, Inc., 53 Md. App. 236, 452 A.2d 1259 (1982).

[63] U.C.C. § 2-313 (1983) provides that an affirmation of fact or any promise constitutes an express warranty.

[64] U.C.C. § 2-314 (1983) provides an implied warranty of merchantability; U.C.C. § 2-315 provides an implied warranty of fitness for the intended purpose.

[65] *Id.* § 2-316 provides that implied warranties may be excluded or modified, but only if the exclusion is conspicuous and all rules are complied with strictly. *Id.* § 2-319 deals with limitation of remedies. *Id.* § 2-207 relates to "the battle of the forms."

[66] Howard Dodge & Sons, Inc. v. Finn, 391 N.E.2d 638 (Ind. Ct. App. 1979).

[67] S.J. Groves & Son v. Warner Co., 576 F.2d 524 (3d Cir. 1978).

[68] Clevenger & Wright Co. v. A.O. Smith Harvestore Prods., Inc., 625 S.W.2d 906 (Mo. Ct. App. 1981).

[69] Anthony Pools, Div. of Anthony Indus., Inc. v. Sheehan, 455 A.2d 434 (Md. Ct. App. 1983).

[70] Cates v. Morgan Portable Bldg. Corp., 591 F.2d 17 (7th Cir. 1979).

[71] Aluminum Co. of Am. v. Elector-Flo Corp., 451 F.2d 1115 (10th Cir. 1971).

[72] Pittsburgh-Des Moines Steel Co. v. Brookhaven Manor Water Co., 532 F.2d 572 (7th Cir. 1976).

[73] Omaha Pollution Control Corp. v. Carver-Greenfield Corp., 413 F. Supp. 1069 (D. Neb. 1976).

[74] Lincoln Pump & Paper Co. v. Dravo Corp., 436 F. Supp. 262 (D. Me. 1977).

of goods.'' In each case, whether the end product involves a sale of goods under the U.C.C. is a question of fact.[75]

Separate and apart from the application of the U.C.C., in general, building contractors impliedly warrant that work will be done in a workmanlike manner and in accordance with good usage and accepted practices in the community in which the work is done.[76] Additionally, virtually all jurisdictions recognize the existence of implied warranties of workmanlike performance and of habitability made by builder/vendors of new residential homes.[77] Most jurisdictions imply these warranties as a matter of common law, but some have statutes fixing them.[78] Some jurisdictions limit this warranty by providing that it only applies as between the builder/vendor and the original purchaser of the residential home,[79] or by limiting the applicability of the implied warranty of habitability to contracts by builders/vendors for construction of a new dwelling house.[80]

While the implied warranty of workmanlike performance applies to building contracts in most jurisdictions,[81] the warranty does not guarantee that work will be perfect forever; rather, the builder impliedly warrants that what it builds will be constructed in a reasonably skillful and workmanlike manner.[82] The warranty does not expire after a set time but, as time goes on, it becomes more likely that defects which arise are not due to the builder's breach of the implied warranty of workmanlike performance.[83] Many jurisdictions hold that implied warranties extend from the builder to all subsequent purchasers, subject to the limitations imposed by time and other defenses such as waiver. However, it has been held that the *reasonable time* limitation on bringing a suit for implied breach of warranty is tolled during the period in which builders fraudulently conceal defects.[84]

The primary limitations on implied warranties available to construction contractors as defenses, other than the reasonable time limitation described above,

---

[75] *See* U.C.C. § 2-105 (1983). *See also* Space Leasing Assocs. v. Atlantic Bldg. Sys., Inc., 144 Ga. App. 320, 241 S.E.2d 438 (1977). *See generally* Annot., 4 A.L.R.4th 912 (1981).

[76] 13 Am. Jur. 2d *Building and Construction Contracts* § 27 (1964). *See also* Henggler v. Gindra, 191 Neb. 317, 214 N.W.2d 925 (1974); Hodgson v. Chin, 168 N.J. Super. 549, 403 A.2d 942 (App. Div. 1979).

[77] Annot., 25 A.L.R.3d 383, 413 (1969).

[78] *See* D.C. Code Ann. § 45-1801 *et seq.* (1981), especially § 45-1847(b); La. Civ. Code Ann. art. 2762 (West 1952); Md. Real Prop. Code Ann. § 10-201(c)-203 (1981).

[79] *See, e.g.*, Sousa v. Albino, 388 A.2d 804 (R.I. 1978).

[80] Groover v. Magnavox, 71 F.R.D. 638 (D. Pa. 1976).

[81] Kubbe v. Crescent Steel, 105 Ariz. 459, 466 P.2d 753 (1970); Aced v. Hobbs-Sesack Plumbing Co., 55 Cal. 2d 573, 360 P.2d 897, 12 Cal. Rptr. 257 (1961); Vernali v. Centrali, 28 Conn. Supp. 476, 266 A.2d 200 (Super. Ct. 1970); Whirlpool Corp. v. Morse, 222 F. Supp. 645 (D. Minn. 1963), *aff'd*, 332 F.2d 901 (8th Cir. 1964); Mann v. Clowser, 190 Va. 887, 59 S.E.2d 78 (1950); Gosselin v. Better Homes, Inc., 256 A.2d 629 (Me. 1969).

[82] Parsons v. Beaulieu, 429 A.2d 214 (Me. 1981).

[83] *Id.*

[84] Reichelet v. Urban Inv. & Dev. Co., 577 F. Supp. 971 (D. Ill. 1984).

are contractual. The contractor may, by contract, limit or exclude the implied warranties of workmanlike performance or habitability. In order to do so, though, the exclusion must not have the appearance of a contract of adhesion. While the contractor may express a limit or exclude such warranties, it must do so clearly. Clauses excluding implied warranties will be strictly construed against the drafter.[85]

Where a contract for construction contains express warranties relating to the exact same subject matter the implied warranties normally covered by (i.e., the quality of work or habitability of the building), the express warranties control. Thus, it has been held that where express warranties as to the details of construction of a home excluded implied warranties on the same subject matter, the express warranties did not eliminate implied warranties relating to soil conditions.[86]

Express warranties are common in construction contracts, usually representing an agreement by the parties as to the quality of the work which the contractor is to perform. Such warranties are given effect as they are written. For example, a warranty by a builder that its work would be free of defects in labor and materials for one year has been given effect.[87]

Express warranties may be a part of a written contract, or may be made orally subsequent to the making of the contract. One court held that "certification" of steel work in a building constituted a warranty that the building was designed to carry the required loads.[88] Also, oral assurances made by a contractor subsequent to the written agreement between the parties that defects would be cured or warranties supplied was given effect by the court.[89]

A contractor faced with allegations of breach of express warranties in a contract with an owner should consider each of the contractual defenses discussed in this chapter; all, some, or none may be viable alternatives, depending on the facts of the case.

§ 15.8 Damages

While the formula for recovery of damages differs in construction defect cases and in delay cases, the theory for an award of damages remains the same, i.e., to put the nonbreaching party where it would have been had the promise been performed.[90] There are two components in the practical application of this theory: (1) the cost of substitute performance, and (2) any further losses resulting from

---

[85] Casavant v. Campopiano, 327 A.2d 831 (R.I. 1974).

[86] F.&S. Constr. Co. v. Berube, 322 F.2d 782 (10th Cir. 1963).

[87] Austin Co. v. Vaughn Bldg. Corp., 643 S.W.2d 113 (Tex. 1982); Fidelity & Cas. Co. of N.Y. v. J.A. Jones Constr. Co., 200 F. Supp. 264 (D. Ark. 1961), *aff'd,* 325 F.2d 605 (8th Cir. 1963).

[88] Industrial Dev. Bd. v. Fequa Indus., Inc., 523 F.2d 1226 (5th Cir. 1975).

[89] Salem Towne Apartments, Inc. v. McDaniel & Sons Roofing Co., 330 F. Supp. 906 (E.D.N.C. 1970).

[90] *See, e.g.,* 525 Main St. Corp. v. Eagle Roofing Co., 34 N.J. 251, 254, 168 A.2d 33 (1961).

the breach that any reasonable person, at the time of entry into a contract, would have foreseen would occur from a breach.

The cost of substitute performance, or the *standard measure* of damages in a situation where an owner alleges breach of contract by a contractor, pertains also to allegations of delay and defect by the owner. Generally, in determining the measure of damages, it is appropriate to consider the time of the alleged breach. Where the contractor's breach is proven before construction begins, the owner's measure of damages is the cost of completion of the work, together with reasonable compensation for unavoidable delay. Where the owner proves a contractor's breach during construction, damages include the cost of completing the project, assuming there is no undue economic waste,[91] and reasonable compensation for unavoidable delay. The contractor, in turn, may be able to claim recovery in *quasi-contract* for its partial performance. Where the contractor completes its performance, but is late, the owner can claim damages for the loss incurred by not being able to use the property on the contractually scheduled completion date. If the damages for this lost use were not foreseeable at the time the contract was entered into, the owner's delay damages are severely limited.[92]

Where there has been substantial performance of the construction contract, the contractor has a widely recognized claim for the contract price, less the cost of remedying defects or omissions. While this is a deviation from the general rule that absolute completion of performance must be proved before the right to payment can be enforced, it has been justified by one court on the basis that where the labor and materials of the contractor go to form an addition affixed to lands, the right of the owner of the land to refuse to accept the work cannot be exercised without doing an injustice to one party or the other.[93]

Lacking substantial performance, an owner's damages for breach by a contractor are limited to the cost of completion, provided there is no undue economic waste. Examples of this principle in operation are best set forth in the *Restatement of Contracts*:

2. A contracts to build a house for B for $100,000 but repudiates the contract after doing part of the work and having been paid $40,000. Other builders will charge B $80,000 to finish the house. B's damages include the $80,000 cost to complete the work less the $60,000 cost avoided or $20,000, together with damages for any loss caused by delay.
3. A contracts to build a house for B for $100,000. When it is completed, the foundations crack, leaving part of the building in a dangerous condition. To make it safe would require tearing down some of the walls and strengthening the foundation at a cost of $30,000, and would increase the market value of the house by $20,000. B's damages include the $30,000 cost to remedy the defects.

---

[91] Restatement (2d) of the Law of Contracts § 348(2) (1981).

[92] *Id.* § 348(1).

[93] Palmieri v. Albanese, 12 N.J. Super. 338, 79 A.2d 699 (App. Div. 1951).

4. A contracts to build a house for B for $100,000 according to specifications that include the use of Reading pipe. After completion, B discovers that A has used Cohoes pipe, an equally good brand. To replace the Cohoes pipe with Reading pipe would require tearing down part of the walls at a cost of over $20,000 and would not affect the market price of the house. In an action by B against A, A gives no proof of any special value that Reading pipe would have to him. B's damages do not include the $20,000 cost to remedy the defects, because that cost is clearly disproportionate to the loss in value to B. B can recover only nominal damages.[94]

In addition to a contractor's potential defense that a particular remedy may involve undue economic waste, an owner must also prove that the losses it allegedly suffered as a result of the contractor's breach were certain in their nature and not speculative. It has been held that in determining the "consequences" of a breach, damages "must be reasonably certain and definite . . . as distinguished from the mere quantitative uncertainty."[95] Applying this rule to a claim of delay damage, the fact of delay in completion of the project must be clearly established, although the value of that delay need not be proved with such certainty. Of course, without certain proofs of the value of the delay, an owner may limit its damage recovery.[96]

Likewise, an owner confronted with a breach by a contractor is under a duty to mitigate damages by reasonable efforts.[97] The term *reasonable* is overriding in the sense that a owner must make all reasonable attempts to mitigate its damages. The evaluation of that duty is measured by whether or not the plaintiff acted reasonably under the circumstances.[98] The burden of proof, however, is on the defendant-contractor to show actual or potential mitigation and the amount of damages for which the owner itself must account.[99]

§ 15.9 When Owner Retains Duty to Coordinate

At common law, an owner is charged with certain coordinating responsibilities as a result of its duty to furnish access to the construction site. This duty in turn arises from the implied provision in every contract that the other party promises to cooperate and that failure to do so is a breach.[1]

---

[94] Restatement (2d) of the Law of Contracts § 348, illustrations 2, 3, and 4 (1981).

[95] Tessmar v. Grosner, 23 N.J. 193, 203, 128 A.2d 467 (1957).

[96] Restatement (2d) of the Law of Contracts § 348(1) (1981).

[97] McGraw v. Johnson, 42 N.J. Super. 267, 273, 126 A.2d 203 (App. Div. 1956).

[98] McDonald v. Mianecki, 79 N.J. 275, 299, 398 A.2d 1283 (1979).

[99] Sandler v. Lawn-A-Mat, 141 N.J. Super. 437, 455, 358 A.2d 805 (App. Div.), *cert. denied,* 71 N.J. 503, 366 A.2d 658 (1976).

[1] 11 S. Williston, A Treatise on the Law of Contracts §§ 1296, 1316 (3d ed. 1968). *See also* George A. Fuller Co. v. United States, 69 F. Supp. 409 (Ct. Cl. 1947); Glassman Constr. Co., Inc. v. Maryland City Plaza, Inc., 371 F. Supp. 1154 (D. Md. 1974), *aff'd*, 530 F.2d 968 (4th Cir. 1975).

A restatement of these principles as they stand today was presented by the New Jersey Supreme Court in the case of *Broadway Maintenance Corp. v. Rutgers*:

> If no one were designated to carry on the overall supervision, the reasonable implication would be that the owner would perform those duties. In so doing the owner impliedly assumes the duty to coordinate the various contractors to prevent unreasonable delays on the project. That is a reasonable assumption because the contracting authority has the power to use its superior position and to invoke its contractual rights to compel cooperation among contractors. The owner is impliedly obligated to act in good faith and to do that which it reasonably can to ensure that the other contractors adhere to the time schedules established for the project. An owner's failure to take action in the face of unnecessary and unreasonable delays by one of the contracting parties would ordinarily evidence bad faith and constitute a breach of its implied duty to coordinate.[2]

As indicated above, in the absence of a contractual provision to the contrary, an owner is assumed to have retained its duty to coordinate the construction project among the primes.

This duty is expressly noted in American Institute of Architects Document A201.[3] That portion of the A.I.A. contract operates to allow multiple prime contractors at the owner's option:

ARTICLE 6
WORK BY OWNER OR BY SEPARATE CONTRACTORS

6.1 OWNER'S RIGHT TO PERFORM WORK AND TO AWARD SEPARATE CONTRACTS

6.1.1 The Owner reserves the right to perform work related to the Project with his own forces, and to award separate contracts in connection with other portions of the Project or other work on the site under these or similar Conditions of the Contract. If the Contractor claims that delay or additional cost is involved because of such action by the Owner, he shall make such claim as provided elsewhere in the Contract Documents.

6.1.2 When separate contracts are awarded for different portions of the Project or other work on the site, the term Contractor in the Contractor Documents in each case shall mean the Contractor who executes each separate Owner-Contractor Agreement.

6.1.3 The Owner will provide for the coordination of the work of his own forces and of each separate contractor with the Work of the Contractor, who shall cooperate therewith as provided in Paragraph 6.2.

---

[2] 90 N.J. 253, 265, 447 A.2d 907 (1982).

[3] American Institute of Architects Document A201, General Conditions of the Contract for Construction, Art. 6.1.3 (13th ed., Aug. 1976).

It is frequently difficult in a multiple prime construction project, however, to determine in which instance an owner has retained the duty to coordinate, in the absence of an express provision such as that in the A.I.A. contract.

In the *Broadway Maintenance* case, the electrical and plumbing prime contractors brought actions against Rutgers for damages caused by delays in construction of the defendant's medical school. The prime contractor for general construction, Briscoe, was brought in as a third-party defendant by Rutgers and settled prior to trial. Each of the contracts between Rutgers and the individual primes contained the following language with respect to the coordinating responsibilities of the general contractor, Briscoe:

G-4-N.1 UNIQUE ROLE OF RESPONSIBILITY: STAFFING:

The General Contractor has the responsibility for being the supervisor, manager, overseer, coordinator and expediter of all of the Contractors and of the total construction process and all of its parts in accordance with the Contract. In executing the duties incurred by these responsibilities, the General Contractor shall provide sufficient executive and supervisory staff in the field to enable efficient and expeditious handling of these matters. There shall be at least one full-time project Manager assigned by the General Contractor to his home office, as well as the field staff referred to above; the Project Manager shall attend each Progress Meeting at the site.

G-4-N.2 OWNER'S RELIANCE UPON GENERAL CONTRACTOR:

The Owner relies upon the organization, management, skill, cooperation and efficiency of the General Contractor to supervise, direct, control and manage the General Construction work and the efforts of the other Contractors so as to deliver the intended building conforming to the Contract and within the scheduled time.

G-4-N.3 OTHER CONTRACTORS' RELIANCE UPON GENERAL CONTRACTOR:

All other Contractors shall rely upon the organization, management, skill, cooperation and efficiency of the General Contractor to supervise, direct, control and manage the general construction work and the efforts of the other Contractors so as to deliver the intended building conforming to the Contract and within the scheduled time.

G-4-N.4 PROJECT PLANNING, SCHEDULING AND CONTROL:

(''CRITICAL PATH METHOD''): As an aid to the General Contractor, and all other Contractors, to bring the completion of the project within the time allocated, the Owner has contracted the services of a Critical Plan Method Scheduling Consultant for project planning, scheduling, and control, and the General Contractor shall incorporate and enforce the combined schedule as his own. Each Contractor

> agrees to cooperate and coordinate his own operations in order to meet effectively all scheduled task deadlines.[4]

Despite this language delegating coordination responsibilities to Briscoe, the main means of enforcing coordination remained within the owner's control, pursuant to the following contract provision, also contained in each of the separate prime contracts:

> f. In case the Contractor, by his own acts or the acts of any person or persons in his employ, shall unnecessarily delay, in the opinion of the Architect, the work of the Owner or other Contractors, by not properly cooperating with them or by not affording them efficient opportunity of facility to perform work as may be specified, the Contractor shall, in that case, pay all costs and expenses incurred by such parties due to any such delays and he hereby authorizes the Owner to deduct the amount of such costs and expenses from any monies due or to become due the Contractor under this Contract, based on the investigations and recommendations of the Architect.

The heart of the court's opinion rejecting any cause of action by the primes against Rutgers for failure to coordinate the work was the reasoning that the power to impose sanctions and the duty to coordinate were distinct.[5]

In contrast to *Broadway Maintenance*, other courts have found the right to impose sanctions on a delaying contractor so significant as to become tantamount to a duty to coordinate. By way of example, in *Tippets-Abbett-McCarthy-Stratton v. New York State Thruway Authority*,[6] an engineer sought compensation from the state for additional supervision services extended because of a contractor's delay. The New York Court of Claims stated:

> The Attorney General further argues that it was the [engineer]'s duty to insure that the job was progressed satisfactorily. But the real power to impose sanctions upon an inefficient or dilatory contractor was retained by the Thruway Authority. [The engineer] could not remove the contractor from the job; could not hold up payment; could not cancel the contract nor avail itself of any penalty for the contractor's delay.

As can be readily seen from the *Broadway Maintenance* and *Tippets* cases, courts will use a factual balancing process in determining whether or not an owner in a multiple prime construction contract has retained its common-law duty to coordinate. The balance is delicate and consideration of who has the power to impose

---

[4] Edwin J. Dobson, Jr., Inc. v. Rutgers State Univ., 157 N.J. Super. 357, 384 A.2d 1121 (Law Div. 1978).

[5] Broadway Maintenance Corp. v. Rutgers, 90 N.J. 253, 266-68, 447 A.2d 907 (1982).

[6] 27 Misc. 2d 522, 523 (N.Y. Ct. Cl. 1961), *rev'd on other grounds*, 18 A.D.2d 402, *aff'd*, 13 N.Y.2d 1091 (1963).

sanctions is a factor to be considered. More important, however, may be the actual conduct of the owner throughout the project; whether, for instance, the owner has exercised the power to terminate a prime contractor or has taken over the work of any prime. In those instances, although the owner may have expressed its intent in the contract to have someone else supervise the work, conduct in the field may have altered the expressed intent.[7]

Of course, an owner who has retained the duty to coordinate shares in the responsibility to see that the project is completed on time. Its failure to do so could support not only an affirmative defense, but perhaps a counterclaim of the contractor against the owner for delay damages.

§ 15.10 Liquidated Damages

In general, damages for breach of contract by either an owner or a contractor may be liquidated in an agreement between them, but only at an amount that is reasonable in light of the anticipated or actual damages caused by the breach, and considering the difficulty of proof of the loss.[8] If the amount fixed by the contract as liquidated damages is unreasonably large in light of the anticipated or actual damages, a liquidated damages provision is unenforceable, since it is regarded as a penalty.[9] Whether an amount fixed by a contract as damages is a valid liquidated damages clause or an invalid penalty is a question of law.[10]

Reasonableness of the amount fixed as damages is determined as of the time of the contract. The question to be determined by the court is: what damages were contemplated as likely by the parties at that time in the event of delay in completion of the project?[11] Thus, where the contractor promises to complete performance by a specified date or to pay $1,000 a day for every day's delay in completing it, and the contractor takes 10 days beyond the date stipulated to complete the project, then the contractor must pay the $10,000 if $1,000 a day is not unreasonable in the light of the anticipated loss. Additionally, the actual loss to the

---

[7] *See, e.g.,* Natkin & Co. v. George A. Fuller Co., 347 F. Supp. 17, 35-36 (W.D. Mo. 1972).

[8] Restatement (2d) of the Law of Contracts § 356(1) (1981); 5 S. Williston, A Treatise on the Law of Contracts § 775A (3d ed. 1961); Leasing Serv. Corp. v. Justice, 673 F.2d 70 (2d Cir. 1982); Dairy Farm Leasing Co., Inc. v. Hartley, 395 A.2d 1135 (Me. 1978); Wilson v. Clarke, 470 F.2d 1218 (1st Cir. 1972).

[9] Restatement (2d) of the Law of Contracts § 356(1) (1981); 5 S. Williston, A Treatise on the Law of Contracts § 776 (3d ed. 1961).

[10] 5 S. Williston, A Treatise on the Law of Contracts § 778 (3d ed. 1961).

[11] *Id.* § 777; E.C. Ernst, Inc. v. Manhattan Constr. Co., 551 F.2d 1026 (5th Cir. 1977) (subsequent history omitted); United Order of Am. Bricklayers & Stone Masons Union No. 21 v. Thorleif Larsen & Son, Inc., 519 F.2d 331 (7th Cir. 1975).

owner must be difficult to prove. In these circumstances, the amount fixed is not a penalty.[12]

With regard to particular liquidated damages clauses, it has been held that a liquidated damages clause in a contract is not to be considered a penalty simply because the amount of damages assessed by the contract escalates with the period of delay.[13] However, where the amount fixed by the liquidated damages clause is not *reasonably proportionate* to the damages actually sustained, the courts may refuse to enforce the clause as being against public policy.[14] Also, a court may refuse to enforce such a clause where it determines that the amount fixed could not have been based upon the damages which would likely have resulted from breach of the contract.[15]

After satisfying the court that the liquidated damages are not a penalty, a party seeking to enforce a liquidated damages clause must show that the contract was actually breached and, more importantly, that it itself strictly complied with the terms of the contract.[16] Courts following this rule in effect place the burden of proof upon the party (usually the owner) seeking to enforce a liquidated damages clause automatically upon delayed completion of the project. In this respect, liquidated damages claims are similar to suits for actual damages; that is, the party seeking recovery by way of a contract provision must still prove that it is entitled to recover. Where a party seeking to enforce a clause which fixes damages for delay has itself contributed to that delay, there is some authority for the proposition that it cannot enforce a liquidated damages clause at all.[17]

The better view, though, is that the person seeking recovery by way of a clause alloting damages for delay on a per diem basis can only recover such damages

---

[12] Restatement (2d) of the Law of Contracts § 356, comment B, illustration 3 (1981); United States v. Bethlehem Steel Co., 205 U.S. 105 (1907); Dahlstrom Corp. v. State Highway Comm'n, 590 F.2d 614 (5th Cir. 1979); Gustav Hirsch Org., Inc. v. East Ky. Rival Elec. Co-op Corp., 201 F. Supp. 809 (D. Ky. 1962).

[13] Grenier v. Compratt Constr. Co., 189 Conn. 144, 454 A.2d 1289 (1983).

[14] San Ore-Gradner v. Missouri Pac. R.R. Co., 496 F. Supp. 1337 (E.D. Ark. 1980); Hungerford Constr. Co. v. Florida Citrus Exposition, Inc., 410 F.2d 1229 (5th Cir.), *cert. denied,* 396 U.S. 928 (1969).

[15] 218-220 Market St. Corp. v. Krich-Radisco, Inc., 124 N.J.L. 302, 11 A.2d 109 (1940); Muller v. Light, 538 S.W.2d 487 (Tex. Civ. App. 1976); Hammaker v. Schleigh, 157 Md. 652, 147 A. 790 (1929); Rye v. Public Serv. Mut. Ins. Co., 34 N.Y.2d 470, 315 N.E.2d 458, 358 N.Y.S.2d 391 (1974).

[16] Utica Mut. Co. v. DiDonato, 187 N.J. Super. 30, 453 A.2d 559 (App. Div. 1982); Vines v. Orchard Hills, Inc., 181 Conn. 501, 435 A.2d 1022 (1980); Dairy Farm Leasing Co., Inc. v. Hartley, 395 A.2d 1135 (Me. 1978); Wilson v. Clarke, 470 F.2d 1218 (1st Cir. 1972); General Ins. Co. of Am. v. Commerce Hyatt House, 5 Cal. App. 3d 460, 85 Cal. Rptr. 317 (1970).

[17] Psaty & Fuhrman, Inc. v. Housing Auth., 76 R.I. 87, 68 A.2d 32 (1949); Gillioz v. State Highway Comm'n, 348 Mo. 211, 153 S.W.2d 18 (1941).

as are not attributable to its own acts.[18] Only delay attributable to a contractor's fault can be the subject of recovery under a liquidated damages clause.[19] It would be inherently unfair to impose liquidated damages upon a contractor for a period of time for which delay was in fact caused by another contractor or the owner.

In sum, the contractor must remember, in defending against the claim of an owner seeking to enforce a liquidated damages clause, that the owner has the burden of proving in most instances that: (1) the amount of damages fixed is a reasonable estimate of the harm caused by the alleged breach; (2) the harm caused by the alleged breach is one incapable or very difficult of accurate estimate; and (3) the delay was not attributable to the owner or another contractor. Absent proof of these three elements, any liquidated damages assessed should be deemed a penalty and unenforceable.

§ 15.11 Substantial Performance

Where a contractor has, in good faith, rendered performance which turns out to be defective or incomplete, it may not be in breach of its contract if it has *substantially performed* the terms of the contract.[20] If the owner sues the contractor for breach based on an allegation of defective performance by the contractor, the contractor may plead *substantial performance* as a defense. If the contractor is successful in establishing the defense, the owner may recover only the difference between the contract price paid to the contractor and the value of the work "as is," or the cost of repairing the defect, if such cost would not be too great.[21] Absent substantial performance, a breach by the contractor may result in complete forfeiture, or the owner may recover in full the costs of remedying the defective performance in order to have it comply with the contract specifications.

Thus, where a contractor's performance under a contract is grossly incomplete or ineffective, it will probably be liable to the owner for the full cost of repair.[22] The same rule of damages applies where the contractor renders defective, though possibly substantial, performance in bad faith.[23]

---

[18] 13 Am. Jur. 2d *Building and Construction Contracts* § 87 (1964).

[19] Bedford-Carthage Stone Co. v. Ranney, 34 S.W.2d 387 (Tex. Civ. App. 1930).

[20] 13 Am. Jur. 2d *Building and Construction Contracts* § 41 (1964); 5 S. Williston, A Treatise on the Law of Contracts § 805 (3d ed. 1961).

[21] 13 Am. Jur. 2d *Building and Construction Contracts* §§ 42, 81 (1964); Farnsworth, Contracts § 8.12 (1982); 5 S. Williston, A Treatise on the Law of Contracts § 805 (3d ed. 1961).

[22] 13 Am. Jur. 2d *Building and Construction Contracts* § 41 (1964); Tolstoy Constr. Co. v. Minter, 78 Cal. App. 3d 665, 143 Cal. Rptr. 570 (1978).

[23] Fidelity & Deposit Co. v. Steel, 607 S.W.2d 17 (Tex. Civ. App. 1980).

The rule of payment to the contractor upon substantial performance is based upon quantum meruit principles: where the owner has received something of value, he or she must pay for it according to its value, whether it was what the owner contracted for or not.[24] The following definition of *substantial performance* may be helpful: "There is substantial performance of a contract where all the essentials necessary to the full accomplishment of the purposes for which the thing contracted for has been constructed are performed with such approximation to complete performance that the Owner obtains substantially what is called for by the Contract."[25] Nonetheless, "the rule does not apply where the deviations from the contract are such that an allowance out of the contract price would not give the other party essentially what he contracted for."[26]

Thus, in a famous case where a contract for construction of a house called for a certain brand of pipe to be used, the builder built the house exactly according to specifications except that he used a different brand of pipe, essentially equal in quality to the brand specified in the contract. The court held that the doctrine of substantial performance limited the owner's possible recovery to the difference in value of the two kinds of pipe, rather than the full cost of tearing out the old pipe and putting in new pipe.[27] Another court held that a slight deviation in the color of brick siding on a house did not amount to a failure to comply with the substantial provisions of a contract to apply the siding.[28] Even where there are significant deviations from plans, courts have found substantial performance where the plans or other circumstances indicated that such deviations were of a nonessential character.[29]

Notwithstanding this, one court has held that a failure to install a roof of uniform color under a roofing contract rendered performance *not* substantial.[30] It has also been said that in construction contracts there can be no substantial performance where the defect is "structural," in the sense that it affects the soundness of the building.[31]

Sometimes courts measure substantial performance in terms of cost required to remedy the defect relative to the entire contract price. Where the cost of remedy

---

[24] 13 Am. Jur. 2d *Building and Construction Contracts* § 41 (1964); 5 S. Williston, A Treatise on the Law of Contracts § 805 (3d ed. 1961).

[25] 13 Am. Jur. 2d *Building and Construction Contracts* § 43 (1964); Jardine Estates, Inc. v. Donna Brook Corp., 42 N.J. Super. 332, 126 A.2d 372 (App. Div. 1956); Foeller v. Heintz, 137 Wis. 169, 118 N.W. 543 (1908).

[26] 5 S. Williston, A Treatise on the Law of Contracts § 805 (3d ed. 1961).

[27] Jacob & Youngs v. Kent, 230 N.Y. 239, 129 N.E. 889 (1921).

[28] S.D. & D.L. Cota Plastering Co. v. Moore, 247 Iowa 972, 77 N.W.2d 475 (1956).

[29] Plante v. Jacobs, 10 Wis. 2d 567, 103 N.W.2d 296 (1960); Dixon v. Nelson, 79 S.D. 44, 107 N.W.2d 505 (1961).

[30] D.W. Grun Roofing & Constr. Co. v. Lope, 529 S.W.2d 258 (Tex. Civ. App. 1975).

[31] Farnsworth, Contracts § 8.12 (1982); Spence v. Harm, 163 N.Y. 220, 57 N.E. 412 (1900).

is great compared to the contract price, courts are hesitant to find that there has been substantial performance.[32] More recently, however, one court noted that "the question whether the building contract has been substantially performed is not to be decided upon a percentage basis . . . ."[33] In the end, as Justice Holmes said, "the question is one of degree, to be answered, if there is doubt, by the triers of fact."[34]

§ 15.12 Accord and Satisfaction

By definition, an *accord and satisfaction* consists of two elements. First is the *accord,* or agreement, whereby one of the parties undertakes to give or perform, and the other to accept, in satisfaction of a claim, something other than or different from that to which the second party is or considers itself to be entitled. Second is the *satisfaction*: the execution or performance of the accord or agreement.[35] Until performance of the accord, the original duty is suspended unless there is such a breach of the accord as to discharge a new duty of the obligee to accept the performance in satisfaction. If there is such a breach, the obligee may enforce either the original duty or any duty under the accord.[36]

The accord entitles the obligor to a chance to render the substitute performance in satisfaction of the original duty. Under the rules set forth in the *Restatement*, the obligee's right to enforce that duty is suspended subject to the terms of the accord until the obligor has had that chance.[37]

An accord is like the making of a new contract. Generally, there is no accord and satisfaction without offer by one party and acceptance by the other of sub-

[32] Nees v. Weaver, 222 Wis. 492, 269 N.W. 266 (1936); Manthey v. Stock, 133 Wis. 197, 113 N.W. 443 (1907).

[33] Jardine Estates, Inc. v. Donna Brook Corp., 42 N.J. Super. 332, 126 A.2d 372 (App. Div. 1956); *accord,* Plante v. Jacobs, 10 Wis. 2d 567, 103 N.W.2d 296 (1960).

[34] Jacob & Youngs v. Kent, 230 N.Y. 239, 243, 129 N.E. 889 (1921).

[35] Restatement (2d) of the Law of Contracts § 281(1) (1981); Long v. Weiler, 395 S.W.2d 234 (Mo. Ct. App. 1965); 15 S. Williston, A Treatise on the Law of Contracts § 1838 (3d ed. 1972).

[36] Restatement (2d) of the Law of Contracts § 281(2) (1981); W.F. Constr. Co., Inc. v. Kalik, 103 Idaho 713, 652 P.2d 661 (Ct. App. 1982) (original contract called for plaintiff to be paid 15% of the direct construction costs; defendant refused to pay, claiming that plaintiff rendered defective performance. Parties agreed that plaintiff would remedy the defects and defendant would pay plaintiff 10% of the direct costs. After plaintiff performed according to the new contract, defendant still refused to pay; *held,* for plaintiff for amount stated in original contract (15%)); Stinson v. Mueller, 449 A.2d 329 (D.C. App. 1982) (document, signed by contractor, promising repayment of money owed as refund on renovation project, did not constitute accord and satisfaction barring suit by owner against contractor for breach of original renovation contract, since contractor failed to perform fully the repayment required by the document).

[37] Restatement (2d) of the Law of Contracts § 281, comment B (1981).

stituted performance in full settlement.[38] Thus, it was held that a contractor and a homeowner did not reach accord and satisfaction regarding a damage claim for defects in a garage floor when there was evidence that the homeowners were aware of the cracks around the time that they paid for the garage and no evidence indicated that they communicated this knowledge to the contractor.[39] Further, neither the original bill nor the reduced bill, which the homeowners paid, referred on its face, to the claim for damages.

It is sometimes said that determining whether accord and satisfaction has been reached between the parties turns on the intent of the parties.[40] One case held that mutual assent of the parties to settlement of the dispute was a requirement for accord and satisfaction, and that the creditor must fully understand that the amount tendered was conditioned upon its being accepted as a full disposition of the underlying obligation.[41]

The contractor-defendant should be aware, when considering raising the defense of accord and satisfaction, that it is an affirmative defense on which the defendant bears the burden of proof.[42]

§ 15.13 Arbitration Defense

A contract between an owner and a contractor often contains an agreement to arbitrate some or all of the disputes which may arise between the parties; yet an owner may bring a claim against a contractor in a court of law. Prior to answering the complaint of the owner in that forum, the contractor should consider raising as a defense the agreement to arbitrate.

Arbitration is a remedy favored by the courts, and they will look to enforce any such agreements to arbitrate that can be read to govern a particular dispute.[43] Many jurisdictions now have statutory schemes allowing a party to an action brought in any court upon an issue arising out of an agreement providing for arbitration

---

[38] *Id.* § 281, comment A; Laganas v. Installation Specialties, Inc., 291 A.2d 187 (D.C. App. 1972).

[39] Parsons v. Beaulieu, 429 A.2d 214 (Me. 1981).

[40] Publicker Indus., Inc. v. Romar Ceramics Corp., 603 F.2d 1065 (3d Cir. 1979); Tuskegee Alumni Hous. Found., Inc. v. National Homes Constr. Corp., 450 F. Supp. 714 (S.D. Ohio 1978), *aff'd,* 624 F.2d 1101 (5th Cir. 1980).

[41] Flowers v. Diamond Shamrock Corp., 693 F.2d 1146 (5th Cir. 1982).

[42] Delhomme Indus., Inc. v. Houston Beechcraft, Inc., 669 F.2d 1049 (5th Cir. 1982); Larson v. Erickson, 549 F.2d 1136 (8th Cir. 1977); Studiengesellschaft Kohle M&H v. Novamont Corp., 485 F. Supp. 471 (S.D.N.Y. 1980).

[43] Long Branch Sewerage Auth. v. Molnar, Keppler & Terhune, 88 N.J. Super. 455, 212 A.2d 683 (App. Div. 1965); Ench Equip. Corp. v. N.K. Foods, Inc., 43 N.J. Super. 500, 503, 129 A.2d 313 (App. Div. 1957); Collingswood Hosiery Mills v. American Fed. Hosiery Works, 28 N.J. Super. 605, 101 A.2d 372 (Ch. Div. 1953).

to obtain a stay of the action pending arbitration.[44] Those same statutory schemes often include other provisions for enforcement of an agreement to arbitrate, including procedures for confirmation of an arbitration award, allowing the award the force and effect of a judgment rendered by a court of law.[45]

Even if the contractor answers a complaint in an action filed in a court of law, an owner's allegation that the contractor has thereby waived its right to arbitration will not be readily acknowledged by the courts.[46] Likewise, an owner's allegation that a contractor is guilty of laches by answering a complaint filed in a court of law, notwithstanding an agreement to arbitrate, is also subject to challenge by the contractor. To constitute a valid defense of laches, generally the delay must not only be unexplained and inexcusable, but must have visited prejudice upon the party asserting the delay.[47]

An example of an arbitration clause which clearly sets forth arbitration as a remedy for all disputes between an owner and a contractor appears in A.I.A. Document A201, which provides:

> All claims, disputes and other matters in question between the Contractor and Owner arising out of, or relating to, the Contract Documents or the breach thereof . . . shall be decided by arbitration in accordance with the Construction Industry Arbitration Rules of the American Arbitration Association then obtaining unless the parties mutually agree otherwise.[48]

While the A.I.A. provision does not refer to the location of the arbitration or the law to be applied, it is often wise for the contractor to consider, at the time of its agreement with the owner, whether to add specific provisions providing for arbitration to take place at a convenient location pursuant to the rules of a friendly jurisdiction. Other items a contractor may wish to consider include preserving legal remedies in certain areas and providing for a particular means of arbitrator selection.

Most arbitration clauses in construction contracts provide for arbitration pursuant to the Construction Industry Arbitration Rules. In general, those rules generate a more loosely structured and faster dispute resolution process than can be obtained in the courts. For example, the Construction Industry Arbitration Rules do not

---

[44] United States Arbitration Act, 9 U.S.C. § 1 *et seq.*; Cal. Civ. Proc. Code § 1208 *et seq.* (West 1982); Fla. Stat. Ann. § 682.01 *et seq.* (West 1966); Ind. Code Ann. § 34-4-2-1 *et seq.* (West 1983); Mo. Ann. Stat. § 435.350 *et seq.* (Vernon 1952); N.J. Stat. Ann. § 2A:24-1 *et seq.* (West 1953); N.Y. Civ. Prac. Law Ann. § 7501 *et seq.* (Law. Co-op. 1978); Ohio Rev. Code Ann. § 2711.01 *et seq.* (Page 1981); Pa. Stat. Ann. tit. 5, § 161 *et seq.* (Purdon 1963).

[45] This is true of all the state statutes cited *supra* n.44.

[46] *See, e.g.,* Hilti v. Oldach, 392 F.2d 368 (1st Cir. 1968).

[47] *See, e.g.,* Mitchell v. Alfred Hoffman, 48 N.J. Super. 396, 137 A.2d 569 (App. Div. 1958), citing Stroebel v. Jefferson Trucking & Rigging Co., 125 N.J.L. 484, 487, 15 A.2d 805 (E&A 1940).

[48] American Institute of Architects Document A201, General Conditions of the Contract for Construction, Art. 7.9.1 (13th ed., Aug. 1976).

provide for any discovery (i.e., deposition of witnesses prior to the hearings, or interrogatories—written questions—which could be served on a party prior to the hearing). Additionally, the Rules provide that "conformity to legal rules of evidence shall not be necessary."[49]

In addition to the considerations set forth above, lawyers and contractors look to arbitration for dispute resolution because the arbitrators (either one or a panel of three) are from within the construction industry and have knowledge of that industry beyond that which can be expected from the average judge. This is particularly important to note: the arbitrator comes to the hearing with a background in the construction industry, *not* the "blank slate" one can reasonably expect from a jury. However, as can be seen from a review of the various statutory schemes for arbitration in different jurisdictions, an award based upon an error in fact or law *cannot* be overturned.[50]

Careful consideration should be given to all these factors as applied to the particular facts at hand before determining to seek relief through arbitration, when it is provided for in the contract. In any event, a contractor sued by an owner in a court of law, who has entered into an arbitration agreement with the owner in any form, should assert that arbitration clause as a defense in its answer to the complaint.

§ 15.14 Termination Defense

In the absence of a provision clearly delineating procedures and conditions for termination of a contractor during the course of a construction project, an owner can only terminate a contractor who has materially breached the terms of the contract between them. What constitutes a *material breach of the contract* or a *material failure* understandably varies from one situation to the next. However, a *material breach* is fairly characterized as a substantial failure of the agreement between the parties which justifies the other party in not performing its obligations under the agreement any further. A contractor's failure to provide labor forces for the project may amount to a material breach, depending upon several variable factors, including the time period that labor forces were not available to the project, the extent of the work remaining on the project, the effect of the lack of labor forces on other phases of the work, and other considerations, such as whether the owner was making timely payments under the contract for the contract work completed.

---

[49] American Arbitration Association, Construction Industry Arbitration Rules, Rule 31. Copies of these rules may be obtained from the American Arbitration Association, 140 West 51st Street, New York, New York 10020.

[50] *See* statutes cited *supra* n. 44.

The *Restatement (Second) of Contracts* sets forth the following rule bearing on the contractor's primary defense to termination by an owner—failure of an owner to satisfy a material requirement of the agreement between the parties. It provides, in pertinent part: "[I]t is a condition of each party's remaining duties to render performances to be exchanged under an exchange of promises that there be no uncured material failure by the other party to render any such performance due at an earlier time."[51]

The illustration in the *Restatement* indicated the operation of this principle in a typical situation where a contractor would have a defense to termination by the owner. The illustration reads as follows:

1. A contracts to build a house for B for $50,000.00, progress payments to be made monthly in an amount equal to eighty-five per cent of the price of the work performed during the preceeding month, the balance to be paid on the architect's certificate of satisfactory completion of the house. Without justification, B fails to make a $5,000.00 progress payment. A thereupon stops work on the house and a week goes by. A's failure to continue the work is not a breach and B has no claim against A. B's failure to make the progress payment is an uncured material failure of performance which operates as a non-occurrence of a condition of A's remaining duties of performance under the exchange. If B offers to make the delayed payment and in all the circumstances, it is not too late to cure the material breach, A's duties to continue the work are not discharged. A has a claim against B for damages for partial breach because of the delay.[52]

In this instance, if the contractor (A) were terminated by the owner (B), the contractor would have a defense that the prior material breach by the owner excused any further performance of the contractor. If this were found to be a material breach, the contractor would have been wrongfully terminated and would not only have a defense to the termination by the owner, but also an affirmative claim against the owner. Generally, the failure of an owner to make a progress payment within the time set forth in the contract for payment is an uncured material failure of performance excusing the contractor who stops work on the project. As can be seen from the illustration, it is critical to determine which party caused the first material breach. Obviously, if the contractor stops work and *then* the owner fails in making a progress payment, the contractor will not be protected by the law.

A separate defense that a contractor may have after termination by an owner is that the owner may in some instances be deemed to have waived a material failure on the part of the contractor. For instance, if a contractor is terminated by an owner based upon the material failure to use equipment which conforms to the specifications of the owner, the contractor may claim that the owner waived this material failure in certain instances. In order to properly claim this waiver

[51] Restatement (2d) of the Law of Contracts § 237 (1981).

[52] *Id.* § 237, illustration 1.

defense, the contractor must show that the owner was aware of the material failure at some earlier point and that the contractor was allowed to continue despite this failure.

Contract language often provides the contractor with relatively clear guidance as to what specific acts on its part will entitle the owner to terminate it. For example, the widely used A.I.A. Document A201 provides:

> 14.2.1 If the Contractor is adjudged bankrupt, or if he makes a general assignment for the benefit of his creditors, or if a receiver is appointed on account of his insolvency, or if he persistently or repeatedly refuses or fails, except in cases for which extension of time is provided, to supply enough properly skilled workmen or proper materials, or if he fails to make prompt payment to subcontractors or for materials or labor or persistently disregards laws, ordinances, rules or regulations or orders of any public authority having jurisdiction or otherwise is guilty of a substantial violation of a provision of the Contract documents, then the Owner, upon certification by the Architect that sufficient cause exists to justify such action, may without prejudice to any right or remedy and after giving the Contractor and his surety, if any, seven days' written notice, terminate the employment of the Contractor and take possession of the site and of all materials, equipment, tools, construction equipment and machinery thereon owned by the Contractor and may finish the Work by whatever method he may deem expedient. In such case, the Contractor may not be entitled to receive any further payment until the work is finished.[53]

While such a provision clearly establishes conditions and procedures for termination of a contractor, a contractor nonetheless has available to it the defenses discussed above, as well as, in certain instances, many of the other defenses discussed in this chapter.

---

[53] American Institute of Architects Document A201, General Conditions of the Contract for Construction, Art. 14.2.1 (13th ed., Aug. 1976).

CHAPTER 16

PRICING THE LOSS

James T. Schmid

James T. Schmid is Regional Manager of Construction Litigation Services for Coopers & Lybrand. Mr. Schmid has worked extensively throughout his career helping clients evaluate, present, and negotiate construction claims. In addition to his construction claims experience, Mr. Schmid has also provided extensive operations-related assistance to construction companies in the areas of financial management, contract administration, systems, and scheduling. Mr. Schmid, a member of the American Arbitration Association's Commercial Panel, holds an M.B.A. degree from the University of Michigan and a B.S. degree from the Oakland University School of Engineering. He has seven years of experience as a consultant and three years as an industrial engineer.

PRELIMINARIES

§ 16.1 Introduction

Pricing the loss in a construction claim refers to determining the amount of money due the contractor because of actions or inactions of the opposing parties during the construction project. The analysis relies heavily on an understanding of the business practices and records of the contractor. This chapter begins with a discussion of some of the business considerations which might help avoid becoming involved in a construction claim, followed by a discussion of the critical business capabilities that should be in place to help calculate and prove damages should a claim arise.[1] The focus then turns to selecting an approach to calculating damages and some of the basic concepts involved in the different approaches. The discussion of alternative damage approaches is by no means exhaustive, since most claims have unique characteristics which require special consideration. The discussion does, however, present the basic guidelines and major analysis options.[2] The final sections review some of the key features that should be incorporated into a damage analysis when it is presented to the court.[3]

§ 16.2 Protecting the Contractor

Construction claims are simply demands for money. They are constituted of sets of facts which are intended to prove both entitlement and the extent to which the contractor should be compensated. These facts are then incorporated into a litigation process, the outcome of which is frequently difficult for the contractor to manage or forecast.

The uncertainty of the litigation process dictates that it is in the contractor's best interests to avoid it if possible. Although this is not always reasonable, the contractor can minimize the risk of being involved in a lawsuit by:

---

[1] See **§§ 16.2–16.5.**

[2] See **§§ 16.6–16.18.**

[3] See **§§ 16.19–16.20.**

1. Using thorough construction contracts
2. Anticipating problems
3. Facing issues
4. Always being prepared to prove damages.

The contractor should consider the litigation risk associated with every construction project pursued. The construction contract should be thorough enough to address all potential problems identified by the contractor through the use of contract clauses which focus on changed conditions, quantification of variations, responsibility for delay, and other potential problems. Characteristics such as unique technical requirements, owner/architect capabilities, clarity of work scope as defined in the specifications and drawings, and owner financial strength should be assessed in detail. The contractor should also look inward to ensure that the project is consistent with its organization's technical and financial capabilities and that sufficient staff will be available when required.

The contractor should be alert to the early warning signs of potential litigation throughout the entire life of the contract. If a problem develops, the contractor would be wise to consult an attorney as soon as possible to identify the major issues connected with a potential claim. The attorney can help isolate those areas where the contractor should mitigate damages during the pendency of the claim in order to avoid unreimbursed costs. The attorney and contractor should also work together with a construction damages expert to develop an appropriate strategy for calculating the value of the claim. This is particularly critical if the project is still under construction; otherwise, important opportunities to capture key cost data may be missed.

The contractor should also conduct an early review of the contract documents with an attorney. The goal of this review is to assess how the contract terms affect the contractor's ability to pursue specific remedies. It is important to know who assumes responsibility for various contract risks and to identify any contract clauses which may limit the remedies that may be pursued. A thorough understanding of the contract provisions is a critical factor in the development of the contractor's damage claim.

§ 16.3 Be Prepared

The best way to avoid costly construction litigation is to be prepared to win the case. Adherence to this rule and implementation of all of the business policies it implies will aid in reaching a negotiated settlement with the adverse parties at the earliest possible opportunity. Contractor efforts to resolve a dispute should not be postponed. The longer a dispute exists, the less likely it is that a satisfactory conclusion can be achieved and the more likely it is that ongoing project costs will increase.

Factual information provided by the contractor's accounting and administrative systems should possess four key characteristics which will significantly enhance its value if used to support a claim. The information should be:

1. Objective in content
2. Consistent with other factual information
3. Obtained from reliable sources
4. Unambiguous in interpretation.

Five critical business capabilities are key to enhancing the contractor's ability to be prepared for a potential lawsuit. These include:

1. *Accounting capabilities.* Once entitlement is proved, the contractor's accounting capabilities will be the single most important factor in determining whether all costs incurred are actually reimbursed. Although courts sometimes award damages in a case where liability is proved even if the actual cost cannot be precisely calculated, reliance on this option jeopardizes the claim's prospects and credibility. The risk of being unable to prove the size of the loss without good accounting capabilities is so significant that it should be avoided if possible.

   Accounting capabilities are also a key element of the contractor's overall project management process. Poor accounting practices and controls may expose the contractor to risks which are even more significant than potential undercompensation. Adverse parties may attempt to prevail in the entire claim by asserting that the inadequate accounting capabilities are a symptom of poor project management controls and that a failure of project management caused the claimed extra costs.

   The contractor should implement a cost accounting system which is sophisticated enough to support a detailed claim assertion. Ideally, this system should tie to the original estimate on a line-item basis, as well as identify costs by time period and major project component. In so doing, it should also be flexible enough to identify costs associated with change orders and extra work orders. Such a system should be properly documented, reliable, and should reconcile to the source documents and the general ledger.
2. *Competent staff.* A knowledgeable and well-trained staff is another critical element of good project controls. The contractor should recognize the importance of hiring competent staff, particularly if temporary project management personnel are used, such as in the case of a joint venture or in a situation where the contract requires local participation. The contractor should assess the control risks associated with the use of short-term local talent, and should avoid using them to perform key project roles such as quality control or funds disbursement. Every staff member should be trained in the overall procedures that will be used throughout the project and should

be familiar with the contract provisions which affect their area of responsibility so they can recognize when potential problems develop.

3. *Administrative controls.* Construction claims rely heavily on many of the documents used throughout the project. Documents such as bid estimates, cost records, project schedules, progress reports, correspondence, change orders, design specifications, field reports, quality control reports, photo logs, and pay draws are examples of the project management documents that may be needed. It is in the contractor's best interest to be sure that these documents are properly filed and controlled and that they provide reliable, detailed information regarding planned and actual activities. Use of transmittal forms, correspondence logs, and submittal logs, which indicate the dates of key events, provide excellent support in this area.
4. *Change order controls.* The contractor should implement a change order control system which establishes a file for every potential change order. All relevant documentation, correspondence, cost information, and engineering notes should be accumulated in the file. If the change order is not approved for the activity in question, then this file may serve as the starting point for preparing the claim.
5. *Schedule controls.* Delay-related costs are frequently a major component of construction claims. Proving delay relies primarily on comparing changes between the planned and as-built construction schedules. Differences between these two schedules must then be associated with the cost records to determine the financial impact. Control and maintenance of the contract schedules are critical tasks of the contractor if a delay claim is to be asserted. The contractor should establish a single person or group of people to assume responsibility for schedule documents. The schedule should be updated in a timely manner for actual performance and new schedules should be prepared which reflect the impact of required changes. Correspondence, meeting minutes, and other documentation which initiate required schedule changes should be referenced and maintained in the scheduling files.

Numerous other financial and administrative practices exist which will assist ongoing project management activities as well as support a claim. Some of these are common to all construction projects, while many are unique to a specific provision of the contract. Project characteristics such as fast-track, cost plus, lump-sum, unit price, shared incentive, or special project organization features all affect the composition of the contractor's ideal financial and administrative control systems. The main objective when designing these systems is to ensure that they can provide the information needed to manage the business and its construction projects, while simultaneously generating the information needed to support a potential claim. The contractor is well advised to solicit the help of an expert to ensure that all reasonable options are considered in the context of the contractor's unique business and project management needs.

When asserting a claim, the contractor will usually be in a better position to control the process than the adverse parties. The contractor should use this position to maximum advantage. The damage analysis, with detailed and factual documentation, should be prepared and thoroughly reviewed well in advance of asserting the formal claim. This analysis should be consolidated into a clear and simple presentation format that summarizes the costs in a manner corresponding to the major entitlement issues.

§ 16.4 Preliminary Project Review

Once the contractor and attorney conclude that a potential construction claim exists, a damages expert should be retained. One of the expert's first tasks should be to conduct a preliminary review of both the contract and the project costs to date. The primary goal of this review is to prepare an initial assessment of the damages incurred and to help develop a strategy for the cost analysis. This review should include an analysis of the contract, the project's cost system, and its supporting documentation, as well as an examination of the pertinent management, quality control, administrative, and other documents associated with the project.

This initial review should generate financial work products that demonstrate the extent of the damages. These include:

1. An assessment of the reliability of the financial systems
2. An analysis of the project's total cost and a comparison to the revenue received
3. A comparison of the original bid to the actual costs incurred by: (a) type of cost, such as labor, subcontracts, material and overhead; and (b) work item or physical location
4. A reconciliation of the original contract amount with the actual contract revenue.

An important element of this preliminary project review is to identify and exclude costs which are inappropriate, i.e., the costs that cannot be attributed to the adverse parties. These include costs within the scope of the contract as well as costs stemming from the contractor's failure to mitigate damages. Exclusion of these inappropriate costs will eliminate potential weaknesses which could undermine the valid claims.

Another important element of the preliminary project review is the early identification of issues which will require expert testimony. Early engagement of experts speeds up the claim development and reduces wasted effort.

The preliminary review should initiate several important work projects which will be relied upon and further developed throughout the course of litigation. These include:

1. *Claim data base:* a data base used to accumulate cost information. It should include source document references and should be able to cross-reference to the claim's summary cost schedules
2. *Proposed interrogatory questions:* suggested questions to be submitted to the opposing side to assist in development of the damage analysis
3. *Proposed requests to admit:* a list of uncontroversial facts which will help to support the financial analysis
4. *Proposed deposition listing:* people who need to be deposed regarding key cost or management issues
5. *Proposed deposition questions:* suggested questions to be directed to key personnel in order to obtain a better understanding of the critical cost or management issues
6. *Document request:* a compilation of the documents required from opposing parties or third parties in order to obtain cost information needed to develop the damage calculation.

A final product of the preliminary project review is the identification of information capable of supporting other expert analyses. Engineers, schedulers, and other construction claims experts frequently rely on the financial information provided by the construction damages expert. Consequently, the construction damages expert must be alert to the types of information that will be of importance to the testimony of other experts.

§ 16.5 Types of Claims

There are seven types of claims which contractors most frequently assert in construction projects. These involve:

1. Termination of construction activities
2. Changes in conditions
3. Changes in project scope
4. Delay in start-up or in ongoing project activities
5. Acceleration of project activities
6. Disruption or inefficiency
7. Owner abuse of the payment process.

Construction claims may involve one or a combination of the preceding claim types. Each of these claim types may be traced to one or more source events which precipitated their occurrence. **Table 16-1** illustrates the relationship between sample events and the types of claims they could potentially cause. Obviously, this presentation of claim types and source events is not exhaustive; nevertheless, it serves as a good starting point for structuring the claim's analysis.

Table 16-1

CLAIMS LIKELY TO RESULT FROM COMMON SOURCES

Source Events	Type of Claim						
	Termination	Changes in Conditions	Changes in Scope	Delay	Acceleration	Disruption	Payment Process
Owner financial difficulty	X		X	X		X	X
Delayed payments						X	X
Increased retainage							X
Work suspension or stoppage	X			X			X
Schedule interference				X	X	X	
Delayed owner decisions or quality approvals				X	X	X	
Inaccurate bid specifications		X	X	X	X	X	
Specification or drawing errors		X	X	X	X	X	
Late drawings				X	X		
Limited site access				X	X	X	
Interfering work crews				X	X	X	
Strikes				X	X	X	
Inappropriate site conditions		X	X				
Excessive inspection or quality control requirements			X	X	X	X	
Construction method interference		X		X	X	X	

ALTERNATE APPROACHES TO CALCULATING DAMAGES

§ 16.6 Deciding Which Approach to Use

The approach used to calculate the contractor's damage claim can vary significantly from one case to the next. Three key factors which determine the appropriate approach are: (1) the point in the construction process at which the contractor identifies the likelihood of a potential litigation; (2) the quality of the contractor's cost, project management, contract administration, and other information systems; and (3) the unique factual circumstances of the dispute and the nature of the events from which it arose. Depending upon these factors and the type of potential claims identified, one or more basic approaches may be utilized to calculate the damages incurred. These approaches include:

1. *Cost forecast analysis:* comparing forecast cost to contract revenue (used in wrongful termination cases)
2. *Total cost analysis* (or some modification thereof): comparing total cost to total revenue
3. *Discrete cost analysis:* studying in detail the excess costs associated with each event in dispute.

Construction claims calculations sometimes use both the total cost and discrete cost approaches. In theory, either approach should arrive at the correct answer, although the total cost approach starts with total cost and works down, while the discrete cost approach starts with no cost and builds up. Calculating damages two different ways enhances the credibility of the claim by demonstrating an inherent consistency in the amount asserted. Nevertheless, given a choice between these two options, the court will probably focus on the discrete cost analysis because this approach is more directly related to the entitlement issues of the case. A discussion of each of the three basic approaches follows in §§ **16.7-16.18**.

§ 16.7 Cost Forecast Approach

Calculating damages in wrongful termination cases frequently involves estimating the contractor's expected profit by subtracting a forecast of total project cost from the established contract revenue. This approach can be used if the construction activities have not yet begun or if they are only partially completed.

If construction has not yet begun, the forecast of total project cost may be drawn from the original bid estimates, assuming these estimates were done in sufficient detail and with reasonable accuracy. Otherwise, the contractor may have to prepare a revised forecast of total project cost based on a more detailed analysis of the project requirements and anticipated construction costs.

If the project is partially complete, the analysis should center on the expected cost to complete the remainder of the contract. If possible, this should be based on extrapolating the contractor's cost performance from the date at which actual data is no longer available. Although this method involves more effort, it should ultimately be more accurate than using the bid estimate because the method takes into account the performance of the contractor under actual working conditions.

When developing a forecast of total cost based on actual costs prior to termination, care should be taken to ensure that disruptive events and excess costs which were not the contractor's responsibility are accounted for properly. These events should be identified, costed, and excluded from the base statistics used to forecast total project cost. Reimbursement for these costs can then be pursued as a separate cause of action based on the specific disruptive events, rather than as part of the wrongful termination claim.

For partially complete contracts, there are two other basic methods for analyzing the damages in wrongful termination cases. These techniques rely more heavily on the original bid estimate to forecast completion costs, and may ignore the contractor's actual cost performance. Hence, these methods run the risk of arriving at conclusions which are inconsistent with the contractor's actual performance prior to termination, thus impairing the claim's credibility. The contractor should therefore use caution when using these alternative approaches to ensure that the factual circumstances support their use and that the conclusions are reasonable. The first of these methods is to forecast total contract cost based on the actual cost for activities performed to date, plus the cost estimated in the bid for the activities that remain to be completed. The second method is to forecast the total contract cost by relying entirely on the original bid estimate and ignoring any actual cost experience.

A final item that should be addressed when analyzing wrongful termination claims is the proper identification of costs unique to mid-project demobilization. For example, if the construction method used cost more to demobilize during the project than it would have cost to demobilize after the project had been completed, the contractor should include the excess demobilization costs as part of the claim. Claim considerations of this type clearly depend on the nature of the project and a thorough understanding of the contractor's actual costs.

§ 16.8 Total Cost Approach

The total cost approach is generally the least complicated method of calculating the damages associated with a completed contract. This approach compares the contractor's actual costs, plus fee, to the revenue received and claims the difference as damages. Courts often refuse to allow this technique because it does not necessarily relate the costs claimed to the entitlement issues. Nevertheless, this approach may be the only viable option if the contractor lacks detailed historical cost information, or if the nature of the dispute is such that the costs cannot be calculated by any other reasonable method.

In its simplest form, the total cost approach involves calculating the contractor's direct costs (labor, material, subcontract, and other), indirect costs (site overhead, home office interest, and other) and fee. The actual revenue received is then subtracted from total costs and fee to obtain the amount to be claimed. To be successful in using this method, the contractor must be able to:

1. Prove that the total costs claimed are accurate and that they were developed based on the contractor's actual accounting records
2. Prove that the contractor's acts did not contribute to extra costs claimed
3. Prove that the contractor's price was based on a properly developed bid estimate
4. Prove that either the contractor's cost records or the nature of the claim precludes the contractor from employing a different approach.

If the contractor is unable to prove any of the preceding four points, then use of the total cost approach may be disallowed by the court. It should be noted that the total cost approach can involve significant risk, because the defense need only identify one disruptive or delaying event for which the contractor was partially responsible to bring the entire claim into question.

To reduce this risk, a contractor may choose to use a modification of the total cost approach which adjusts for excess costs incurred due to the contractor or third parties. In this approach, the total cost is reduced by the costs incurred which were not the fault of the adverse parties. Fee is then added to the adjusted total cost and revenue received is subtracted to derive the claim amount.

§ 16.9 Discrete Cost Approach

The discrete cost approach calculates the cost of damages incurred on an item-by-item basis for specific excess cost events. Generally, this involves more effort than the total cost approach because it is performed on a more detailed level. The primary advantage of this technique is the ease with which the specific entitlement issues can be directly related to the elements of the damage cost calculation. An analysis using the discrete cost approach might include the following nine major elements:

1. Extra work costs
2. Delay costs
3. Escalation costs
4. Acceleration costs

5. Disruption costs
6. Cost of capital
7. Home office costs
8. Fee
9. Opportunity-related costs.

The specific elements that will be employed when using the discrete cost approach will depend on the nature of the entitlement issues and the cost information available for analysis. Since this method is more complicated than other approaches, care should be taken to ensure that all costs are counted and that no costs are double-counted. The contractor and the damages expert must clearly understand the theory associated with each of the cost elements used. The expert must then be able to structure an analysis and describe the approach in simple terms for the jury.[4]

§ 16.10 —Extra Work Costs

Extra work costs include both the direct and indirect costs of work performed for which the owner refuses to pay. Extra work costs can result from performing tasks such as rebuilding a site access road or excavating an extra 30 yards. The contractor must be alert to the needs of the project versus the contract plans and specifications, so that work outside the scope of the contract can be identified on an ongoing basis.

Direct costs of extra work can include labor, material, subcontracts, and equipment. Contractors frequently document the direct costs associated with extra work claims by establishing separate cost account codes, implementing detailed equipment logs, and by implementing force account documentation. *Force account documentation* consists of special daily cost reports, which the contractor submits to the owner or the engineer for signature, confirming that the costs were in fact incurred. If the costs of extra work are documented in this way, then direct cost damages are much easier to prove.

Indirect costs of extra work apply to home office costs[5] and site overhead. *Site overhead costs* are the costs of temporary facilities, clerical and supervisory workers, office supplies, and other site-related expenditures. These usually cannot be associated with specific direct work activities; instead, they are incurred in support of the overall work effort and generally fluctuate with the overall level of project activity. Contractors therefore often include charges for site overhead based on an average project site overhead rate incurred during the period of extra work. This rate is calculated by dividing total site overhead by total direct project cost. The rate is then applied to the extra work direct costs claimed to derive the extra

---

[4] See § **16.20.**

[5] See § **16.16.**

work site overhead cost to be claimed. Other variations of this approach may also be asserted, depending on the nature of the claim or unique characteristics of the project's costs. These variations involve the use of different time periods or different allocation statistics.

§ 16.11 —Delay Costs

Delay costs are the excess costs the contractor incurs because of an interruption to project activities. Most often these are the time-related costs of site overhead incurred during the delay period. If the project is delayed a month, for example, then the extra rental cost of the site trailer is a legitimate delay cost.

One problem often encountered when asserting a delay claim is that the delay may actually consist of a series of delays which occurred throughout the project. In this situation, it is difficult to isolate the site overhead costs associated with each delay period. A further complication is the fact that site overhead costs will probably fluctuate throughout the project, following a bell-shaped curve. An alternative frequently used to resolve these problems is to calculate an average daily site overhead cost rate for the project and apply it to the total delay days claimed.

Sometimes direct costs may also be asserted in a delay claim if these costs are incurred to retain assets, such as key workers or equipment, which would be economically unfeasible to remove from the project. This generally applies if the delay period is expected to be relatively short. These costs, which would normally be considered direct costs during the project, actually assume characteristics similar to fixed overhead costs.

Delays, in and of themselves, cause the contractor to incur other unique costs which may also be included in the claim. These include costs for extra demobilization, remobilization, and extra storage costs. Costs for refurbishing construction materials, such as removing surface rust, fall into this category. These costs are actually extra costs which result solely from the delay.

§ 16.12 —Escalation Costs

Escalation costs are the incremental inflation costs associated with postponing an expenditure due to a delay, error in design specification, or other reason. Contractors can incur escalation costs for labor, material, subcontract, overhead, or other expenditures. Some escalation costs are easy to calculate, such as the case of a subcontract which is bid at one price but executed at a higher price because of a delay in the issuance of mobilization orders. However, the calculation of escalation costs must usually rely on other techniques involving economic statistics, the project schedule, or the original cost estimate. The challenge is to establish the baseline costs that would have been incurred had construction expenditure not been postponed.

§ 16.13 —Acceleration Costs

Acceleration means performing scheduled work quicker than originally planned. Overtime premiums, reduced labor efficiency, excess supervision, and other productivity-related costs are often incurred because of acceleration. An acceleration cost claim should address only work activities which are within the contract scope. If the activity is an extra work activity, then associated acceleration costs are not an issue because the entire cost of the activity will have been included in an extra work claim.

Determining the productivity-related components of costs usually requires the testimony of an expert who develops a comparison of actual costs to either an engineering estimate or to a sufficiently detailed and reliable bid estimate. Engineering estimates of the percent productivity impact may also be developed and asserted as the cost calculation basis. The difficulty of this task will be significantly reduced if the contractor develops productivity-related documentation or engages the services of an expert during the period in which the accelerated work is being performed.

An accurate acceleration claim should deduct certain cost savings if they are achieved. This may be appropriate, for example, if the entire project duration is shortened. In this instance, a credit would stem from savings in those project overhead costs which are a function of time, such as site trailer rental costs.

§ 16.14 —Disruption Costs

Disruption, or inefficiency, applies to the excess productivity-related costs of routine project activities which are affected by the extra work activities claimed. If a work crew's labor efficiency is adversely affected because of limited site access or the sharing of work space with other crews performing extra work, this inefficiency cost should be included in the contractor's claim. Establishing disruption costs involves challenges similar to those encountered in establishing acceleration costs. One method of analyzing disruption is to prepare a comparison of actual cost to an engineering estimate or the original bid. Another method is to estimate the percent inefficiency, based on an engineering survey or sample, and apply this percentage to the work affected. This latter approach may be the only reasonable approach if there are many work activities. The contractor is well advised to prepare thorough documentation and to retain the services of an expert during the period of disruption.

§ 16.15 —Cost of Capital

The *cost of capital* is the cost arising from the excess financial burden created by the extra costs claimed. These costs, when claimed, typically are based on the rationale either that extra funds had to be borrowed by the contractor to finance the extra costs required by the project (*actual cost rationale*), or that the funds used to finance a delay or extra work costs came from working capital which could have earned interest if it had not been required by the project (*opportunity cost rationale*).

Claims made for the cost of capital must be carefully prepared to ensure that they have sound support and that they do not overlap costs claimed elsewhere. For example, if the cost of capital is calculated using an opportunity cost rationale, then care should be taken to ensure that interest charges, which may be included under home office costs, do not flow into the claim.

§ 16.16 —Home Office Costs

Home office costs are the general and administrative costs incurred which should have been absorbed by the project due to delays or extra work costs. Costs claimed in this area are generally calculated using some form of allocation.

The *Eichleay formula* is one method of allocating home office costs to a delay period.[6] This method uses the sales revenue of the project versus total company sales revenue for the period of the project to perform an initial allocation of home office costs to the project. This amount is then converted to a daily rate, based on total project days, and used to assign home office costs to the delay period. The *Eichleay* formula received wide acceptance until recently, when cases appeared wherein it was argued that the formula has several problems:

1. The formula may be inappropriate for corporations with significant non-construction activities such as manufacturing operations
2. The formula allocates overhead costs based on sales dollars, although these costs may actually fluctuate more closely with direct project cost
3. The method may significantly understate the daily rate if the delay period is long. This occurs because delay-period days are included in the project period from which both home office costs (the pool of costs being allocated) and total company sales (used to allocate these costs between the different projects) are obtained.

Other techniques of allocating home office overhead to a delay period involve the same basic approach as the *Eichleay* formula, except that they do not use sales dollars as the allocation statistic. Instead, they use allocation statistics such as total direct cost or total direct labor hours. These techniques also rely on certain assumptions which could affect their validity; foremost among these is that the mix of cost types within the contractor's projects does not vary significantly and that the delay period is not so long that it distorts the resulting daily rate.

---

[6] It is named after the case in which it was first formulated, Eichleay Corp., 60-2 B.C.A. (CCH) ¶ 2688 (A.S.B.C.A. 1960).

Another alternative is to analyze each home office cost account individually and to assign it the allocation statistic which is expected to track the closest with that account's cost behavior. This approach is probably the most time-consuming method and may not justify the required investment in effort.

Recently courts tend to demand greater justification for the use of a daily rate to allocate home office costs to a delay claim. Two areas frequently questioned are: (1) whether the delay caused some increase in home office costs; and (2) whether the company was unable to obtain other work during the delay to help absorb some of the unabsorbed home office costs now being claimed.

Calculating home office costs that should have been absorbed by extra work generally involves determining the company's average overhead rate (home office costs "per" project costs for all projects) during a specific time period. This cost is then applied to the extra costs claimed to derive the corresponding home office claim. Variations on this basic approach involve different allocation statistics (e.g., man-hours instead of direct cost) or different time periods (e.g., total project period or period during which extra work occurred). Justification of these various options will depend on the cost behavior of the project in dispute.

§ 16.17 —Fee

Fee is generally viewed as the profit the contractor should have obtained for the extra costs being claimed. Fee claims often encounter strong resistance and may be unsuccessful even when the associated extra cost claims are won. Part of the reason for this is because they are sometimes viewed as speculative and because of the numerous rebuttal issues which are often asserted. Rebuttals to fee claims frequently involve areas such as how the profit percentage was selected, how the profit percentage's base was selected, and whether the fee claim potentially overlaps with other cost claims. One approach to calculating the fee rate is to base it on the company's average historical rate, in total or for a specific line of business. Other approaches include basing the fee on the industry average or on the contract rate specified for change orders.

§ 16.18 —Opportunity-Related Costs

Contractors frequently find that a job fraught with difficulties has an adverse financial impact on other projects or future projects. Problem projects which consume limited resources, such as equipment or cash, may adversely affect the cost performance of the projects from which the resources were diverted. Problem projects may also affect a company's reputation or ability to obtain bonding, thereby limiting future sales performance. Unfortunately, these costs are frequently difficult to quantify. Nevertheless, every prudent contractor should attempt to calculate the losses related to opportunity costs and to include them in the claim if appropriate. This will probably require the services of an expert capable of preparing an analysis geared to the unique characteristics of the contractor's business.

PRESENTING THE CLAIM

§ 16.19 Use of Computer Models

Modeling of damage claims has become an important factor in construction litigation. The declining cost of microcomputers, their increased capacity, and the availability of inexpensive, user-friendly software packages such as spreadsheets, have contributed to their widespread use. The primary reason, however, for the proliferation of microcomputer modeling in construction litigation is that it provides tremendous analytical advantages, including:

1. Computational accuracy
2. Enhanced data maneuverability and instantaneous recalculation capability, thus increasing user opportunities to test variable sensitivity, test different analysis approaches, perform higher-level, more sophisticated analyses, and reduce the cost of developing the damage analysis
3. Improved traceability and repeatability of the logical approach followed in the analysis
4. Typically, generation of higher credibility among jurors and courts than manual techniques
5. Ability to easily generate accurate graphic exhibits.

Microcomputer models also introduce a unique set of disadvantages or concerns which should be assessed whenever they are being used or when their use is being considered. Primary among these concerns is that the use of computers can lead the analysis into too much detail and omit simplifying assumptions which could enhance the claim's presentation in the courtroom. The computer model also introduces a new layer of technical issues such as quality control, user expertise, expertise continuity, system compatibility, and program documentation.

§ 16.20 Presenting the Damage Analysis

The format for presenting the contractor's damage claim is probably the most critical component of the entire damage analysis effort. An extremely detailed analysis may win all the technical battles, but lose the war because it was not understood by the jury. The average juror may not appreciate the complexities of a sophisticated damage analysis unless it is presented in an easily digestible manner.

Throughout the development of the damage analysis, the damages expert must keep this goal in mind.

A good damages claim presentation incorporates several elements which aid in communicating the contractor's position to the jury. These include:

1. *Simplicity:* the use of simple analogies and graphic presentations are particularly helpful
2. *Support:* although the analysis should be presented in simple terms, the jurors should be aware that the information presented is supported by extensive cost records
3. *Summary:* after presenting each component of the cost claim, a summary should be presented which consolidates all the costs claimed in a manner consistent with the major facts supporting the contractor's assertions
4. *Presentation:* the damage analysis should be presented by an expert witness who is knowledgeable in the subject area, credible, and (most importantly) capable of presenting the concepts in a manner which the jury can comprehend. A key task of this expert should be to present the summary of the damage claim at some point near the end of the trial
5. *Documentation:* it is advantageous to enter into evidence a summary or other documentation to which the jury may refer when deliberating. This is particularly helpful in a long, complicated trial because it may refresh the jurors' memories regarding issues important to the contractor.

Combining these attributes into the final damage presentation is essential for its success. The contractor should critically assess the capability, objectivity, and credibility of potential experts and attorneys when selecting the team to prepare and present the damage claim. The contractor must always remember that being right does not guarantee success if the facts supporting that position are not effectively communicated.

CHAPTER 17

TAX CONSIDERATIONS OF CONSTRUCTION LITIGATION

Robert T. Evans, Jr.

Robert T. Evans, Jr. is a certified public accountant and a tax manager in the San Jose, California office of Coopers & Lybrand, an international firm of certified public accountants. He graduated with a bachelor's degree in accounting from San Jose State University in 1978, and obtained a master's degree in taxation from Golden Gate University in 1985. Mr. Evans is involved in many areas of taxation, and specializes in the taxation of real estate, partnerships, and corporations. He has had particular experience and concentration in the taxation of real estate and corporate consolidations, mergers and acquisitions. He has spoken at various meetings, seminars and functions, and has authored numerous articles and publications on many areas of taxation. He also serves as an adjunct professor of taxation at Golden Gate University.

WHEN INCOME FROM A JUDGMENT OR SETTLEMENT IS RECOGNIZED FOR TAX PURPOSES

TAX TREATMENT AND CHARACTERIZATION OF AMOUNTS PAID OR INCURRED AS A RESULT OF A JUDGMENT OR SETTLEMENT

WHEN LITIGATION EXPENSES ARE DEDUCTED FOR TAX PURPOSES

OTHER CONSIDERATIONS

INTRODUCTION

§ 17.1 Overview

It is of course the primary goal of the construction contractor to successfully prevail in the prosecution or defense of a claim or settlement. The economic consequences to the contractor as a result of any judgment or settlement may, however, be either greatly enhanced or significantly reduced by the resultant income tax consequences. Proper structuring of the character and timing of proceeds received from litigation or the settlement of a dispute will increase the after-tax economic benefits to the contractor. Similarly, proper structuring of the character and timing of amounts paid or incurred as a result of an unfavorable judgment or settlement may mitigate the after-tax economic loss to the contractor.

This chapter gives a general overview and discussion of such income tax consequences to the contractor and the tax treatment of amounts received or paid as a result of litigation. It is intended to identify and familiarize the reader with the various tax issues involved in the litigation or resolution of disputes; also, where possible, it discusses general tax planning opportunities. The income tax consequences of litigation to the construction contractor are generally the same as for any other taxpayer; nevertheless, this chapter discusses these tax consequences from the perspective of the construction contractor.

This chapter does not purport to be a complete and ultimate authority on the tax consequences of construction litigation; rather, it identifies the general tax issues involved in construction litigation. For a more detailed discussion, the reader is advised to perform further, more in-depth research, or consult a professional specializing in the taxation of construction litigation.

The chapter is divided into two major parts. First, the chapter discusses the tax consequences and treatment of amounts received from judgments and settlements. Next, it addresses the tax consequences and treatment of amounts paid or incurred as a result of litigation or settlement of a dispute. Each part is divided again into two areas: characterization of the amounts, and timing of the recognition of the amounts.

TAX TREATMENT AND CHARACTERIZATION OF AMOUNTS RECEIVED FROM JUDGMENTS AND SETTLEMENTS

§ 17.2 Introduction

The proceeds received by a contractor from a judgment or settlement of a claim against an owner, developer, subcontractor, or other party may produce significant tax consequences to the contractor. The proceeds may be fully or partially taxable.[1] Alternatively, the amounts received may be considered a nontaxable return of capital,[2] or in some instances, simply nontaxable.

§ 17.3 How Judgments and Settlements Are Taxed

The characterization of amounts received from litigation or the resolution of a dispute is important, as it directly affects the amount of federal and state income taxes which the contractor will be required to pay. If the proceeds received are considered fully or partially taxable as ordinary income, the contractor must include the taxable amounts in gross income when required by the contractor's tax accounting method.[3] Ordinary income is generally taxed at graduated rates which currently may reach as high as 50 percent for noncorporate taxpayers and 46 percent for corporate taxpayers.[4]

Amounts received which are considered to be a return of capital reduce the contractor's tax basis in the property or the investment to which the amounts relate.[5] Thus, they are not currently taxed to the contractor unless they exceed the basis of the property or investment.[6] Any excess of the proceeds received over the basis of the property or investment is generally taxed as ordinary income, not as capital gain (taxed at lower, more beneficial rates), discussed below, because a settlement or judgment is not considered the equivalent of a sale or exchange of property.[7]

A reduction in basis not exceeding the original basis of the property or investment, although not currently taxed, will eventually be taxed when the property or investment is sold. The reduced basis of the property or investment will produce a greater gain or smaller loss upon the sale or disposition as a result.[8]

Gain or loss upon the sale or disposition of property or investment will usually be taxed differently from ordinary income. Such gain or loss is often treated as *capital* gain or loss. Capital gain from the sale or disposition of property or investment acquired on or after June 23, 1984 and held for more than six months,

---

[1] I.R.C. § 61. All references in this chapter to the Internal Revenue Code (I.R.C.) refer to the Internal Revenue Code of 1954, as amended, Title 26 of the United States Code.

[2] I.R.C. § 1061(a).

[3] I.R.C. §§ 61, 451. See **§§ 17.12-17.17**.

[4] I.R.C. §§ 1, 11.

[5] *Id.* § 1016(a).

[6] Raytheon Prods. Corp. v. Commissioner, 144 F.2d 110 (1st Cir. 1944).

[7] Harwick v. Commissioner, 133 F.2d 110 (8th Cir. 1943); I.R.C. § 1222.

[8] I.R.C. §§ 1001(a), 1011(a), 1016(a).

or acquired before June 23, 1984 and held for more than one year (long-term capital gain) is effectively taxed at rates which are significantly less than ordinary income tax rates. For noncorporate taxpayers, 60 percent of long-term capital gain is excluded from gross income.[9] For corporate taxpayers, long-term capital gain is taxed at rates not exceeding 28 percent.[10]

The tax benefits of capital losses, in contrast to capital gains, are limited. For noncorporate taxpayers, the excess of long-term capital losses over capital gains in any taxable year is reduced by 50 percent.[11] The net capital losses are then only deductible against ordinary income to the extent of $3,000 in any taxable year,[12] although any unused net capital loss may be indefinitely *carried over* to succeeding taxable years.[13] For corporate taxpayers, capital losses may only be applied to reduce capital gains.[14] Unapplied capital losses may be *carried back* and applied against capital gains in each of the three taxable years preceding the year of loss, and carried forward to the succeeding five taxable years.[15]

Special, more favorable rules apply to gains and losses from the sale or disposition of property used in the contractor's trade or business, such as furniture, machinery or equipment.[16] In general, these gains and losses, known as § 1231 gains and losses, generated in any taxable year are aggregated. Net gain receives favorable treatment as capital gain.[17] Net loss is treated as ordinary loss, deductible in full against ordinary income.[18] Some or all of the § 1231 net gain may instead be treated as ordinary income, however, if the contractor has taken depreciation deductions over the life of the property.[19] The tax provisions in this area are very complex, and a complete discussion of them is beyond the scope of this chapter.

In most cases, it is more advantageous for tax purposes to characterize proceeds received from a judgment or settlement as a return of capital rather than as ordinary income. If treated as a return of capital, the proceeds in most cases are not currently taxed, but are deferred until the property or investment which was the subject of the litigation or dispute is sold or otherwise disposed of. Further, gain recognized upon the eventual sale or disposition may be taxed at favorable capital gain rates. Finally, loss recognized upon the sale or disposition of property used in the contractor's trade or business may be treated as an ordinary loss under certain circumstances.

---

[9] *Id.* § 1202(a).

[10] *Id.* § 1201(a).

[11] *Id.* § 1212(b)(1)(B).

[12] *Id.* § 1211(b).

[13] *Id.*

[14] I.R.C. § 1211(a).

[15] *Id.* § 1212(a)(1).

[16] *Id.* § 1231.

[17] *Id.* § 1231(a)(1).

[18] *Id.* § 1231(a)(2).

[19] *Id.* §§ 1245, 1250.

§ 17.4 Characterizing Proceeds Received from Litigation or a Dispute

The characterization for tax purposes of proceeds received from a judgment or settlement of a dispute depends primarily upon the origin and nature of the claim or litigation.[20] The nature of the loss for which damages have been awarded will usually determine the tax treatment of the award.[21]

§ 17.5 –Proceeds Generally Treated as Ordinary Income

Damages awarded or settlements received as recovery of lost profits are generally taxed as ordinary income.[22] Proceeds received by a contractor in settlement for amounts which the contractor previously deducted against taxable income are also generally treated as ordinary income.[23] This is an application of the *tax benefit theory,* which treats such proceeds as reimbursements or recoveries of previously deducted expenditures.[24] Thus, proceeds received by a contractor in reimbursement of cost overruns are taxed as ordinary income. Similarly, amounts received representing further contract proceeds for change orders or other additional work performed are taxed as ordinary income. Recovery of amounts paid to subcontractors for their portion of the performance of a construction contract will also be taxed as ordinary income.

§ 17.6 –Proceeds Generally Treated as a Return of Capital

Damages awarded for loss of capital or to compensate for loss of property represent a return or recovery of that capital or property, and are not taxed unless they exceed the basis of the capital.[25] Thus, amounts received by a contractor from a subcontractor or an owner for damages to the contractor's machinery or equipment used in the performance of a construction contract are considered a return of capital.

---

[20] United States v. Gilmore, 372 U.S. 39 (1963).

[21] Raytheon Prods. Corp. v. Commissioner, 144 F.2d 110 (1st Cir. 1944).

[22] Swastika Oil & Gas Co. v. Commissioner, 123 F.2d 382 (6th Cir. 1941).

[23] Treas. Reg. § 1.111-1(a). All references in this chapter to the Treasury Regulations or Proposed Regulations refer to the regulations on income tax codified at Title 26 of the Code of Federal Regulations.

[24] Treas. Reg. § 1.111-1(a).

[25] Raytheon Prods. Corp. v. Commissioner, 144 F.2d 110 (1st Cir. 1944); Durkee v. Commissioner, 162 F.2d 184 (6th Cir. 1947), *rev'g & remanding* 6 T.C. 773 (1946).

Amounts received as compensation for injury to goodwill may also be treated as a return of capital.[26] Defamatory statements made by a subcontractor or owner regarding the contractor's performance or capabilities may have damaged the goodwill attached to the contractor's business. Awards to compensate this damage may thus be considered as return of capital. However, compensation may often be treated as ordnary income, as discussed in **§ 17.7**.

§ 17.7 Damages for Injury to Reputation

Damages received to compensate losses for *personal* injuries are completely nontaxable, unless the losses were deducted previously.[27] This includes compensation for personal embarrassment, mental and physical strain, and injury to personal reputation.[28] However, awards for damages to *business* reputation are generally taxed as ordinary income.[29] This includes damages for slander or libel of the contractor's business or business product.[30]

When does compensation received for damages to busness reputation constitute a nontaxable return of capital, and when does it constitute ordinary income? The weight of judicial interpretation generally supports treatment as ordinary income, under the theory that damage to business reputation is actually an impairment of business profits. Therefore, compensatory proceeds more closely resemble compensation for lost profits. However, an argument may be made to treat amounts received for damage to business reputation as received for damage to goodwill, and therefore as a nontaxable return of capital. This argument has some judicial support,[31] and is further strengthened if the contractor has in fact a tax basis in the goodwill. Goodwill is an intangible asset, and may result, for example, from the contractor's purchase of an existing construction business where the amount paid exceeded the fair market value of the business's net assets.

§ 17.8 Punitive Damages

Awards received for punitive damages are generally taxable as ordinary income.[32] By definition, *punitive damages* are not compensation for loss of profits or capital;[33]

---

[26] Durkee v. Commissioner, 162 F.2d 184 (6th Cir. 1947), *rev'g & remanding* 6 T.C. 773 (1946).

[27] I.R.C. § 104(a)(2).

[28] Seay, 58 T.C. 32 (1972).

[29] Wolfson, T.C. Memo 1978-445.

[30] Agar v. Commissioner, 290 F.2d 283 (2d Cir. 1961); Rev. Rul. 58-418, 1958-2 C.B. 18.

[31] Raytheon Prods. Corp. v. Commissioner, 144 F.2d 110 (1st Cir. 1944).

[32] Commissioner v. Glenshaw Glass Co., 18 T.C. 860 (1952), *aff'd,* 211 F.2d 928 (3d Cir. 1954), *rev'd,* 348 U.S. 426 (1955).

[33] *Taxation of Damage Recoveries from Litigation,* 40 Cornell L.Q. 345 (1955).

rather, they are awarded by operation of law to the prevailing party for the unlawful conduct of the other party. The Internal Revenue Service, until recently, had issued conflicting revenue rulings which held that awards of punitive damages were alternatively taxable[34] and nontaxable.[35] Recently, however, the Service revoked its revenue ruling holding that punitive damages were nontaxable and clarified its earlier position that punitive damages were taxable as ordinary income.[36] Some courts have also held that proceeds received for punitive damages are nontaxable,[37] but most of those decisions were reversed on appeal, and the majority of cases hold punitive damages to be fully taxable as ordinary income.[38]

§ 17.9 Allocating between Ordinary Income and Return of Capital

Awards received by a contractor in a judgment or settlement often represent compensation for both lost profits and damages to capital. The contractor must prove that the awards represent compensation for loss of capital to avoid current taxation as ordinary income.[39] Where no allocation is made, or if the contractor does not have sufficient evidence supporting the allocation of the award, the court, in a subsequent tax dispute between the contractor and the Internal Revenue Service, will make an allocation.[40] The contractor will then bear the burden of proving the court's allocation wrong if the contractor wishes it changed.[41] The courts often provide for awards which represent compensation both for lost profits and for loss of capital.[42] Accordingly, the contractor is well advised to document the nature of the original claim adequately and to structure the written documents concerning the litigation properly, in order to justify the desired tax treatment and characterization of the claim and resultant award. Otherwise, the contractor risks leaving the allocation in a subsequent tax dispute to the judgment of the courts, which may be unsympathetic to the contractor's arguments, or may not understand the contractor's business.

---

[34] Rev. Rul. 58-418, 1958-2 C.B. 18.

[35] Rev. Rul. 75-45, 1975-1 C.B. 47, *revoked by* Rev. Rul. 84-108, 1984-29 I.R.B. 5.

[36] Rev. Rul. 84-108, 1984-29 I.R.B. 5.

[37] Commissioner v. Glenshaw Glass Co., 18 T.C. 860 (1952), *aff'd,* 211 F.2d 928 (3d Cir. 1954), *rev'd,* 348 U.S. 426 (1955); Commissioner v. Obear-Nester Glass Co., 217 F.2d 56 (7th Cir. 1954), *aff'g* 20 T.C. 1102 (1953).

[38] Commissioner v. Glenshaw Glass Co., 18 T.C. 860 (1952), *aff'd,* 211 F.2d 928 (3d Cir. 1954), *rev'd,* 348 U.S. 426 (1955); Commissioner v. Obear-Nester Glass Co., 217 F.2d 56 (7th Cir. 1954), *aff'g* 20 T.C. 1102 (1953); William Goldman Theatres, Inc. v. Commissioner, 19 T.C. 637 (1953), *aff'd* 211 F.2d 928 (3d Cir. 1954), *rev'd,* 348 U.S. 426 (1955).

[39] Raytheon Prods. Corp. v. Commissioner, 144 F.2d 110 (1st Cir. 1944).

[40] Collins, T.C. Memo 1959-174.

[41] Phoenix Coal Co. v. Commissioner, 231 F.2d 420 (2d Cir. 1956), *aff'g* T.C. Memo 1955-28.

[42] *Taxation of Damage Recoveries from Litigaton,* 40 Cornell L.Q. 345 (1955).

§ 17.10 Professional Fees and Other Expenditures Incurred in Obtaining a Judgment or Settlement

The characterization for tax purposes of litigation expenditures generally follows the characterization of the proceeds received as a result of the litigation. Litigation expenditures incurred to recover lost profits are generally deductible against ordinary income, if they are incurred in the ordinary course of the owner's trade or business,[43] for the production or collection of income,[44] or for the management, conservation, or maintenance of property held for the production of income.[45] Litigation expenditures incurred to recover lost capital are generally nondeductible;[46] rather, they increase the basis of the investment or property which is the subject of litigation.[47] A more detailed discussion of the tax treatment of litigation expenditures is presented in §§ **17.22-17.27**.

WHEN INCOME FROM A JUDGMENT OR SETTLEMENT IS RECOGNIZED FOR TAX PURPOSES

§ 17.11 General Rules

The timing, for tax purposes, of the recognition of income from litigation or the settlement of a dispute or claim is important, as it directly influences when federal and state income taxes will have to be paid. Generally, amounts received which are treated as gross income for tax purposes are recognized according to the tax method of accounting elected by the contractor to report income and expense from operations.[48] The emphasis on *tax* method of accounting is important. The contractor may elect to use a method of accounting for taxes which differs from the method of accounting it uses to recognize income and expenses for internal management purposes or to report financial information to stockholders, investors, creditors, or other outsiders.[49] In fact, many contractors, like most taxpayers, use tax methods of accounting which defer taxable income to the greatest extent possible, while using methods of accounting for internal management or financial statement purposes which accelerate income.

---

[43] I.R.C. § 162.

[44] *Id.* § 212(1).

[45] *Id.* § 212(2).

[46] *Id.* § 263.

[47] *Id.* § 1016(a).

[48] I.R.C. §§ 445, 451.

[49] *Id.* §§ 451(a), 461(a).

At first glance, the Internal Revenue Code appears to require a taxpayer to use tax accounting methods which conform to the accounting methods the taxpayer uses for *book* or financial accounting.[50] However, the Treasury Regulations and case law overwhelmingly support the use of different tax and book accounting methods if the taxpayer maintains adequate records and prepares schedules reconciling taxable income to book income.[51] Thus, the contractor has the flexibility to choose tax accounting methods which allow it to in large part influence and plan for the payment of income taxes.

Generally, the contractor must formally elect to adopt one of the permissible tax accounting methods on the first tax return for which income or expenses are reported using the desired tax accounting method.[52] The contractor must then consistently use the elected tax accounting method for all future taxable years.[53] In order to make use of or change to a different tax accounting method, the contractor must generally receive permission from the Commissioner of the Internal Revenue Service before such a change can be made.[54]

§ 17.12 Methods of Accounting for Tax Purposes

There are two general tax accounting methods from which the owner may choose to report income from operations, which would include reporting income from the resolution of litigation or a dispute. They are the *cash receipts and disbursements method* and the *accrual method.*[55] In some cases, the contractor may use a hybrid method of tax accounting if the hybrid method clearly reflects income and is consistently applied.[56]

The contractor may elect to use one of two other tax accounting methods for income earned from the performance of *long-term contracts.*[57] They are the *completed contract method* and the *percentage of completion method.*[58] A *long-term contract* is defined in the Treasury Regulations as a "building, installation, construction or manufacturing contract which is not completed within the taxable year in which it is entered into."[59] Further, the contract must involve the manufacture or construction of items which are not normally considered inventory of the contractor, or which normally require more than 12 months to complete.[60] As their

---

[50] *Id.* § 446(a).

[51] Treas. Reg. § 1.446-1(a)(4), (c); Patchen v. Commissioner, 258 F.2d 544 (5th Cir. 1958).

[52] Treas. Reg. § 1.446-1(e)(1).

[53] *Id.* § 1.446-1(c)(2)(ii).

[54] I.R.C. § 446(e).

[55] I.R.C. § 446(c).

[56] Treas. Reg. § 1.446-1(c)(2)(ii).

[57] *Id.* § 1.451-3; Prop. Reg. § 1.451-3.

[58] Treas. Reg. § 1.451-3(a)(1); Prop. Reg. § 1.451-3(a)(1).

[59] Treas. Reg. § 1.451-3(b)(1)(i); Prop. Reg. § 1.451-3(b)(1)(i).

[60] Treas. Reg. § 1.451-3(b)(1)(ii); Prop. Reg. § 1.451-3(b)(1)(ii).

major source of income is the performance of construction contracts, most contractors should consider adopting and using one of these two long-term contract tax accounting methods for reporting the income from construction contracts.

Income earned by a contractor from a judgment or settlement of a dispute or claim is generally recognized under the tax accounting method used by the contractor for the business activity to which the litigation or dispute relates. Thus, income earned from litigation connected with the performance of a long-term construction contract is recognized using the long-term contract tax method of accounting adopted by the contractor.[61] On the other hand, income earned from litigation connected with the contractor's general business activity is recognized using the tax method of accounting adopted by the contractor with respect to general operations (usually either the cash or accrual method).

§ 17.13 —Cash Method versus Accrual Method of Tax Accounting

Under the cash method of accounting, income is recognized when it is actually or constructively received.[62] Under the accrual method, income is recognized when the contractor becomes entitled to it.[63]

The use of the accrual method generally results in earlier recognition of income from a judgment or settlement, as the right to receive the proceeds of litigation or a settlement usually precedes its actual receipt by the contractor. The accrual method of accounting may not be as unfavorable in comparison to the cash method as it first appears, however, because the accrual method also allows the contractor to deduct expenses, including litigation expenses, when they are due. The contractor using the accrual method thus may deduct expenses for tax purposes before they are actually paid. The choice of the cash or the accrual method should be considered in light of the contractor's operations as a whole.

§ 17.14 —Percentage of Completion

Under the percentage of completion method of tax accounting, gross income from a long-term construction contract is recognized as the contract progresses toward completion. The portion of the gross contract price which corresponds to the percentage of the entire construction contract completed during any taxable year is included in the contractor's gross income for that year.[64] The cumulative percentage for each construction contract at the end of any year may be calculated by using one of two methods:

---

[61] Treas. Reg. § 1.451-3(a)(1); Prop. Reg. § 1.451-3(a)(1).

[62] Treas. Reg. § 1.446-1(c)(1)(i).

[63] *Id.*

[64] Treas. Reg. § 1.451-3(c)(1).

1. By comparing the actual contract costs incurred to date with the estimated total contract costs
2. By comparing the work performed on the contract to date with the estimated total work to be performed.[65]

This cumulative percentage is applied to the gross contract price, giving the cumulative gross contract income recognized through the current year. The amount of gross contract income recognized in prior years is then subtracted from the current year's cumulative gross contract income to give the amount of gross contract income recognized in the current taxable year.

The costs incurred in performing the construction contract are deducted against gross contract income using the accrual method of accounting.[66] Generally, these costs include expenses incurred in litigating or otherwise pursuing a judgment, settlement, or claim.

§ 17.15 —Completed Contract

Under the completed contract method of tax accounting, gross income from a construction contract is not recognized until the contract has been completed and accepted by the person for whom the contract has been performed.[67] The direct and many of the indirect costs incurred in performing the construction contract are aggregated and deducted against the contract gross income in the taxable year that the construction contract is completed.[68] Generally, these contract costs include any litigation expenses directly related to the construction contract.

Special rules apply to determine which indirect or *period* contract costs must be aggregated and deducted when the construction contract is completed, and which may instead be deducted when paid or incurred, regardless of the contract's completion.[69] These rules vary depending upon the length of the construction contract and its gross income.[70] The special rules are very complicated, and a discussion of them is beyond the scope of this chapter.

§ 17.16 —Percentage of Completion and Completed Contract Methods Contrasted

Generally, the use of the completed contract method of tax accounting results in a longer deferral of taxable income than the percentage of completion method.

---

[65] *Id.* § 1.451-3(c)(2).

[66] *Id.* § 1.451-3(c)(3).

[67] Treas. Reg. § 1.451-3(d)(1); Prop. Reg. § 1.451-3(d)(1).

[68] Treas. Reg. § 1.451-3(d)(1); Prop. Reg. § 1.451-3(d)(1).

[69] Treas. Reg. § 1.451-3(d)(5)(ii); Prop. Reg. § 1.451-3(d)(5)(ii).

[70] Prop. Reg. § 1.451-3(d)(6).

Under the percentage of completion method, the net profit on a well-managed, successful construction contract will be recognized in increments over the life of the contract. In contrast, under the completed contract method, recognition of the contract net profit is deferred until completion of the construction contract. By using the completed contract method, the contractor may more easily plan for the completion of the construction contract, and also may often be able to influence the contract's completion date and resultant tax liability.

§ 17.17 Recognizing Income from a Judgment or Settlement under Long-Term Contract Tax Accounting Methods

Proceeds from a judgment, settlement, or claim are generally recognized as income when awarded to the contractor under the percentage of completion method of accounting. The work performed on the construction contract which was the subject of the litigation or dispute will most likely have been completed prior to or concurrently with the award or settlement. Thus, the percentage of completion method effectively places the contractor on the accrual method of tax accounting with respect to income from a judgment or settlement.

Under the completed contract method of accounting, income from a judgment or settlement earned prior to completion of the construction contract will generally not be recognized until the contract is completed.[71] Awards or settlements earned after completion of the contract are generally treated as separate from the construction contract to which the litigation relates.[72] Thus, the net profit from the construction contract is recognized upon completion of the contract, and the proceeds from the litigation award or settlement are subsequently and separately recognized as income when earned, presumably under the accrual method of tax accounting.[73]

In some cases, because of a dispute or litigation involving a significant amount of money, the contracting party for whom the construction is being performed will refuse to accept the project which is the subject of the contract. A construction contract which has been finished but not yet accepted is not considered complete for purposes of recognizing income under the completed contract method of tax accounting. The contract thus will not be considered completed until the dispute or litigation is resolved.[74] An amount in dispute significant enough that the contractor cannot determine whether the construction contract will result in a net profit or loss may also hold the construction contract open until the dispute is resolved.[75]

---

[71] Treas. Reg. § 1.451-3(d); Prop. Reg. § 1.451-3(d).

[72] Treas. Reg. § 1.451-3(d)(3)(ii).

[73] *Id.* § 1.451-3(d)(3)(iii).

[74] *Id.* § 1.451-3(b)(2); Prop. Reg. § 1.451-3(b)(2).

[75] Treas. Reg. § 1.451-3(d)(2)(ii); Prop. Reg. § 1.451-3(d)(2)(ii).

§ 17.18 Contested or Contingent Claims

To a contractor using the cash basis method of tax accounting, awards from judgments or settlements are not recognized as income until they are received or made available to the contractor.[76] In contrast, amounts from a judgment or settlement are generally not recognized as income to an accrual basis contractor until the litigation or dispute is settled, the amount of the award or settlement is fixed, and the right of the contractor to receive the proceeds is uninhibited.[77] If a judgment is appealed, the income is generally not recognized until the appeal has been decided.[78] In some cases, where the contractor is reasonably assured of prevailing and the amount of the potential award can be determined with reasonable accuracy, the award must be accrued to income.[79]

A disputed amount may be placed in trust or escrow prior to the outcome of the litigation or dispute, to be distributed to the prevailing party. Such escrowed funds are not recognized as income to a prevailing contractor until the litigation or dispute is settled, as the contractor receives no beneficial right to the income during the period of escrow.[80]

Amounts accrued or received by a contractor held under a claim of right without restrictions are generally held to constitute income to the contractor upon accrual or receipt, in accordance with the contractor's tax accounting method.[81] A claim filed against the contractor to recover amounts after they have been properly recognized as income by the contractor will not cause the amounts to be removed from taxable income. If the contractor must refund some or all of the income, the contractor will then be allowed to deduct the refunded amounts when paid or accrued, according to the contractor's tax accounting method.

§ 17.19 Tax Planning

By properly structuring the timing of the proceeds from a judgment or settlement, the contractor may further mitigate any tax liability generated by award of the proceeds. Recall that a cash basis taxpayer is not taxed on income until it is received or made available to the taxpayer.[82] An accrual basis taxpayer is not taxed until all the events have occurred which fix the right to receive the income and the amount of the income can be determined with reasonable accuracy.[83] If the proceeds of

---

[76] Treas. Reg. §§ 1.446-1(c)(1)(i), 1.451-1.

[77] Breeze Corps., Inc. v. United States, 117 F. Supp. 404 (Ct. Cl. 1954); Jamaica Water Supply Co. v. Commissioner, 125 F.2d 512 (2d Cir. 1942), aff'g 42 B.T.A. 359 (1940).

[78] United States v. Safety Car Heating & Lighting Co., 297 U.S. 88 (1936).

[79] Boston Elevated Ry. v. Commissioner, 16 T.C. 1084 (1951), B.T.A. Memo (Nov. 29, 1938).

[80] McLaughlin v. Commissioner, 113 F.2d 611 (7th Cir. 1940).

[81] Hexter v. Commissioner, 8 B.T.A. 888 (1927), *acq.*, VII-1 C.B. 14 (1927).

[82] Treas. Reg. §§ 1.446-1(c)(1)(i), 1.451-1(a).

[83] *Id.* §§ 1.446-1(c)(1)(ii), 1.451-1(a).

a judgment or settlement are paid or made available to the construction contractor in a lump sum, they are immediately taxable.

The proceeds of a judgment or settlement may instead be paid to the contractor in installments over a period of time. If properly structured, the contractor would then be taxed as the installment payments are received or accrued. The contractor would be able to plan for the receipt of income from the judgment or settlement, and where possible, offset the income with available deductions, which may lower the effective rate at which the income will be taxed.

Income from a judgment or settlement will be taxed as the installment proceeds are received or accrued *only if* the contractor has no right to or control over the income prior to its receipt or accrual.[84] If the contractor may choose between a lump-sum settlement and a settlement payable in installments, or if the contractor can in any way control or influence the amount of the installment payments or the timing of their receipt, the contractor is said to be in *constructive receipt* of the judgment or settlement proceeds. Constructive receipt often includes the ability of the contractor to pledge the future judgment or settlement proceeds due as collateral for a loan.[85] To avoid constructive receipt, the manner in which the proceeds are administered and distributed must be determined *prior to* the judgment or settlement. The contractor may negotiate the manner in which the expected proceeds are to be administered and distributed, but must have no control over the proceeds once the litigation is settled.

The proceeds may be administered in several ways. The party against whom the litigation or dispute was settled may be charged with responsibility to pay the proceeds to the contractor in installments. However, the contractor must have no security interest in the undistributed proceeds if the contractor is to avoid taxation on the undistributed amounts. Many contractors will therefore avoid this method of administration, as they will be considered unsecured creditors of the losing party charged with paying the amounts; that is, without any security in the judgment or settlement proceeds. Alternatively, the judgment or settlement may provide for the proceeds to be irrevocably entrusted to and deposited with an independent third-party trustee or escrowee responsible for administering and distributing the proceeds.[86] Finally, the judgment or settlement may call for the purchase of an annuity providing for installment payments over a period of time. Again, the contractor may have no right to or control over the undistributed proceeds if the contractor is to avoid current taxation of the undistributed amounts.

The undistributed judgments or settlement proceeds will most likely earn interest while in trust or escrow. An annuity will by design distribute over the life of the annuity an aggregate amount greater than the purchase price of the annuity, the excess of which is considered taxable income to the contractor as it is received or accrued.

---

[84] Hart, T.C. Memo 1983-364.

[85] Watson, Jr. v. Commissioner, 613 F.2d 594 (5th Cir. 1980), *aff'g* 69 T.C. 544 (1978).

[86] Reed v. Commissioner, 723 F.2d 138 (1st Cir. 1983), *rev'g* T.C. Memo 1982-734.

§ 17.20 Bad Debts

What are the tax consequences to the contractor when the owner or other contracting party fails or is otherwise unable to pay amounts owing the contractor? The answer is simple for a contractor using the cash basis method of tax accounting: because income is not recognized until received, the cash basis contractor will never recognize as income the unpaid amounts.[87]

For a contractor using the accrual or one of the long-term contract tax accounting methods, tax treatment of a bad debt depends on whether the amount is considered collectible at the time of its accrual to income. An amount will generally not be accrued as income if, at the time of accrual, serious and reasonable doubt exists as to its collectibility.[88] An extreme example of this principle is bankruptcy or insolvency of the debtor. However, mere financial difficulty of the debtor, or the fact that the contractor does not expect to be paid timely by the contracting party does not constitute doubtful collectibility.[89]

If an amount, though unpaid, is considered collectible at the time of accrual, the contractor must recognize the amount as income.[90] Relief is provided, however, as the contractor is allowed a deduction against ordinary income for bad debts.[91] The contractor may deduct bad debts under one of two methods: the *specific write-off method* or the *reserve method.*[92]

Under the specific write-off method, bad debts are deducted when determined to be uncollectible.[93] Under the reserve method, the contractor accrues and deducts an amount each year representing that portion of accrued income expected eventually to become uncollectible.[94] The deduction each year adds to a *reserve.* As bad debts become uncollectible, they are subtracted from the reserve. The annual addition to the reserve must be reasonable, consistent, and based on objective criteria.[95] Most additions to bad debt reserves are either based on a percentage of the contractor's accounts receivable at the end of the year, the contractor's gross income for the year,[96] or a combination of these. The percentage applied is largely dependent on the contractor's historical bad debt experience.

Uncollectible amounts receivable by the contractor not generated in the ordinary course of the contractor's business, or amounts which have become uncollectible and have not previously been adequately provided for through deductions to reserve,

---

[87] Treas. Reg. § 1.446-1(c)(1)(i).

[88] Corn Exch. Bank v. United States, 37 F.2d 34 (2d Cir. 1930).

[89] Harmont Plaza, Inc. v. Commissioner, 542 F.2d 414 (6th Cir. 1975), *aff'g* 64 T.C. 632 (1975).

[90] Treas. Reg. § 1.446-1(c)(1)(ii); Johnson, T.C. Memo 1983-517.

[91] I.R.C. § 166.

[92] Treas. Reg. § 1.166-1(a).

[93] *Id.* § 1.166-1(a)(1).

[94] *Id.* § 1.166-4(a).

[95] *Id.* § 1.166-4(b).

[96] *Id.* § 1.166-4(b)(1).

may be deducted in the year they become uncollectible. These amounts are deductible in addition to amounts deducted through the reserve.[97]

The contractor must formally elect to use one of the two methods for deducting bad debts on the tax return for the first year in which the contractor incurs a bad debt.[98] The method elected must be consistently used for all subsequent taxable years. If the contractor wishes to change methods, it must formally obtain permission in advance from the Commissioner of the Internal Revenue Service.[99]

Generally, the reserve method of accounting for bad debts is more favorable, for tax purposes, to the contractor than the specific write-off method, as tax deductions for bad debts may be obtained in advance of the actual determination of uncollectibility.

When a debt becomes worthless or uncollectible depends upon the facts and circumstances surrounding the debt, the debtor, and the possibility of repayment. Important criteria include the financial condition of the debtor and the value of any collateral.[1] Bankruptcy of the debtor is generally an indication of the worthlessness of at least part of an unsecured or unpreferred debt.[2] However, bankruptcy does not necessarily by and of itself establish uncollectibility; in some cases, uncollectibility can be established only after a settlement has been reached in bankruptcy, which may not occur until some time after the debtor files for protection under the Bankruptcy Act.[3]

To establish uncollectibility, it is recommended that the contractor take positive, documented action to attempt collection.[4] Written correspondence requesting payment of overdue amounts, and documented negotiations with the debtor provide support of the contractor's attempts to collect unpaid amounts. Legal action, although excellent evidence of the worthlessness of a debt, is not necessarily required. Where the surrounding circumstances indicate that a debt is worthless and uncollectible, and that legal action to enforce payment would in all probability not result in satisfaction of the debt, a showing of these facts will be sufficient evidence of the worthlessness of the debt.[5]

If an account receivable which was previously deducted as a bad debt is subsequently paid by the debtor (*recovered* in whole or in part), the payment becomes income in the year of recovery to a contractor using the specific write-off method.[6] A contractor using the reserve method may either recognize income or reduce

---

[97] *Id.* § 1.166-4(b)(2).

[98] *Id.* § 1.166-1(b)(1).

[99] *Id.*

[1] Treas. Reg. § 1.166-2(a).

[2] *Id.* § 1.166-2(c)(1).

[3] *Id.* § 1.166-2(c)(2).

[4] A. Finkenberg's Sons, Inc., 17 T.C. 973 (1951).

[5] Treas. Reg. § 1.166-2(b).

[6] *Id.* § 1.166-1(f).

the reserve in the year of recovery, which generally decreases the amount of the deduction to the reserve for that year.[7]

§ 17.21 Summary

Proper structuring of the character and the timing for tax purposes of amounts received from judgments and settlements may save significant tax dollars for the contractor. Obviously, tax considerations should not take precedence over economics or other issues critical to obtaining a favorable recovery by the contractor. However, proper tax planning may save or defer income taxes which would otherwise reduce the amount of the award or settlement to the contractor.

The contractor is well advised for many reasons, including tax planning, to document and support the nature of a claim or dispute. Characterizing a loss and the resultant claim as a loss of or damage to property or capital, where applicable, may result in an award which is currently nontaxable.

The contractor should also consider the timing of an award or settlement by electing proper tax accounting methods which defer taxation of the income from a judgment or settlement. In some cases, the award or settlement may also be structured, where applicable, so as to be taxed in future taxable years or over time. In this way, the resultant tax liability may be deferred, thereby allowing the contractor time to pay the taxes and the opportunity to effectively plan to mitigate or reduce them.

TAX TREATMENT AND CHARACTERIZATION OF AMOUNTS PAID OR INCURRED AS A RESULT OF A JUDGMENT OR SETTLEMENT

§ 17.22 Introduction

The contractor is faced with obvious undesirable economic consequences when a judgment or settlement is reached in favor of the other contracting party. The contractor may be required to physically pay out amounts to the prevailing party. Alternatively, the judgment or settlement may reduce the amount of the proceeds due the contractor from the original construction contract. Additionally, the contractor will more than likely have incurred costs of legal, accounting, engineering, or other professional fees in pursuing or defending against the litigation or dispute. Finally, the contractor may also be assessed court costs, fees, and fines. For purposes of the following discussion, all of the above expenditures of the contractor will be referred to as *litigation expenditures*.

---

[7] *Id.* § 1.166-4(b)(2).

The effect of these undesirable economic consequences may be mitigated through proper tax planning and structuring. For example, amounts paid or incurred as a result of litigation or a dispute which represent tax-deductible expenditures will reduce the contractor's income tax liability.[8] The government could be viewed as thus having subsidized or reimbursed a portion of the contractor's loss. The contractor's litigation or settlement costs may alternatively be considered nondeductible capital expenditures,[9] personal or otherwise nondeductible expenditures without tax benefit,[10] or a combination of the above.

§ 17.23 How Litigation Expenditures Are Taxed

The tax consequences to the contractor of expenditures paid or incurred as a result of a judgment or settlement are generally the same as for amounts received, discussed in **§§ 17.2-17.21**. Amounts paid or incurred which are treated as deductible expenditures reduce, dollar for dollar, the contractor's taxable income as determined by the contractor's tax accounting method.[11] Since ordinary income is taxed at graduated rates which may reach as high as 50 percent, deductible expenditures may, on an after-tax basis, cost the contractor only half the amount of the expenditures.[12]

Amounts paid or incurred as a result of litigation or a dispute which are characterized for tax purposes as capital expenditures are not currently deductible. Rather, they increase the basis of the property or investment which was the subject of the litigation or dispute.[13] These *capitalized* expenditures produce a tax benefit when the property or investment is sold or otherwise disposed of. Because the amount of gain or loss upon sale or disposition is measured by the excess of the sales proceeds over the basis of the property or investment,[14] an expenditure increasing basis therefore results in a reduction of gain or an increase in loss.[15] The gain or loss upon sale or disposition may be treated as a capital gain or loss, depending on the character of the property or investment.[16] Capital gain is taxed at significantly lower rates than ordinary income,[17] while capital loss is generally limited in its deductibility.[18] (A more complete discussion of gains and losses appears in § **17.3**.)

---

[8] I.R.C. § 62.

[9] *Id.* § 263.

[10] *See, e.g.*, I.R.C. §§ 162(f), 261, 262.

[11] I.R.C. §§ 62, 461.

[12] *Id.* §§ 1, 11.

[13] *Id.* §§ 263, 1016(a).

[14] *Id.* § 1001(a).

[15] *Id.* §§ 1001(a), 1011(a), 1016(a).

[16] *Id.* § 1222.

[17] *Id.* §§ 1201(a), 1202(a).

[18] *Id.* § 1211.

Generally, from a tax standpoint, it is more beneficial to the contractor to characterize amounts paid or incurred from litigation or a dispute as deductible expenses. The contractor is able to generate tax benefits directly reducing ordinary income which would otherwise be taxed at high rates. Capitalized expenditures generally produce tax benefits at capital gain rates, which are lower than ordinary income tax rates. Further, as noted above, capital losses are limited in their deductibility.

Current tax deductions generated by deductible expenses are also more valuable than deferred deductions generated by capitalized expenditures. Money is worth more today than it is later. Thus, the present value of a current deduction is more valuable than a deferred deduction.

§ 17.24 Expenditures Treated as Deductible Expenses

Unlike the tax treatment of certain expenditures such as taxes,[19] interest,[20] and bad debts,[21] tax treatment of litigation expenditures is not specifically addressed by the Internal Revenue Code. Generally, litigation expenditures are treated as deductible expenses for tax purposes if they are paid or incurred in connection with the ordinary operation of a trade or business.[22] Most litigation expenditures paid or incurred by a contractor are treated as currently deductible expenses, since litigation usually arises either from or relating to work being done by the contractor during the ordinary course of performing a construction contract, or in connection with the general, everyday operation of the contractor's business.

Specifically, in order to be treated as currently deductible expenses, litigation expenditures must be ordinary and necessary, and paid or incurred during the taxable year.[23] In addition, the expenditures must be connected with one of two general activities: (1) the carrying on of a trade or business,[24] or (2) the production or collection of income, or the management, conservation, or maintenance of property held for the production of income.[25]

The term *ordinary* has been applied by the courts to differentiate deductible expenses from those expenditures which must be capitalized.[26] Expenditures are *ordinary* if they are reasonably common or customary to the continuing operation of a trade or business, or if they are paid or incurred within usual business practice.[27] An expenditure is considered *necessary* if it is appropriate and helpful for the continuing development or maintenance of the taxpayer's business.[28]

---

[19] I.R.C. § 164.

[20] *Id.* § 163.

[21] *Id.* § 166.

[22] *Id.* § 162(a).

[23] *Id.* §§ 162(a), 212.

[24] *Id.* § 162(a).

[25] *Id.* § 212(1), (2).

[26] Commissioner v. Tellier, 383 U.S. 687 (1966).

[27] Welch v. Helvering, 290 U.S. 111 (1933).

[28] *Id.*

§ 17.25 Business versus Personal Expenditures

Both the *trade or business* and the *production or collection of income* provisions discussed in **§ 17.24** require that expenditures, in order to be currently deductible, must be paid or incurred in a business or profit-seeking activity.[29] Expenditures paid or incurred for personal purposes are normally not deductible.[30] Thus, expenditures made to prosecute a claim for personal injuries are not deductible;[31] likewise, litigation expenditures paid or incurred in connection with the settlement of a personal injury action are not deductible.[32]

On the other hand, litigation which may at first appear to give rise to nondeductible personal expenditures may instead produce deductible business expenditures if the dispute or charges arise out of or directly involve the contractor's business. For example, litigation expenditures paid or incurred in a libel or slander action are deductible if the contractor's trade or business is directly involved in or affected by the litigation or dispute.[33]

Generally, because a corporation is considered organized for the purpose of carrying on a trade or business, most litigation expenditures paid or incurred by a contractor operating in corporate form are considered deductible business expenses.[34] However, litigation expenditures paid or incurred by a corporation on behalf of or solely benefiting a shareholder are not deductible.[35] Damages paid or incurred on behalf of shareholders or employees charged with unlawful or tortious conduct arising from activities associated with the corporation's business are deductible, however.[36]

The following litigation expenditures have been held by various courts to constitute currently deductible expenses considered made in connection with a trade or business activity: the costs of defending against accusations which would harm the person's business or profession;[37] fees paid by an employer in defending an employee against criminal charges arising out of a crime in connection with the employee's business duties;[38] fees paid to defend a suit for damages based on fraud;[39] litigation expenditures in bankruptcy actions and claims;[40] litigation

---

[29] Kornhauser v. United States, 276 U.S. 145 (1928).

[30] I.R.C. § 262.

[31] Murphy, 48 T.C. 569 (1967).

[32] McCaa, Sr., T.C. Memo 1967-152.

[33] Dyer v. Commissioner, 36 T.C. 456 (1961).

[34] R.C. Walthall, *Deductibility of Legal and Accounting Fees, Bribes and Illegal Payments,* 342 Tax Mgmt (BNA) at A-2.

[35] Ecco High Frequency Corp. v. Commissioner, 167 F.2d 583 (2d Cir. 1948).

[36] Central Coat, Apron & Linen Serv., Inc. v. United States, 298 F. Supp. 1201 (S.D.N.Y. 1969).

[37] Draper v. Commissioner, 26 T.C. 201 (1956).

[38] Union Inv. Co. v. Commissioner, 21 T.C. 659 (1954).

[39] Matson Navigation Co. v. Commissioner, 24 B.T.A. 14 (1931).

[40] International Shoe Co. v. Commissioner, 38 B.T.A. 81 (1938); Harvey v. Commissioner, 12 T.C.M. (CCH) 1358 (1953).

expenditures in connection with the collection of accounts receivable or payable;[41] damages paid or incurred to avoid adverse publicity or controversy;[42] and expenditures to determine compensatory damages to employees.[43]

§ 17.26 Expenditures Which Must Be Capitalized

Not all business-related litigation expenditures are currently deductible. Expenditures which increase the value of property or prolong its useful life must instead be capitalized.[44] These expenditures include costs paid or incurred in acquiring, constructing, developing, or improving property; defending or perfecting title to property; recovering property; and disposing of property.[45]

Expenditures which must be capitalized are generally associated with property or an investment. The contractor will usually be involved in disputes relating to the performance of a construction contract, which is not considered property or an investment in the hands of the contractor. Thus, the contractor will not often be involved in litigation or a dispute the costs of which must be capitalized.

The contractor may nevertheless be involved in a dispute or litigation where litigation expenditures are paid or incurred to acquire, reacquire, or protect title to machinery and equipment used in the performance of construction contracts; these costs must generally be capitalized to the basis of the machinery and equipment involved.[46] Litigation expenses paid or incurred to acquire or protect title to patents, copyrights, processes, or trade secrets must also be capitalized,[47] as must expenditures incurred by a corporation to acquire the stock of a dissenting shareholder.[48]

Sometimes a contractor may incur expenditures which benefit the contractor's business as a whole, and these expenses may have to be capitalized. For example, a contribution to the construction of a state-owned bridge which facilitates the operations of the contractor must be capitalized.[49] However, if the expenditures merely facilitate the contractor's performance of a single construction contract, the

---

[41] Koerner v. Commissioner, 10 T.C.M. (CCH) 867 (1951).

[42] International Shoe Co. v. Commissioner, 38 B.T.A. 81 (1938).

[43] Guttman v. United States, 181 F. Supp. 290 (D. Pa. 1960).

[44] Treas. Reg. § 1.263(a)-1(b).

[45] *Id.* § 1.263(a)-2.

[46] *Id.* § 1.263(a)-2(a).

[47] *Id.* § 1.263(a)-2(b).

[48] Commerce Photo Print, T.C. Memo (Apr. 11, 1947).

[49] Rev. Rul. 58-373, 1958-2 C.B. 125.

expenditures would generally be deductible, since they would be considered paid or incurred in the normal course of the contractor's business.[50]

Although capitalized litigation expenditures are not currently deductible, they may nevertheless produce ordinary tax deductions in future periods. Litigation expenditures which must be capitalized to the tax basis of equipment, machinery, or buildings may produce tax benefits over time through depreciation, amortization, or cost recovery deductions.[51] Any undepreciated or unamortized basis reduces gain or increases loss upon sale or disposition of the property, as discussed in **§ 17.23**.

§ 17.27 Expenditures Which Are Neither Deductible nor Capitalizable

Some litigation expenditures are neither currently deductible nor capitalizable, and thus never produce a tax benefit. These include expenditures paid or incurred in connection with purely personal claims or disputes, discussed in **§ 17.25**.[52] Other nondeductible and noncapitalizable expenditures include expenditures made to produce or collect tax-exempt income,[53] and fines and penalties.[54]

Fines and penalties are amounts made to or assessed by federal, state, or local governments or their agencies, including the courts.[55] They include amounts paid or incurred pursuant to conviction or a plea of guilty or nolo contendere to a crime (felony or misdemeanor) in a criminal proceeding.[56] Also included are amounts paid or incurred as civil penalties imposed by federal, state, or local law,[57] or payments made in settlement of an actual or potential liability for civil or criminal fines or penalties.[58]

Litigation expenditures paid or incurred in defense of a prosecution or civil action arising from an alleged violation for which fines and penalties would be assessable are not included as fines and penalties. Thus, legal and other professional fees, court costs, and other expenditures paid or incurred in such a defense would be deductible or capitalizable under criteria discussed in **§§ 17.24-17.26**.

---

[50] I.R.C. §§ 162, 212.

[51] *Id.* §§ 167, 168, 178.

[52] I.R.C. § 262.

[53] *Id.* § 265.

[54] *Id.* § 162(f).

[55] Treas. Reg. § 1.162-21(a).

[56] *Id.* § 1.162-21(b)(1)(i).

[57] *Id.* § 1.162-21(b)(1)(ii).

[58] *Id.* § 1.162-21(b)(1)(iii).

WHEN LITIGATION EXPENSES ARE DEDUCTED FOR TAX PURPOSES

§ 17.28 Cash Method versus Accrual Method

As discussed in §§ **17.12-17.16**, regarding proceeds received from judgments or settlements, litigation expenses paid or incurred are deducted according to the tax method of accounting elected by the contractor.[59] The cash basis contractor recognizes deductions for litigation expenses as they are paid;[60] the accrual basis contractor recognizes deductions for litigation expenses as they become due and payable.[61]

§ 17.29 Litigation Expenses Deducted under Long-Term Contract Methods

A detailed description of the two tax methods of accounting for long-term contracts was presented in §§ **17.14-17.16**. Under the percentage of completion method of tax accounting, all contract costs incurred during the taxable year must be deducted in the taxable year.[62] The contractor is thus required to use the accrual method to deduct litigation expenses attributable to construction contracts under the percentage of completion method.

Under the completed contract method of tax accounting, all costs properly allocable to a construction contract must be deferred and deducted when the contract is completed.[63] Indirect costs not properly allocable to the contract are not deferred. Rather, they are currently deductible as *period costs* under the tax method of accounting elected by the contractor with respect to general operations, excluding long-term contracts.[64] Thus, the timing of deductions for litigation expenditures under the completed contract method depends on whether the dispute or litigation directly relates to the performance of the construction contract. If the litigation expenses are directly related, the deductions must be deferred until the contract is completed.

On the other hand, litigation expenses paid or incurred with respect to the contractor's general business would not be deferred. General and administrative costs attributable to the performance of services which affect the contractor's business

---

[59] I.R.C. §§ 446(a), 461(a).

[60] Treas. Reg. § 1.461-1(a)(1).

[61] *Id.* § 1.461-1(a)(2).

[62] Treas. Reg. § 1.451-3(c)(3).

[63] *Id.* § 1.451-3(d)(1); Prop. Reg. § 1.451-3(d)(1).

[64] Treas. Reg. § 1.451-3(d)(5); Prop. Reg. § 1.451-3(d)(5), (6).

operations as a whole are considered period costs not directly related to any construction contract.[65] Similarly, costs incurred with respect to guarantees, warranties, maintenance, or other service agreements are not considered directly related to any construction contract.[66]

§ 17.30 —Effect of Litigation Expenses on Timing of Income under Completed Contract Method

A claim or dispute may affect the completion date of a construction contract accounted for tax purposes under the completed contract method. If the amount of a dispute or claim is large enough that the contractor is unable to determine whether the contract will eventually result in a profit or a loss, the contract will generally not be considered complete until the dispute is settled.[67] If the contractor is assured of a profit regardless of the outcome of the litigation or dispute, the contract will be considered complete without regard to the dispute or litigation.[68] If the contract will result in a loss regardless of the outcome of the litigation or dispute, the contract will be considered closed and net profit reduced (but not below zero) by a reasonable estimate of the litigation expenses.[69] The balance of the litigation expenses is generally deducted, upon settlement or judgment, in accordance with the contractor's general method of accounting. However, if the litigation expenses represent additional work to be performed, they are not deductible until the additional work is completed.[70]

In some cases, the contracting party may not accept the subject of the disputed construction contract because additional work or rework is requested of the contractor. Although the Treasury Regulations generally require the amount represented by the dispute to be segregated from the original contract,[71] some courts have taken a different position. Courts in the First,[72] Sixth,[73] and Ninth[74] Circuits have held that a contract is considered closed only when *completely* finished and *accepted* by the contracting party. The contractor is advised to consider the positions taken by these courts, in light of the conflicting position taken by the Treasury Regulations, before attempting to hold a contract open for additional work or rework.

---

[65] Treas. Reg. § 1.451-3(d)(5)(iii)(E); Prop. Reg. § 1.451-3(d)(5)(iii)(E).

[66] Treas. Reg. § 1.451-3(d)(5)(iv); Prop. Reg. § 1.451-3(d)(5)(iv).

[67] Treas. Reg. § 1.451-3(d)(2)(iii).

[68] *Id.* § 1.451-3(d)(2)(iv).

[69] *Id.* § 1.451-3(d)(2)(v).

[70] *Id.* § 1.451-3(d)(2)(vi).

[71] See **§ 17.17**.

[72] Rice, Barton & Fales, Inc. v. Commissioner, 41 F.2d 339 (1st Cir. 1930).

[73] Thompson-King-Tate, Inc. v. United States, 296 F.2d 290 (6th Cir. 1961), *rev'g & remanding* 185 F. Supp. 748 (D. Ky. 1955).

[74] E.E. Black, Ltd. v. Alsup, 211 F.2d 879 (9th Cir. 1954).

§ 17.31 Effect of Contingencies or Restrictions on Timing of Litigation Expense Deductions

An accrual basis contractor may not deduct litigation expenses for tax purposes until they are in fact due and payable.[75] Thus, an accrual basis contractor cannot accrue deductions for estimated amounts,[76] for amounts which have not yet been assessed,[77] amounts contingent upon the happening of a future event,[78] or amounts which the contractor is contesting and does not yet have a legal obligation to pay.[79] It is common and prudent financial statement accounting practice for a contractor to establish *reserves* for liabilities reasonably expected to occur in the future. Because these reserves represent estimates of future liabilities which have not yet occurred, they are not deductible for tax purposes until the liabilities become fixed and determinable.

In certain instances, litigation expenses may be accrued and deducted for tax purposes if they can be determined with reasonable accuracy, even if the exact amount of the liability and its date of payment are not yet fixed.[80] This exception to general accrual method tax accounting provisions normally applies only to expenses which are recurring and customary to the contractor's general operations, such as insurance, repairs, and maintenance. Litigation expenses may be currently deductible, even though they are being contested, if the contractor irrevocably transfers money or property to provide for satisfaction of the asserted liability.[81] The contractor must give up control of the money or property transferred. This may be accomplished by transferring funds or property to the person asserting the claim or to an independent third party under an agreement to distribute the money or property in accordance with the eventual resolution of the dispute.[82] Independent third parties include a trustee under an irrevocable trust or escrow arrangement, a government agency, and a court with jurisdiction over the litigation. To support a deduction for such a transfer, it is strongly advisable to have both parties to the dispute formally agree to the plan of transfer, as opposed to a unilateral transfer by the contractor.[83]

---

[75] Treas. Reg. §§ 1.446-1(c)(ii); 1.461-1(a)(2).

[76] *Id.* § 1.461-1(a)(2).

[77] United States v. Anderson, 269 U.S. 422 (1926).

[78] Lustman, T.C. Memo 1960-116, *aff'd,* 322 F.2d 253 (3d Cir. 1953).

[79] Shepherd Constr. Co., Inc., 51 T.C. 890 (1969), *acq.,* 1969-2 C.B. 25.

[80] Lucas v. American Code Co., Inc., 280 U.S. 445 (1930).

[81] I.R.C. § 461(f).

[82] Treas. Reg. § 1.461-1(c).

[83] Poirier & McLane Corp. v. Commissioner, 547 F.2d 161 (2d Cir. 1976), *rev'g & remanding* 63 T.C. 570 (1975), *nonacq.*

§ 17.32 Tax Planning for Litigation Expenses to Be Paid

By properly structuring the payment of litigation expenses, an accrual basis contractor may generate tax deductions in advance of their actual payment. The accrual basis contractor may generate a current tax deduction for the total amount of a liability for damages in a judgment or settlement even though the amounts may be payable in installments over a period of time, as long as the contractor's liability is fixed and the amount of the liability can be determined with reasonable accuracy.[84] By structuring an unfavorable settlement as payable in installments, the accrual basis contractor may generate immediate tax benefits while paying for them over time. As the time value of money is important to good cash management, the contractor is well advised to consider such structuring.

A cash basis contractor may generate tax deductions only as the litigation expenses are actually paid. Thus, a settlement structured in installments will not benefit a cash basis contractor from a tax standpoint as it would an accrual basis contractor.

OTHER CONSIDERATIONS

§ 17.33 Tax Law Changes

There is a widespread, growing concern with the complexity of the tax laws, and a general agreement that they should be simplified. In the past 10 years, since 1975, there have been no less than five major general tax acts, several technical corrections acts, and a host of other acts aimed at revising specific areas of taxation. The Treasury is currently working on a backlog of regulations for over 100 different amendments to the Internal Revenue Code made as a result of the above acts.

At the time of this writing, various proposals have been made calling for basic tax reform. Various bills have been introduced in Congress, and the Treasury Department, with the general support of the President, has introduced its own proposals, all of which call for reductions in tax rates and a general elimination of deductions, exclusions, and credits. These proposals would generally create an income taxation system which would apply lower-than-current tax rates to gross income, reduced by very few if any deductions, and have become known in colloquial terms as *flat tax* proposals. At the time of this writing, it is not known if any of these proposals or a combination of them will ultimately be enacted. It is widely felt, however, that some form of a simplified tax system could be adopted within the next few years.

---

[84] Treas. Reg. § 1.461-1(a)(2).

If a simplified tax system is enacted, many, if not most, of the tax deductions currently allowed may be eliminated. Additionally, the ability to defer taxation of income by the use of special tax provisions, such as long-term contract methods of accounting or installment payouts, may be severely restricted or eliminated. The contractor is cautioned to consult both current and proposed tax law when planning for and structuring judgments and settlements.

§ 17.34 Conclusion

The contractor should be aware of tax consequences and planning opportunities when involved in construction litigation or disputes. Proper tax planning for the effects of a judgment or settlement may increase the after-tax benefits of proceeds received from judgments and settlements; likewise, proper planning may mitigate the after-tax cost of litigaton expenditures.

The contractor is well advised to consult counsel and tax professionals when initiating a claim, planning litigation, and structuring settlements. With proper consideration and appropriate use of the tax laws, the contractor will assure favorable tax results, and above all, save money.

CHAPTER 18

USING EXPERTS TO CONTAIN LITIGATION COSTS

Janice Anderson Gram

Janice Anderson Gram is Regional Director for Coopers & Lybrand's Business Analysis Group, where she is responsible for working with trial attorneys to put together financial, actuarial, economic and/or other consulting analyses for use in the resolution of claims and litigation. Ms. Anderson Gram has appeared as a guest speaker on a variety of issues connected with construction claims and litigation, including the preparation of expert testimony and damage claims, as well as the use of computer systems to manage documents and to create exhibits. She has published articles in *Trial Magazine*, *The National Law Journal*, and the *Software Newsletter* and she authored a chapter for *Construction Litigation: Representing the Owner*, published by Wiley Law. Ms. Anderson Gram graduated from the University of California, Berkeley and subsequently obtained a National Association of Securities Dealers License. She serves on the Board of Directors of several nonprofit organizations and is listed in *Who's Who in California* and *Who's Who of California Executive Women.*

§ 18.1 Introduction

This chapter is about how to contain the cost of construction litigation by spending money on experts. It is about *who* the experts are and *when* to hire them: should they be hired at the outset of the dispute, or is it more cost effective to wait and hire them right before trial? It is about *how many* experts to hire: will one good generalist suffice, or will several be needed? Finally, it is about *where* to find experts, and once located, how to make sure their time is spent in the most efficient and cost-effective manner possible.

Construction litigation can be a complex, expensive maze, regardless of whether the dispute is over the construction of a residence or a nuclear power plant. There are many interrelated facts, and few straightforward answers to questions like, "Who caused the job to come in over budget?" or "How much should the contractor be paid for lost opportunity costs resulting from delays?" Moreover, it sometimes takes a scorecard to keep track of all the claims and counterclaims that can be filed by contractors, architects, engineers, owners, and suppliers.

Construction claims may, in fact, be inevitable. A large percentage of construction projects seem to end up in some kind of a dispute, whether over additional compensation resulting from acceleration, delay, change in the work sequence, or quality of the workmanship on the job. Putting together the facts to prepare or defend a construction claim is like putting together a puzzle with 10,000 pieces. Both require a great deal of thought, concentration, and care. To organize and understand, and ultimately resolve, a construction dispute, the contractor will need to seek assistance from experts.

Attorneys reading the word *expert* probably think immediately of *expert witnesses.* Contractors may think of *technical experts* such as structural engineers. Use of the word in this chapter conforms to the definition found in *Webster's Dictionary*:

ex.pert/'ek,spert,ik-'/adj 1 obs:EXPERIENCED 2: having, involving, or displaying special skill or knowledge derived from training or experience syn see PROFICIENT ant amateurish

ex.pert/'ek-,spert/n: one who has acquired special skill in or knowledge of a particular subject: AUTHORITY[1]

Proving the factual basis of the claim requires individuals with skill, experience, and a special knowledge of the legal, technical, managerial, graphic, and testimonial aspects of a construction case.

§ 18.2 Recruiting Experts

Imagine the plight of a contractor who was on a job where sloped roofs were specified. Just before the roofs were to be installed, the owner decided that flat roofs would be less expensive. The architectural and engineering drawings arrived late and were incomplete. The new material was partially defective, and the installer did a lousy job. Everyone involved on the job sued everyone else. This contractor needs the advice of experts to proceed.

The first and most important expert to hire is someone with legal expertise. The attorney will be responsible for the single most important aspect of the case, that of persuading the jury of the correctness of the contractor's arguments. The contractor who takes the time to interview several candidates for this position will reap the benefit of that time when the verdict is delivered.

Some things to consider during the interviews with attorneys are:

1. **Choose an individual, not a law firm**. The jury will neither know, nor will they care, about the law firm affiliation of the attorney trying the case. What makes them decide in the contractor's favor is whether or not the attorney is a powerful and compelling communicator. The jury knows that Mark Twain was correct when he said, "the difference between the right word and the *almost* right word is the difference between lightning and the lightning bug." The object is to win, and words will be the way to accomplish that goal.

   An attorney whose speech is hesitant and replete with "um, ah, I guess," or who fails to establish eye contact during the interview, is not the best choice for a strong advocate. If the attorney seems arrogant and disorganized to the contractor, he or she will appear that way to the jury too, and they will tune him or her out. Ten minutes into the interview, ask "If I were a juror, what would I think of this person?" Based on the response, either walk out the door, or hire the attorney on the spot.

2. **Choose a leader who is also a team player**. The qualities needed in an attorney are like those of the captain of a football team. This is the in-

[1] Webster's New Collegiate Dictionary (150th anniversary ed. 1973).

dividual who will be responsible for hiring all the other members of the team, and whose job it will be to lead and motivate them. The skills sought are not those of an individualist; the contractor is not looking for someone who insists on overcoming all obstacles singlehandedly.

3. **Choose someone who is organized and who listens.** There will be a myriad of facts bearing on the claim which must be interrelated and organized into a simple, understandable argument. Determining which theories are worth pursuing and which are not requires someone with exceptional organizational skills.

   Some people are good listeners, some are not. It takes time and patience to listen to the experts so carefully that a case-winning point imbedded in a long technical dissertation is not missed. Hold out for an individual who demonstrates this skill.

4. **Choose someone with trial experience.** Ideally, the contractor will find an attorney who is first and foremost a trial attorney. This is as opposed to someone who has worked on a number of construction cases, but who rarely gets into the courtroom. One of the ways to keep the cost of pursuing the dispute down is to hire an attorney who knows, based on experience, how dangerous it is to become too enamored of details, thereby forgetting to focus on the most important aspect of the case, namely, how the case will sound to the jury. This requires the proficiency of an experienced trial attorney.

§ 18.3 How Many Experts Are Needed?

Using the guidelines set out in **§ 18.2**, together with intuition and common sense, the perfect attorney is hired. The next step is for that attorney to put together a team of people to uncover the proof that will enable the contractor to prevail at trial. It is unlikely that the attorney will find all of the training, skills, and personality traits needed to help win the dispute in just one individual. If it took an architect, engineers, contractors, and suppliers to design and install the roof in the first place, the attorney will probably need help from people who have the same skills in order to determine what really happened on the project. If a product defect is alleged or suspected, other technical experts and testing facilities may be needed. All claims involve damages, so an accountant or financial analyst will have to be hired to prepare or defend the claim for damage.

In addition to legal, technical, and financial exerts, construction litigation requires a person with managerial expertise. Someone must keep track of and coordinate the day-to-day work of the experts. This person must have excellent communication skills and a temperament which remains unruffled by the crisis-oriented environment of a construction case. In some instances, management tasks are a full-time job, and cannot be handled efficiently by a trial attorney. If this is true, the role needs to be delegated to a litigation specialist with management credentials.

While all these people are experts, they are *not* expert witnesses. They are fact-finders and fact gatherers. They are advisors and consultants.

§ 18.4 When to Hire Experts

Experts should be hired at the outset of litigation. In the case of a plaintiff, it is wise to employ experts prior to filing the complaint to help give an objective view of the allegations, and to provide an initial assessment of damages. The sooner the experts start to work, the sooner the attorney has a base of knowledge to use in weighing the relative merits of the various allegations for purposes of settlement.

There is a temptation *not* to do this, not to bother hiring experts until the last minute, especially if there is a good chance for settlement. This can be a costly mistake. The attorney will not be able to evaluate what is realistic or reasonable for purposes of settlement without having hired the experts to examine the issues surrounding both liability and damages. The more the attorney knows of the facts, the stronger the settlement position he or she can assume.

The author's experience on a construction matter illustrates this point. The structural engineering expert suggested that the attorney hire two additional experts to perform some tests, because he was convinced that the tests would prove that the plaintiff's contentions were severely overstated. The two experts, plus the tests, would have cost approximately $6,000. The decision was made to forgo this extra work because there was a "good chance of settlement," and, in fact, the case did settle. The plaintiff was awarded approximately $60,000 on a $120,000 claim. This seemed pretty good. Later, however, because of circumstances surrounding other issues in the case, the settlement was abandoned, the additional experts were hired, and the tests were performed. They proved that there was *no* damage. The settlement was renegotiated and the amount was substantially reduced.

It would be simplistic to say that it was better to pay no money than it would have been to have paid $60,000. Of course, it was also possible that the tests could have shown damage of $240,000. The point is that to be convincing during settlement one must deal from a position of strength, and it is the work of the experts that lends that credibility and strength to the position. Hiring experts very early in the process may mean the difference between dealing in supportable fact or dealing with suppositions.

§ 18.5 What to Look for in an Expert

The qualities to be sought in an expert are the same, regardless of whether the case looks like it will settle early, or whether it will be decided by an arbitration panel, a judge, or a jury. The characteristics to look for in a technical expert, however, are different from those found in a witness who will be used to present testimony at trial. Technical experts need to be good, but not necessarily great communicators. They should be thorough and detail-oriented. They should have an in-depth understanding of their technical area and a wealth of experience to back up their point of view. They must also be comfortable with taking the position of an advocate. The attorney should beware the purist who says, for example, ''All flat roofs are bad'' when the roof on the complex in question is flat, or whose overwhelming preference is for techniques which would produce a Taj Mahal when a low-cost development was what was originally intended.

People with a technical specialty tend to be modest about their ability to provide expert advice, and, when asked, may decline the opportunity to work on the dispute. If instead they are asked to *consult* on some problems, they may willingly sign onto the team. Later, if a technical expert becomes comfortable in the role of a witness, and the attorney believes that he or she will be able to present the facts in a comprehensible, convincing, and understandable manner, the consultant may turn into the witness.

§ 18.6 Locating Experts

There are a number of sources which can be explored to try to find just the right experts, including the following:

1. Ask the contractor client for suggestions
2. Call the national or regional association for the trade or specialty needed
3. Look in the American Bar Association's list of construction experts
4. Call a large accounting firm and ask to speak to a litigation specialist
5. Look in the yellow pages of the local telephone book under the specialty needed
6. Ask the first expert hired for suggestions as to other experts who could be used.

To save time, make a list of all the potential areas and issues which may need to be addressed. Then, when talking to people in the above organizations, ask them to recommend other sources for experts. The person contacted in the accounting firm may very well be able to recommend or suggest names of design professionals and engineers, as well as accountants.

§ 18.7 Getting Started

All or most of the experts have been hired. Next, before meeting with the experts, the attorney must formulate what Judge Herbert J. Stern calls ''one central theme'' for the case. This is because ''jurors (and experts) look for that one explanation, that one central theme that best reconciles the greatest number of discrepancies . . . . [W]hatever's life's problem, we all want to think there is some simple solution to it somewhere.''[2] Without this central focus, experts are likely to go off on a variety of tangents, pursuing at will what they think is the most important aspect of the case.

Once this capsulized view of the lawsuit is developed, what should the attorney do to make sure that the experts' time is spent in the most productive manner?

> The first step in dealing with an expert is a preliminary conference. Any lawyer who does not avail himself of such a conference is both lazy and negligent, for this can be one of the most important elements in preparing a lawsuit. . . . The most important element of the preliminary conference is communicating to the experts your objectives. . . . He should know exactly how his opinions are to be used in constructing the proof of the case.[3]

Experts will want this discussion to include a synopsis of the potential weak points in the case as well.

§ 18.8 —Provide Summaries

The initial interview is a good time to provide the expert with a brief written narrative containing all the relevant facts surrounding the case. This summary may be contained in the complaint, counterclaim, and/or answers to interrogatories. If it is not, the summary might include information like:

1. Dates of all important events affecting the allegations
2. Names of all companies involved and their specialty
3. Names of individuals who are central to the case and their positions
4. Information unique to the case
5. Names of substances or processes which were used to install allegedly faulty products
6. Date of the close of discovery
7. Tentative dates for dispositions, arbitration, or trial
8. Amount of damages sought.

Having this information in one place will save the expert time; otherwise, each expert will end up compiling the list himself or herself. Further, summaries help the expert to give uniform and consistent information about the case to his or her staff.

---

[2] H.J. Stern, *Winning Trial Advocacy Techniques,* Law & Business Inc. (Harcourt/Brace/Jovanovich 1984).

[3] This material is reprinted from Hullverson, *Pretrial Preparation of the Expert Witness,* in Experts in Litigation 11-12 (1973), published by the Institute of Continuing Legal Education, Ann Arbor, Michigan.

If the attorney does not yet have some of the information needed for the summary, consider including the information that is available and leaving blanks for the missing data:

Roof installer: EZ Roof Installers (Document #123)
Key personnel: ______________________ (Foreman) (Document ?)
Contract signed: 3/29/83 by Joe Krepps (Document #456)
First date on site: ______________________ 1983 (Document ?)
Job complete: 10/? ?/85 (Document #789)

This serves to alert the experts to the basic data which is important, but which is still unknown. If in the course of their work the experts come across the missing information, they should be instructed to call the attorney, who then can update the summary.

The time spent creating high-level summaries can be leveraged when there is a need to bring a new expert or attorney up to speed quickly on the facts of the case (and this saves the client's money). If the case is dormant for a while, summaries provide a good means of refreshing people's memories as to the dates, persons, and sequence of events which are at issue.

General checklists are a way to broaden the horizons of the expert's work. Normally the attorney selects the documents which he or she believes are relevant to the expert's specialty. The author saw one situation where this was not done; instead, the attorney put together an extensive list which included all the types of documents which were to be produced in the litigation. Using the list as a guide, the expert was asked to do two things: (1) determine if there were any other types of documents which should be added to the list; and (2) select from the list the documents the expert believed would be helpful in pursuing his investigation. The advantage to this approach is that the *expert,* based on his or her knowledge of the technical area, selected the documents relevant to the analysis, rather than the attorney who knew very little about the relative importance of one kind of document over another.

That list of the kinds of documentation produced could include the following:

1. The contract and amendments
2. Inspection certificate reports
3. Documents regarding insurance coverage
4. Shop drawings and as-built drawings with all revisions
5. Invoices for materials and equipment
6. Time sheets and payroll records
7. Daily logs
8. Bonding information
9. Purchase orders
10. Change orders
11. Punch lists
12. Installation information

13. Maintenance information
14. Testing procedures
15. Critical path method (CPM) and scheduling information
16. Product brochures
17. Photographs and/or videotapes
18. Bid estimates
19. Cost ledgers and the chart of accounts
20. Correspondence
21. Identity and description of products used
22. Architectural and engineering plans and specifications.

A conscientious expert will refer to this list at various points during the investigation to make sure that all pertinent documents are reviewed for purposes of the analysis.

§ 18.9 —Encourage Expert Candor

The experts will need to be encouraged to take issue with the attorney if there is a point with which they disagree. The author has seen an expert sit through a meeting with an attorney and agree to follow a particular strategy, walk out of the meeting, and say "I don't think his approach makes any sense, but he must know what he is doing." Experts tend to assume that the attorney, because he or she is more intimately involved with all aspects of the case, has superior knowledge about what steps should be taken next in order to answer questions raised in the lawsuit. Further, the attorney handling the lawsuit may have a more assertive personality than the expert; therefore, it is necessary for the attorney to *actively* seek the honest opinions of the experts.

Ask the experts to point out what they believe to be the strongest and weakest points of their own analyses. They will probably not volunteer this information. Experts need to be *instructed* to tell the attorney about each and every inconsistency they find in the data, regardless of how mundane any single observation may seem to the expert. Facts that appear obvious to an expert may be case-winners for the attorney. In one recent case the accounting expert noticed that the depreciation schedule used for the damage claim was different from the one the construction company normally used. The expert thought that this point was so obvious that he did not point it out to the attorney. In refuting the damage claim, no mention was made of this potentially significant point. When the case was concluded, the accountant wondered out loud to the attorney why this fact was not used. The attorney was totally taken aback, since he of course had not noticed the discrepancy; if he had been told, his client could have received a significantly better result at trial. Had the attorney asked the accountant to tell him about *all* the accountant's observations, perhaps this fact would have been revealed.

§ 18.10 —Keep the Experts Informed

Withholding information from the experts, either consciously or unconsciously, is wasteful and a disincentive to the members of the litigation team. On the subject of teamwork, Akio Morita, the cofounder and chairman of the Sony Corporation, has said that "in Japan, management does not treat labor as a tool, but as a partner. We share a common fate."[4] This philosophy is a good one for a trial attorney to follow when working with experts.

It can directly affect the result of the analysis if experts are kept informed of *all* events which impact the litigation. Unfortunately, many attorneys tell the experts only the barest details surrounding the progress of the case. It is not uncommon for an expert to be engaged, perform work under very tight deadlines, and then suddenly not hear anything further from the attorney for three to six months (or maybe never!). It is extremely demotivating for the expert to find out in a chance meeting in a restaurant that the case settled, or that the case is on hold for the time being. Keeping experts up to date on the case does take time, but it can pay dividends in the end.

A case in point involved a construction claim which required the services of both a structural engineer and an electrical engineer. Both experts were thoroughly informed about the progress of the case; they were given copies of all legal documents filed by both sides, and were regularly told of the pitfalls being encountered in the other expert's area. While reviewing documents, the electrical engineer found some information that he knew the structural engineer had been unable to locate. He called the attorney immediately. Discovery of these documents saved the structural engineer several weeks of work.

Status meetings are not a waste of time. If it seems impossible to schedule these meetings, consider arranging for a telephone or video conference call with the experts. If the experts know in advance that they can count on speaking to the attorney at a specific time, they may be able to hold their questions and observations until then. This saves everyone's time and cuts down on the inherent frustration of trying to talk to an attorney who is frequently unavailable. Another benefit of this group electronic conference is that people have a greater tendency to be brief when they know that others are listening. Status meetings are a good time to make sure that everyone clearly understands the current priorities.

§ 18.11 —Maintain Contact

It should go without saying that the expert's telephone calls need to be returned promptly. The following case is a dramatic example of why this rule should be followed. An expert, during destructive tests, uncovered a hazardous electrical situation. He tried numerous times to notify the attorney to get further direction

[4] Playboy (Aug. 1982).

as to who should be informed of the situation, but the attorney failed to return the phone calls because he was in trial on another case. The building burned down.

§ 18.12 —Establish Guidelines for Written and Computer-Generated Materials

Most experts have never heard of the attorney work product privilege. Giving the experts guidelines on what to document and how to document their findings in writing may save the future embarrassment of damaging documentation being introduced at trial.

Special consideration should be given to information created on computers. Electronic spreadsheets make it faster and easier to add and/or delete information. Additionally, the software automatically and accurately recalculates the figures affected by any change. Some experts routinely save all superceded electronic worksheets, others merely write over on the disk the changes and/or corrections, and do not keep a record of the original data. Whatever the attorney's decision regarding guidelines for keeping information stored on magnetic media, the policy should be consistent and clear.

Since the ability to respond quickly to changing events is essential during litigation, the versatility and speed of a computer can mean the difference between an expert's timely response to an attorney's request (''Oh, by the way, would you also analyze the housing industry trends?'') and an inability to perform the analysis at all.

Computers also assist in organizing complex information, and so, in the interest of saving costs, their use by experts should be encouraged. In litigation, the attorney typically does not have all of the documents needed at the outset of the case. Computers aid in the quick assimilation of new information as it becomes available.

§ 18.13 Preparing to Win

To receive one's money's worth from the efforts of the experts, the attorney must help them structure their work and set specific goals to be accomplished. Fans of Perry Mason have the impression that trial work is a fast-paced, exciting life. On television, witnesses appear at the last minute, just in time to make a startling revelation that wins the case. In reality, winning—or successfully settling—a case requires methodical preparation and attention to detail, and the people who provide the details are the experts. Encourage them, however, to focus on the big picture rather than on overworking the minutia. Most technical experts are used to working from the bottom up, from the smallest detail up to the development of the final conclusion. This is the reverse of the way an attorney works. The attorney needs to know as soon as possible the conclusions which will be reached, and only then will he or she want to fill in the details. The attorney should be aware, however, that most experts will be reluctant and uncomfortable about verbalizing a conclusion which may only be partially supported or researched.

If several experts are hired, the goal should be to have each of their work products support one another. Unless the attorney is constantly watching over the experts' work, each expert will develop a separate, distinct, and unconnected collection of data. "To be most memorable, information must be heard more than once . . . it must be repeated."[5] It is the attorney's responsibility to see that each expert's work ties in with and supports or enhances that of the other experts. Otherwise, the jury will get lost trying to put together disjointed points, and will not understand the case the attorney is trying to make.

By asking the expert to first draw tentative, overview conclusions, the attorney is helping the expert to understand the priorities. For instance, the expert may be requested to structure his or her work so that he or she can answer the following questions:

1. What is the existing condition? ("The roof leaks.")
2. What caused it? ("It appears that either the roof was installed wrong, or possibly the material was defective.")
3. What are the alternative ways to fix it? ("Patch the roof, or tear it completely off.")
4. How much will the method cost?
5. How would you assess the case from the other side's point of view? (This can be very revealing and may point out biases on the part of the expert.)

At this point, ask the expert what work will be required to obtain the next level of detail ("Send a sample of the roof to a testing laboratory"), and ask what the approximate cost will be to get that detail. The attorney using this top-down technique will always be better prepared to discuss settlement at an early stage, since part of the information elicited from the expert was the potential range of damage.

§ 18.14 How to Use the Expert

People with technical expertise can be used for more than technical evaluations. They can assist the attorney to:

1. Review and summarize, possibly even code, documents
2. Write and answer interrogatories
3. Write requests for document production
4. Analyze deposition transcripts

---

[5] S. Hamlin, *What Makes Juries Listen,* Law & Business Inc. (Harcourt/Brace/Jovanovich 1984).

5. Prepare for deposition of the opposing expert
6. Prepare for cross-examination of the opposing expert.

A financial expert's summary of the deposition of a controller will probably bear no resemblance to the summary prepared by a paralegal. True, the expert's time is more costly, but the resulting product will be more meaningful and valuable.

§ 18.15 Controlling the Cost of Destructive Tests

Many construction disputes require that destructive tests be performed. For guidance in this area the attorney must rely on the advice of experts to arrange for and assess the findings of these tests. This can be a very expensive endeavor. To control this effort, the attorney should understand in advance the answers to the following questions:

1. What specifically is to be tested?
2. How is the test to be done?
3. What will the test show or prove?
4. Is there an alternate way to determine the same or similar information?
5. Which contractors, subcontractors, or laboratory will be used? (In one case, the answer to this question indicated that the expert was planning on using the other side's testing laboratory)
6. What is the approximate cost of the test?
7. What is the approximate cost of the repairs?
8. How long will it take?
9. What is the approximate cost of the expert's time to oversee the process?

This information provides a concrete basis for determining whether the test is, in fact, justified. It also gives the attorney the ability to evaluate the cost-benefit of the proposed test.

§ 18.16 Use Visual Aids to Gain Consensus

A study by the Wharton School's Applied Research Center,[6] which was commissioned by the audio/visual division of 3-M, found that using visual aids can reduce the length of business meetings by 28 percent, proving that a picture *is* worth a thousand words. If the use of visual aids or exhibits reduces the time it takes to

[6] 3-M Corporation, St. Paul, Minnesota.

present a case, the early use of exhibits can save the client money. Not only do visual aids cut down on the time it takes to make a point, but this study showed that:

1. There was a 79 percent rate of group consensus when visuals were used
2. When a presentation was made without visuals, there was only a 58 percent rate of consensus
3. If visual aids were used, decisions were reached at the time of the presentation 64 percent of the time
4. Decisions were delayed 52 percent of the time, because people could not agree on a course of action, when the presentation was made without visual aids.

In addition, the study showed that presenters who used visual aids were considered more persuasive and interesting than presenters who did not. These results argue strongly for encouraging experts to use exhibits, graphs, and charts, and, since "80 percent of jurors make up their minds regarding liability after opening statement, and never change their minds,"[7] it may be a good idea to use them at the very beginning of trial.

Consider getting a sample of the roof, as built, or have a picture of the roof drawn, and have the expert explain his or her opinion of what happened using the visual aid. It will cut down on the time it takes to understand the expert's ex-

PRESENTING INFORMATION

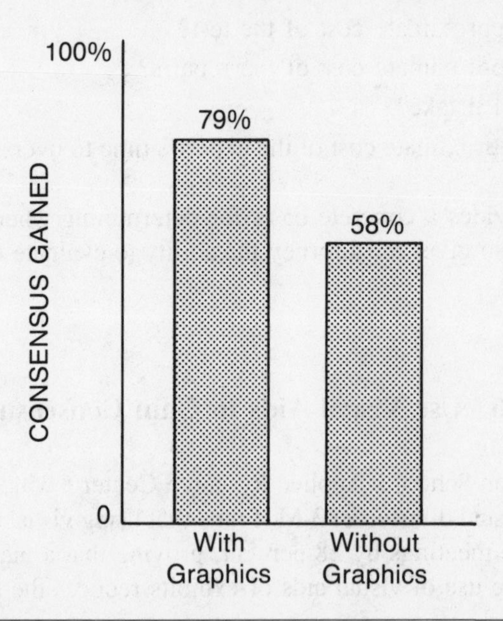

[7] Colley, *Friendly Persuasion: Gaining Attention, Comprehension, and Acceptance in Court,* Trial 42, 44 (Aug. 1981).

planation of the problem. It also gives the expert a chance to become comfortable explaining the facts using a visual aid.

If a construction project has many interrelated structures which are in dispute, consider having a scale model of the whole project built. Then, as the facts come together, the model can be revised to reflect the results of the investigation. Sometimes scale models are built to assist in obtaining building permits, to entice financial backers, or to convince planning commissions of the appropriateness of a project. If these models have already been constructed, they can be used to assist in explaining the conclusions reached about the cause and effect of the dispute.

Using a visual aid or a model also reduces the time needed to introduce new experts and attorneys to the facts of the case. They can see, and therefore understand more readily, what everyone is talking about. Use of models and exhibits during settlement conferences is an indication that the contractor's approach to the litigation is going to be thorough and well-prepared, thereby giving a him or her a psychological edge over the opponent.

Another kind of visual aid which can be used from the outset is a *critical path chart* showing the anticipated progress of the construction project over time. Again, these may have been used to plan and track the progress of the construction. If a critical path chart does not exist, one can be built after the fact by using the timeline of events identified in the summary of the litigation. Contrasting the anticipated timeline with the actual timeline, as in **Illustration 18-1**, can be very revealing and can assist in supporting contentions of damage as a result of acceleration or delay.

Illustration 18-1

CRITICAL PATH CHART

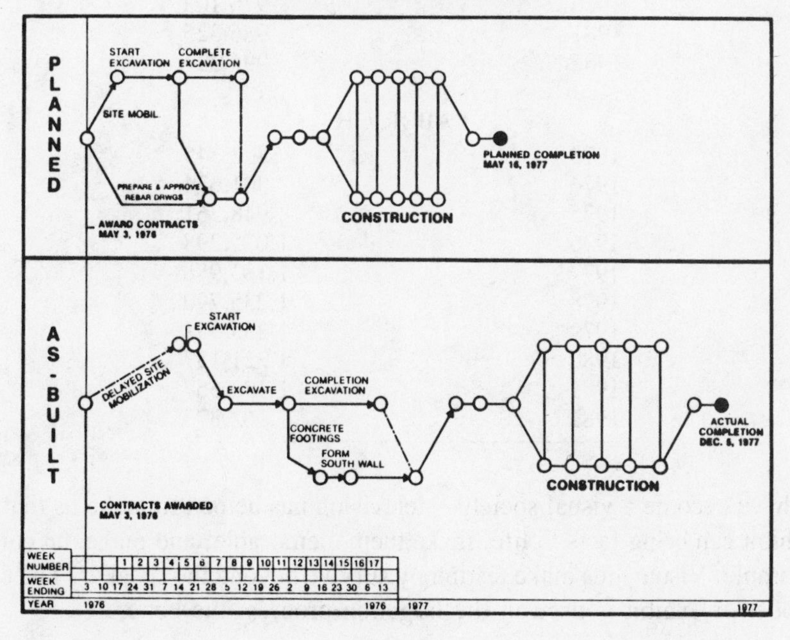

All numerical information is more easily understood if presented in a bar, line, or pie-chart format. People can read information presented this way five times faster than an attorney can talk. Look at the difference between reading the information as presented in **Illustration 18-2**, in tabular form, and as presented in **Illustration 18-3**, in line-chart form.

Illustration 18-2

LABOR COST ANALYSIS

OVERHEAD	
1973	$412,224
1974	421,443
1975	422,132
1976	445,785
1977	463,221
1978	470,000
1979	478,746
1980	495,111
1981	512,593
1982	519,842
LABOR	
1973	$910,592
1974	918,765
1975	927,669
1976	935,168
1977	953,398
1978	954,287
1979	964,078
1980	976,103
1981	986,325
1982	995,321
LABOR OH	
1973	$852,345
1974	901,674
1975	948,361
1976	1,075,398
1977	1,183,988
1978	1,235,790
1979	1,255,351
1980	1,221,157
1981	1,172,567
1982	1,162,873

We have become a visual society— television has helped to make us that way. An exhibit can bring facts to life, make them memorable, and make the complex seem simple. Visual aids make testimony survive beyond the speaker. The earlier the model or exhibit is used in the litigation process, the better.

Illustration 18-3

LABOR COST ANALYSIS

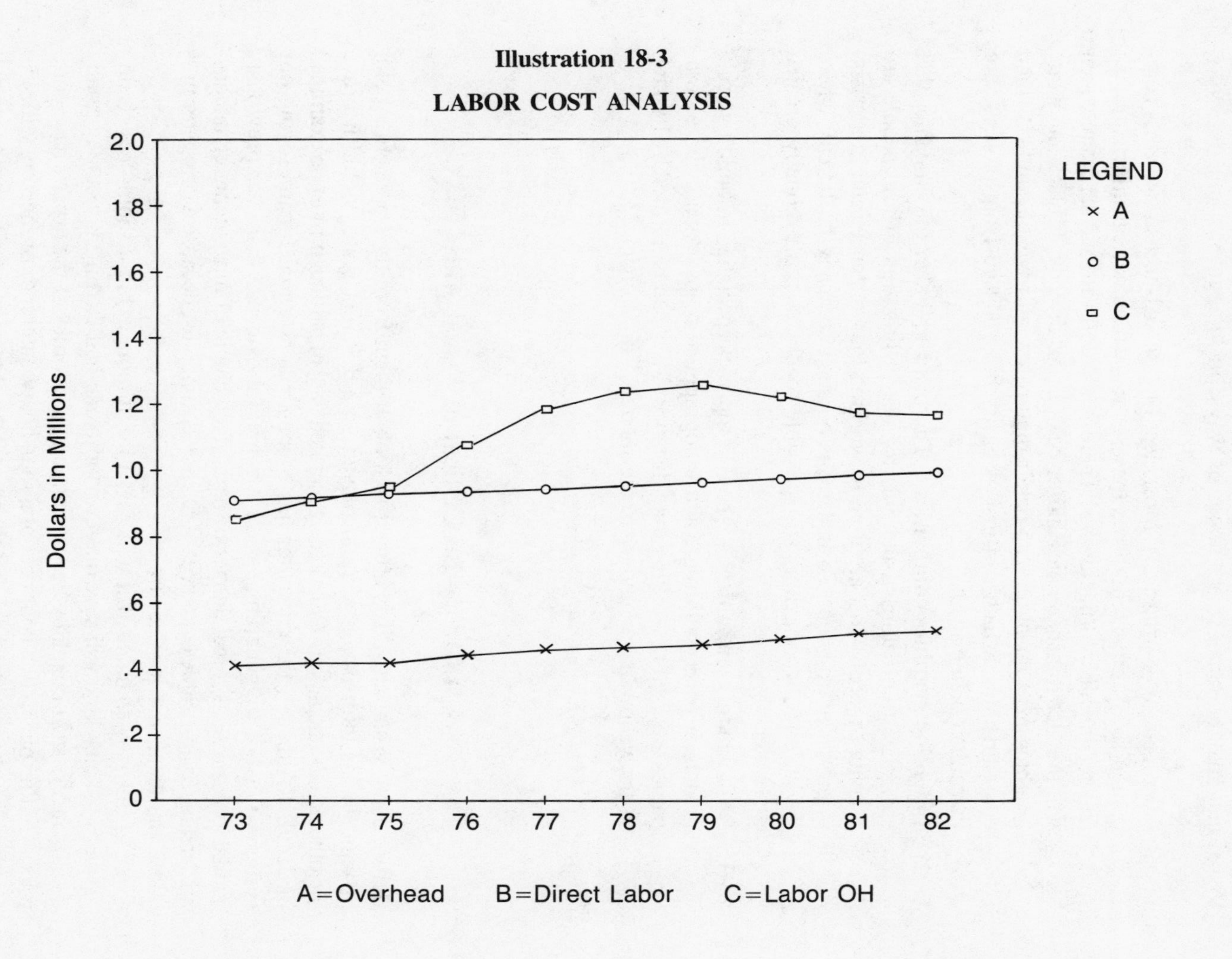

§ 18.17 —Creating Visual Aids

Visual aids may be made in a number of ways, including:

1. **Microcomputer programs**. Most electronic spreadsheet software has companion software which allows the user to feed numerical information in and get graphic information out. There is also software and hardware which will produce transparencies and/or color slides directly. If black and white graphics are a quantum leap better than rows and rows of numbers, then color graphics, which can just as easily be produced from plotters, are that much better again.
2. **Litigation graphics companies**. There are a number of companies that specialize in creating visual aids for use in litigation. These people are graphic experts. Most are used to working long distance, and in a crisis-oriented environment, so if there is no such company in the city where the attorney works, it is still possible to obtain professionally created graphics.
3. **Scale model companies**. These companies specialize in building scale models of parts, buildings, and topographical maps. Again, these companies are used to working long distance. They can be located by looking in the yellow pages under "scale models" or "models."

§ 18.18 —Be Critical of Visual Aids

Before using visual aids, it is crucial to look at them from the opponent's point of view. If the information were displayed on a different scale, would it prove the other side's allegation? Computer-generated data in particular must be examined in detail. "Although the computer has tremendous potential for improving our system of justice by generating more meaningful evidence than was previously available, it presents a real danger of being the vehicle of introducing erroneous, misleading or unreliable evidence."[8] Do not assume, just because a computer produced the data, that it is accurate.

All of this may sound as if it will cost a lot of money, but it should not. Using graphics of some kind will save one-quarter to one-half the time it would normally take to understand or explain the information the attorney or the experts have compiled. Cutting one to two hours out of a four-hour conference or meeting will save hundreds (perhaps thousands) of dollars, and the chance of gaining consensus will be increased by approximately 20 percent.

---

[8] T. Harper, *Computer Evidence is Coming*, 70 A.B.A. J. 82 (1984).

§ 18.19 Hiring Expert Witnesses

Once the facts are in, the attorney and his or her client need someone who can look the other side in the eye and translate those facts to a jury in a believable, concise manner; in other words, an expert witness. First, a word about timing. The best time to look for this paragon is prior to settlement conferences. The worst time is two weeks before trial.

It is hard to find people who are comfortable being put in a position where the other side's objective is to make them look like a fool. Americans fear public speaking more than sickness, heights, or even death. No wonder it is so difficult to find good expert witnesses.

Genius or brilliance is not enough in an expert witness. Dr. Albert Mehrabian did research[9] to determine what influences us to trust and believe a speaker. He found that:

1. Fifty-five percent of the message comes from what we see
2. Thirty-eight percent comes from what we hear
3. Seven percent comes from the actual words or data that are presented.

The first criterion, then, for selecting an expert witness is the way the person looks and acts, and the second is the person's ability to communicate information in an understandable fashion. Finally, one should determine the witness's technical competence. It is easy to forget this. The author recently sat with an attorney who was interviewing a prospective accounting witness for a construction lawsuit. When the interview was finished, the author turned to the lawyer and said, ''Well, what do you think?'' He replied, ''He has great credentials and I guess he must know what he's doing because I didn't understand a word he said!'' The best reason to choose a witness is because he or she is likeable, understandable, and believable, and last of all because that person has good credentials and is competent. It can be assumed that, if you have checked his or her references, the witness being interviewed is competent. The worst reason to choose a particular witness is because he or she is the best technical expert around.

A good technique for interviewing a prospective witness is to use videotape. This gives the attorney the ability to play back the interview and really concentrate on what impressions the person makes. The attorney can sit back and listen to the answers as if he or she were a juror. If there is any question about the expert, the attorney can ask others in the office to view the tape and give their opinions about the witness. A witness who cannot express his or her point of view well in an initial interview will not be good at trial.

Videotaped depositions are becoming more and more commonplace, and have added another element to what may be required from a witness. Not everyone who comes across well in person looks good on tape. Videotaping the interview

[9] A. Mehrabian, Silent Messages: Implicit Communication of Emotions and Attitudes (2d ed. 1981).

eliminates this element of risk because the attorney can immediately see how the prospective witness appears on tape.

Another testing method is to inform the expert that the attorney will be asking for a statement of credentials during the initial interview as if the expert were in court. Amazingly, some experienced experts will still stumble over their own credentials. Since this is the most straightforward aspect of the witness's testimony, and since first impressions are crucial, it would be better not to hire this person.

§ 18.20 Preparing the Expert to Be Convincing

A judge noted, "Witnesses decide trials, not lawyers; the impression *witnesses* make on the trier of fact is decisive."[10] Before trial it is helpful to talk to the witness about some of the *dos* and *don'ts* of testifying. That conversation might go something like this:

DO:	Listen to the questioner.	**DON'T**:	Interrupt.
DO:	Be brief and to the point.	**DON'T**:	Get caught on a long tangent.
DO:	Look at the questioner.	**DON'T**:	Show annoyance.
DO:	Use the question.	**DON'T**:	Be evasive.
DO:	Answer honestly.	**DON'T**:	Guess.
DO:	Be accommodating; be cooperative.	**DON'T**:	Volunteer information.
DO:	Pause with a tough question	**DON'T**:	Look away or look up or down at length.
DO:	Be dignified.	**DON'T**:	Forget to be human.
DO:	Know the meaning of all technical terms and acronyms.	**DON'T**:	Posture.
DO:	Be impartial.	**DON'T**:	Be overbearing.
DO:	Be objective.	**DON'T**:	Argue or get angry.
DO:	Be careful and be precise.	**DON'T**:	Talk down to the jurors.

By the time the trial is set, the attorney should be comfortable that the individuals chosen to testify will intuitively follow these guidelines. If, however, an expert does not seem to know when to stop talking or loses control when questioned too carefully, it is not too late to rethink that witness choice. Many an attorney has

[10] C.E. Goff, *Common Errors Made by Less-Experienced Trial Lawyers: Tips from the Other Side*, Brief/case (Nov. 1981).

regretted sticking with a witness, who loses his or her credibility on the stand because of his or her personality. It is worth the time and trouble to re-educate and train a new person on the issues. Expert witnesses can make or break a construction case.

A good expert witness is a performer, an actor who is articulate, believable, confident, courteous, and quick-witted. To ensure that experts will come across to the jury in this manner, it is necessary to carefully prepare witnesses so that all efforts to discredit their testimony or diminish the impact of their testimony on cross-examination will be foiled. This preparation may include the following:

1. A detailed review of all work on which the expert may rely. Some or all of this work may have been prepared by other experts or by the witness's staff. The witness must be so familiar with the work product that he or she will be able to explain it as if he or she had created it all
2. A discussion about the probable testimony of all other witnesses on both sides
3. A review of any articles written or testimony previously given by the witness
4. Several practice sessions at explaining exhibits that may be introduced, and any analogies that may be used
5. Organization of all documents which provide the basis for testimony (documents are critical to construction disputes, so take the time to get them organized)
6. An outline of the testimony to be given
7. An outline of the questions that are likely to be asked on cross-examination
8. Practice stating the conclusions reached, and practice explaining in a concise fashion any technical concepts. (As Thomas Jefferson wisely observed, "The most valuable of talents is never using two words when one will do")
9. An explanation of courtroom procedures, including, if possible, a visit to the courtroom
10. Practice saying two possibly crucial phrases: "*I don't know*" and "*It's my opinion*"
11. A frank discussion of any disadvantages in the witness's style
12. A discussion of what the appropriate dress should be. First impressions count—do not leave this to chance.

Finally, there is the chance to practice both direct and cross-examination. If it is at all possible, try videotaping this session so that the examination can be critiqued in minute detail. Videotaping quickly points out how clear the explanations really are. If, for some reason, videotaping is not possible, use a tape recorder or dictating machine to record the practice session. All of this takes time, but practice is the only assurance of prevailing at trial.

§ 18.21 Conclusion

Because construction litigation is extraordinarily complex, it is impossible to prepare a construction case without the assistance of experts. The skills of these individuals are diverse, and must be managed carefully in order to contain the cost of their time. The author's advice is, when in doubt, hire an expert, even if it is only for an hour or a day. There is little to lose and possibly a lot to gain. Experts should be hired early. The plaintiff should consider hiring them prior to filing the complaint to help evaluate the strength of the claim and to quantify damages. The author recently evaluated a claim that a subcontractor wanted to submit to the general contractor. The subcontractor thought the claim was worth $750,000; documentation was found to support a claim for twice that amount.

It is the attorney's responsibility to make sure that a minimum amount of time is spent educating the experts about the facts of the case, while at the same time giving them all the tools and freedom they need to do an effective job. Someone must take responsibility for managing the case. It is imperative that the attorney insures that the experts are kept informed about the progress of the case and that he or she is available to the experts in order to answer questions and review their work.

The use of visual aids in construction litigation can save time and money. They increase the chance of being understood and make the message more compelling. It is a good idea to hire an expert to help put the visual aids together. Graphs, charts, and models should be used in the initial stages of the dispute, and then modified to reflect any changes. The sooner they are prepared, the more the benefit which can be derived from them.

The outcome of construction litigation is highly dependent on how convincing the witnesses are. A significant amount of time should be spent preparing the witnesses so that testimony is delivered in a compelling fashion. Someone someday will probably do a study showing that the more time spent preparing an expert, the greater the chances of victory.

CHAPTER 19

TRIAL AND APPEAL

William G. Schaefer*

William G. Schaefer is a partner in the firm of Sidley & Austin in Washington, D.C. Mr. Schaefer has a litigation and general business practice that involves various aspects of construction litigation, including private and government contracts, antitrust, safety, and employment matters. Mr. Schaefer was formerly Vice President and General Counsel of DeKalb AgResearch Inc. He is a member of the bars of the District of Columbia, Maryland and Illinois.

* The author would like to express his appreciation to his colleague, Per A. Ramfjord, for his substantial assistance, and to his partners, George W. McBurney in Los Angeles and Thomas F. Ryan in Chicago.

INTRODUCTION

§ 19.1 Overview

The lawyer today faces unprecedented challenges in litigating construction disputes. The size and complexity of projects has increased, and there is increasing complexity in the relationships among the parties responsible for planning and construction. The lawyer responsible for the trial of construction disputes must therefore understand relationships that do not fit traditional patterns, and master and apply new legal theories to cases that involve increasingly complex fact situations and growing numbers of documents.

This chapter attempts to provide some guidance for the attorney and client who have some experience with construction law or litigation, but who are not experienced in developing the strategy and focus necessary for the effective trial of complex construction cases.

INITIAL ASSESSMENT

§ 19.2 General

The initial assessment of the case is among the most important steps in managing any complex construction dispute. When faced with litigation, every lawyer makes some kind of assessment, but habits of a routine practice or the pressures of a difficult one may result in the lawyer failing to make the kind of rigorous assessment that larger cases require. As a result, many lawyers lose control of complex cases and end up merely reacting, ineffectively and inefficiently, to the adversary's more focused initiatives or the deadlines imposed by the court.

To take the initiative and avoid this pattern of reaction, the lawyer and client should make a detailed assessment of the case as soon as possible. The assessment is most useful if it is set out in writing. In a complex case, if the lawyer does not prepare an assessment, the client should demand it. Where counsel is retained by an insurer to represent the insured, the assessment should be prepared for both the insurer and the insured.

The form of assessment will vary according to the nature of the dispute, but a few deceptively simple questions should be asked at the start of every case. The following questions give both lawyer and client a framework for efficient and effective trial preparation.

§ 19.3 What Facts Are Most Important to the Case?

Almost no construction case is as simple as it first appears. Consequently, there are substantial risks in approaching a trial, or even discovery or settlement negotiations, without having critically examined the available facts.

The first step in thoroughly examining the facts of any case is typically the same—an extended interview with the client. The lawyer should begin by having the client identify the parties and relate the facts with as little prompting as possible. At the same time, the lawyer should ask the client to include his or her opinions of the legal issues involved. These discussions will bring to the surface any misperceptions of fact the lawyer may have, and any legal misunderstandings the client may have.

The lawyer can then focus on the relevant facts in more detail and obtain from the client the names of other individuals who may be able to provide more complete information and later testify at trial, if necessary. Significant issues that deserve consideration at this point include relevant contract provisions and their background, applicable professional standards, and identification of all potential parties. Weaker points in the case must also be discussed, and the lawyer should obtain the names of individuals who might provide further information concerning them. Finally, the lawyer should ask the client what documents might be relevant to the claim. If the client does not already have a reliable system for organizing and preserving key documents, it will be important to establish a proper system immediately.

It is often useful, after the first interview, for the client to make a detailed written account of the facts. This written account will ensure that important facts are not lost as the client's memory fades over the extended period it may take to resolve the dispute. Concentrating more closely on relevant events and reviewing backup documents in more detail may also help the client to remember significant but otherwise forgotten details. The client's written account, and all other writings prepared in anticipation of trial, should be marked to indicate that they are privileged and confidential.

Soon after interviewing the client, counsel should visit the job site. By making this visit early—as close to the time the dispute arose as possible—the lawyer gets a feel for the job that will provide an invaluable base for developing the client's case and conveying an accurate picture to the judge and jury. The lawyer's site visit may also be a good time to arrange for photographs of certain key portions of the job site for later use in trial. In addition, the lawyer may take advantage of the visit to conduct preliminary interviews of the client's key employees. Beginning these interviews early insures that valuable testimony is not lost, and contributes substantially to the lawyer's understanding of the case.

§ 19.4 What Legal Arguments Support Each Side's Claim?

Close scrutiny of relevant contract provisions is usually the first step in determining the parties' respective legal rights and duties. Most lawyers with construction law experience will have little difficulty in assessing the strengths and weaknesses of the various contract claims presented. Relevant legal doctrines are not always flagged by the terms of the contract, however, and the lawyer must also consider other possible sources of liability,[1] including:

1. Whether any parties may be held liable for negligence
2. Whether any parties may be held liable under strict liability tort theories
3. Whether any parties may be held liable under third-party beneficiary theories
4. Whether any express warranties were made
5. Whether the court might apply any implied warranties, such as the judicially created warranty of good workmanship, or the U.C.C. warranty of merchantability or fitness

[1] For more detailed information on noncontractual theories of liability, *see* Grandoff, *Step Three: Consider Theories of Legal Liability Apart From The Contract,* in Seven Steps to Resolving Construction Contract Claims 3-1 (Section of Litigation of the American Bar Association 1983).

6. Whether there has been misrepresentation or fraud
7. Whether a theory of tortious interference with contractual duties might apply
8. Whether the owner is liable for improperly accelerating the contract or changing the specifications
9. Whether violations of statutes may have created liability
10. Whether any insurers or sureties might be held liable.

In identifying applicable legal theories, it is also important to consider how various potential claims conflict with one another. In this regard, it is necessary to focus on whether a possible successful legal position might jeopardize more important client interests. For example, in representing a contractor, it may be possible to win a dispute with a subcontractor by proving that the subcontractor acted negligently, only to have the owner use that very evidence to obtain a larger judgment against the contractor under a negligent entrustment theory.

§ 19.5 What Are the Equities of Each Side's Case?

In complex construction cases, the lawyer must go beyond articulating technical legal principles and strive to portray the case in terms of equitable considerations that will demonstrate why the client should prevail. If these considerations are developed early, they can provide strong leverage during settlement negotiations, facilitating speedy and economical resolution of the case. Moreover, even if settlement is not possible, a refined view of the equities can be used profitably to influence the judge during pretrial proceedings. Finally, during trial, the equities become critical, and in a complex case they may provide the most important basis for the judge or jury's decision. The effective lawyer will capitalize on this potential, using the equities as recurring themes and supporting them with evidence throughout the trial.

With equities, as with all issues in litigation, it is important that the lawyer examine both sides of the case. The client's position cannot be developed and presented effectively unless counsel sees clearly the equitable appeal of the opposing case, and works from the outset to deflect or minimize that appeal.

§ 19.6 What Are the Client's Objectives?

Construction trials typically involve disputes over large amounts of money, and maximizing or minimizing the amount paid is the most obvious objective. Nonetheless, monetary objectives are not the only ones to be identified. For one client, the most important goal may be avoiding adverse publicity. Another may be concerned about harmful precedent or even collateral estoppel. The insurer, for example, may be less concerned about the amount of the award than about litigation expenses, or how the outcome of a case will affect other contractors it insures.

Counsel should also remember that not all litigants, even in major cases, are intractable enemies. Owners, for example, may have financial or political obligations that require them to pursue a claim. In such a context, the primary goal of a contractor-defendant may be preserving its goodwill with the owner or its general reputation for competence and fair dealing. The lawyer in such a case might determine that an informal *mini-trial* or other alternative form of dispute resolution would be more likely than litigation to produce a desirable result.[2]

§ 19.7 How Should Damages Be Measured?

Ascertaining the amount of damages in construction cases is complicated by the difficulties of selecting the damage theories that best fit the client's objectives, and by the frequent necessity of sifting through thousands of pages of accounting documents to support the amounts claimed. For these reasons, it is essential that the lawyer promptly identify the appropriate damage theories and establish an organized method for calculating the amount sought. Such early organization will guarantee that the best possible damage evidence is available for trial. If the client wishes to pursue settlement, solid damage figures established early in the case will also act as a strong catalyst for successful negotiations.

There are a number of traditional damage theories in construction law, including the basic theories of expectancy and quantum meruit. The lawyer should begin his or her evaluation by analyzing the traditional theories against the facts. The creative lawyer, however, will then determine whether any newer theories might permit recovery of such items as indirect expenses, consequential damages or economic losses, litigation expenses, or even punitive damages.[3]

After identifying applicable damage theories, the attorney must decide upon a means for calculating the amounts due. In some instances, it will be possible to establish an accounting system that permits maintenance of separate records for dispute-related costs. If such a *damage account* can be set up immediately after the damages begin to accrue, it will provide a convenient means of proving damages at trial.

In many instances, however, establishing separate contemporaneous accounting is not practical; either the dispute is not recognized early enough to establish

---

[2] See § **19.10**.

[3] Construction damage theories are treated more thoroughly in **ch. 14**. *See also* Martell & Brenner, *Engineering Trends in Construction Damages: Direct, Indirect, Consequential and Punitive, and ''No Damage for Delay'' Clauses and Legislative History,* in Emerging Trends in Construction Law (Presentation of the American Bar Association, Forum Committee on the Construction Industry 1984); Bruner, *Construction Damages,* in Construction Litigation 251 (New York Law Journal Seminars-Press 1979); Currie, *Winning Strategies in Construction Negotiations, Arbitration and Litigation,* in Construction Contracts 1984, at 737, 819-25 (J. Pierce ed. 1984).

an account that completely measures damages, or the damages involve a complex variety of expenses that cannot be accurately or conveniently grouped into appropriate accounts. In larger cases involving such difficulties, it will often be necessary to consult accounting experts who can do the auditing work necessary to reach a reasonable damage figure. Obtaining such expertise early may be extremely beneficial. First, the expert can identify the documents that must be preserved or discovered to measure damages. Second, the expert may provide an edge in establishing solid damage figures for use in settlement or alternative dispute resolution.

§ 19.8 What Are the Potential Conflicts of Interest?

Unfortunately for all concerned, conflicts of interest are increasingly present in construction litigation. For example, as more projects use multiple contractors, the lawyer is more frequently faced with the potential conflict of defending several contractors who may be jointly and severally liable.[4] Recently developed *wrap-up* insurance policies entail similar conflicts, as insurers under such plans are bound to represent all defendants on a given project claim.[5] Finally, particularly because of legal theories that allow contractor suits directly against architects, conflicts may pose serious problems for attorneys who represent both contractors and architects.[6]

Because these conflicts are increasingly common and not always obvious, the lawyer should make a special effort to evaluate potential conflicts in the initial assessment. Moreover, this effort should be sustained throughout the trial, because conflicts are likely to become more apparent as the lawyer's understanding of the case develops.

---

[4] The lawyer representing more than one defendant may not have adequate incentive to press the defenses of the defendants against each other. The problem is exacerbated in states where joint tortfeasors have a right of contribution against one another. In such circumstances the potential conflict must be fully disclosed to the clients. *See* Aetna Cas. & Sur. Co. v. United States, 570 F.2d 1197, 1200-02 (4th Cir.), *cert. denied,* 439 U.S. 821 (1978); Kerry Coal Co. v. United Mine Workers, 470 F. Supp. 1032 (W.D. Pa. 1979); A.B.A Code of Professional Responsibility DR 5-105(B) (1980).

[5] In this connection, however, it should also be noted that the comprehensive general liability insurance policy usually contains complex coverage exemptions which increase the chances that insured and insurer will disagree over the actual coverage of the policy. Where these conflicts are not resolved, the insurance company is increasingly required to pay for separate counsel chosen by the insured to conduct the defense. United States Fidelity & Guar. Co. v. Louis A. Roser Co., 585 F.2d 932 (8th Cir. 1978); Storm Drilling Co. v. Atlantic Richfield Corp., 386 F. Supp. 830 (E.D. Cal. 1974).

[6] As the traditional roles of architect and contractor have changed, the law has shifted an increasing share of liability onto architects. Most prominent has been the courts' abandoning the privity requirement for suits by contractors against architects. Note, *Architectural Malpractice: A Contract-Based Approach,* 92 Harv. L. Rev. 1075, 1081-83 (1979).

§ 19.9 How Can Trial Preparation Be Planned?

Because modern construction cases typically involve complex facts and large numbers of documents, it is essential that the lawyer develop a plan at the outset to focus the necessary trial preparation. A preliminary plan should therefore be included in the written assessment and discussed with the client.

A successful plan is an evolving document that centers upon obtaining the evidence necessary to establish or defend against the basic elements of the contractual breach, tort, or statutory violation. Yet preparation should not be focused so narrowly that counsel will not discover valuable evidence beyond the basic facts. In particular, the context or background of a dispute may provide valuable information and evidence. For example, if the claim is for a subcontractor's delay due to lack of diligence, the subcontractor's general incompetence or inexperience may be as important to the case as the fact of its absence from the job at key times.

Planning the case also requires putting aside claims that are not valuable to the client's case. For example, as noted earlier, a contractor may decide against pursuing an attractive negligence claim against a subcontractor, because it might leave the contractor itself liable to the owner. If this possibility is not flagged early in the case, valuable time and money will be wasted. In this regard, the lawyer should also make an assessment of which parties are best suited to compensating the injured party. Even the best case or claim is useless against a party who cannot pay a judgment. Therefore, although keeping litigation options open may require compiling a laundry list of allegations or defenses against every potential party, counsel and the client must both be able to keep their focus on those allegations and defenses most likely to produce a successful prosecution or defense.

Focused preparation also avoids overpreparation, which few clients can afford, or underpreparation, which no client should tolerate. If, for example, eventual trial of the case requires several experts on different issues, counsel who has assessed and focused the case may be able to involve one key expert at the outset and leave the others to a later and more cursory role. The client who understands the case may itself be able to do much of the early work. In some instances, where the case is resolved through settlement, early involvement of the client or a single expert may be all that is necessary.

The traditional trial notebook is particularly useful in focusing trial preparation in complex cases. Its initial preparation and continual development by the trial lawyer imposes an invaluable discipline. The notebook can be divided into sections reflecting each major aspect of the case. As additional information is acquired, and as the trial strategy grows more refined, the notebook is updated. Ultimately, it becomes a comprehensive guidebook for the trial, with all the key exhibits and anticipated questions for each witness.

§ 19.10 Should Alternatives to Litigation Be Pursued?

In addition to the situations where arbitration is specified in the contract, or where the dispute requires special government administrative procedures, instances will arise where a traditional trial is not the most appropriate means of dispute resolution. In some cases, clients simply cannot afford the money and management time required for a full trial. In other cases, the client may be too concerned about preserving a valuable business relationship to risk the pressures created by full discovery and trial. Finally, the client may be involved in other related disputes that may affect trial of the current dispute through collateral estoppel. In these, and many other situations, both lawyer and client should give special attention to the potential for settling the dispute or deciding it by some alternative form of dispute resolution.

Settlement, in particular, should always be considered. There are numerous times during the course of a case when settlement may be profitably discussed, and trial counsel should identify those times for the client as part of the initial assessment. In many instances, the optimal time will be before the complaint has been filed. At this early stage, the parties' views have not yet been polarized by confrontation, and the most flexible agreements can be reached. There is also a strong motivation to settle at this early stage in order to avoid the expense and time associated with filing a claim and conducting discovery. By settling early, however, the parties run the risk of an inequitable agreement based on insufficient knowledge of each other's claims.

In some cases, a settlement may best be obtained after suit has commenced and some portion of discovery is completed. At this point, the parties will have a better developed knowledge of the case, but may want to settle to avoid the burdens or embarrassment of further discovery. In still other cases, the parties may settle just before or during trial—either with or without the judge's assistance. Pursuing or deflecting such settlement possibilities is an important part of the trial of complex construction cases.

When parties who want to settle are unable to reach agreement without assistance, they may consider using an alternative means of dispute resolution. Often parties who want to settle are unable to focus on central issues or reconcile radically different views of the facts. In such cases, the presence of a mediator, arbitrator, or some figure combining the roles of both mediator and arbitrator may be the best way to reach an accord.[7] The procedures involving mediation or arbitration can be tailored to the circumstances to maximize the potential for agreement.

---

[7] For further information on mediation and noncontractual arbitration of construction disputes, *see* Howell, *Putting Arbitration in Perspective,* 5 Construction Litigation Rep. 314 (1984); Olsen, *Dispute Resolution: An Alternative for Large Case Litigation,* 6 Litigation 22 (1980).

Some creative lawyers have also recently begun settling disputes through procedures that do not fit the traditional mediation-arbitration models. The *mini-trial* is one example. In the mini-trial, the lawyers make abbreviated presentations of the case to a small group of key representatives of the parties, either alone or in the presence of a neutral advisor. The businessmen then meet and try to hammer out an agreement. It they are unsuccessful, the neutral advisor can draft a recommended agreement which is submitted to the businessmen as a catalyst for further negotiations.[8]

Where alternative dispute resolution is pursued, the role of trial counsel will change, but will not be diminished. Counsel's careful assessment, thorough investigation, and concentrated effort to focus the case and develop the equities remain key to a successful presentation.

TRIAL PREPARATION

§ 19.11 What Constitutes Trial Preparation?

There is no formal demarcation between the initial assessment and trial preparation. Generally, however, it is useful to consider trial preparation as encompassing primarily those activities that are undertaken with a specific view to what will happen in the courtroom. As with the initial assessment, trial preparation will vary with the dispute, but answering a few basic questions can be useful.

§ 19.12 What Is the Contractor's Case?

The first principle of effective trial preparation was articulated effectively by a project manager who was meeting with his company's attorneys prior to a major trial. Exhibits were reviewed, expected testimony of expert and lay witnesses was discussed in detail, and the prospects for success were evaluated. The project manager's response was simple and straightforward: ''What is our case?'' He understood the evidence and the anticipated legal issues, but he also saw the fatal flaw in counsel's preparation: the evidence would not present a clear story that would engage and convince the judge and the jury.

To be effective, the trial lawyer must distill the complex facts and intricate legal arguments of a large construction case into a focused and compelling presentation. This process demands simplifying the case into the critical facts that clearly

---

[8] Additional information on the mini-trial may be found in Olsen, *Dispute Resolution: An Alternative for Large Case Litigation,* 6 Litigation 22 (1980); *see also,* Green, *Growth of the Mini-Trial,* 9 Litigation 12 (1982); Franklin, *The Mini-Trial: What it is and What it isn't and What it Can Do,* Construction Law. (Fall, 1983) at 1.

demonstrate why the client should prevail. If simple and clear enough, the lawyer's presentation and the trial evidence will convey a picture or story that allows the judge or jury to decide the case confidently, even without full knowledge or understanding of all the underlying complexities.

§ 19.13 Should the Case Be Tried to a Judge or a Jury?

Much attention has been given to the difficulties juries face in deciding complex cases.[9] The length of the trial, the number of parties, the number of claims and cross-claims, witnesses, and documents, combined with unfamiliar factual and legal settings, may make a complex case incomprehensible to lay juries. As a result, at least one court has held that the Fifth Amendment due process requirement supersedes the Seventh Amendment right to a jury trial in highly complex cases.[10]

Certainly, much of the concern over jury inability to handle complex cases is justified. Accordingly, there will be some instances where the lawyer should try to reach an agreement with opposing counsel to permit a bench trial. This is especially true where the lawyer knows the judge is competent and conscientious. Prior to making a final decision to seek a bench trial, however, the lawyer should consider a few of the less obvious drawbacks of such trials. First, experienced lawyers have noted that a judge sitting without a jury may be willing to apply a more expanded concept of liability than he or she would incorporate in a jury charge, thinking that he or she can later limit damages to an amount that seems appropriate. Second, judges, who must hear evidence to pass on its admissibility, may be influenced by material they would not allow to go to a jury.[11] They are also generally more willing to admit arguably inadmissible evidence.[12] Finally, judges may organize a jury trial better to avoid undue protraction.[13]

---

[9] *See, e.g.*, Redish, *Seventh Amendment Right to a Jury Trial, A Study in the Irrationality of Rational Decision Making,* 70 NW. L. Rev. 486 (1975); Note, *Unfit for Jury Determination, Complex Civil Litigation and the Seventh Amendment Right of Trial by Jury,* 20 B.C.L. Rev. 511 (1979); Note, *The Right to a Jury Trial in Complex Civil Litigation,* 92 Harv. L. Rev. 898 (1979).

[10] *In re* Japanese Elecs. Prods. Antitrust Litig., 631 F.2d 1069 (3d Cir. 1980).

[11] Lempert, *Civil Juries and Complex Cases: Let's Not Rush to Judgement,* 80 Mich. L. Rev. 68, 94 (1981).

[12] Appellate courts are less likely to reverse a trial judge for admission of incompetent evidence than for exclusion of proper evidence. C. McCormick, Evidence § 60, at 153 (3d ed. 1984).

[13] [A] jury trial creates an urgency about deadlines that may be absent when a judge sits alone, and there is strong pressure to minimize interruptions in the presentation of evidence to a jury. While an efficient and energetic judge might be able to accomplish more in less time if unconstrained by a jury, judges who are not outstanding in these respects may benefit from the discipline that a jury imposes. It is, for example, unlikely that either the

If the lawyer is unable to obtain a jury waiver, or decides that a jury would be a better arbiter of the client's case than a particular judge, there remain a variety of techniques that may be used to assist the jury:

1. Sever the trial into smaller, more manageable sets of issues. Even the division of liability and damages may result in substantially clearer cases for the jury to decide
2. Ask the judge to use special verdicts to simplify issues
3. Use pretrial conferences to limit the testimonial and documentary evidence that will be admitted at trial
4. Ask the judge to provide the jury with preliminary jury instructions at the start of the trial.[14] The jury can use these instructions as a framework to focus upon and remember the most relevant evidence
5. Guarantee that the jury instructions given at either the beginning or the end of the trial are in plain language, understandable to the jury. Too often, instructions are copied from form books and are plagued with arcane language unintelligible to the jury[15]
6. Use summaries, charts, or calculations that reduce voluminous evidence to an understandable form[16]
7. Ask the judge to allow the jurors to take notes, or to review the trial transcripts periodically.[17] This increases both the jurors' sense of participation in the trial and their ability to remember key facts
8. Ask the judge to allow the jury to question the witnesses. The questions may be in written form and submitted to the judge, who may decide which questions do not call for inadmissible evidence.

> counsel or the judge in the *IBM* case . . . would have allowed the matter to proceed as it has if trial had been to a jury.

Lempert, *Civil Juries and Complex Cases: Let's Not Rush to Judgment,* 80 Mich. L. Rev. 68, 93 (1981).

[14] A number of cases have upheld the use of preliminary jury instructions. *See, e.g.,* Stonehocker v. General Motors Corp., 587 F.2d 151, 157 (4th Cir. 1978); Cavendish v. Sunoco Serv. of Greenfield, Inc., 451 F.2d 1360, 1364-66 (7th Cir. 1971).

[15] Research has shown that traditional jury instructions are difficult for jurors to comprehend and can be rephrased to increase their clarity. A. Elwork, B. Sales & J. Alfini, Making Jury Instructions Understandable (1982).

[16] Fed. R. Evid. 1006 provides in part: "The contents of voluminous writings, recordings, or photographs which cannot conveniently be examined in court may be presented in the form of a chart, summary or calculation."

[17] An increasing number of courts allow jurors to take notes in appropriate cases. *See, e.g.,* United States v. Braverman, 522 F.2d 218, 224 (7th Cir.), *cert. denied,* 423 U.S. 985 (1975)(jurors allowed to take notes while jury instructions were being played back to them from a tape recording); SCM Corp. v. Xerox Corp., 77 F.R.D. 10, 15 (D. Conn. 1977) (jurors allowed to take notes during trial); *see also* United States v. Marquez, 449 F.2d 89 (2d Cir. 1971), *cert. denied,* 405 U.S. 963 (1972).

Most judges would not permit use of all these techniques, but even if only a few are used, the jury is likely to comprehend the case more clearly and reach a more rational verdict.

§ 19.14 How Can Documents Be Managed Effectively?

The complexity of construction litigation makes it imperative for the attorney to use pretrial discovery tools for systematic investigation and development of trial evidence on key issues. Commercial arbitration and administrative proceedings demand similar investigation, although the formal procedures differ. Whatever the dispute or forum, however, many of the relevant facts will be set out or reflected in documents and, as a result, much of the most important evidence may be in document form.

Comprehensive review of the multitude of documents involved in most large construction cases requires thorough planning and organization. Initially, a list of the various types of documents available should be prepared. Typical construction projects involve the following types of documents:

1. Contracts between the respective parties, including the owner, the contractor, and the architect, and various subcontractors
2. Insurance policies
3. Specifications and drawings, including addenda and engineering change notices
4. Correspondence and internal memoranda
5. Bid estimates of the contractor, subcontractors and, possibly, budget estimates by the designer
6. Daily work reports and payroll records
7. Progress payment requests and related payment orders
8. Change orders and related documentation
9. Progress schedules including, in some cases, construction project management charts, bar charts, and updates to such charts
10. Equipment use logs
11. Job meeting notes
12. Private or governmental inspection reports
13. Accounting records.

Prior to submitting a document request, the lawyer must narrow this list to the key documents necessary for supporting the client's case. In most instances, these will be the documents specifying what standards the parties were obligated to meet, how the parties actually performed, and what damages resulted from any breaches.

Identifying these key documents should progress naturally from the work done in preparing the initial assessment. Consultation with the client and experts may also be required, however, and in some cases the lawyer may even serve a preliminary set of interrogatories to obtain the information necessary to narrow discovery appropriately.

Before actually reviewing any significant number of documents, it is also critical to develop a plan for managing the documents and insuring that they remain accessible before and during trial. For the very large case, a computer data base may be the only practical way to achieve the necessary organization. A data base is often used for access to documents through the recording of basic data (subject, date, author, address, etc.), though including a summary of the text of important documents is also possible and may be useful. Numerous professional services are available to provide the necessary classification, software or hardware systems, and the manpower to run them. In some instances, clients themselves may have the required capability. Finally, there are a growing number of document organization software programs that can be used economically in conjunction with personal computers already owned by the lawyer's firm.

Whether it is desirable to create a computer data base of document information or only a careful index, a few basic principles should be applied to any document management plan:

1. *Organization.* No one system will fit all or even most cases. In some cases, documents are best coded or filed by site, in others by chronology, in others by subject matter. The paramount purpose of organization is attorney access during discovery, trial preparation, and trial.
2. *Numbering.* Individual documents or groups of documents should be numbered for future reference. A document left unnumbered may be lost forever. Numbering codes can be developed to help manage and identify the source and original file location of documents.
3. *Summaries.* Key documents often must be summarized for ease of reference, but summaries prepared by incompetent personnel are useless.
4. *Lawyer control.* No document organization scheme is a subsitute for personal review of documents by trial counsel. Similarly, no document organization scheme is appropriate unless trial counsel understands it and feels completely comfortable with it.
5. *Use at trial.* The mechanics of rapid document retrieval must be both planned and tested prior to trial. Locating documents, especially for effective cross-examination, can be a nightmare for the unprepared.

While documents are being reviewed, the lawyer should also be utilizing interrogatories and depositions to gain further information in preparation for trial.[18]

---

[18] For a very thorough treatment of all discovery problems, *see* M. Callahan & B. Bramble, Discovery in Construction Litigation (1983).

§ 19.15 How Should Witnesses for Trial Be Selected?

The lawyer preparing for trial should use both interviews and depositions to locate and test potential witnesses. During the interview or deposition, counsel has the unique opportunity of probing the limits of the witness's knowledge, veracity, and effectiveness on the stand, without the risks inherent in courtroom testimony. Fully capitalizing on this opportunity, however, requires careful planning.

Counsel must begin by identifying every potential witness likely to be called by any party; ideally, each one should be either interviewed or deposed prior to trial. All documents relating to each witness's potential testimony must be thoroughly reviewed, both to insure that counsel obtains any possibly relevant information the potential witness can provide, and so that counsel has the background necessary to evaluate and test the potential witness's knowledge and veracity. Finally, it is advisable to mix formal and informal approaches in both interviews and depositions. Interviews should be planned as methodically as depositions, and where possible, deposition witnesses should be engaged with the same informal tactics used in interviews.

After depositions and interviews, the universe of potential witnesses will generally be clear, but actual selection of witnesses to testify at trial may require additional consideration. Final selection is commonly based on such traditional factors as demeanor, knowledge, and vulnerability to cross-examination. Yet, in large construction cases, it is also important to select witnesses capable of explaining technically complex matters in plain and clear language. Lower-level workers directly involved in the project are frequently better witnesses in this respect than more articulate management personnel who are too distant from the job to be able to explain the facts in detail. In other cases, however, the reverse may be true.

§ 19.16 When and How Should Expert Witnesses Be Used?

Most complex construction cases require both expert assistance and expert testimony.[19] There are a variety of contexts in which such expertise may be necessary. For example, a challenge to the adequacy of job specifications may require engineering and architectural experts to evaluate the sufficiency of the design, a metallurgist to testify to the adequacy of the specified steel supports, and geologists or hydrologists to attest to whether the specifications reasonably accounted for job-site soil conditions. Usually, it is wise to begin searching for the proper experts early. Particularly in complex cases, the expert will need time to develop an opinion and prepare for trial. The expert is also likely to be of assistance in discovery and initial investigation of the case. Finally, the expert who is contacted

---

[19] See generally **ch. 14** regarding the use of experts.

early will be able to develop preliminary conclusions useful during any settlement negotiations.

Before interviewing and selecting experts, counsel must understand the dispute. Gaining this understanding should be part of the initial assessment, and requires assistance from the client. It may also be necessary to consult secondary sources that provide basic information on the construction technology involved in the dispute. From this, counsel can judge what type of expert and what level of expertise will be necessary to convince the trier of fact.

In some instances, only a minimally qualified witness will be necessary, and it may be possible to use someone from the client's own in-house staff. For example, a contractor's field inspector may have sufficient expertise to identify errors in certain engineering work, and a highly-trained, highly-paid engineering consultant may not be required. If the errors are neither highly complex nor central to proving liability, the field inspector may be the best witness. The lawyer should exercise caution, however, to avoid in-house experts who have anything in their background that might detract from their stature or necessary objectivity.

Where greater expertise is required, the lawyer should consider a number of factors in selecting a suitable expert. If the expert seems generally suitable and does not hold views adverse to the client's position, counsel should begin by careful scrutiny of the expert's credentials. Obviously, the expert should be formally trained in the substantive area on which he or she will testify, but it is not always best to select the expert with the best "paper" credentials. For example, training at a school well-known in the forum state may impress judge or jury more than a degree from a more prestigious university. Likewise, an expert with impeccable academic credentials may have little practical experience to back them up.

In addition to educational background, counsel should consider which professional societies or organizations the expert belongs to. Membership in such organizations is not merely another credential with which to impress the jury; it is an indication to the lawyer that the expert is likely to be well versed in recent developments in his or her field. Information on changes in construction-related technology is disseminated primarily through these organizations, and in some cases it is particularly important that the expert be familiar with the most current developments and practices. In addition, membership in professional organizations may convey to the judge or jury that the expert is well-recognized in his or her field.

The expert's experience in analyzing related situations is also vital. The lawyer should seek experts who have worked on the same type and size of projects as that involved in the dispute, both to insure complete understanding of the potential complexities of the case and to show the judge or jury that the expert's opinion is well grounded in experience.

Finally, the expert must be articulate. In particular, the expert must be able to avoid the technical jargon of his or her field and testify in plain and simple terms that are understandable to the jury. Professors are frequently the best experts in this respect. Years of simplifying complexities for their students, combined with frequent public speaking, often make them excellent communicators. It is important, however, that a professor have practical, as well as academic, experience. In addition, a professor must be capable of handling the type of hostile and domineering cross-examination common in the courtroom but not in the classroom.

There are a number of ways to locate an appropriate expert. Counsel should begin by consulting with the client, and then talking to other lawyers who have handled similar disputes requiring similarly qualified experts. Other lawyers can be a particularly valuable source of information, because they can provide a candid evaluation of the expert's courtroom performance in addition to his or her basic knowledge.[20]

Once selected, the expert must be prepared for deposition or trial testimony. In particular, counsel must insure that the expert's opinion is grounded on a thorough examination of all available data and is solid enough to withstand vigorous cross-examination. The lawyer should also familiarize the expert with the rules of evidence, both with regard to testimony and in relation to authentication of any exhibits the expert has prepared. Finally, to insure that the expert is fully prepared, the lawyer should conduct thorough mock direct and cross-examination.

MOTIONS AND FILINGS BEFORE TRIAL

§ 19.17 Pretrial Motions

Lawyers often fail to pay sufficient attention to pretrial motions, viewing them as irritating encumbrances. In fact, carefully prepared pretrial motions may be vehicles for substantial strategic advantage.

Motions to dismiss or strike can be an important trial tool in the construction case. Issues of indemnity, punitive damages, or even general liability are among the matters that often can be resolved before trial, frequently even before discovery and other trial preparation. For example, an owner may be able to strike a contractor's delay claims based upon a contract containing a no damages for delay clause.[21] Likewise, any claims brought after a relevant statute of limitations has run may be dismissed peremptorily. The lawyer must exercise some caution in filing such early motions, however, as they may precipitate adverse opinions before the judge fully understands the case.

Motions for summary judgment can also be an effective method for disposing of a case, narrowing the issues, or educating the judge. Many cases can be fully presented on uncontested facts or simply do not have enough substance to justify

---

[20] A useful list of experts recommended by lawyers has been published by the ABA Section of Litigation: Register of Expert Witnesses in the Construction Industry (1984).

[21] *See* Unicon Mgmt. Corp. v. City of Chicago, 404 F.2d 627 (7th Cir. 1968); Peter Kiewit Sons' Co. v. Iowa S. Utils. Co., 355 F. Supp. 376 (S.D. Iowa 1973).

trial. Judges also recognize that summary disposition of appropriate cases or issues reduces court backlogs and permits others to have their timely day in court. Finally, even if the court denies summary judgment, it may grant partial judgment or at least give the parties guidance as to what issues it believes remain for trial.

Requests for admissions can be used to lay the groundwork for pretrial motions by narrowing the legal issues to be tried. They are also a powerful tool for simplifying presentation of a case by limiting the amount of evidence required at trial. Finally, they can facilitate admission of documentary evidence by allowing counsel to circumvent tedious authentication procedures.

All of these advantages would be significant in normal litigation, but they are especially important in complex construction cases where simplification is at a premium. Yet requests for admissions followed by thorough summary judgment motions are often not used, perhaps because counsel have come to expect routine denial of admission requests arising from a general reluctance to resolve disputed issues prior to completing discovery. Unfortunately, such reluctance may also reflect counsel's failure to develop the clear focus that is necessary to determine what admissions would be significant, and to prepare requests that cannot easily be evaded.

§ 19.18 Pretrial Conferences

Formal pretrial procedures are provided in the rules of virtually every court, government board, or arbitration body. Each judge or presiding officer usually has an additional set of formal and informal procedures for bringing a case to trial.

The most important aspect of pretrial procedures is the pretrial conferences. Although such conferences vary widely in form and substance, most judges use initial and final conferences to attempt to identify and narrow the issues for discovery and trial. In these conferences, aggressive and prepared counsel has available formal and informal procedures that may be used to significant advantage. In addition, counsel has a unique opportunity to educate the judge in a setting less formal than argument on motions and far less restrained than the trial courtroom.

§ 19.19 Pretrial Filings

As the final pretrial conference approaches, trial counsel should give particular attention to the following major filings: pretrial or trial briefs, proposed findings, jury instructions, motions to exclude or limit evidence, and lists of witnesses and exhibits.

The pretrial or trial brief requires the most sophisticated advocacy. Its form may be dictated by local rule or practice, but, whatever the form, it is probably counsel's most comprehensive statement of the case. At its best, the trial brief is a favorable and complete presentation of all the significant issues, assertions of fact, legal arguments, and anticipated evidence. However, it must remain simple and compelling, and portray the equities so as to create the appropriate sympathetic appeal. Finally, it should provide the judge with a road map for favorably deciding critical issues both prior to and during trial.

Special attention should be paid to the choice and balance of trial brief subjects. Because the complexity of modern construction can so easily lead either to misunderstanding or to overwhelmed inattention, the trial brief must include clear explanations of the relevant facts. If one or more parties are seeking damages for delay, for example, the trial brief may have to explain the relevant aspects of contract administration, the significance of arcane contract provisions, and the implications of critical path scheduling. Similarly, the trial brief may have to explain pedestrian realities of the job site, even if they do not appear complex. For example, if safety is an issue, the brief must explain who had responsibility for that safety, what was the scope of responsibility, and what was the practical chain of command. When explaining such details, however, the trial brief must remain simple and clear. In particular, care should be taken not to let explanation of details become so protracted that it detracts from the main issues.

The jury instructions and the proposed findings of fact and conclusions of law, if they are submitted prior to trial, flow naturally from the trial brief. Again, the practice varies from judge to judge, but the objective is universal: the judge should be offered a set of instructions or findings and conclusions that are comprehensive and compelling. Counsel can start with a conservatively worded set, most likely to be adopted if the judge is at all sympathetic, and then weave in paragraphs reaching for the limits of the law and facts. Merely seeking the findings formally sufficient to uphold a judgment on appeal is not sufficient; the lawyer should try to obtain findings that will assure appellate sympathy and account for the appellate prerogative to expand or change the law.

For example, consider a case involving design problems with structural steel on a project. It might be sufficient for the judge to make a simple finding that the architect-engineer exercised due care in preparation of the drawings and specifications, but why should the advocate stop there? If the evidence shows that the plans were reviewed by three senior engineers, each with 30 years of experience in structural steel design of large-scale commercial projects, why not ask for a specific finding setting out those facts in detail?

Jury instructions provide fewer opportunities for advocacy on the facts, but they are a useful vehicle for focus and emphasis. This is especially true, as already noted, where judges permit a copy of the instructions to be given to the jury for use during its deliberations. Counsel should also consider the desirability of requesting special interrogatories to the jury. In addition to their role in providing structure for the jury's thinking, they can also play a significant role on appeal by exposing irrational verdicts and supporting proper verdicts.

Other seemingly commonplace steps in conjunction with the final pretrial conference also offer important opportunities. Exhibit and witness lists should be scrutinized with the same care as when the exhibits are offered or the witness is testifying. Successful motions to limit evidence can avoid courtroom controversy over unduly prejudicial testimony. In addition, they can eliminate entire areas of evidence which may be only marginally relevant, but would consume substantial trial time, require extensive rebuttal, and unduly confuse the jury if admitted.

Optional trial procedures should also be considered for recommendation at the pretrial conference, even if they go beyond normal practice in the jurisdiction. As already noted, complex construction litigation may benefit materially from such techniques as note-taking by jurors, creative use of site models, site visits, or videotaped exhibits, and submission of technical testimony in written form. Written testimony can be particularly useful in expediting bench trials and in presenting complex engineering information in a clear form. Controversial procedures such as permitting jurors to ask questions during the trial may also be useful in confusing cases. Counsel should request these innovative procedures at the pretrial conference, where judges are often more flexible than they will be once the trial has begun.

Finally, as noted earlier, the pretrial conference presents an excellent opportunity to press for settlement. As the judge focuses on the details of the anticipated trial, he or she can be engaged as an invaluable ally in either supporting bona fide demands or attacking inflated ones.

TRIAL

§ 19.20 Opening Statement

The opening statement is particularly important in complex construction trials. The effective opening statement presents a simple and compelling picture of the facts and law that triers of fact can use to organize their perception of the entire trial. To achieve this level of effectiveness, however, the opening statement must be rooted in a clear understanding of the case and a well-defined trial plan. Accordingly, it requires careful preparation and, in a case of any substance, it should be reviewed in advance with the client. The statement may be pitched differently for a judge than for a jury, but the framework it provides remains important and counsel should not forego making a statement in a bench trial unless required to do so. This is particularly true where the judge may have prior experience with the construction industry that leaves him or her with either outdated ideas or inapplicable prejudices.

Much has been written on what should be included in the opening statement.[22] Most commentators agree, however, that the overall goal should be providing the

---

[22] For a conventional viewpoint on what should be included in the opening statement, *see* J. Appleman, Preparation & Trial 189 (1967); I S. Schweitzer, Cyclopedia of Trial Practice § 172, at 474 (1970).

judge or jury with a brief road map explaining how the client's case will be proved, including what witnesses will be called and what their testimony will be. Counsel should refrain from including any excess detail that might detract from the clarity and simplicity of this outline. Accordingly, facts should be explained in a brief and logical manner. Visual aids—especially photographs—should be used to give the trier of fact a feel for the overall project and to convey necessary background information more quickly and compellingly.[23] The bases for significant equitable considerations should be woven in throughout the statement to invoke sympathy and emotional appeal, although counsel must avoid jeopardizing fundamental credibility by making overstatements that the evidence will not support.

§ 19.21 Presentation of Evidence

Effective presentation of the evidence requires blending traditional principles of trial advocacy with the special demands of complex construction cases. Because every case has its own group of strategical problems requiring individual analysis, the most this chapter can do is identify some of the recurring considerations that should be part of presenting the evidence in any complex construction case.

1. **Start slowly.** Traditional wisdom calls for presenting the strongest witness first. In complex litigation, however, counsel may wish to compromise this principle and select a first witness who is not substantively important, but is capable of clearly describing the parties and relevant facts. Proving liability in complex trials cannot even begin until counsel has educated the trier of fact in the intricacies of the dispute, and this should normally be the function of the first witness.
2. **Present the case in practical terms.** The complexities of large construction trials, especially those involving multiple expert witnesses, make it easy to lose the trier of fact in technological details. Trial counsel must recognize that no matter how many experts testify, the final judgments are made by judges and juries who will base their decisions on whatever practical understanding they can form from the relevant facts. Counsel must therefore continually refocus on the most relevant and important facts to insure that they will be plainly understood.
3. **Emphasize people.** The competent completion of a construction contract is a matter of people, not just engineering and management. The trier of fact will know this, and thus treat the contractor sympathetically, only if the people on the job explain what they did and why they did it. In this respect, the inarticulate practical witness is often more effective than the glib managerial one.

[23] Counsel must remember, however, that photographs, maps, diagrams, and other exhibits may not be used in the opening statement without prior approval of the court. R. Figg, R. McCullough & J. Underwood, Civil Trial Manual (1980).

4. **Use construction jargon only where its meaning is clear.** Terms of art and professional jargon tend most often to obscure the facts. For example, consider the many meanings of apparently simple terms like *inspection* or *engineering* or *overhead*. Accordingly, such terms should be explained and used only where they are clearly understood.
5. **Consider submitting complex testimony in written form.** Courts are experimenting increasingly with written direct testimony for complex matters.[24] There are obvious limitations and complications, but they can be reduced by carefully considered procedures. The testimony can be reviewed in advance by opposing counsel for objections. After objections are resolved, the testimony can be admitted and given to the judge or jury for review. The witness can then be made available for additional direct testimony and cross-examination. The practice has most obvious appeal in bench trials, but it has been suggested for jury trials as well. If written testimony is appropriate in jury trials, the witness might read the prepared testimony to the jury.

   Written testimony has been used for years in administrative proceedings, where experience suggests that there are often substantial benefits to be gained from counsel and the witness working together to formulate a clear and comprehensive statement.
6. **Make clear the significance of each piece of evidence.** No document or fact that the lawyer presents at trial is without some connection to the overall understanding of the dispute. For counsel, who has studied the dispute extensively, these connections are obvious, but for the uninitiated judge or jury member, the connection may remain invisible unless it is revealed by the lawyer.
7. **Make sure the trier of fact understands everything about key documents.** Key documents—contract clauses, inspection reports, memoranda, etc.—inevitably get special attention from the judge and jury, even though the actual practice of construction contracting seldom comports with the documents. The minimum that must be done is assuring that nothing in the key documents is left unexplained. The record, for the trier of fact and for the appeal, must be complete.
8. **Select the best possible company representative to sit at counsel table.** The assistance of someone who can understand the facts and the issues is essential. Any individual selected for this purpose must also make a good impression in the courtroom, and should be coached on appropriate conduct.

[24] *See* Richey, *A Modern Management Technique for Trial Courts to Improve the Quality of Justice: Requiring Direct Testimony to be Submitted in Written Form Prior to Trial,* 72 Geo. L.J. 73 (1983); *see also* Local Rule 1-15 of the United States District Court for the District of Columbia; Local Rule 9.11.2 of the United States District Court for the Central District of California.

§ 19.22 Cross-Examination

The basic cross-examination techniques of attacking and impeaching a witness's testimony are as applicable in construction cases as in any type of litigation. There are, however, a few aspects of cross-examination that are especially important in the construction setting. First, extra care should be taken in deciding whether to cross-examine each witness. In large construction cases there will be many opposing witnesses who do not present any real threat to the client's case. If the trier of fact is likely to recognize the partisan role of these witnesses and is not likely to place great reliance upon their testimony, the lawyer may not be prejudiced by failing to cross-examine them. And, of course, cross-examination may backfire. A witness antagonized by cross-examination may make a damaging remark, and the lawyer risks detracting from the power of his cross-examination of more important witnesses.

Second, the lawyer should be particularly careful to use clear language in framing questions. Once again, although counsel is likely to have become versed in construction terminology while preparing the case, the judge or jury is not likely to understand or appreciate the significance of lines of questioning couched in technical jargon.

Counsel should also take into account some of the special considerations involved in cross-examining expert witnesses. Most important, the lawyer must research both the expert's particular views and more general authorities relating to the issues in question. With this background information, the lawyer may be able to impeach the expert's opinion, either by showing that it is contrary to the general weight of authority, or that it contradicts the expert's own published statements. Counsel may also attack the expert's testimony by revealing that the expert's conclusions are based on improper testing or lack of detailed knowledge of the actual construction methods used. To aid in detecting such deficiencies, the lawyer may wish to have the client's own expert present during direct examination of the opposing expert.

§ 19.23 Closing Argument

Closing arguments are so much a part of the lore and art of trial practice that little can be added to the volumes of available advice. A few points, however, cannot be overemphasized. Most importantly, the closing argument must do more than summarize the evidence and law relating to the case; it must tell a story. The lawyer must gather up the facts and equitable consideration of the case and paint a picture that compels the judge or jury to find in his or her client's favor.

Unfortunately, creating such a clear image of the case demands a level of preparation that lawyers frequently cannot, or are not willing to, attain. In the most complex cases, such as large construction cases, the burdens of preparation may be particularly onerous, yet it is in just such complex cases that a clear picture of the case is most essential. In some cases, it may be the only real basis for the ultimate decision, as the underlying dispute is too difficult for the judge or jury to comprehend adequately. Therefore, before the trial even starts counsel should prepare a detailed written outline of the closing argument. In some bench trials, it may even be possible to include a list of principal cases, and offer the outline to the court and opposing counsel just prior to the argument. Such an outline can be a compelling road map for both the judge and the law clerk, and the time spent preparing it will insure a clearer argument.

APPEAL

§ 19.24 Planning an Appeal

The appeal can best be considered in three parts: reassessment of the case, briefing, and oral argument. Each of these steps must reflect a recognition that in some respects, a case on appeal is closed (e.g., some findings of fact may be virtually irreversible), in some respects, it is entirely new (e.g., the appellate court may reverse on grounds not ruled on by the trial court), and in most respects the door is open only a crack. However, so long as the door is open at all, counsel and client must again collaborate to focus and present the case.

§ 19.25 Reassessment of the Case

In many ways, reassessment at the start of an appeal is easier than the assessment made at the start of the case. The factual universe is defined by the record in the lower court. The issues have generally been defined as they have been preserved for appeal. Also, because during the appeal there will be no discovery, few motions, and no cumbersome trial preparation, the legal fees and associated costs are easier to estimate.

In other respects, the appellate reassessment is very difficult. First, the same questions must be asked about the claims, interests, objectives, facts, equities, and focus.[25] They must be asked with the record in mind, but counsel must recognize that the record will look very different to appellate judges than it did to the trial court. Counsel's most important task may be developing the appellate court's understanding of what is in the record and what its significance is. Once again, the context and the equities can be as important as the evidence specifically related to liability.

---

[25] See **§§ 19.2-19.10, 19.12**.

The most difficult part of the reassessment may be presented by the wide range of possible results. Each must be identified and evaluated by counsel and client. The difficulty of this process is illustrated by the case on appeal from a trial court's granting of a motion for summary judgment. On large projects with recurring litigation, such dismissals may become almost routine, and often no appeal is taken, especially where other ''deep-pocket'' defendants are available. Yet, if plaintiff does pursue an appeal, it may result in unforeseen consequences. Even though the record is likely to be skimpy and the equities undeveloped, the appellate court may overstep its traditionally limited role and interpret the contract out of context. The result may be a serious misinterpretation of the parties' relationships.

As with the trial, the assessment should have a clear focus. Lawyers who fail to develop a focus often find irresistible the temptation to brief and argue every possible issue. Of course, the risk of exhaustive argument is exhausted judges (or their overworked clerks) who miss the best points and are alienated by the worst.

§ 19.26 Appellate Brief

Preparing the appellate brief starts with research encompassing both the law and the record. The same basic considerations apply to developing the issues on appeal as apply at the trial level. Thus, attention should be centered on those arguments that best serve the client's overall objectives. This focus allows the lawyer to keep control of the appeal and avoid needless overpreparation.

Rules of the specific appellate court may cause the form of the brief to vary, but the basic format is similar in almost every appellate forum. For example, most appellate courts require a statement of each issue presented for review. The statement must be clear, and it should be worded to suggest the desired resolution.

A statement of the case is also generally required. It is essential to the judges' understanding of the basic issues. The statement must show how the controversy arose, what roles the parties played in the construction work at issue, and how the trial court incorporated these facts into its decision. Encompassing all of these details may require that the lawyer make the statement longer than otherwise appropriate and, in particularly complex cases, the lawyer may break down various areas of the statement into different sections. For example, in a job-site safety case where relationships and duties are important, the statement of the case might be organized under the following subheadings:

1. Overall organization of the project
2. The relationship between the contractor and the owner
3. Safety responsibilities on the project
4. The decision of the district court.

The statement of the case should not be argument, but, like the statement of the issues, its wording and focus should clearly and persuasively suggest the desired result.

The core of the brief is, of course, the argument. Each argument will be developed differently, with the emphasis reflecting the overall position counsel wishes to convey to the judges. The argument section should be organized with headings and subheadings that state succinctly but positively the points being made. The argument section is and should appear entirely different from the statement of the case, where an appearance of objectivity is important.

It is often useful to organize the argument section of an appellate brief into two major parts. First, the contractor's arguments; second, the contractor's answers to the opponent's arguments. The major headings may simply be "The District Court Correctly Found That . . ." and "Appellant's Arguments that . . . Are Without Merit." The subheadings can then be one-sentence summaries of the argument made in each subsection. A carefully organized argument with effective captions should always result in a table of contents that is itself readable and persuasive.

It is easy to overlook the importance of record excerpts and appendices. In addition to the formal record on appeal, the Federal Rules of Appellate Procedure and many state appellate rules require an appendix to the briefs.[26] The appendix includes any parts of the record to which the parties wish to direct the attention of the court. While the filing of an appendix does not preclude reliance on parts of the record not included, it typically becomes the handbook for the judges and their clerks, both when the briefs are being considered and during oral argument. Selection of its contents is therefore a very important piece of counsel's advocacy.

§ 19.27 Oral Argument

Oral argument to the appellate court parallels both the opening statement and the closing argument at the trial level. The judges must be oriented to the contractor's case and then must be convinced that the facts and the law support a decision for the contractor. As part of planning the argument, it is important to find out as much as possible about the judges. What are their biases? How carefully will they prepare? How have they decided similar cases or issues? What types of questions do they typically raise during argument?

Oral argument is often interrupted by questions from the judges. Sometimes the questions are so numerous or so insistent that virtually none of the planned argument is delivered. If this happens, counsel should be delighted, not dismayed. The most effective oral arguments can be those which become a dialogue between counsel and the court. Confusion may be distressing, but it is better elicited during argument than left to surface in a misinformed adverse opinion.

---

[26] *See, e.g.*, Fed. R. App. Proc. 30.

To handle such a dialogue, counsel must know both the record and the law. In addition, counsel will occasionally be called on for facts that are not in the record, especially where the case is before the court on a pretrial appeal. An informed answer which includes all the facts important to the client's position is the best response to such a question. Counsel's informed representation is entirely appropriate in such instances, so long as the judges are informed that the facts are not in the record.

Because questioning is likely, preparation for it is essential. At the least, counsel must think through the probable questions from the court and then consider how to answer them, both from the perspective of providing the desired information and furthering the overall argument. Mock questioning by a colleague is especially useful. If counsel has not previously argued before the court, an hour's observing in the courtroom during preparation is time well spent.

At the oral argument, the complexity of construction litigation once again influences the strategy. It is especially important to use the argument as a time to identify and correct any misperceptions. It can be especially useful to tell the court at the outset what will be covered in the oral argument. This permits the judges to indicate particular areas of interest, matters they do not fully understand, or areas in which they do not have further interest.

Finally, counsel should recall that postargument filings may be possible, even if only with special leave of the court. An error, a missing record citation, a new relevant decision, or failure to have briefed a case, or an issue raised during the argument may make it necessary for counsel to request leave to make a supplemental filing. Where this is done, local rules and informal procedures should be carefully reviewed, the motion for leave should be made promptly, and the filing should normally accompany it.

TABLE OF CASES

Case	*Book §*
Alyeska Pipeline Serv. v. Aurora Air Serv., Inc., 604 P.2d 1090 (Alaska 1979)	§ 10.27
Amelco Window Corp. v. Federal Ins. Co., 127 N.J. Super. 342, 317 A.2d 398 (1974)	§ 10.15
American Auto Ins. Co. v. United States *ex rel.* Luce, 269 F.2d 406 (1st Cir. 1959)	§ 11.10
American Bridge Co., Inc. v. State, 245 A.D. 535, 283 N.Y.S. 577 (1935)	§ 5.33
American Cas. Co. v. Shattuck, 228 F. Supp. 834 (W.D. Okla. 1964)	§ 10.24
American Cas. Co. v. Timmons, 352 F.2d 563 (6th Cir. 1968)	§ 9.8
American Cas. Co. of Reading v. Irvin, 426 F.2d 647 (5th Cir. 1970)	§ 11.5
American Cas. Co. of Reading v. Southern Materials Co., 261 F.2d 197 (4th Cir. 1958)	§ 11.12
American Fidelity Fire Ins. Co. v. Construcciones Werl, 407 F. Supp. 164 (V.I. 1975)	§ 10.32
American Fidelity Fire Ins. Co. v. Pavia-Byrne Eng'g Corp., 393 So. 2d 830 (La. Ct. App.), *cert. denied,* 397 So. 2d 1362 (La. 1981)	§§ 10.24, 10.33
American Masons' Supply Co. v. F.W. Brown Co., 394 A.2d 378 (Conn. 1978)	§ 11.13
American Star Ins. Co. v. American Employers Ins. Co., 165 Cal. App. 3d 728, 210 Cal. Rptr. 836 (1985)	§ 9.5
American Sur. Co. v. Wheeling Structural Steel Co., 114 F.2d 237 (4th Cir. 1940)	§ 10.23
American Sur. Co. of N.Y. v. Westinghouse Elec. Mfg. Co., 296 U.S. 133 (1935)	§ 11.17
Anderson Constr. Co., Inc. v. Lyon Metal Prods., Inc., 370 So.2d 935 (Miss. 1979)	§ 8.2
Anthony Pools, Div. of Anthony Indus., Inc. v. Sheehan, 455 A.2d 434 (Md. Ct. App. 1983)	§ 15.7
Apache Powder Co. v. Ashton Co., Inc., 264 F.2d 417 (9th Cir. 1959)	§ 11.11
Apex Siding & Roofing Co. v. First Fed. Sav. & Loan Ass'n, 301 P.2d 352 (Okla. 1956)	§§ 7.9, 7.15
Appalachian Ins. Co. v. Superior Court (Union Carbide Corp.), 162 Cal. App. 3d 427, 208 Cal. Rptr. 627 (1984)	§ 9.2
Applegreen v. Milbank Mut. Ins. Co., 268 N.W.2d 114, (N.D. 1978)	§ 9.9
Arc Elec. Constr. Co. v. George A. Fuller Co., 299 N.Y.S.2d 129 (N.Y. 1969)	§ 3.16
Arcon Constr. Co. v. South Dakota Cement Plant, 349 N.W.2d 407 (S.D. 1984)	§ 3.13
Argonaut Ins. Co. v. Commercial Standard Ins. Co., 380 So. 2d 1066 (Fla. Dist. Ct. App. 1980), *review denied,* 389 So. 2d 1108 (Fla. 1980)	§ 10.33
A.R. Moyer, Inc. v. Graham, 285 So. 2d 397 (Fla. 1973)	§§ 5.16, 5.17, 5.22, 5.41, 6.2
Armstrong & Armstrong, Inc. v. United States, 514 F.2d 402 (9th Cir. 1975)	§ 4.5

Case	*Book §*
Berkel & Co. v. Providence Hosp., 454 So. 2d 496 (Ala. 1984)	§ 3.1
Berley Indus., Inc. v. City of N.Y., 45 N.Y.2d 683, 385 N.E.2d 281, 412 N.Y.S.2d 589 (1978)	§§ 3.13, 14.40
Bernard Johnson, Inc. v. Continental Constructors, Inc., 630 S.W.2d 365 (Tex. Civ. App. 1982)	§§ 6.2, 15.2
Bescor, Inc. v. Chicago Title & Trust Co., 446 N.E.2d 1209 (Ill. App. Ct. 1983)	§ 7.24
Biebel Bros. v. United States Fidelity & Guar. Co., 522 F.2d 1207 (8th Cir. 1975)	§ 9.8
Bigelow-Liptak Corp. v. Continental Ins. Co., 417 F. Supp. 1276 (E.D. Mich. 1976)	§ 9.9
Bird v. Pan W. Corp., 261 Ark. 56, 546 S.W.2d 417 (1977)	§ 5.7
B.K. Instrument, Inc. v. United States, 715 F.2d 713 (2d Cir. 1983)	§ 4.5
Blackfield v. Underwriters at Lloyd's, London, 245 Cal. App. 2d 271, 53 Cal. Rptr. 838 (1966)	§ 9.12
Blackhawk Heating & Plumbing Co. v. Seaboard Sur. Co., 534 F. Supp. 309 (N.D. Ill. 1982)	§ 10.35
Black Lake Pipe Line Co. v. Union Constr. Co., Inc., 538 S.W.2d 80 (Tex. 1976)	§ 3.13
Blake Constr. Co. v. C.J. Coakley Co., Inc., 431 A.2d 569 (D.C. 1981)	§§ 3.10, 5.35, 8.6
Blinderman Constr. Co. v. United States, 695 F.2d 552 (Fed. Cir. 1982)	§ 1.25
Bloomsburg Mills, Inc. v. Sordoni Constr. Co., 164 A.2d 201 (Pa. 1960)	§§ 5.19, 6.3
Board of Educ. v. Sever-Williams Co., 258 N.E.2d 605 (Ohio 1970)	§ 2.30
Board of Trustees of Union College v. Kennerly, Slomanson & Smith, 400 A.2d 850 (N.J. Super. Ct. Law Div. 1979)	§ 6.3
Boise Jr. College Dist. v. Mattefs Constr. Co., 450 P.2d 604 (Idaho 1969)	§ 2.30
Bond Bros., Inc. v. Robinson, 471 N.E.2d 1332 (Mass. 1984)	§ 9.8
Bondumin-McClure, Inc. v. Snow, 183 Cal. App. 58, 6 Cal. Rptr. 52 (1960)	§ 5.6
Bonebrake v. Corx, 499 F.2d 951 (8th Cir. 1974)	§ 1.19
Boomer v. Abbett, 121 Cal. App. 2d 449, 263 P.2d 476 (1953)	§ 3.2
Borell v. Municipality of Metropolitan Seattle, 698 F.2d 1229 (9th Cir. 1983)	§ 14.29
Borman's Inc. v. Lake State Dev. Co., 60 Mich. App. 175, 230 N.W.2d 363 (1975)	§§ 5.6, 6.3
Bor-Son Bldg. Corp. v. Employees Commercial Union Ins. Co., 323 N.W.2d 58 (Minn. 1982)	§ 9.9
Boston Elevated Ry. v. Commissioner, 16 T.C. 1084 (1951), B.T.A. Memo (Nov. 29, 1938)	§ 17.18
Boston Ins. Co. v. Grable, 352 F.2d 368 (5th Cir. 1965)	§ 9.9
Boston Plate & Window Glass Co. v. John Bowen Co., 335 Mass. 697, 141 N.E.2d 715 (1957)	§ 15.6
Bowden v. United States *ex rel.* Malloy, 239 F.2d 572 (9th Cir. 1956), *cert. denied,* 353 U.S. 957 (1957)	§ 11.11
Bradley Envtl. Contractors v. Village of Sylvan Beach, 98 A.D.2d 973, 470 N.Y.S.2d 214 (4th Dep't 1983)	§ 14.17
Brady v. Board of Educ., 222 A.D. 504, 226 N.Y.S. 707 (1926)	§ 5.33

Case	*Book §*
Carlo C. Gelardi Corp. v. Miller Brewing Co., 502 F. Supp. 637 (D.N.J. 1980)	§ 10.27
Carnicle v. Swann, 314 N.W.2d 311 (S.D. 1982)	§ 3.14
Caron v. Andrews, 133 Cal. App. 2d 402, 284 P.2d 544 (1955)	§ 10.22
Carpenter v. Suffolk Franklin Sav. Bank, 346 N.E.2d 892 (Mass. 1976)	§ 10.37
Carroll-Boone Water Dist. v. M.&P. Equip. Co., 280 Ark. 560, 661 S.W.2d 560 (1984)	§ 5.6
Casavant v. Campopiano, 327 A.2d 831 (R.I. 1974)	§ 15.7
Case Prestressing Corp. v. Chicago College of Osteopathic Medicine, 455 N.E.2d 811 (Ill. App. Ct. 1983)	§§ 5.15, 5.27
Casey v. Employers Nat'l Ins. Co., 539 S.W.2d 181 (Tex. Civ. App. 1976)	§ 9.15
Casson Constr. Inc., GSBCA Nos. 4884 *et al.*, 83-1 B.C.A. (CCH) ¶ 16,523 (1983)	§ 5.27
Castaldo v. Pittsburgh-Des Moines Steel Co., 376 A.2d 88 (Del. 1977)	§ 6.3
Cates v. Morgan Portable Bldg. Corp., 591 F.2d 17 (7th Cir. 1979)	§ 15.7
Cavendish v. Sunoco Serv. of Greenfield, Inc., 451 F.2d 1360 (7th Cir. 1971)	§ 19.13
C.E. Frazier Constr. Co. v. Campbell Roofing & Metal Works, Inc., 373 So. 2d 1036 (Miss. 1979)	§§ 1.24, 8.2
Central Coat, Apron & Linen Serv., Inc. v. United States, 298 F. Supp. 1201 (S.D.N.Y. 1969)	§ 17.25
Central Mechanical Constr., ASBCA Nos. 29431, 29432, 29433 85-2 B.C.A. (CCH) ¶ 18.061 (1985)	§ 3.3
Central Mut. Ins. Co. v. Del Mar Beach Club Owners Ass'n, 123 Cal. App. 3d 916, 176 Cal. Rptr. 895 (1981)	§ 9.1
Ceramic Tiles of Fairfield, Inc. v. Aetna Cas. & Sur. Co., 39 Conn. Supp. 477, 466 A.2d 348 (1983)	§ 9.9
Charles E. Brohawn & Bros., Inc. v. Board of Trustees, 304 A.2d 819 (Md. 1973)	§§ 4.2, 4.6, 4.9
Charles H. Siever, PSBCA No. 460, 80-2 B.C.A. (CCH) ¶ 14,740 (1980)	§ 3.3
Chevron Chem. Corp. v. United States, 5 Cl. Ct. 807 (1984)	§ 4.4
Chicago Bonding & Sur. Co. v. Augusta Savannah Navigation Co., 250 F. 616 (7th Cir.), *cert. denied,* 247 U.S. 509 (1918)	§ 10.23
Chicago College of Osteopathic Medicine v. George A. Fuller Co., 719 F.2d 1335 (7th Cir. 1983)	§ 3.10
Chlan v. K.D.I. Sylvan Pools, Inc., 53 Md. App. 236, 452 A.2d 1259 (1982)	§ 15.7
C.H. Leavell & Co. v. Fireman's Fund Ins. Co., 372 F.2d 784 (9th Cir. 1967)	§ 9.18
C.H. Leavell & Co. v. Grafe & Assocs., Inc., 414 P.2d 873 (Idaho 1966)	§ 8.3
Chouteau v. United States, 95 U.S. 61 (1877)	§ 15.5
Christensen v. Idaho Land Developers, Inc., 660 P.2d 70 (Idaho 1983)	§ 10.28
Cincinnati Elecs. Corp. v. Kleppe, 509 F.2d 1080 (6th Cir. 1975)	§ 4.5
Claiborne Parish School Bd. v. Fidelity & Deposit Co. of Md., 40 F.2d 577 (5th Cir. 1930)	§ 10.32
Clardy v. Barco Constr. Co., 208 A.2d 793 (Pa. Super. Ct. 1965)	§ 7.9

Case	*Book §*
Clear v. Patterson, 80 N.M. 654, 459 P.2d 358 (1969)	§ 15.3
Clevenger & Wright Co. v. A.O. Smith Harvestore Prods., Inc., 625 S.W.2d 906 (Mo. Ct. App. 1981)	§ 15.7
Clifford F. MacEvoy Co. v. United States *ex rel.* Calvin Tomkins Co., 332 U.S. 102 (1944)	§§ 10.37, 11.6
Coac, Inc. v. Kennedy Eng'rs, 136 Cal. Rptr. 890 (Ct. App. 1977)	§ 6.2
Cochran v. Ozark Country Club, Inc., 339 So. 2d 1023 (Ala. 1976)	§ 5.7
Coco Bros. Inc. v. Pierce, 741 F.2d 675 (3d Cir. 1984)	§§ 2.16, 4.5
Coleman Engineering Co. v. North American Aviation, Inc., 65 Cal. 2d 396, 55 Cal. Rptr. 1 (1966)	§ 3.2
Collingswood Hosiery Mills v. American Fed. Hosiery Works, 28 N.J. Super. 605, 101 A.2d 372 (Ch. Div. 1953)	§ 15.13
Collins, T.C. Memo 1959-174	§ 17.9
Colonial Refrigerated Transp., Inc. v. Worsham, 705 F.2d 821 (6th Cir. 1983)	§ 9.8
Columbus, City of v. Clark-Dietz & Assocs.-Eng'rs, Inc., 550 F. Supp. 610 (N.D. Miss. 1982)	§§ 6.2, 6.5, 6.6
Commerce Photo Print, T.C. Memo (Apr. 11, 1947)	§ 17.26
Commercial Contractor, Inc. v. United States Fidelity & Guar. Co., 524 F.2d 944 (5th Cir. 1975)	§ 15.3
Commercial Standard Ins. Co. v. Bank of Am., 57 Cal. App. 3d 241, 129 Cal. Rptr. 91 (1976)	§ 7.24
Commercial Union Assurance Cos. v. Gollan, 118 N.H. 744, 394 A.2d 839 (1978)	§ 9.8
Commissioner v. Glenshaw Glass Co., 18 T.C. 860 (1952), *aff'd,* 211 F.2d 928 (3d Cir. 1954), *rev'd,* 348 U.S. 426 (1955)	§ 17.8
Commissioner v. Obear-Nester Glass Co., 217 F.2d 56 (7th Cir. 1954), *aff'd* 20 T.C. 1102 (1953)	§ 17.8
Commissioner v. Tellier, 383 U.S. 687 (1966)	§ 17.24
Commonwealth v. Eastern Paving Co., 288 Pa. 571, 136 A. 853 (1927)	§ 4.6
Comptroller of Treasury v. Paintz, 297 A.2d 289 (Md. 1972)	§ 4.9
Condenser Serv. & Eng'g Co., Inc. v. American Mut. Liab. Ins. Co., 58 N.J. Super. 179, 155 A.2d 789 (1959)	§ 9.9
Conforti & Eisele, Inc. v. John C. Morris Assocs., 175 N.J. Super. 341, 418 A.2d 1290 (1981)	§§ 5.16, 5.22, 5.41, 6.2, 15.2
Connie's Constr. Co. v. Continental Ins. Co., 227 N.W.2d 204 (Iowa 1975)	§ 9.19
Connor v. Great W. Sav. & Loan Ass'n, 69 Cal. 2d 850, 477 P.2d 609, 73 Cal. Rptr. 369 (1969)	§ 7.7
Continental Cas. Co. v. Gilborne Bldg. Co., 391 Mass. 143, 461 N.E.2d 209 (1984)	§ 9.9
Continental Cas. Co. v. Hartford Accident & Indem. Co., 243 Cal. App. 2d 565, 52 Cal. Rptr. 533 (1966)	§ 10.15
Continental Heller Corp., GSBCA No. 6812, 84-2 B.C.A. (CCH) ¶ 17,275 (1984)	§ 3.7
Continental Realty Corp. v. Andrew J. Crevolin Co., 380 F. Supp. 246 (D.W. Va. 1974)	§ 10.23
Contracting & Material Co. v. City of Chicago, 20 Ill. App. 3d 684, 314 N.E.2d 598 (1974), *rev'd on other grounds,* 64 Ill. 2d 21, 349 N.E.2d 389 (1976)	§§ 14.26, 14.27

Case	*Book §*
Cook v. Citizens Sav. & Loan Ass'n, 346 So.2d 370 (Miss. 1977)	§ 7.24
Coombs v. Beede, 36 A. 104 (Me. 1896)	§ 10.33
Cooper v. Jevne, 129 Cal. Rptr. 724 (Ct. App. 1976)	§ 6.2
Corinno Civetta Constr. Corp. v. City of N.Y., N.Y.L.J., Jan. 22, 1985	§ 14.17
Corn Exch. Bank v. United States, 37 F.2d 34 (2d Cir. 1930)	§ 17.20
County Asphalt, Inc. v. State, 40 A.D.2d 26, 337 N.Y.S.2d 415 (3d Dep't. 1972)	§ 14.20
Cowles v. City of Minneapolis, 151 N.W. 184 (Minn. 1915)	§ 6.3
Craig v. Everett M. Brooks Co., 351 Mass. 497, 222 N.E.2d 752 (1967)	§ 5.11
Craviolini v. Scholer & Fuller, 357 P.2d 611 (Ariz. 1961)	§ 6.7
Creely v. Leisure Living, Inc., 437 So. 2d 816 (La. 1983)	§ 10.28
Cunningham v. Cunningham, 157 F.2d 859 (D.C. Cir. 1946)	§ 10.23
Cunningham Bros., Inc. v. City of Waterloo, 117 N.W.2d 46 (Iowa 1962)	§§ 5.35, 14.13
Cuthrell v. Milwaukee Mechanics Ins. Co., 234 N.C. 137, 66 S.E.2d 649 (1951)	§ 9.17
Dahlstrom Corp. v. State Highway Comm'n, 590 F.2d 614 (5th Cir. 1979)	§ 15.10
Dairy Farm Leasing Co., Inc. v. Hartley, 395 A.2d 1135 (Me. 1978)	§ 15.10
Dakota Block Co. v. Western Cas. & Sur. Co., 81 S.D. 213, 132 N.W.2d 826 (1965)	§ 9.6
Dale Benz, Inc. Contractors v. American Cas. Co., 303 F.2d 80 (9th Cir. 1962), *reh'g denied,* 305 F.2d 641 (9th Cir. 1962)	§ 11.10
D'Angelo v. State, 41 A.D.2d 77, 341 N.Y.S.2d 84 (3d Dep't 1973), *modified,* 46 A.D.2d 983, 355 N.Y.S.2d 377 (1974)	§ 14.42
Daniels v. Big Horn Fed. Sav. & Loan Ass'n, 604 P.2d 1046 (Wyo. 1980)	§ 7.24
Darner Motor Sales, Inc. v. Universal Underwriters Ins. Co., 682 P.2d 388 (Ariz. 1984)	§ 10.37
Davidson & Jones, Inc. v. County of New Hanover, 255 S.E.2d 580 (N.C. Ct. App. 1979)	§§ 6.2, 6.5
Deer Park Bank v. Aetna Ins. Co., 493 S.W.2d 305 (Tex. 1973)	§ 10.24
Delancey v. Insurance Co., 52 N.H. 581 (1873)	§ 9.2
Delaware Dredging Co. v. Tucker Stevedoring Co., 25 F.2d 44 (3d Cir. 1928)	§ 11.7
Delhomme Indus., Inc. v. Houston Beechcraft, Inc., 669 F.2d 1049 (5th Cir. 1982)	§ 15.12
Della Ratta, Inc. v. American Better Community Developers, Inc., 380 A.2d 627 (Md. Ct. Spec. App. 1977)	§ 3.12
Del Mar Beach Club Owners Ass'n, Inc. v. Imperial Contracting Co., Inc., 176 Cal. Rptr. 886 (Ct. App. 1981)	§ 6.3
Demharter v. First Fed. Sav. & Loan Ass'n of Pittsburgh, 194 A.2d 214 (Pa. 1963)	§ 7.9
Depot Constr. Corp. v. State, 278 N.Y.S. 2d 363 (1st Dep't 1969)	§ 14.29
Detweiler Bros., Inc. v. John Graham & Co., 412 F. Supp. 416 (E.D. Wash. 1976)	§§ 5.22, 5.41, 6.2, 6.7

Case	*Book §*
Ecco High Frequency Corp. v. Commissioner, 167 F.2d 583 (2d Cir. 1948)	§ 17.25
E.C. Ernst, Inc. v. City of Tallahassee, 527 F. Supp. 1141 (D. Fla. 1981)	§ 12.14
E.C. Ernst, Inc. v. Koppers Co., 626 F.2d 324 (3d Cir. 1980)	§§ 14.5, 14.35
E.C. Ernst, Inc. v. Manhattan Constr. Co., 551 F.2d 1026 (5th Cir.), *cert. denied,* 434 U.S. 1067 (1977)	§§ 5.33, 6.2, 6.7, 6.9, 14.15, 14.21, 15.10
Economy Lumber Co. v. Insurance Co. of N. Am., 157 Cal. App. 3d 641, 203 Cal. Rptr. 135 (1984)	§ 9.6
Ed Sparks & Sons v. Joe Campbell Constr. Co., 578 P.2d 681 (Idaho 1978)	§ 3.15
Edward Joy Co. v. Noise Control Prods., Inc., 443 N.Y.S.2d 361 (N.Y. 1981)	§ 8.2
Edwin J. Dobson, Jr., Inc. v. Rutgers State Univ., 157 N.J. Super. 357, 384 A.2d 1121 (1978)	§§ 5.37, 15.9
E.E. Black, Ltd. v. Alsup, 211 F.2d 879 (9th Cir. 1954)	§ 17.30
E.H. Leavell & Co. v. Glantz Contracting Corp. of La., 322 F. Supp. 779 (E.D. La. 1971)	§ 10.33
E.H. Morrill Co. v. State, 65 Cal. 2d 787, 423 P.2d 551 (1967)	§ 3.9
Eichleay Corp., 60-2 B.C.A. (CCH) ¶ 2688 (A.S.B.C.A. 1960)	§§ 3.13, 14.40, 16.16
Eichler Homes, Inc. v. Underwriters at Lloyd's, London, 238 Cal. App. 2d 532, 47 Cal. Rptr. 843 (1965)	§ 9.12
Ekco Enters., Inc. v. Remi Fortin Constr., Inc., 382 A.2d 368 (N.H. 1978)	§ 10.38
Elcar Mobil Home, Inc. v. D.K. Baxter, Inc., 66 N.J. Super. 478, 169 A.2d 509 (1961)	§ 9.9
Ell-Dorer Contracting Co. v. State, 484 A.2d 356 (N.J. Super. Ct. App. Div. 1984)	§ 3.6
Ely v. Bottini, 179 Cal. App. 2d 287, 3 Cal. Rptr. 756 (1960)	§ 3.12
Emond v. Tyler Bldg. & Constr. Co., Inc., 438 So. 2d 681 (La. Ct. App. 1983)	§ 10.33
Empire Constr. Co., Inc., ASBCA No. 27540, 84-3 B.C.A. (CCH) ¶ 17,531 (1984)	§ 3.6
Employer Ins. of Wausau v. Cajun Contractors & Eng'rs, Inc., 459 So. 2d 610 (La. Ct. App. 1984)	§ 10.34
Employers Cas. Co. v. Holm, 393 S.W.2d 363 (Tex. 1965)	§ 9.20
Ench Equip. Corp. v. N.K. Foods, Inc., 43 N.J. Super. 500, 129 A.2d 313 (App. Div. 1957)	§ 15.13
Endres Plumbing Corp. v. State, 95 N.Y.S.2d 574 (Ct. Cl. 1950)	§ 5.33
Equitable Cas. & Sur. Co. v. Helena Wholesale Grocery Co., 60 F.2d 380 (8th Cir. 1932)	§ 11.7
Equitable Fire & Marine Ins. Co. v. Allied Steel Constr. Co., 421 F.2d 512 (10th Cir. 1965)	§ 9.20
Eric A. Carlstrom Constr. Co. v. Independence School Dist. No. 77, 256 N.W.2d 479 (Minn. 1977)	§§ 1.11, 1.25
Essex House v. St. Paul Fire & Marine Ins. Co., 404 F. Supp. 970 (D.C. Ohio 1975)	§§ 9.16, 9.20
Estrin Constr. Co., Inc. v. Aetna Cas. & Sur. Co., 612 S.W.2d 413 (Mo. Ct. App. 1981)	§ 9.9

Case	*Book §*
Fitzgerald v. LaPorte, 40 S.W. 61 (Ark. 1897)	§ 10.34
525 Main St. Corp. v. Eagle Roofing Co., 34 N.J. 251, 168 A.2d 33 (1961)	§ 15.8
Fleisher Eng'g & Constr. Co. v. United States *ex rel.* Hallenbeck, 311 U.S. 15 (1940)	§ 11.11
Florida *ex rel.* Consolidated Pipe & Supply Co. v. Houdaille Indus., 372 So. 2d 1177 (Fla. Dist. Ct. App. 1979)	§ 10.37
Flowers v. Diamond Shamrock Corp., 693 F.2d 1146 (5th Cir. 1982)	§ 15.12
Foeller v. Heintz, 137 Wis. 169, 118 N.W. 543 (1908)	§ 15.11
Fontainebleau Hotel Corp. v. United Filigree Corp., 298 So. 2d 455 (Fla. Dist. Ct. App. 1974)	§ 9.8
Fort Wayne, City of, (Protest of Davis Excavating, Inc.) (EPA Region V, Mar. 24, 1981)	§ 2.10
Foster Constr. C.A. & Williams Bros. Co. v. United States, 435 F.2d 873 (Ct. Cl. 1970)	§ 3.6
Fostic v. United Home Improvement Co., 428 N.E.2d 1351 (Ind. Ct. App. 1981)	§ 14.43
Framlau Corp. v. United States, 568 F.2d 687 (Ct. Cl. 1977)	§ 14.41
Frank Briscoe Co., Inc. v. Georgia Sprinkler Co., 713 F.2d 1500 (11th Cir. 1983)	§ 9.23
Frank Corrente v. Conforti & Eisele Co., Inc., 468 A.2d 920 (R.I. 1984)	§ 10.34
Franklin Contracting Co. v. State, 365 A.2d 952 (N.J. Super. Ct. 1976)	§ 5.32
Frank T. Hickey, Inc. v. Los Angeles Jewish Community Council, 128 Cal. App. 2d 676, 276 P.2d 52 (Dist. Ct. App. 1954)	§ 15.5
Frederick v. Electro-Coal Transfer Corp., 548 F. Supp. 83 (E.D. La. 1982)	§ 10.37
Fresard v. Michigan Millers Mut. Ins. Co., 414 Mich. 686, 327 N.W.2d 286 (1982), *aff'g but overruling as to this issue* 97 Mich. App. 584, 296 N.W.2d 112 (1982)	§ 9.8
Fretz Constr. Co. v. Southern Nat'l Bank, 626 S.W.2d 478 (Tex. 1982)	§§ 7.12, 7.15
Friedman v. Edward L. Bakewell, Inc., 654 S.W.2d 367 (Mo. Ct. App. 1983)	§ 10.27
Friedman v. Hampden County, 204 Mass. 494, 90 N.E. 851 (1910)	§ 10.1
Friestad v. Travelers Indem. Co., 393 A.2d 1212 (Pa. Super. Ct. 1978)	§ 9.15
Fritz-Rumer Cooke Co. v. United States, 279 F.2d 200 (6th Cir. 1960)	§ 14.25
F.&S. Constr. Co. v. Berube, 322 F.2d 782 (10th Cir. 1963)	§ 15.7
Fuchs v. Parsons Constr. Co., 172 Neb. 719, 111 N.W.2d 727 (1961)	§ 15.2
Garcia v. Truck Ins. Exch., 35 Cal. 3d 426, 682 P.2d 1100, 204 Cal. Rptr. 435 (1984)	§ 9.2
Gardner Displays Co. v. United States, 346 F.2d 585 (Ct. Cl. 1965)	§ 14.42
Gateway Erectors v. Lutheran Gen. Hosp., 102 Ill. App. 3d 300, 430 N.E.2d 20 (1981)	§§ 5.16, 5.17, 5.23, 5.28

Case	*Book §*
Governor v. Madrazo, 26 U.S. (1 Pet.) 110 (1828)	§ 10.32
Granite Constr. Co., IBCA No. 1-1442-3-81, 82-2 B.C.A. (CCH) ¶ 15,975 (1982)	§ 5.30
Grant Constr. Co. v. Burns, 92 Idaho 408, 443 P.2d 1005 (1968)	§ 14.16
Gravely v. Providence Partners, 549 F.2d 958 (4th Cir. 1977)	§ 5.6
Graves Constr. Co., Inc. v. Rockingham Nat'l Bank, 28 U.C.C. Rep. Serv. (Callaghan) 588 (Va. 1980)	§ 10.31
Gray v. Zurich Ins. Co., 65 Cal. 2d 263, 419 P.2d 168 (1966)	§ 9.1
Great Am. Ins. Co. v. United States, 492 F.2d 821 (Ct. Cl. 1974)	§ 10.32
Great Lakes Constr. Co. v. Republic Creosoting Co., 139 F.2d 456 (8th Cir. 1943)	§ 1.11
Green Constr. Co. v. William Form Eng'g Corp., 506 F. Supp. 173 (W.D. Mich. 1980)	§ 5.6
Grenier v. Compratt Constr. Co., 189 Conn. 144, 454 A.2d 1289 (1983)	§§ 15.6, 15.10
Griners' & Shaw, Inc. v. Federal Ins. Co., 234 F. Supp. 753 (D.S.C. 1964)	§ 11.16
Groover v. Magnavox, 71 F.R.D. 638 (D. Pa. 1976)	§ 15.7
Gruman Ecosys. Corp. v. Gainesville-Alachua 402 F. Supp. 582 (N.D. Fla. 1975)	§ 2.21
Gulf M. & O. Ry. Co. v. Illinois Cent. R.R. Co., 128 F. Supp. 311 (N.D. Ala. 1954), *aff'd,* 225 F.2d 816 (5th Cir. 1955)	§ 1.17
Gunderson's, Inc. v. James H. Tull, 678 P.2d 1061 (Colo. Ct. App. 1984)	§ 3.12
Gurtler, Hebert & Co., Inc. v. Weyland Mach. Shop, Inc., 405 So. 2d 660 (La. Ct. App. 1981)	§§ 6.2, 6.7, 10.33
Gustav Hirsch Org., Inc. v. East Ky. Rival Elec. Co-op Corp., 201 F. Supp. 809 (D. Ky. 1962)	§ 15.10
Guttman v. United States, 181 F. Supp. 290 (D. Pa. 1960)	§ 17.25
Guy James Constr. Co. v. Trinity Indus., Inc. 644 F.2d 525 (5th Cir. 1981)	§ 3.13
G-W-L, Inc. v. Robichaux, 643 S.W.2d 392 (Tex. 1982)	§ 15.7
Hadley v. Baxendale, 156 Eng. Rep. 145 (1854)	§ 14.35
Hammaker v. Schleigh, 157 Md. 652, 147 A. 790 (1929)	§ 15.10
Hans v. Louisiana, 134 U.S. 1 (1890)	§ 10.32
Harmont Plaza, Inc. v. Commissioner, 542 F.2d 414 (6th Cir. 1975), *aff'g* 64 T.C. 632 (1975)	§ 17.20
Harrington v. Oliver, 459 So. 2d 111 (La. Ct. App. 1984)	§ 10.28
Harrison Plumbing & Heating, Inc. v. New Hampshire Ins. Group, 37 Wash. App. 621 P.2d 875 (1984)	§ 9.8
Harry Harris, Inc. v. Quality Constr. Co. of Benton, Ky., Inc., 593 S.W.2d 872 (Ky. 1979)	§ 8.2
Hart, T.C. Memo 1983-364	§ 17.19
Hartford Accident & Indem. Co. v. Boise Cascade Corp., 489 F. Supp. 855 (N.D. Ill. 1980)	§ 10.32
Hartford Accident & Indem. Co. v. Millis Roofing & Sheet Metal, Inc., 418 N.E.2d 645 (Mass. Ct. App. 1981)	§ 10.27
Hartford Accident & Indem. Co. v. Shelby Mut. Ins. Co., 208 So. 2d 465 (Fla. Dist. Ct. App. 1968)	§ 9.9
Harvey v. Commissioner, 12 T.C.M. (CCH) 1358 (1953)	§ 17.25

Case	*Book §*
Harvey's Wagon Wheel, Inc. v. MacSween, 96 Nev. 215, 606 P.2d 1095 (Nev. 1980)	§ 9.23
Harwick v. Commissioner, 133 F.2d 110 (8th Cir. 1943)	§ 17.3
Hauenstein v. St. Paul-Mercury Indem. Co., 242 Minn. 354, 65 N.W.2d 122 (1954)	§§ 9.5, 9.6
Haugan v. Home Indem. Ins. Co., 86 S.D. 406, 197 N.W.2d 18 (1972)	§ 9.8
Hawthorne v. Kober Constr. Co., 640 P.2d 467 (Mont. 1982)	§ 5.6
Hayes Int'l Corp. v. McLucas, 509 F.2d 247 (5th Cir.), *cert. denied,* 423 U.S. 864 (1975)	§ 4.5
Heheman v. E.W. Scripps Co., 661 F.2d 1115 (6th Cir. 1981)	§ 10.27
Helfeldt v. Robinson, 290 S.E.2d 896 (W. Va. 1981)	§ 9.8
Hendrix v. New Amsterdam Cas. Co., 390 F.2d 299 (10th Cir. 1968)	§ 9.17
Hendrix Elec. Co., Inc. v. Casualty Reciprocal Exch., 297 So. 470 (La. Ct. App. 1974)	§ 9.9
Henggler v. Gindra, 191 Neb. 317, 214 N.W.2d 925 (1974)	§ 15.7
Henningsen v. United States Fidelity & Guar. Co., 208 U.S. 404 (1908)	§§ 10.31, 10.32
Hensler v. City of Los Angeles, 124 Cal. App. 2d 71, 268 P.2d 12 (1954)	§ 3.2
Herbison v. Employers Ins. Co. of Ala., 593 S.W.2d 923 (Tenn. Ct. App. 1979)	§ 9.9
Herman H. Neumann, 70-1 B.C.A. (CCH) ¶ 8216 (1970)	§ 14.24
H.E. Wiese, Inc. v. Western Stress, Inc., 407 So. 2d 464 (La. Ct. App. 1981)	§ 9.9
Hexter v. Commissioner, 8 B.T.A. 888 (1927), *acq.*, VII-1 C.B. 14 (1927)	§ 17.18
Higgins v. City of Fillmore, 639 P.2d 192 (Utah 1981)	§§ 1.25, 3.13
Highland Constr. Co. v. Union Pac. R.R. Co., 683 P.2d 1042 (Utah 1984)	§ 3.13
Hill v. Polar Pantries, 64 S.E.2d 885 (S.C. 1951)	§ 5.6
Hilti v. Oldach, 392 F.2d 368 (1st Cir. 1968)	§ 15.13
Himmel Corp. v. Stade, 52 Ill. App. 3d 294, 367 N.E.2d 411 (1977)	§ 5.13
Hionis v. Northern Mut. Ins. Co., 230 Pa. Super. 511, 327 A.2d 363 (1974)	§ 9.2
H.O. Bragg Roofing, Inc. v. First Fed. Sav. & Loan Ass'n, 226 Cal. App. 2d 24, 37 Cal. Rptr. 775 (1964)	§§ 7.15, 7.16
Hodgins v. Marquette Iron Mining Co., 503 F. Supp. 88 (D. Mich. 1980)	§ 3.17
Hodgson v. Chin, 168 N.J. Super. 549, 403 A.2d 942 (App. Div. 1979)	§ 15.7
Hogan v. Postin, 695 P.2d 1042 (Wyo. 1985)	§ 6.2
Holman v. Sorenson, 556 P.2d 499 (Utah 1976)	§ 3.12
Home Indem. Co. v. Leo L. Davis, Inc., 79 Cal. App. 3d 863, 145 Cal. Rptr. 158 (1978)	§ 9.9
Home Indem. Co. v. United States, 313 F. Supp. 212 (W.D. Mo. 1970)	§ 10.31
Home Mut. Fire Ins. Co. v. Hosfelt, 233 F. Supp. 360 (D. Conn. 1962)	§ 9.5
Housing Auth. v. Hubbell, 325 S.W.2d 880 (Tex. Civ. App. 1959)	§ 5.34

Case	*Book §*
Houston, City of v. R.F. Ball Constr. Co., Inc., 570 S.W.2d 75 (Tex. Civ. App. 1978)	§ 5.32
Howard Dodge & Sons, Inc. v. Finn, 391 N.E.2d 638 (Ind. Ct. App. 1979)	§§ 10.31, 15.7
Hudik-Ross, Inc. v. 1530 Palisades Corp., 131 N.J. Super. 159 (1974)	§ 12.14
Hughes Masonry v. Greater Clark County School Bldg. Corp., 659 F.2d 836 (7th Cir. 1981)	§§ 5.29, 5.38
Hungerford Constr. Co. v. Florida Citrus Exposition, Inc., 410 F.2d 1229 (5th Cir.), *cert. denied,* 396 U.S. 928 (1969)	§ 15.10
Hurst v. Town of Shelburn, 422 N.E.2d 322 (Ind. App. Ct. 1981)	§ 10.27
Ibex Indus., Inc. v. Coastline Waterproofing, 563 F. Supp. 1142 (D.D.C. 1983)	§ 11.7
Illinois Sur. Co. v. John Davis Co., 244 U.S. 276 (1917)	§ 11.7
Illinois Valley Asphalt, Inc. v. J.F. Edwards Constr. Co., 413 N.E.2d 209 (Ill. 1980)	§ 8.2
INA v. Northhampton Nat'l Bank, 708 F.2d 13 (1st Cir. 1983)	§ 10.31
Indiana Ins. Co. v. DeZutti, 408 N.E.2d 1275 (Ind. 1980)	§ 9.8
Indiana Mortgage & Realty Inv. v. Peacock Constr., 348 So. 2d 59 (Fla. Dist. Ct. App.), *cert. denied,* 353 So. 2d 677 (Fla. Dist. Ct. App. 1977)	§§ 7.15, 7.25
Industrial Dev. Bd. v. Fequa Indus., Inc., 523 F.2d 1226 (5th Cir. 1975)	§ 15.7
Industrial Research Assocs., 68-1 B.C.A. (CCH) ¶ 7069, at 32,685 (1968)	§ 4.10
Ingle v. Jones, 2 Wall. 1, 17 L. Ed. 2d 763 (1864)	§ 9.16
Inglewood-Los Angeles County Civic Center Auth., City of v. Superior Court, 7 Cal. 3d 861, 500 P.2d 601, 103 Cal. Rptr. 689 (1972)	§§ 2.8, 5.9
Insurance Co. of N. Am. v. Forty-Eight Insulations, 633 F.2d 1212 (6th Cir. 1980)	§ 9.5
Integrated, Inc. v. Alec Fergusson Elec. Contractor, 250 Cal. App. 2d 287, 58 Cal. Rptr. 503 (1967)	§§ 3.8, 3.14
International Derrick & Equip. Co. v. Buxbaum, 240 F.2d 536 (3d Cir. 1956)	§ 9.9
International Eng'g Co. v. Richardson, 512 F.2d 573, (D.C. Cir. 1975), *cert. denied,* 423 U.S. 1048 (1976)	§ 4.4
International Shoe Co. v. Commissioner, 38 B.T.A. 81 (1938)	§ 17.25
Jacobson & Co., GSBCA No. 5605, 80-2 B.C.A. (CCH) ¶ 14,521 (1980)	§ 5.37
Jacob & Youngs v. Kent, 230 N.Y. 239, 129 N.E. 889 (1921)	§ 15.11
J.A. Jones Constr. Co. v. Greenbriar Shopping Center, 332 F. Supp. 1336 (N.D. Ga. 1971), *aff'd,* 461 F.2d 1269 (5th Cir. 1972)	§ 14.21
Jamaica Water Supply Co. v. Commissioner, 125 F.2d 512 (2d Cir. 1942), *aff'g* 42 B.T.A. 359 (1940)	§ 17.18
James King & Son, Inc. v. DeSantis Constr. No. 2 Corp., 413 N.Y.S.2d 78 (Sup. Ct. 1977)	§ 8.2
James McKinney & Son, Inc. v. Lake Placid 1980 Olympic Games, Inc., 461 N.Y.S.2d 483 (Sup. Ct. App. Div. 1983)	§§ 5.16, 5.19, 6.2

Case	*Book §*
Japanese Elecs. Prods. Antitrust Litig., *In re,* 631 F.2d 1069 (3d Cir. 1980)	§ 19.13
Jardine Estates, Inc. v. Donna Brook Corp., 42 N.J. Super. 332, 126 A.2d 372 (App. Div. 1956)	§ 15.11
J.A. Ross & Co. v. United States, 115 F. Supp. 187 (Ct. Cl. 1953)	§ 14.11
Jasper Constr. v. Foothill Junior College Dist., 91 Cal. App. 3d 1, 153 Cal. Rptr. 767 (1979)	§ 14.4
J.A. Tobin Constr. v. State Highway Comm'n, 680 S.W.2d 183 (Mo. Ct. App. 1984)	§ 3.7
Jenkins v. Krebs & Sons, Inc., 322 So. 2d 426 (La. Ct. App. 1975)	§ 10.33
J.F. White Eng'g Corp. v. United States *ex rel.* Pittsburgh Plate Glass Co., 311 F.2d 410 (10th Cir. 1962)	§ 11.10
J.G. Plumbing Serv., Inc. v. Coastal Mortgage Co., 329 So. 2d 393 (Fla. Dist. Ct. App.), *cert. denied,* 339 So. 2d 1169 (Fla. Dist. Ct. App. 1976)	§§ 7.20, 7.25
Joba Constr. Co. v. Burns & Roe, Inc., 29 N.W.2d 760 (Mich. Ct. App. 1982)	§§ 5.18, 10.27
Joba Constr. Co., Inc. v. Burns & Roe, Inc., 500 F. Supp. 910 (E.D. Mich. 1980)	§ 5.21
John Burns Constr. Co. v. Interlake, Inc., 433 N.E.2d 1126 (Ill. App. Ct. 1982)	§ 3.14
John E. Green Plumbing & Heating Co., Inc. v. Turner Constr. Co., 742 F.2d 965 (6th Cir. 1984)	§§ 5.25, 5.26, 6.2, 14.35
John E. Green Plumbing & Heating Co., Inc. v. Turner Constr. Co., 500 F. Supp. 910 (E.D. Mich. 1980)	§§ 5.18, 5.24, 5.31
John F. Harkins Co. v. School Dist. of Philadelphia, 460 A.2d 260 (Pa. Super. Ct. 1983)	§ 3.13
John J. Calnan Co. v. Talsma Builders, Inc., 367 N.E.2d 695 (Ill. 1977)	§ 3.14
John R. Dudley Constr., Inc. v. Drott Mfg. Co., 66 A.D.2d 368, 412 N.Y.S.2d 512 (1979)	§ 5.41
Johns-Manville Sales Corp. v. Reliance Ins. Co., 410 F.2d 277 (9th Cir. 1969)	§ 11.17
Johnson, T.C. Memo 1983-517	§ 17.20
Johnson v. City of N.Y., 191 A.D. 205, 181 N.Y.S. 137 (2d Dep't 1920)	§ 14.18
Johnson v. Fenestra, Inc., 305 F.2d 179 (3d Cir. 1962)	§§ 1.11, 15.3
Johnson v. State, 24 N.Y.2d 958, 302 N.Y.S.2d 489 (1969)	§ 14.16
Johnson Controls, Inc. v. National Valve & Mfg. Co., 569 F. Supp. 758 (E.D. Okla. 1983)	§ 14.28
John T. Brady & Co. v. Board of Educ., 222 A.D. 504, 226 N.Y.S. 707 (1st Dep't 1928)	§ 14.19
Jones v. J.H. Hiser Constr., Inc., 484 A.2d 302 (Md. Ct. Spec. App. 1984)	§ 10.28
Jones v. Scofield Bros., 73 F. Supp. 395 (D. Md. 1947)	§ 4.2
Joseph F. Hughes & Co. v. George H. Robinson Corp., 175 S.E.2d 413 (Va. 1970)	§ 11.5
Joseph Pickard's Sons Co. v. United States, 532 F.2d 739 (Ct. Cl. 1946)	§ 14.36
J.S. Watkins v. Department of Highways, 290 S.W.2d 28 (Ky. 1956)	§ 4.6

Case	*Book §*
J.W. Bateson Co. v. United States *ex rel.* Board of Trustees, 434 U.S. 586 (1978)	§§ 11.6, 11.8
J.W. Cooper Constr. Co. v. Public Hous. Admin. *ex rel.* Rio Grande Steel Prods. Co., 390 F.2d 175 (10th Cir. 1968)	§ 11.6
Kalisch-Jarcho, Inc. v. City of N.Y., 58 N.Y.2d 377, 448 N.E.2d 413, 461 N.Y.S.2d 746 (1983)	§§ 3.10, 5.33, 5.35, 14.17, 14.19
Kandalis v. Paul Pet Constr. Co., 210 Md. 319, 123 A.2d 345 (1956)	§ 15.3
Kanters v. Kotick, 102 Wash. 523, 173 P.2d 329 (1918)	§ 10.21
Katz v. Bedford, 77 Cal. 319, 19 P. 523 (1888)	§ 15.3
Keco Indus., Inc. v. United States, 492 F.2d 1200 (Ct. Cl. 1974)	§ 14.30
Keco Indus., Inc. v. United States, 428 F.2d 1233 (Ct. Cl. 1970)	§ 2.16
Keene Corp. v. Insurance Co. of N. Am., 667 F.2d 1034 (D.C. Cir. 1981)	§ 9.5
Kennedy Elec. Co., Inc. v. United States Postal Serv., 508 F.2d 954 (10th Cir. 1974)	§ 11.17
Kerry Coal Co. v. United Mine Workers, 470 F. Supp. 1032 (W.D. Pa. 1979)	§ 19.8
Kessler v. Peck, 266 Ala. 669, 98 So. 2d 606 (1957)	§ 5.44
K&G Constr. Co. v. Harris, 223 Md. 305, 164 A.2d 451 (1960)	§ 15.3
Khabbaz v. Swartz, 319 N.W.2d 279 (Iowa 1982)	§ 7.8
Kinchen v. Gilworth, 454 So. 2d 1130 (La. Ct. App. 1984)	§ 10.28
King Enters., *In re,* 678 F.2d 73 (8th Cir. 1982)	§ 3.13
Klapper v. Hanover Ins. Co., 39 Misc. 2d 215, 240 N.Y.S.2d 284 (1963)	§ 9.9
Klein-Dickert Oshkosh, Inc. v. Frontier Mortgage Corp., 287 N.W.2d 742 (Wis. 1980)	§ 7.25
Knight Constr. Co. v. Barnett Mortgage Trust, 572 S.W.2d 381 (Tex. Civ. App. 1978), *writ refused n.r.e.*	§§ 7.8, 7.9
Knoell Constr. Co. v. Hanson, 287 N.W.2d 435 (Neb. 1980)	§ 3.17
Koerner v. Commissioner, 10 T.C.M. (CCH) 867 (1951)	§ 17.25
Koppers Co., Inc. v. Garling & Langlois, 594 F.2d 1094 (6th Cir. 1979)	§ 7.26
Kornhauser v. United States, 276 U.S. 145 (1928)	§ 17.25
Kosanovich v. Meade, 463 N.E.2d 257 (Ind. 1984)	§ 10.35
Kraemer Bros. Inc. v. United States Fire Ins. Co., 89 Wis. 2d 555, 278 N.W.2d 857 (1979)	§ 9.20
Kreimer v. Second Fed. Sav. & Loan Ass'n of Pittsburgh, 176 A.2d 132 (Pa. Super. Ct. 1961)	§ 7.9
Krichner v. Hartford Accident & Indem. Co., 440 S.W.2d 751 (Mo. Ct. App. 1969)	§ 9.9
Krieger v. S.E. Griner Co., Inc., 282 Md. 50, 832 A.2d 1069 (1978)	§§ 5.10, 5.11
Kubbe v. Crescent Steel, 105 Ariz. 459, 466 P.2d 753 (1970)	§ 15.7
Kvenild v. Taylor, 594 P.2d 972 (Wyo. 1979)	§ 10.27
Laburnum Constr. Corp. v. United States, 325 F.2d 451 (Ct. Cl. 1963)	§ 14.4
Laclede Inv. Corp. v. Kaiser, 596 S.W.2d 36 (Mo. Ct. App. 1980)	§ 7.8

Case	*Book §*
Louisiana, State of *ex rel.* William J. Guste, Jr. v. M/V Testbank, 752 F.2d 1019 (5th Cir. 1985), *appeal pending*	§ 6.2
Lucas v. American Code Co., Inc., 280 U.S. 445 (1930)	§ 17.31
Lumbermen's Mut. Cas. Co. v. Town of Pound Ridge, 362 F.2d 430 (2d Cir. 1966)	§ 9.8
Lustman, T.C. Memo 1960-116, *aff'd,* 322 F.2d 253 (3d Cir. 1953)	§ 17.31
Manchester Bank v. Connecticut Bank & Trust Co., 497 F. Supp. 1304 (D.N.H. 1980)	§ 10.27
Mann v. Clowser, 190 Va. 887, 59 S.E.2d 78 (1950)	§§ 15.2, 15.7
Mannella v. Pittsburgh, 6 A.2d 70 (Pa. 1939)	§ 10.38
Manshul Constr. Co. v. Dormitory Auth. of N.Y., 79 A.D.2d 383, 436 N.Y.S.2d 724 (1st Dep't 1981)	§§ 14.35, 14.40
Manshul Constr. Corp. v. Dormitory Auth. of N.Y., 111 Misc. 2d 209, 444 N.Y.S.2d 792 (1981)	§ 14.43
Manthey v. Stock, 133 Wis. 197, 113 N.W. 443 (1907)	§ 15.11
Martin v. National Sur. Co., 300 U.S. 588 (1937)	§ 10.32
Martinez v. Hawkeye-Security Ins. Co., 576 P.2d 1017 (Colo. 1978)	§ 9.15
Martin Fireproofing Corp. v. Aetna Ins. Co., 346 Mass. 498, 194 N.E.2d 101 (1963)	§ 10.4
Maryland Cas. Co. v. Cunningham, 234 Ala. 80, 173 So. 2d 506 (1937)	§ 10.23
Maryland Cas. Co. v. Farmers Alliance Mut. Ins. Co., 566 P.2d 168 (Okla. Ct. App. 1977)	§ 9.9
Masson v. Reindollar, 69 A.2d 482 (Md. 1949)	§ 4.9
Matson Navigation Co. v. Commissioner, 24 B.T.A. 14 (1931)	§ 17.25
Matthews Constr. Co. v. Brady, 104 N.J.L. 438, 140 A. 433 (1928)	§ 15.5
Mattingly v. Anthony Indus., Inc., 167 Cal. Rptr. 292 (Ct. App. 1980)	§ 10.34
Mayor of Baltimore v. Fidelity & Deposit Co. of Md., 386 A.2d 749 (Md. 1978)	§§ 11.4, 11.5
Mayville-Portland School Dist. v. C.I. Linfoot Co., 261 N.W.2d 907 (N.D. 1978)	§ 15.2
Maxfield v. Simmons, 449 N.E.2d 110 (Ill. 1983)	§ 8.8
Maxwell v. Anderson, 593 P.2d 29 (Mont. 1979)	§ 3.17
McCaa, Sr., T.C. Memo 1967-152	§ 17.25
McCarty Corp. v. United States, 499 F.2d 633 (Ct. Cl. 1974)	§ 14.30
McCloskey & Co. v. Wright, 363 F. Supp. 223 (E.D. Va. 1973)	§ 5.43
McCord, Condron & McDonald, Inc. v. Twin City Fire Ins. Co., 607 S.W.2d 956 (Tex. Civ. App. 1980)	§ 9.9
McCreary Roofing Co., Inc. v. Northern Ins. Co. of N.Y., 275 A.2d 388 (Pa. Super. 1971)	§ 9.9
McDaniel v. Ashton-Mardian Co., 357 F.2d 511 (9th Cir. 1966)	§§ 11.9, 14.14
McDonald V. Mianecki, 79 N.J. 275, 398 A.2d 1283 (1979)	§ 15.8
McGann v. Hobbs Lumber Co., 150 W. Va. 364, 145 S.E.2d 476 (1965)	§ 9.2
McGrath v. American Sur. Co. of N.Y., 307 N.Y. 552, 122 N.E.2d 906 (1954)	§ 10.12
McGraw v. Johnson, 42 N.J. Super. 267, 126 A.2d 203 (App. Div. 1956)	§ 15.8

Case	*Book §*
Montgomery County Bd. of Educ. v. Glassman Constr. Co., 225 A.2d 448 (Md. 1967)	§ 11.4
Montgomery Indus. Int'l, Inc. v. Thomas Constr. Co., 620 F.2d 91 (5th Cir. 1980)	§§ 1.24, 8.2
Moore v. Pavex, Inc., No. 2843 S 1984, slip op. (Ct. of Common Pleas of Dauphin County, Pa., Oct. 4, 1985)	§ 6.2
Moorehead Constr. Co., Inc. v. City of Grand Fortis, 508 F.2d 1008 (8th Cir. 1973)	§ 14.36
Morrison v. MacNamara, 407 A.2d 555 (D.C. 1979)	§ 5.16
Morrison-Knudsen Co., ASBCA No. 16483, 72-2 B.C.A. (CCH) ¶ 9733 (1972)	§ 4.10
Mortgage Assocs., Inc. v. Monona Shores, Inc., 177 N.W.2d 340 (Wis. 1970)	§§ 7.7, 7.15
Mounds View, City of v. Walijarvi, 263 N.W.2d 420 (Minn. 1978)	§§ 5.6, 6.3
M&R Contractors & Builders v. Michael, 215 Md. 340, 138 A.2d 350 (1958)	§ 14.35
M. Steinthal & Co. v. Seamans, 455 F.2d 1289 (D.C. Cir. 1971)	§ 4.5
Muller v. Light, 538 S.W.2d 487 (Tex. Civ. App. 1976)	§ 15.10
Murphy, 48 T.C. 569 (1967)	§ 17.25
Nanuet Nat'l Bank v. Eckerson Terrance, Inc., 47 N.Y.2d 243, 391 N.E.2d 983, 417 N.Y.S.2d 901 (1979)	§ 7.5
Nardi & Co., Inc. v. Allabastro, 314 N.E.2d 367 (Ill. App. Ct. 1974)	§ 10.28
Nat Harrison Assocs., Inc. v. Gulf States Utils. Co., 491 F.2d 578 (5th Cir. 1974)	§ 14.27
National Coop. Refinery Ass'n v. Northern Ordnance, Inc., 258 F.2d 803 (10th Cir. 1956)	§ 10.34
National Investors Fire & Cas. Co. v. Preddy, 248 Ark. 320, 451 S.W.2d 457 (1970)	§ 9.20
National Shawmut Bank of Boston v. New Amsterdam Cas. Co., 411 F.2d 843 (1st Cir. 1969)	§§ 10.24, 10.31
National Sur. Corp. v. United States, 319 F. Supp. 45 (N.D. Ala. 1970)	§ 10.32
National Sur. Corp. v. United States *ex rel.* Way Panama, S.A., 378 F.2d 294 (5th Cir.), *cert. denied,* 389 U.S. 1004 (1967)	§ 11.12
Nationwide Mut. Ins. Co. v. Wenger, 222 Va. 263, 278 S.E.2d 874 (1981)	§ 9.8
Natkin & Co. v. George A. Fuller Co., 347 F. Supp. 17 (W.D. Mo. 1972)	§§ 3.13, 14.27, 15.9
Nees v. Weaver, 222 Wis. 492, 269 N.W. 266 (1936)	§ 15.11
Nelse Mortensen v. Group Health, Inc., 566 P.2d 560 (Wash. Ct. App. 1977)	§ 5.32
Nether Providence Township School Auth. v. Thomas M. Durkin & Sons, Inc., 476 A.2d 904 (Pa. 1984)	§§ 3.3, 14.11
New Amsterdam Cas. Co. v. Lundquist, 198 N.W.2d 543 (Minn. 1972)	§ 10.30
Newman v. National Fire Ins. Co., 152 Miss. 344, 188 So. 295 (1928)	§ 9.17
Newport News, City of v. Doyle & Russell Inc., 179 S.E.2d 493 (Va. 1971)	§ 2.30

Case	*Book §*
Pathman Constr. Co., A.S.B.C.A. No. 22343, 81-1 B.C.A. (CCH) 15010 (1981)	§ 1.25
Pathman Elec. Co. v. Hi-Way Elec. Co., 382 N.E.2d 453 (Ill. App. Ct. 1978)	§ 3.16
Paxton v. County of Alameda, 259 P.2d 934 (Cal. Ct. App. 1953)	§ 6.3
PDM Plumbing & Heating, Inc. v. Findlan, 431 N.E.2d 594 (Mass. App. Ct. 1982)	§ 3.13
Peach State Uniform Serv. Inc. v. American Ins. Co., 507 F.2d 996 (5th Cir. 1975)	§ 9.20
Pearlman v. Reliance Ins. Co., 371 U.S 132 (1962)	§§ 10.31, 10.32, 11.3, 11.17
Peerless Ins. Co. v. Cerny & Assocs., Inc., 199 F. Supp. 951 (D. Minn. 1961)	§ 10.33
Peltz Constr. Co. v. Dunham, 436 N.E.2d 892 (Ind. Ct. App. 1982)	§ 15.7
Pembroke v. Caudill, 37 So. 2d 538 (Fla. 1948)	§ 10.34
Peninsular Supply Co. v. C.B. Day Realty, 423 So. 2d 500 (Fla. Dist. Ct. App. 1982)	§ 3.17
Penn Bridge Co. v. New Orleans, 222 F. 737 (5th Cir. 1915)	§ 15.2
Pennsylvania State Shopping Plazas, Inc. v. Olive, 202 Va. 862, 120 S.E.2d 372 (1961)	§ 15.5
People v. Craig City, 133 N.E. 419 (N.Y. 1921)	§ 5.33
People *ex rel.* Wells & Newton Co. v. Craig, 232 N.Y. 125, 133 N.E. 419 (1921)	§ 14.19
People's Gas, Light & Coke Co. v. United States Postal Serv., 658 F.2d 1182 (7th Cir. 1981)	§ 4.5
Perkins v. Lukens Steel Co., 310 U.S. 113 (1940)	§ 4.5
Peter Kiewit Sons' Co. v. Iowa S. Utils. Co., 355 F. Supp. 376 (S.D. Iowa 1973)	§§ 5.31, 5.35, 6.2, 6.3, 6.8, 14.16, 19.17
Peter Kiewit Sons' Co. v. Summit Constr. Co., 422 F.2d 242 (8th Cir. 1969)	§ 14.7
Peters Township School Auth. v. United States Fidelity & Guar. Co., 467 A.2d 904 (Pa. 1983)	§ 10.36
Petrol Indus., Inc. v. Gearhart-Owen Indus., Inc., 424 So. 1059 (La. Ct. App. 1982)	§ 9.9
Peyronnin Constr. Co., Inc. v. Weiss, 208 N.E.2d 489 (Ind. Ct. App. 1965)	§ 6.2
P.H.A.C. Servs. v. Seaways Int'l, Inc., 403 So. 2d 1199 (La. 1981)	§ 10.28
Philadelphia Facilities Management Corp. v. St. Paul Fire & Marine Ins. Co., 379 F. Supp. 780 (E.D. Pa. 1974)	§ 9.17
Philadelphia Hous. Auth. v. Turner Constr. Co., 343 Pa. 512, 23 A.2d 426 (1942)	§ 15.6
Phil Crowley Steel Corp. v. Sharon Steel Corp., 536 F. Supp. 429 (E.D. Mo. 1982)	§ 10.27
Phillips & Colby Constr. Co. v. Seymour, 91 U.S. 646 (1875)	§ 15.3
Phoenix Coal Co. v. Commissioner, 231 F.2d 420 (2d Cir. 1956), *aff'g* T.C. Memo 1955-28	§ 17.9
Phoenix Contractors, Inc. v. General Motors Corp., 135 Mich. App. 787, 355 N.W.2d 673 (1984)	§ 8.6
Phoenix of Hartford v. Holloway Corp., 268 So. 2d 195 (Fla. Dist. Ct. App. 1972)	§ 9.9

Case	*Book §*
Ralph C. Sutro Co. v. Paramount Plastering Inc., 216 Cal. App. 2d 433, 31 Cal. Rptr. 174 (1963)	§ 7.18
Raytheon Prods. Corp. v. Commissioner, 144 F.2d 110 (1st Cir. 1944)	§§ 17.3, 17.4, 17.6, 17.7, 17.9
R.C. Stanhope, Inc. v. Roanoke Constr. Co., 539 F.2d 992 (4th Cir. 1976)	§ 11.2
Redman Dev. Corp. v. Piedmont Heating & Air Conditioning Co., 197 S.E.2d 167 (Ga. Ct. App. 1974)	§ 1.19
Redna Marine Corp. v. Poland, 46 F.R.D. 81 (D.C.N.Y. 1969)	§ 9.21
Reed v. Commissioner, 723 F.2d 138 (1st Cir. 1983), *rev'g* T.C. Memo 1982-734	§ 17.19
Reichelet v. Urban Inv. & Dev. Co., 577 F. Supp. 971 (D. Ill. 1984)	§ 15.7
Reinharcz v. Chrysler Motors Corp., 514 F. Supp. 1141 (C.D. Cal. 1981)	§ 10.27
Reliance Ins. Co. v. Jones, 296 F.2d 71 (10th Cir. 1961)	§ 9.17
Reliance Ins. Co. v. Trane Co., 194 S.E.2d 817 (Va. 1971)	§ 11.5
Remmer v. Glens Falls Indem. Co., 140 Cal. App. 2d 84, 295 P.2d 19 (1956)	§ 9.5
Renown, Inc. v. Hensel Phelps Constr. Co., 201 Cal. Rptr. 242 (Ct. App. 1984)	§ 10.34
Rice, Barton & Fales, Inc. v. Commissioner, 41 F.2d 339 (1st Cir. 1930)	§ 17.30
Rickel v. Schwinn Bicycle Co., 192 Cal. Rptr. 732 (Ct. App. 1983)	§ 10.27
Ridley v. Pipe Maintenance Servs., Inc., 477 A.2d 610 (Pa. Commw. Ct. 1984)	§ 4.12
Ridley Inv. Co. v. Croll, 192 A.2d 925 (Del. 1963)	§§ 15.2, 15.6
Ridley, Township of v. Haulaway Trash Removal, Inc., 68 Pa. Commw. 176, 448 A.2d 654 (1982)	§ 4.12
Robert Gordon, Inc. v. Ingersoll-Rand Co., 117 F.2d 654 (7th Cir. 1941)	§ 8.3
Roberts v. Sinnot, 55 Mont. 369, 177 P. 252 (1918)	§ 15.2
Roberts Constr. Co., *In re,* 111 N.W.2d 767 (Neb. 1961)	§ 1.25
Robins Dry Dock & Repair Co. v. Flint, 275 U.S. 303 (1927)	§ 10.27
Rockhill v. United States, 418 A.2d 197, (Md. Ct. App. 1980)	§ 7.24
Ronald Adams, Contractor, Inc. v. State, 457 So. 2d 778 (La. Ct. App. 1984)	§ 3.5
Rosos Litho Supply Corp. v. Richard T. Hanson, 123 Ill. App. 3d 290, 462 N.E.2d 566 (1984)	§ 5.41
Royal Indem. Co. v. T.B. Smith, 121 Ga. App. 272, 173 S.E.2d 738 (1970)	§ 9.9
Royal Indem. Co. v. United States, 371 F.2d 462 (Ct. Cl. 1967), *cert. denied,* 389 U.S. 833 (1967)	§ 10.32
R.P. Russo Contractors v. C.J. Pettinato Realty, 482 A.2d 1086 (Pa. Super. Ct. 1984)	§§ 7.9, 7.10
R.S. Noonan, Inc. v. Morrison-Knudsen Co., Inc., 522 F. Supp. 1186 (E.D. La. 1981)	§ 5.23
Rushlight Auto Sprinkler Co. v. City of Portland, 219 P.2d 732 (Or. 1950)	§ 2.30
Ryan v. Morgan Spear Assocs., Inc., 546 S.W.2d 678 (Tex. Civ. App. 1977)	§ 5.6

Case	*Book §*
Security Ins. Co. v. United States *ex rel.* Haydis, 338 F.2d 444 (9th Cir. 1964)	§ 11.14
Security Ins. Co. of Hartford v. Kay Mill Supply, Inc., 211 N.W.2d 519 (Minn. 1973)	§ 9.15
Segovia Dev. Corp. v. Constructora Maza, Inc., 628 F.2d 724 (1st Cir. 1980)	§ 10.25
Seif v. Long Beach, 286 N.Y. 382, 36 N.E.2d 630 (1941)	§ 4.12
Seretto v. Rockland, S.T.& O.H. Ry. Co., 63 A. 651 (Me. 1906)	§ 1.25
Shakin v. Buskirk, 454 Mich. 490, 93 N.W.2d 293 (1958)	§ 14.35
Shepherd Constr. Co., Inc., 51 T.C. 890 (1969), *acq.*, 1969-2 C.B. 25	§ 17.31
Shoffner Indus. v. W.B. Lloyd Constr. Co., 257 S.E.2d 50 (N.C. Ct. App. 1979)	§§ 6.2, 6.6
Sides v. Richard Mach. Works, Inc., 406 F.2d 445 (4th Cir. 1969)	§ 5.16
Siefford v. Housing Auth. of City of Humboldt, 192 Neb. 943, 223 N.W.2d 816 (1974)	§ 14.28
Silverdale Hotel v. Lomas & Nettleton Co., 36 Wash. App. 762, 677 P.2d 773 (1984)	§§ 7.9, 7.15
Singsaas v. Diederich, 238 N.W.2d 878 (Minn. 1976)	§ 9.15
S.J. Groves & Son v. Warner Co., 576 F.2d 524 (3d Cir. 1978)	§ 15.7
Slate Constr. Co. v. Bituminous Cas. Corp., 228 Pa. Super. 1, 323 A.2d 141 (1974)	§ 9.9
Slattery Assocs. v. City of N.Y., 98 A.D.2d 686, 469 N.Y.S.2d 758 (1st Dep't 1983)	§ 14.17
Sloane v. Malcolm Price, Inc., 339 A.2d 43 (D.C. 1975)	§ 3.17
S.L. Rowland Constr. Co. v. St. Paul Fire & Marine Ins. Co., 72 Wash. 2d 682, 434 P.2d 725 (1967)	§ 9.12
Smither & Co. v. Calvin Humphrey Corp., 232 F. Supp. 204 (D.D.C. 1964)	§ 15.3
Smith & Lowe Constr. Co. v. Herrera, 79 N.M. 239, 442 P.2d 197 (1968)	§§ 2.30, 3.14
S.N. Nielsen Co. v. United States, 141 Ct. Cl. 793 (1958)	§ 3.2
Solite Masonry Units Corp. v. Piland Constr. Co., 232 S.E.2d 759 (Va. 1977)	§ 11.4
Son-Shine Grading, Inc. v. ADC Constr. Co., 315 S.E.2d 346 (N.C. Ct. App. 1984)	§ 10.38
Sorenson v. City of N.Y., 202 F.2d 857 (2d Cir. 1953), *cert. denied,* 47 U.S. 951 (1951)	§ 4.12
Sousa v. Albino, 388 A.2d 804 (R.I. 1978)	§ 15.7
Southeastern Sales & Serv. Co. v. T.T. Watson, Inc., 172 So. 2d 239 (Fla. Dist. Ct. App. 1965)	§§ 1.24, 8.3
Southern Bell Tel. & Tel. Co. v. Acme Elec. Contractors, Inc., 418 So. 2d 1187 (Fla. 1983)	§ 3.13
Southern Cal. Edison Co. v. Harbor Ins., 83 Cal. App. 3d 747, 148 Cal. Rptr. 106 (1978)	§ 9.22
Southern Gulf Utils., Inc. v. Boca Ciega Sanitary Dist., 238 So. 2d 458 (Fla. Dist. Ct. App. 1970)	§ 5.34
Southern New England Contracting Co. v. State, 165 Conn. 644, 345 A.2d 550 (1974)	§ 15.2
Space Leasing Assocs. v. Atlantic Bldg. Sys., Inc., 144 Ga. App. 320, 241 S.E.2d 438 (1977)	§ 15.7

Case	*Book §*
Spang Indus., Inc., Fort Pitt Bridge Div., v. Aetna Cas. & Sur. Co., 512 F.2d 365 (2d Cir. 1975)	§ 3.12
Spence v. Harm, 163 N.Y. 220, 57 N.E. 412 (1900)	§ 15.11
Sprayregen v. American Airlines, Inc., 570 F. Supp. 16 (D.C.N.Y. 1983)	§ 5.40
S.S. Silberblatt, Inc. v. East Harlem Pilot Block, 608 F.2d 28 (2d Cir. 1979)	§§ 7.7, 7.18, 7.20
Staley v. New, 56 N.M. 745, 250 P.2d 893 (1952)	§ 15.2
Standard Accident Ins. Co. v. United States *ex rel.* Powell, 302 U.S. 442 (1938)	§ 11.7
Standard Venetian Blind Co. v. American Empire Ins. Co., 469 A.2d 563 (Pa. 1983)	§ 9.9
State v. Omega Painting, Inc., 463 N.E.2d 287 (Ind. Ct. App. 1984)	§§ 3.3, 14.11
State *ex rel.* Fisher Constr. Co. v. Linzell Constr. Co., 101 Ohio App. 219 (1955)	§ 2.23
State *ex rel.* Lane v. Dashiell, 75 A.2d 348 (Md. 1950)	§ 4.9
State *ex rel.* National Sur. Corp. v. Malvaney, 72 So. 2d 424 (Miss. 1954)	§ 10.33
State Highway Admin. v. Transamerica Ins. Co., 367 A.2d 509 (Md. 1976)	§ 11.3
State Sur. Co. v. Lamb Constr. Co., 625 P.2d 184 (Wyo. 1981)	§ 10.34
Statesville, N.C., City of, Grant No. C370395-02 (Protest of DPS Contractors, Inc.) (EPA Region IV, Nov. 17, 1982)	§ 2.9
Stauffer Constr. Co., Inc. v. Tate Eng'g, Inc., 407 A.2d 1191 (Md. Ct. Spec. App. 1979)	§ 11.2
St. Claire Shores, City of v. L.&L. Constr. Co., 363 Mich. 518, 109 N.W.2d 802 (1961)	§ 15.3
Stearns Constr. Corp. v. Carolina Corp., 217 N.W.2d 291 (Wis. 1974)	§ 10.34
Steckmar Nat'l Realty & Inv. Corp., Ltd. v. J.I. Case Co., 99 Misc. 2d 212, 415 N.Y.S.2d 946 (1979)	§ 5.41
Stees v. Leonard, 20 Minn. 494 (1874)	§ 15.6
Stehlin-Muller-Henes Co. v. City of Bridgeport, 117 A. 811 (Conn. 1922)	§§ 1.11, 1.25
Steltz v. Armory Co., 15 Idaho 551, 99 P. 98 (1908)	§ 15.3
Stenerson v. City of Kalispell, 629 F.2d 773 (Mont. 1981)	§ 3.9
Stephens v. Great S. Sav. & Loan Ass'n, 421 S.W.2d 332 (Mo. Ct. App. 1967)	§§ 7.9, 7.10
Stillwater Condominium Ass'n v. American Home Assurance Co., 508 F. Supp. 1075 (D. Mont. 1981), *aff'd,* 688 F.2d 848 (9th Cir. 1982), *cert. denied,* 460 U.S. 1038 (1983)	§ 9.8
Stinson v. Mueller, 449 A.2d 329 (D.C. App. 1982)	§ 15.12
Stonehocker v. General Motors Corp., 587 F.2d 151 (4th Cir. 1978)	§ 19.13
Storm Drilling Co. v. Atlantic Richfield Corp., 386 F. Supp. 830 (E.D. Cal. 1974)	§ 19.8
Storms v. United States Fidelity & Guar. Co., 388 A.2d 578 (N.H. 1978)	§§ 9.2, 9.15
St. Paul Fire & Marine Ins. Co. v. Sears, Roebuck & Co., 603 F.2d 780 (9th Cir. 1979)	§ 9.1
St. Paul Fire & Marine Ins. Co. v. Coss, 80 Cal. App. 3d 888, 145 Cal. Rptr. 836 (1978)	§§ 9.6, 9.8

Case	*Book §*
St. Paul Fire & Marine Ins. Co. v. Murray Plumbing & Heating Corp., 65 Cal. App. 3d 66, 135 Cal. Rptr. 120 (1976)	§ 9.20
Stroebel v. Jefferson Trucking & Rigging Co., 125 N.J.L. 484, 15 A.2d 805 (E & A 1940)	§ 15.13
Strouss v. Simmons, 657 P.2d 1004 (Hawaii 1982)	§§ 7.12, 7.15, 7.16
Strubble v. United Servs. Auto. Assocs., 35 Cal. App. 3d 498, 110 Cal. Rptr. 828 (1973)	§ 9.21
Studiengesellschaft Kohle M&H v. Novamont Corp., 485 F. Supp. 471 (S.D.N.Y. 1980)	§ 15.12
Sturla, Inc. v. Fireman's Fund Ins. Co., 684 P.2d 960 (Hawaii 1984)	§ 9.8
Superior Glass Co. v. First Bristol County Nat'l Bank, 8 Mass. App. 356, 394 N.E.2d 972 (1979), *aff'd,* 406 N.E.2d 672 (1980)	§§ 7.22, 7.24
Swarthaut v. Beard, 33 Mich. App. 395, 90 N.W.2d 373 (1971), *rev'd on other grounds,* 388 Mich. 637, 202 N.W.2d 300 (1972)	§ 5.14
Swastika Oil & Gas Co. v. Commissioner, 123 F.2d 382 (6th Cir. 1941)	§ 17.5
Swinerton & Walberg Co. v. Inglewood, 40 Cal. App. 3d 98 (1974)	§ 2.27
Sylla v. United States Fidelity & Guar. Co., 54 Cal. App. 3d 895, 127 Cal. Rptr. 38 (1976)	§ 9.5
Taramac Dev. Co., Inc. v. Delamater Fround & Assocs., P.A., 675 P.2d 361 (Kan. 1984)	§ 5.19
Teeples v. Tolson, 207 F. Supp. 212 (D.C. Or. 1962)	§ 9.16
Terry v. United States Fidelity & Guar. Co., 196 Wash. 206, 82 P.2d 532 (1938)	§ 10.23
Tessmar v. Grosner, 23 N.J. 193, 128 A.2d 467 (1957)	§ 15.8
Texas E. Transmission Corp. v. Marine Office-Appelton & Cox Corp., 579 F.2d 561 (10th Cir. 1978)	§§ 9.18, 9.21
Third Nat'l Bank in Nashville v. Highlands Ins. Co., 603 S.W.2d 730 (Tenn. 1980)	§ 10.31
Thomas W. Hooley & Sons v. Zurich Gen. Accident & Liab. Ins. Co., 235 La. 289, 103 So. 2d 449 (1958)	§ 9.9
Thompson-King-Tate, Inc. v. United States, 296 F.2d 290 (6th Cir. 1961), *rev'g & remanding* 185 F. Supp. 748 (D. Ky. 1955)	§ 17.30
Tiano v. Aetna Cas. & Sur. Co., 102 Mich. App. 177, 301 N.W.2d 476 (1980)	§ 9.15
Tippets-Abbett-McCarthy-Stratton v. New York State Thruway Auth., 27 Misc. 2d 522 (N.Y. Ct. Cl. 1961), *rev'd on other grounds,* 18 A.D.2d 402, *aff'd,* 13 N.Y.2d 1091 (1963)	§ 15.9
Todd Shipyards Corp. v. Turbine Serv., Inc., 674 F.2d 401 (5th Cir. 1982)	§ 9.12
Tolstoy Constr. Co. v. Minter, 78 Cal. App. 3d 665, 143 Cal. Rptr. 570 (1978)	§ 15.11
Tonawanda, Town of v. Stapell, Mumm & Beals Corp., 270 N.Y.S. 377 (N.Y. 1934)	§ 10.34
Torrey v. Simon-Torrey, Inc., 307 So. 2d 569 (La. 1974)	§ 10.38
T.P.M.P.T. Employees Credit Union v. Charpentier, 376 So. 2d 592 (La. Ct. App. 1979)	§ 10.30
Tranco Indus., Inc., ASBCA No. 26955, 83-1 B.C.A. (CCH) ¶ 16,414 (1983)	§ 3.7

Case	*Book §*
Transamerica Ins. Co. v. Gage Plumbing & Heating Co., 433 F.2d 1051 (10th Cir. 1970)	§ 9.23
Trans-Bay Eng'rs & Builders, Inc. v. Hills, 551 F.2d 370 (D.C. Cir. 1976)	§§ 7.7, 7.21
Travelers Indem. Co. v. United States *ex rel.* Construction Specialities Co., 382 F.2d 103 (10th Cir. 1967)	§ 11.14
Travelers Indem. Co. v. United States *ex rel.* Western Steel Co., 362 F.2d 896 (9th Cir. 1966)	§§ 11.6, 11.10
Travelers Indem. Co. v. Western Ga. Nat'l Bank, 387 F. Supp. 1090 (N.D. Ga. 1974)	§ 10.24
Traylor v. Henkels & McCoy, Inc., 585 P.2d 970 (Idaho 1979)	§ 3.12
Treasure State Indus., Inc. v. Welsh, 173 Mont. 403, 567 P.2d 947 (1977)	§ 10.12
Tri-Cor, Inc. v. United States, 458 F.2d 112 (Ct. Cl. 1972)	§ 14.28
Trinity Universal Ins. Co. v. United States, 382 F.2d 317 (5th Cir. 1967), *cert. denied,* 390 U.S. 906 (1967)	§ 10.32
Trip v. Henderson, 158 Fla. 442, 28 So. 2d 857 (1947)	§ 15.5
Triple A Contractors Inc. v. Rural Water Dist., 603 P.2d 184 (Kan. 1979)	§ 2.30
Turman v. Central Billing Bureau, 279 Or. 443, 568 P.2d 1382 (1977)	§ 7.26
Turner Constr. Co. v. John B. Kelly Co., 442 F. Supp. 551 (E.D. Pa. 1976)	§ 9.23
Tuskegee Alumni Hous. Found., Inc. v. National Homes Constr. Corp., 450 F. Supp. 714 (S.D. Ohio 1978), *aff'd,* 624 F.2d 1101 (5th Cir. 1980)	§ 15.12
Twin City Constr. Co. v. ITT Indus. Credit Co., 358 N.W.2d 716 (Minn. Ct. App. 1984)	§§ 7.9, 7.18
218–220 Market St. Corp. v. Krich-Radisco, Inc., 124 N.J.L. 302, 11 A.2d 109 (1940)	§ 15.10
Unicon Mgmt. Corp. v. City of Chicago, 404 F.2d 627 (7th Cir. 1968)	§ 19.17
Union Bldg. Corp. v. J&J Bldg. & Maintenance Contractors, Inc., 578 S.W.2d 519 (Tex. Civ. App. 1979)	§ 3.13
Union Inv. Co. v. Commissioner, 21 T.C. 659 (1954)	§ 17.25
United Bonding Ins. Co. v. Stein, 273 F. Supp. 929 (D.C. Pa. 1967)	§ 10.30
United Elec. Corp. v. United States, 647 F.2d 1082 (Ct. Cl. 1981)	§ 11.17
United Order of Am. Bricklayers & Stone Masons Union No. 21 v. Thorleif Larsen & Son, Inc., 519 F.2d 331 (7th Cir. 1975)	§ 15.10
United Stage Equip., Inc. v. Charles Carter & Co., 342 So. 2d 1153 (La. Ct. App. 1977)	§ 5.7
United States v. Ames, 24 F. Cas. 784 (C.C.D. Mass. 1845)	§ 4.6
United States v. Anderson, 269 U.S 422 (1926)	§ 17.31
United States v. Bellard, 674 F.2d 330 (5th Cir. 1982)	§ 10.30
United States v. Bethlehem Steel Co., 205 U.S 105 (1907)	§ 15.10
United States v. Braverman, 522 F.2d 218 (7th Cir.), *cert. denied,* 423 U.S. 985 (1975)	§ 19.13
United States v. Brooks-Calloway Co., 318 U.S. 120 (1943)	§ 14.23
United States v. Citizens & S. Nat'l Bank of Atlanta, Ga., 367 F.2d 473 (4th Cir. 1966)	§ 1.11

Case	*Book §*
United States v. Commonwealth of Pa., Dep't of Highways, 349 F. Supp. 1370 (E.D. Pa. 1972)	§ 11.8
United States v. Crosland Constr. Co., 217 F.2d 275 (4th Cir. 1954)	§ 11.8
United States v. Farragut, 89 U.S. (22 Wall.) 406 (1875)	§ 4.6
United States v. Gilmore, 372 U.S. 39 (1963)	§ 17.4
United States v. John C. Grimberg Co., 702 F.2d 1362 (Fed. Cir. 1983)	§§ 2.16, 4.5
United States v. Johnson Controls, Inc., 2 Fed. Procurement Dec. ¶ 15 (Fed. Cir. 1983)	§ 5.1
United States v. King, 395 U.S. 1 (1968)	§ 4.4
United States v. Marquez, 449 F.2d 89 (2d Cir. 1971), *cert. denied,* 405 U.S. 963 (1972)	§ 19.13
United States v. Maryland Cas. Co., 323 F.2d 473, (5th Cir. 1963)	§ 11.8
United States v. Rogers & Rogers, 161 F. Supp. 132 (S.D. Cal. 1958)	§§ 5.22, 10.33
United States v. Safety Car Heating & Lighting Co., 297 U.S. 88 (1936)	§ 17.18
United States v. Spearin, 248 U.S. 132 (1918)	§§ 1.10, 14.12, 15.2
United States v. United Eng'g & Contracting Co., 234 U.S. 236 (1913)	§ 10.34
United States v. William F. Klingensmith, Inc., 670 F.2d 1227 (D.C. Cir. 1982)	§ 14.21
United States *ex rel.* Altman v. Young Lumber Co., 376 F. Supp. 1290 (D.S.C. 1974)	§ 11.14
United States *ex rel.* Angell Bros., Inc. v. Cave Constr., Inc., 250 F. Supp. 873 (D. Mont. 1966)	§ 11.16
United States *ex rel.* Astro Cleaning & Packaging Corp. v. Jamison Co., 425 F.2d 1281 (6th Cir. 1970)	§ 11.10
United States *ex rel.* A.V. DeBlasio Constr., Inc. v. Mountain States Constr. Co., 588 F.2d 259 (9th Cir. 1978)	§ 11.10
United States *ex rel.* Bailey v. Freethy, 469 F.2d 1348 (9th Cir. 1972)	§ 11.11
United States *ex rel.* Bailey-Lewis-Williams of Fla., Inc. v. Peter Kiewit Sons Co., 195 F. Supp. 752 (D.D.C. 1961), *aff'd,* 299 F.2d 930 (D.C. Cir. 1962)	§ 11.16
United States *ex rel.* Benkart Co. v. John A. Johnson & Sons, Inc., 236 F.2d 864 (3d Cir. 1956)	§ 11.6
United States *ex rel.* Billows Elec. Supply Co., Inc. v. E.J.T. Constr. Co., Inc., 517 F. Supp. 1178 (E.D. Pa. 1981), *aff'd,* 688 F.2d 827 (3d Cir.), *cert. denied,* 103 S. Ct. 126 (1982)	§§ 11.10, 11.12, 11.14
United States *ex rel.* Board of Trustees v. J.W. Bateson Co., Inc., 551 F.2d 1284 (D.C. Cir. 1977)	§ 11.6
United States *ex rel.* Bordallo Consol., Inc. v. Markowitz Bros. (Delaware), Inc., 249 F. Supp. 610 (D. Guam 1966)	§ 11.8
United States *ex rel.* Bryant v. Lembke Constr. Co., 370 F.2d 293 (10th Cir. 1966)	§ 11.6
United States *ex rel.* Bryant Elec. Co. v. Aetna Cas. & Sur. Co., 297 F.2d 665 (2d Cir. 1962)	§ 11.16
United States *ex rel.* Caldwell Foundry & Mach. Co. v. Texas Constr. Co., 237 F.2d 705 (5th Cir. 1955)	§ 11.10

Case	*Book §*
Inc., 581 F.2d 239 (9th Cir. 1978)	§ 11.11
United States *ex rel.* Honeywell v. A&L Mechanical Contractors, Inc., 677 F.2d 383 (4th Cir. 1982)	§ 11.11
United States *ex rel.* H.T. Sweeney & Son, Inc. v. E.J.T. Constr. Co., Inc., 415 F. Supp. 1328 (D. Del. 1976)	§ 11.14
United States *ex rel.* Humble Oil & Ref. Co. v. Fidelity & Cas. Co., 402 F.2d 893 (4th Cir. 1968)	§ 11.15
United States *ex rel.* I. Burack, Inc. v. Sovereign Constr. Co., 338 F. Supp. 657 (S.D.N.Y. 1972)	§ 11.12
United States *ex rel.* J.A. Edwards & Co. v. Peter Reiss Constr. Co., 273 F.2d 880 (2d Cir. 1959), *cert. denied,* 362 U.S. 951 (1960)	§ 11.12
United States *ex rel.* Jinks Lumber Co. v. Federal Ins. Co., 452 F.2d 485 (5th Cir. 1971)	§ 11.11
United States *ex rel.* Joseph I. Richardson, Inc. v. E.J.T. Constr. Co., Inc., 453 F. Supp. 435 (D. Del. 1978)	§ 11.14
United States *ex rel.* J.P. Byrne & Co. v. Fire Ass'n of Philadelphia, 260 F.2d 541 (2d Cir. 1958)	§ 11.7
United States *ex rel.* Kelly-Mohrhusen Co. v. Merle A. Patnode Co., 457 F.2d 116 (7th Cir. 1972)	§ 11.11
United States *ex rel.* Kinlau Sheet Metal Works, Inc. v. Great Am. Ins. Co., 537 F.2d 222 (5th Cir. 1976)	§ 11.11
United States *ex rel.* Lank Woodwork Co., Inc. v. CSH Contractors, Inc., 452 F. Supp. 922 (D.D.C. 1978)	§ 11.14
United States *ex rel.* Los Angeles Testing Laboratory v. Rogers & Rogers, 161 F. Supp. 132 (S.D. Cal. 1958)	§ 6.1
United States *ex rel.* Mariana v. Piracci Constr. Co., Inc., 405 F. Supp. 904 (D.D.C. 1975)	§ 11.9
United States *ex rel.* Mississippi Road Supply Co. v. H.R. Morgan, Inc., 542 F.2d 262 (5th Cir. 1976), *cert. denied,* 434 U.S. 828 (1977)	§§ 11.2, 11.7
United States *ex rel.* Nelson v. Reliance Ins. Co., 436 F.2d 1366 (10th Cir. 1971)	§ 11.15
United States *ex rel.* Olmsted Elec., Inc. v. Neosho Constr. Co., Inc., 599 F.2d 930 (10th Cir. 1979)	§ 11.12
United States *ex rel.* Pittsburgh-Des Moines Steel Co. v. Dauphin Steel & Eng'g Co., 53 F.R.D. 382 (S.D.N.Y. 1971)	§ 11.14
United States *ex rel.* Pittsburgh-Des Moines Steel Co. v. MacDonald Constr. Co., 281 F. Supp. 1010 (E.D. Mo. 1968)	§ 11.9
United States *ex rel.* Pool Constr. Co. v. Smith Rd. Constr. Co., 227 F. Supp. 315 (D. Okla. 1964)	§ 11.11
United States *ex rel.* Powers Regulator Co. v. Hartford Accident & Indem. Co., 376 F.2d 811 (1st Cir. 1967)	§ 11.6
United States *ex rel.* Pre-Fab Erectors, Inc. v. A.B.C. Roofing & Siding, Inc., 193 F. Supp. 465 (S.D. Cal. 1961)	§ 11.14
United States *ex rel.* Schaefer Elec., Inc. v. O. Frank Heinz Constr. Co., 300 F. Supp. 396 (S.D. Ill. 1969)	§ 11.10
United States *ex rel.* Sherman v. Carter, 353 U.S 210 (1957)	§§ 11.8, 11.10
United States *ex rel.* Soda v. Montgomery, 253 F.2d 509 (3d Cir. 1958)	§ 11.13
United States *ex rel.* State Elec. Supply. Co. v. Hesselden Constr. Co., 404 F.2d 774 (10th Cir. 1968)	§ 11.12

Case	*Book §*
United States *ex rel.* Statham Instruments, Inc. v. Western Cas. & Sur. Co., 359 F.2d 521 (6th Cir. 1966)	§§ 11.13, 11.14
United States *ex rel.* T. Square Equip. Corp. v. Gregor J. Schaefer Sons, Inc., 272 F. Supp. 962 (E.D.N.Y. 1967)	§ 11.14
United States *ex rel.* Vermont Marble Co. v. Roscoe-Ajax Constr. Co., 246 F. Supp. 439 (N.D. Cal. 1965)	§ 11.16
United States *ex rel.* Victory Elec. Corp. v. Maryland Cas. Co., 213 F. Supp. 800 (E.D.N.Y.), *further proceedings,* 215 F. Supp. 700 (E.D.N.Y. 1963)	§ 11.17
United States *ex rel.* Warren v. Kimrey, 489 F.2d 339 (8th Cir. 1974)	§ 10.15
United States *ex rel.* Wellman Eng'g Co. v. MSI Corp., 350 F.2d 285 (2d Cir. 1965)	§ 11.6
United States *ex rel.* West v. Peter Kiewit & Sons' Co., 235 F. Supp. 500 (D. Alaska 1964)	§ 11.8
United States *ex rel.* Westinghouse Elec. Supply Co. v. Endebrock-White Co., 275 F.2d 57 (4th Cir. 1960)	§ 11.7
United States *ex rel.* Wheeling-Pittsburgh Steel Corp. v. Algernon Blair, Inc., 329 F. Supp. 1360 (D.C.S.C. 1971)	§ 10.36
United States Fed. Eng'g & Mfg., ASBCA No. 19909, 75-2 B.C.A. (CCH) ¶ 11,578 (1975)	§ 4.11
United States Fidelity & Guar. v. Nevada Cement Co., 561 P.2d 1335 (Nev. 1977)	§ 9.6
United States Fidelity & Guar. Co. v. Louis A. Roser Co., 585 F.2d 932 (8th Cir. 1978)	§ 19.8
United States Fidelity & Guar. Co. v. Orlando Utils. Comm'n, 564 F. Supp. 962 (M.D. Fla. 1983)	§§ 10.32, 10.35
United States Fidelity & Guar. Co. v. Triborough Bridge Auth., 74 N.E.2d 226 (N.Y. 1974)	§ 10.32
United States Fidelity & Guar. Co. v. United States, 676 F.2d 622 (Ct. Cl. 1982)	§ 10.35
United States Fidelity & Guar. Co. v. United States, 475 F.2d 1377 (Ct. Cl. 1973)	§ 11.17
United States Fidelity & Guar. Co. v. Virginia Eng'g Co., 213 F.2d 109 (4th Cir. 1954)	§ 9.8
United States Fire Ins. Co. v. Colver, 600 P.2d 1 (Alaska 1979)	§ 9.8
United States Propellers v. Zenith Plastics Co., 187 Cal. App. 2d 780, 10 Cal. Rptr. 137 (1960)	§ 3.16
United States Steel Corp. v. Missouri Pac. R.R. Co., 668 F.2d 435 (8th Cir.), *cert. denied,* 459 U.S. 836 (1982)	§§ 5.35, 14.16
Unity Tel. Co. v. Design Serv. Co., Inc., 201 A.2d 177 (Me. 1964)	§ 10.33
Universal By-Prods., Inc. v. City of Modesto, 43 Cal. App. 3d 145 (1974)	§ 2.27
Universal Minerals, Inc. v. C.A. Hughes & Co., 669 F.2d 98 (3d Cir. 1981)	§ 4.4
University of Md. v. Maas, 197 A. 123 (Md. 1938)	§ 4.9
Uptagraff v. United States, 315 F.2d 200 (4th Cir.), *cert. denied,* 375 U.S. 818 (1963)	§ 10.24
U.S. Indus., Inc. v. Blake Constr. Co., 671 F.2d 539 (D.C. Cir. 1982)	§§ 3.10, 14.18
Utica Mut. Ins. Co. v. DiDonato, 187 N.J. Super. 30, 453 A.2d 559 (App. Div. 1982)	§ 15.10

Case	*Book §*
Valley Landscape Co., Inc. v. Rollader, 218 Va. 257, 237 S.E.2d 120 (1977)	§ 5.17
Vanasik v. Hirsch & Stevens, 65 Wis. 2d 1, 221 N.W.2d 815 (1974)	§ 5.11
Varia v. Southwick, 81 Idaho 68, 337 P.2d 374 (1959)	§ 15.6
Vernali v. Centrali, 28 Conn. Supp. 476, 266 A.2d 200 (Super. Ct. 1970)	§ 15.7
Vernon Williams & Son Constr., Inc. v. Continental Ins. Co., 591 S.W.2d 760 (Tenn. 1979)	§ 9.8
Vines v. Orchard Hills, Inc., 181 Conn. 501, 435 A.2d 1022 (1980)	§ 15.10
Virginia Iron, Coal & Coke Co. v. Graham, 124 Va. 692, 98 S.E. 659 (1919)	§ 15.6
Visant Elec. Contractors v. Aetna Cas. & Sur. Co., 530 S.W.2d 76 (Tenn. 1975)	§ 9.9
Waldinger Corp. v. Ashbrook-Simon-Hartley, Inc., 564 F. Supp. 970 (C.D. Ill. 1983)	§§ 5.15, 5.18, 5.24, 6.2, 6.7
Wallace Process Piping Co. v. Martin-Marietta Corp., 251 F. Supp. 411 (E.D. Va. 1965)	§ 14.27
Warner v. Cox, 487 F.2d 1301 (5th Cir. 1974)	§ 4.4
Warner Constr. Corp. v. City of Los Angeles, 2 Cal. 3d 285, 466 P.2d 996, 85 Cal. Rptr. 444 (1970)	§ 14.20
Warren Bros. Roads Co. v. United States, 173 Ct. Cl. 714 (1965)	§ 2.16
Watson, Jr. v. Commissioner, 613 F.2d 594 (5th Cir. 1980), *aff'g* 69 T.C. 544 (1978)	§ 17.19
Weedo v. Stone-E-Brick, Inc., 81 N.J. 233, 405 A.2d 788 (1979)	§ 9.8
Weeshoff Const. Co. v. Los Angeles County Flood Control Dist., 88 Cal. App. 3d 579, 152 Cal. Rptr. 19 (1979)	§ 3.4
W.E. Garrison Grading Co. v. Piracci Constr. Co., Inc., 221 S.W.2d 512 (N.C. Ct. App. 1975), *cert. denied,* (Feb. 3, 1976)	§ 10.38
Weinberg v. Wilensky, 26 N.J. Super. 301, 97 A.D.2d 707 (App. Div. 1953)	§ 15.3
Welch v. Helvering, 290 U.S. 111 (1933)	§ 17.24
Wes-Julian Constr. Corp. v. Commonwealth, 351 Mass. 588, 223 N.E.2d 72 (1967)	§ 14.14
Westerhold v. Carroll, 419 S.W.2d 73 (Mo. 1967)	§ 10.33
Western Cas. & Sur. Co. v. Biggs, 217 F.2d 163 (7th Cir. 1954)	§ 10.30
Western Employers Ins. Co. v. Arciero & Sons, 146 Cal. App. 3d 1027, 194 Cal. Rptr. 688 (1983)	§ 9.12
Western Eng'rs Inc. v. State Road Comm'n, 20 Utah 2d 294, 437 P.2d 216 (1968)	§§ 5.32, 14.14
Western World Ins. Co. v. Travelers Indem. Co., 358 So. 2d 602 (Fla. Dist. Ct. App. 1978)	§ 10.33
W.F. Constr. Co., Inc. v. Kalik, 103 Idaho 713, 652 P.2d 661 (Ct. App. 1982)	§ 15.12
Wheelabrator Corp. v. Chafee, 455 F.2d 1306 (D.C. Cir. 1971)	§ 2.16
Wheeler & Lewis v. Slifer, 195 Colo. 291, 577 P.2d 1092 (1978)	§ 5.14
Whirlpool Corp. v. Morse, 222 F. Supp. 645 (D. Minn. 1963), *aff'd,* 332 F.2d 901 (8th Cir. 1964)	§ 15.7
White v. Berenda Mesa Water Dist., 7 Cal. App. 3d 894 (1970)	§ 2.29
White v. Mitchell, 123 Wash. 630, 213 P. 10 (1923)	§ 15.5

Case	*Book §*
W.H. Lyman Constr. Co. v. Village of Guerney, 403 N.E.2d 1325 (Ill. Ct. App. 1980)	§§ 6.2, 6.7
Wickham Contracting Co., ASBCA No. 19069, 75-1 B.C.A. (CCH) ¶ 11,248 (1975)	§ 4.10
Wierzbicki v. Alaska Mut. Sav. Bank, 630 P.2d 998 (Alaska 1981)	§ 7.7
William F. Wilke, Inc. v. Department of the Army, 485 F.2d 180 (4th Cir. 1973)	§ 4.5
William Goldman Theatres, Inc. v. Commissioner, 19 T.C. 637 (1953), *aff'd* 211 F.2d 928 (3d Cir. 1954), *rev'd,* 348 U.S. 426 (1955)	§ 17.8
William Moors, Inc. v. Pine Lake Shopping Center, Inc., 253 N.W.2d 658 (Mich. Ct. App. 1977)	§ 3.17
Williams v. Aetna Cas. & Sur. Co., 376 A.2d 562 (N.J. Super. Ct. App. Div. 1977)	§ 9.15
Williams v. Kerns, 265 S.E.2d 605 (Ga. Ct. App. 1980)	§ 3.12
Willow Beach Resort, Inc. v. United States, 5 Cl. Ct. 241 (1984)	§ 4.4
Wilson v. Clarke, 470 F.2d 1218 (1st Cir. 1972)	§ 15.10
Wingler v. Niblack, 374 N.E.2d 252 (Ill. App. Ct. 1978)	§ 3.17
Winnebago Homes, Inc. v. Sheldon, 139 N.W.2d 606 (Wis. 1966)	§ 7.9
W.M. Schlosser Co. v. United States, 705 F.2d 1336 (Fed. Cir. 1983)	§ 4.4
W.N. Leslie, Inc. v. Travelers Ins. Co., 215 S.E.2d 448 (S.C. 1975)	§ 9.15
Wolfson, T.C. Memo 1978-445	§ 17.7
Woodard v. North Carolina Farm Bureau Mut. Ins. Co., 261 S.E.2d 43 (N.C. Ct. App. 1979)	§ 9.15
Worsham Constr. Co. v. Reliance Ins. Co., 687 P.2d 988 (Colo. Ct. App. 1984)	§ 9.8
W. Southward Jones, ASBCA No. 6321, 61-2 B.C.A. (CCH) ¶ 3182 (1961)	§ 4.11
Wunderlich v. State, 65 Cal. 2d 777, 423 P.2d 545 (1967)	§ 3.9
Wyatt v. Ruck Constr. Co., Inc., 571 P.2d 683 (Ariz. Ct. App. 1977)	§ 10.27
Wyoming Sawmills v. Transportation Ins. Co., 282 Or. 401, 578 P.2d 1253 (1978)	§ 9.6
Yakima Cement Prods. Co. v. Great Am. Ins. Co., 590 P.2d 371 (Wash. Ct. App. 1979), *rev'd on other grounds,* 608 P.2d 254 (Wash. 1980)	§ 9.13
Yarborough Realty Co. v. Barar, 67 So. 2d 853 (Ala. 1953)	§ 10.34
Zandri Constr. Co. v. Firemen's Ins. Co., 81 A.S.2d 106, (N.Y. App. Div.), *aff'd,* 54 N.Y.2d 999, 430 N.E.2d 922, 446 N.Y.S.2d 45 (1981)	§ 9.8
Zirin Laboratories Int'l v. Mead-Johnson & Co., 208 F. Supp. 633 (E.D. Mich. 1962)	§ 14.35
Zobel & Dahl Constr. v. Crotty, 356 N.W.2d 42 (Minn. 1984)	§ 3.8

TABLE OF FEDERAL STATUTES

INDEX